Two Manuals Combined into One Book

Military Mountaineering

And

Medical Evacuation

Department of Defense

Pentagon
Publishing

PREFACE

Mountains exist in almost every country in the world and almost every war has included some type of mountain operations. This pattern will not change; therefore, soldiers will fight in mountainous terrain in future conflicts. Although mountain operations have not changed, several advancements in equipment and transportation have increased the soldiers' capabilities. The helicopter now allows access to terrain that was once unreachable or could be reached only by slow methodical climbing. Inclement weather, however, may place various restrictions on the capabilities of air assets available to a commander. The unit must then possess the necessary mountaineering skills to overcome adverse terrain to reach an objective.

This field manual details techniques soldiers and leaders must know to cope with mountainous terrain. These techniques are the foundation upon which the mountaineer must build. They must be applied to the various situations encountered to include river crossings, glaciers, snow-covered mountains, ice climbing, rock climbing, and urban vertical environments. The degree to which this training is applied must be varied to conform to known enemy doctrine, tactics, and actions. This FM also discusses basic and advanced techniques to include acclimatization, illness and injury, equipment, anchors, evacuation, movement on glaciers, and training.

This field manual is a training aid for use by *qualified* personnel in conjunction with FM 3-97.6, Mountain Operations, which is used for planning operations in mountainous terrain. Personnel using FM 3-97.61 should attend a recognized Department of Defense Mountain Warfare School for proper training. **Improper use of techniques and procedures by untrained personnel may result in serious injury or death.** Personnel should be certified as Level I, Basic Mountaineer; Level II, Assault Climber; or Level III, Mountain Leader before using FM 3-97.61 for training (see Appendix A).

The measurements in this manual are stated as they are used in training (either metric or standard). Appendix B contains a measurement conversion chart for your convenience.

The proponent of this publication is HQ TRADOC. Submit changes for improving this publication to doctrine@benning.army.mil or on DA Form 2028 (Recommended Changes to Publications and Blank Forms) and forward to the Commander, U.S. Army Infantry School, ATTN: ATSH-RBO, Fort Benning, GA 31905-5593.

Unless otherwise stated, whenever the masculine gender is used, both men and women are included.

FIELD MANUAL
No. 3-97.61

HEADQUARTERS
DEPARTMENT OF THE ARMY
WASHINGTON, DC, 26 August 2002

MILITARY MOUNTAINEERING

CONTENTS

DISTRIBUTION RESTRICTION: Approved for public release; distribution is unlimited.

*This publication supersedes TC 90-6-1, dated 26 April 1989.

CHAPTER 1

MOUNTAIN TERRAIN, WEATHER, AND HAZARDS

Commanders must consider the effects terrain and weather will have on their operations, mainly on their troops and logistics efforts. Weather and terrain combine to challenge efforts in moving supplies to forward areas. Spring storms, which may deposit a foot of snow on dry roads, combined with unprepared vehicles create hazardous situations. Helicopters are a valuable asset for use in moving men and supplies, but commanders should not plan to use them as the only means of movement and resupply. Alternate methods must be planned due to the variability of weather. Units scheduled for deployment in mountainous terrain should become self-sufficient and train under various conditions. Commanders must be familiar with the restraints that the terrain can place on a unit.

Section I. MOUNTAIN TERRAIN

Operations in the mountains require soldiers to be physically fit and leaders to be experienced in operations in this terrain. Problems arise in moving men and transporting loads up and down steep and varied terrain in order to accomplish the mission. Chances for success in this environment are greater when a leader has experience operating under the same conditions as his men. Acclimatization, conditioning, and training are important factors in successful military mountaineering.

1-1. DEFINITION

Mountains are land forms that rise more than 500 meters above the surrounding plain and are characterized by steep slopes. Slopes commonly range from 4 to 45 degrees. Cliffs and precipices may be vertical or overhanging. Mountains may consist of an isolated peak, single ridges, glaciers, snowfields, compartments, or complex ranges extending for long distances and obstructing movement. Mountains usually favor the defense; however, attacks can succeed by using detailed planning, rehearsals, surprise, and well-led troops.

1-2. COMPOSITION

All mountains are made up of rocks and all rocks of minerals (compounds that cannot be broken down except by chemical action). Of the approximately 2,000 known minerals, seven rock-forming minerals make up most of the earth's crust: quartz and feldspar make up granite and sandstone; olivene and pyroxene give basalt its dark color; and amphibole and biotite (mica) are the black crystalline specks in granitic rocks. Except for calcite, found in limestone, they all contain silicon and are often referred to as silicates.

1-3. ROCK AND SLOPE TYPES

Different types of rock and different slopes present different hazards. The following paragraphs discuss the characteristics and hazards of the different rocks and slopes.

a. **Granite.** Granite produces fewer rockfalls, but jagged edges make pulling rope and raising equipment more difficult. Granite is abrasive and increases the danger of ropes or accessory cords being cut. Climbers must beware of large loose boulders. After a rain, granite dries quickly. Most climbing holds are found in cracks. Face climbing can be found, however, it cannot be protected.

b. **Chalk and Limestone.** Chalk and limestone are slippery when wet. Limestone is usually solid; however, conglomerate type stones may be loose. Limestone has pockets, face climbing, and cracks.

c. **Slate and Gneiss.** Slate and gneiss can be firm and or brittle in the same area (red coloring indicates brittle areas). Rockfall danger is high, and small rocks may break off when pulled or when pitons are emplaced.

d. **Sandstone.** Sandstone is usually soft causing handholds and footholds to break away under pressure. Chocks placed in sandstone may or may not hold. Sandstone should be allowed to dry for a couple of days after a rain before climbing on it—wet sandstone is extremely soft. Most climbs follow a crack. Face climbing is possible, but any outward pull will break off handholds and footholds, and it is usually difficult to protect.

e. **Grassy Slopes.** Penetrating roots and increased frost cracking cause a continuous loosening process. Grassy slopes are slippery after rain, new snow, and dew. After long, dry spells clumps of the slope tend to break away. Weight should be distributed evenly; for example, use flat hand push holds instead of finger pull holds.

f. **Firm Spring Snow (Firn Snow).** Stopping a slide on small, leftover snow patches in late spring can be difficult. Routes should be planned to avoid these dangers. Self-arrest should be practiced before encountering this situation. Beginning climbers should be secured with rope when climbing on this type surface. Climbers can glissade down firn snow if necessary. Firn snow is easier to ascend than walking up scree or talus.

g. **Talus.** Talus is rocks that are larger than a dinner plate, but smaller than boulders. They can be used as stepping-stones to ascend or descend a slope. However, if a talus rock slips away it can produce more injury than scree because of its size.

h. **Scree.** Scree is small rocks that are from pebble size to dinner plate size. Running down scree is an effective method of descending in a hurry. One can run at full stride without worry—the whole scree field is moving with you. Climbers must beware of larger rocks that may be solidly planted under the scree. Ascending scree is a tedious task. The scree does not provide a solid platform and will only slide under foot. If possible, avoid scree when ascending.

1-4. ROCK CLASSIFICATIONS

Rock is classified by origin and mineral composition.

a. **Igneous Rocks.** Deep within the earth's crust and mantle, internal heat, friction and radioactive decay creates magmas (melts of silicate minerals) that solidify into igneous rocks upon cooling. When the cooling occurs at depth, under pressure, and over time, the minerals in the magma crystallize slowly and develop well, making coarse-grained plutonic rock. The magma may move upward, propelled by its own lower density, either melting and combining with the overlying layers or forcing them aside. This results in an intrusive rock. If the melt erupts onto the surface it cools rapidly and the minerals form little or no crystal matrix, creating a volcanic or extrusive rock.

(1) ***Plutonic or Intrusive Rocks.*** Slow crystallization from deeply buried magmas generally means good climbing, since the minerals formed are relatively large and interwoven into a solid matrix. Weathering develops protrusions of resistant minerals, which makes for either a rough-surfaced rock with excellent friction, or, if the resistant

crystals are much larger than the surrounding matrix, a surface with numerous knobby holds. Pieces of foreign rock included in the plutonic body while it was rising and crystallizing, or clusters of segregated minerals, may weather differently than the main rock mass and form "chicken heads."

(a) Intrusions are named according to location and size. Large (100 square kilometers or larger) masses of plutonic rock are called "batholiths" and small ones "stocks." Most plutonic rock is in the granite family, differing only in the amounts of constituent minerals contained. A core of such batholiths is in every major mountain system in the world. In the Alps, Sierras, North Cascades, Rockies, Adirondacks, and most other ranges this core is at least partly exposed.

(b) Small plutonic intrusions are stocks, forced between sedimentary strata, and dikes, which cut across the strata. Many of these small intrusive bodies are quickly cooled and thus may look like extrusive rock.

(2) ***Volcanic or Extrusive Rocks.*** Explosive eruptions eject molten rock so quickly into the air that it hardens into loose aerated masses of fine crystals and uncrystallized glass (obsidian). When this ash consolidates while molten or after cooling, it is called "tuff," a weak rock that breaks down quickly and erodes easily. Quieter eruptions, where widespread lava flows from large fissures, produce basalt. Basaltic rocks are fine-grained and often sharp-edged.

(3) ***Jointing Rocks.*** In plutonic rocks, joints or cracks are caused by internal stresses such as contraction during cooling or expansion when overlying rock erodes or exfoliates. Some joints tend to follow a consistent pattern throughout an entire mountain and their existence can often be predicted. Therefore, when a ledge suddenly ends, the joint—and thus the ledge—may begin again around the corner. When molten rock extrudes onto the surface as a lava flow or intrudes into a cold surrounding mass as a dike or sill, the contraction from rapid cooling usually causes so much jointing that climbing can be extremely hazardous. Occasionally, this jointing is regular enough to create massed pillars with usable vertical cracks such as Devil's Tower in Wyoming.

b. **Sedimentary Rocks.** Sedimentary rocks are born high in the mountains, where erosion grinds down debris and moves it down to rivers for transportation to its final deposition in valleys, lakes, or oceans. As sediments accumulate, the bottom layers are solidified by pressure and by mineral cements precipitated from percolating groundwater. Gravel and boulders are transformed into conglomerates; sandy beaches into sandstone; beds of mud into mudstone or shale; and shell beds and coral reefs into limestone or dolomite.

(1) Though in general sedimentary rocks are much more friable than those cooled from molten magmas, pressure and cementing often produce solid rocks. In fact, by sealing up internal cracks cementing can result in flawless surfaces, especially in limestone.

(2) Most high mountain ranges have some sedimentary peaks. Ancient seafloor limestone can be found on the summits of the Himalayas and the Alps. The Canadian Rockies are almost exclusively limestone. With the exception of the Dolomites, in general sedimentary rocks do not offer high-angle climbing comparable to that of granite.

c. **Metamorphic Rocks.** These are igneous or sedimentary rocks that have been altered physically and or chemically by the tremendous heat and pressures within the earth. After sediments are solidified, high heat and pressure can cause their minerals to recrystallize. The bedding planes (strata) may also be distorted by folding and squeezing. Shale changes to slate or schist, sandstone and conglomerate into quartzite, and limestone to marble. These changes may be minimal, only slightly altering the sediments, or extensive enough to produce gneiss, which is almost indistinguishable from igneous rock.

(1) Metamorphic rocks may have not only joints and bedding, but cleavage or foliation, a series of thinly spaced cracks caused by the pressures of folding. Because of this cleavage, lower grades of metamorphic rocks may be completely unsuitable for climbing because the rock is too rotten for safe movement.

(2) Higher degrees of metamorphism or metamorphism of the right rocks provide a solid climbing surface. The Shawangunks of New York are an excellent example of high-grade conglomerate quartzite, which offers world class climbing. The center of the Green Mountain anticline contains heavily metamorphosed schist, which also provides solid climbing.

1-5. MOUNTAIN BUILDING

The two primary mechanisms for mountain-building are volcanic and tectonic activity. Volcanoes are constructed from lava and ash, which begin within the earth as magma. Tectonic activity causes plates to collide, heaving up fold mountains, and to pull apart and crack, forming fault-block mountain ranges.

a. **Plate Tectonics.** The massive slabs composing the outer layer are called tectonic plates. These plates are made up of portions of lighter, granitic continental crust, and heavier, basaltic oceanic crust attached to slabs of the rigid upper mantle. Floating slowly over the more malleable asthenosphere, their movement relative to each other creates earthquakes, volcanoes, ocean trenches, and mountain ridge systems.

b. **Mountain Structure.** The different horizontal and vertical stresses that create mountains usually produce complex patterns. Each type of stress produces a typical structure, and most mountains can be described in terms of these structures.

(1) ***Dome Mountains.*** A simple upward bulge of the crust forms dome mountains such as the Ozarks of Arkansas and Missouri, New York's Adirondacks, the Olympics of Washington, and the High Uintahs of Utah. They are usually the result of the upward movement of magma and the folding of the rock layers overhead. Erosion may strip away the overlying layers, exposing the central igneous core.

(2) ***Fault-Block Mountains.*** Faulting, or cracking of the crust into large chunks, often accompanies upwarp, which results in fault-block mountains. Many forms are created by the motion of these chunks along these faults.

(a) The ranges of the desert country of California, Nevada, and Utah provide the clearest display of faulting. The breakage extends to the surface and often during earthquakes—caused by slippage between the blocks—fresh scarps many feet high develop.

(b) Sometimes a block is faulted on both sides and rises or falls as a unit. More often, however, it is faulted on one side only. The Tetons of Wyoming and the Sierra Nevada display this—along the single zone of faults the range throws up impressive steep scarps, while on the other side the block bends but does not break, leaving a gentler slope from the base of the range to the crest. An example of a dropped block is California's Death Valley, which is below sea level and could not have been carved by erosion.

(3) ***Fold Mountains.*** Tectonic forces, in which continental plates collide or ride over each other, have given rise to the most common mountain form—fold mountains. Geologists call folds geosynclines. Upward folded strata are anticlines and downward folds are synclines. When erosion strips down the overburden of rock from folded mountain ranges, the oldest, central core is all that remains. The Alps and the Appalachians are examples of fold mountains. When the squeezing of a range is intense the rocks of the mountain mass first fold but then may break, and parts of the rocks are pushed sideways and override neighboring formations. This explains why older rocks are often found perched on top of younger ones. Isolated blocks of the over thrust mass may form when erosion strips away links connecting them with their place of origin. Almost every range of folded mountains in the world exhibits an over thrust of one sort or another.

(4) ***Volcanic Mountains.*** Along convergent plate boundaries volcanic activity increases. As it is forced underneath an overriding neighbor, continental crust melts and turns to magma within the mantle. Since it is less dense than the surrounding material it rises and erupts to form volcanoes.

(a) These volcanoes are found in belts, which correspond to continental margins around the world. The best known is the "Ring of Fire" encircling the Pacific Ocean from Katmai in Alaska through the Cascades (Mount Rainier and Mount Saint Helens) down through Mexico's Popocatepetl to the smokes of Tierra del Fuego. This belt then runs west down the Aleutian chain to Kamchatka, south to the volcanoes of Japan and the Philippines, and then east through New Guinea into the Pacific. Smaller volcanic belts are found along the Indonesian-SE Asian arc, the Caucasus region, and the Mediterranean.

(b) Volcanic activity also arises at boundaries where two plates are moving away from each other, creating deep rifts and long ridges where the crust has cracked apart and magma wells up to create new surface material. Examples of this are the Mid-Atlantic Ridge, which has created Iceland and the Azores, and the Rift Valley of East Africa with Kilimanjaro's cone.

(5) ***Complex Mountains.*** Most ranges are complex mountains with portions that have been subject to several processes. A block may have been simply pushed upward without tilting with other portions folded, domed, and faulted, often with a sprinkling of volcanoes. In addition, these processes occur both at the macro and the micro level. One massive fold can make an entire mountain peak; however, there are folds measured by a rope length, and tiny folds found within a handhold. A mountain front may be formed from a single fault, but smaller faults that form ledges and gullies may also be present.

1-6. ROUTE CLASSIFICATION

Military mountaineers must be able to assess a vertical obstacle, develop a course of action to overcome the obstacle, and have the skills to accomplish the plan. Assessment of a vertical obstacle requires experience in the classifications of routes and understanding the levels of difficulty they represent. Without a solid understanding of the difficulty of a chosen route, the mountain leader can place his life and the life of other soldiers in extreme danger. Ignorance is the most dangerous hazard in the mountain environment.

a. In North America the Yosemite Decimal System (YDS) is used to rate the difficulty of routes in mountainous terrain. The YDS classes are:

- Class 1—Hiking trail.
- Class 2—Off-trail scramble.
- Class 3—Climbing, use of ropes for beginners (moderate scrambling).
- Class 4—Belayed climbing. (This is moderate to difficult scrambling, which may have some exposure.)
- Class 5—Free climbing. (This class requires climbers to be roped up, belay and emplace intermediate protection.)

b. Class 5 is further subdivided into the following classifications:

(1) Class 5.0-5.4—Little difficulty. This is the simplest form of free climbing. Hands are necessary to support balance. This is sometimes referred to as advanced rock scrambling.

(2) Class 5.5—Moderate difficulty. Three points of contact are necessary.

(3) Class 5.6—Medium difficulty. The climber can experience vertical position or overhangs where good grips can require moderate levels of energy expenditure.

(4) Class 5.7—Great difficulty. Considerable climbing experience is necessary. Longer stretches of climbing requiring several points of intermediate protection. Higher levels of energy expenditure will be experienced.

(5) Class 5.8—Very great difficulty. Increasing amount of intermediate protection is the rule. High physical conditioning, climbing technique, and experience required.

(6) Class 5.9—Extremely great difficulty. Requires well above average ability and excellent condition. Exposed positions, often combined with small belay points. Passages of the difficult sections can often be accomplished under good conditions. Often combined with aid climbing (A0-A4).

(7) Class 5.10—Extraordinary difficulty. Climb only with improved equipment and intense training. Besides acrobatic climbing technique, mastery of refined security technique is indispensable. Often combined with aid climbing (A0-A4).

(8) Class 5.11-5.14—Greater increases of difficulty, requiring more climbing ability, experience, and energy expenditure. Only talented and dedicated climbers reach this level.

c. Additional classifications include the following.

(1) Classes are further divided into a, b, c, and d categories starting from 5.10 to 5.14 (for example, 5.10d).

(2) Classes are also further divided from 5.9 and below with +/- categories (for example, 5.8+).

(3) All class 5 climbs can also be designated with "R" or "X," which indicates a run-out on a climb. This means that placement of intermediate protection is not possible on portions of the route. (For example, in a classification of 5.8R, the "R" indicates periods of run-out where, if a fall was experienced, ground fall would occur.) Always check the local guidebook to find specific designation for your area.

(4) All class 5 climbs can also be designated with "stars." These refer to the popularity of the climb to the local area. Climbs are represented by a single "star" up to five "stars;" a five-star climb is a classic climb and is usually aesthetically pleasing.

d. Aid climb difficulty classification includes:

(1) A0—"French-free." This technique involves using a piece of gear to make progress; for example, clipping a sling into a bolt or piece of protection and then pulling up on it or stepping up in the sling. Usually only needed to get past one or two more difficult moves on advanced free climbs.

(2) A1—Easy aid. The placement of protection is straight forward and reliable. There is usually no high risk of any piece of protection pulling out. This technique requires etriers and is fast and simple.

(3) A2—Moderate aid. The placement of protection is generally straight forward, but placement can be awkward and strenuous. Usually A2 involves one or two moves that are difficult with good protection placement below and above the difficult moves. No serious fall danger.

(4) A3—Hard aid. This technique requires testing your protection. It involves several awkward and strenuous moves in a row. Generally solid placements which will hold a fall and are found within a full rope length. However, long fall potential does exist, with falls of 40 to 60 feet and intermediate protection on the awkward placements failing. These falls, however, are usually clean and with no serious bodily harm.

(5) A4—Serious aid. This technique requires lots of training and practice. More like walking on eggs so none of them break. Leads will usually take extended amounts of time which cause the lead climber to doubt and worry about each placement. Protection placed will usually only hold a climber's weight and falls can be as long as two-thirds the rope length.

(6) A5—Extreme aid. All protection is sketchy at best. Usually no protection placed on the entire route can be trusted to stop a fall.

(7) A6—Extremely severe aid. Continuous A5 climbing with A5 belay stations. If the leader falls, the whole rope team will probably experience ground fall.

(8) Aid climbing classes are also further divided into +/- categories, such as A3+ or A3-, which would simply refer to easy or hard.

e. Grade ratings (commitment grades) inform the climber of the approximate time a climber trained to the level of the climb will take to complete the route.

- I—Several hours.
- II—Half of a day.
- III—About three-fourths of a day.
- IV—Long hard day (usually not less than 5.7).
- V—1 1/2 to 2 1/2 days (usually not less than 5.8).
- VI—Greater than 2 days.

f. Climbing difficulties are rated by different systems. Table 1-1 shows a comparison of these systems.

- YDS (Yosemite Decimal System)—Used in the United States.
- UIAA (Union des International Alpine Association)—Used in Europe.
- British—The British use adjectives and numbers to designate the difficulty of climbs. This system can be confusing if the climber is not familiar with it.
- French—The French use numbers and letters to designate the difficulty of climbs.
- Brazilian—Brazil uses Roman Numerals and adjectives to designate difficulty.
- Australian—Australia uses only numbers to designate difficulty.

YDS	UIAA	BRITISH	FRENCH	BRAZIL	AUSTRALIA
Class 1	I	easy (E)			
Class 2	II	easy (E)			
Class 3	III	easy (E)	1a, b, c		
Class 4	III-	moderate (MOD)	1a, b, c		
5.0	III	moderate (MOD)	2a, b		4
5.1	III+	difficult (DIFF)	2a, b		5
5.2	IV-	hard difficult	2c, 3a		6
5.3	IV	very difficult	3b, c, 4a		7
5.4	IV+	hard very difficult	3b, c, 4a	II	8, 9
5.5	V-	mild severe	3b, c, 4a	IIsup	10, 11
5.6	V	severe, hard severe, 4a	4a, b, c	III	12, 13
5.7	V+	severe, hard severe, 4b	4a, b, c	IIIsup	14
5.8	VI-	hard severe, hard very severe, 4c	5a, b	IV	15
5.9	VI	5a	5b, c	IVsup	16, 17
5.10a	VII-	E1, 5b	5b, c	V	18
5.10b	VII	E1, 5b	5b, c	Vsup	19
5.10c	VII	E1, 5b	5b, c	VI	20
5.10d	VII+	E1/E2, 5b-5c	5b, c	VIsup	21
5.11a	VIII-	E3, 6a	6a, b, c	VII	22
5.11b	VIII	E3/E4, 6a	6a, b, c	VII	23
5.11c	VIII	E4, 6b	6a, b, c	VIIsup	24
5.12a	IX-	E5, E6/7, 6c	7a	VIII	26
5.12b	IX	E5, E6/7, 6c	7a	VIIIsup	27
5.12c	IX	E5, E6/7, 6c	7a		28
5.12d	IX+	E6/7, 7a	7a		29

Table 1-1. Rating systems.

g. Ice climbing ratings can have commitment ratings and technical ratings. The numerical ratings are often prefaced with WI (waterfall ice), AI (alpine ice), or M (mixed rock and ice).

(1) ***Commitment Ratings.*** Commitment ratings are expressed in Roman numerals.

- I—A short, easy climb near the road, with no avalanche hazard and a straightforward descent.
- II—A route of one or two pitches within a short distance of rescue assistance, with little objective hazard.
- III—A multipitch route at low elevation, or a one-pitch climb with an approach that takes about an hour. The route requires anywhere from a few hours to a long day to complete. The descent may require building rappel anchors, and the route might be prone to avalanche.
- IV—A multipitch route at higher elevations; may require several hours of approach on skis or foot. This route is subject to objective hazards, possibly with a hazardous descent.
- V—A long climb in a remote setting, requiring all day to complete the climb itself. Requires many rappels off anchors for the descent. This route has sustained exposure to avalanche or other objective hazards.

- VI—A long ice climb in an alpine setting, with sustained technical climbing. Only elite climbers will complete it in a day. A difficult and involved approach and descent, with objective hazards ever-present, all in a remote area.
- VII—Everything a grade VI has, and more of it. Possibly days to approach the climb, and objective hazards rendering survival as questionable. Difficult physically and mentally.

(2) ***Technical Ratings.*** Technical ratings are expressed as Arabic numerals.

- 1—A frozen lake or stream bed.
- 2—A pitch with short sections of ice up to 80 degrees; lots of opportunity for protection and good anchors.
- 3—Sustained ice up to 80 degrees; the ice is usually good, with places to rest, but it requires skill at placing protection and setting anchors.
- 4—A sustained pitch that is vertical or slightly less than vertical; may have special features such as chandeliers and run-outs between protection.
- 5—A long, strenuous pitch, possibly 50 meters of 85- to 90-degree ice with few if any rests between anchors. The pitch may be shorter, but on featureless ice. Good skills at placing protection are required.
- 6—A full 50-meter pitch of dead vertical ice, possibly of poor quality; requires efficiency of movement and ability to place protection while in awkward stances.
- 7—A full rope length of thin vertical or overhanging ice of dubious adhesion. An extremely tough pitch, physically and mentally, requiring agility and creativity.
- 8—Simply the hardest ice climbing ever done; extremely bold and gymnastic.

1-7. CROSS-COUNTRY MOVEMENT

Soldiers must know the terrain to determine the feasible routes for cross-country movement when no roads or trails are available.

a. A pre-operations intelligence effort should include topographic and photographic map coverage as well as detailed weather data for the area of operations. When planning mountain operations, additional information may be needed about size, location, and characteristics of landforms; drainage; types of rock and soil; and the density and distribution of vegetation. Control must be decentralized to lower levels because of varied terrain, erratic weather, and communication problems inherent to mountainous regions.

b. Movement is often restricted due to terrain and weather. The erratic weather requires that soldiers be prepared for wide variations in temperature, types, and amounts of precipitation.

(1) Movement above the timberline reduces the amount of protective cover available at lower elevations. The logistical problem is important; therefore, each man must be self-sufficient to cope with normal weather changes using materials from his rucksack.

(2) Movement during a storm is difficult due to poor visibility and bad footing on steep terrain. Although the temperature is often higher during a storm than during clear weather, the dampness of rain and snow and the penetration of wind cause soldiers to chill quickly. Although climbers should get off the high ground and seek shelter and

warmth, if possible, during severe mountain storms, capable commanders may use reduced visibility to achieve tactical surprise.

c. When the tactical situation requires continued movement during a storm, the following precautions should be observed:

- Maintain visual contact.
- Keep warm. Maintain energy and body heat by eating and drinking often; carry food that can be eaten quickly and while on the move.
- Keep dry. Wear wet-weather clothing when appropriate, but do not overdress, which can cause excessive perspiration and dampen clothing. As soon as the objective is reached and shelter secured, put on dry clothing.
- Do not rush. Hasty movement during storms leads to breaks in contact and accidents.
- If lost, stay warm, dry, and calm.
- Do not use ravines as routes of approach during a storm as they often fill with water and are prone to flash floods.
- Avoid high pinnacles and ridgelines during electrical storms.
- Avoid areas of potential avalanche or rock-fall danger.

1-8. COVER AND CONCEALMENT

When moving in the mountains, outcroppings, boulders, heavy vegetation, and intermediate terrain can provide cover and concealment. Digging fighting positions and temporary fortifications is difficult because soil is often thin or stony. The selection of dug-in positions requires detailed planning. Some rock types, such as volcanic tuff, are easily excavated. In other areas, boulders and other loose rocks can be used for building hasty fortifications. In alpine environments, snow and ice blocks may be cut and stacked to supplement dug-in positions. As in all operations, positions and routes must be camouflaged to blend in with the surrounding terrain to prevent aerial detection.

1-9. OBSERVATION

Observation in mountains varies because of weather and ground cover. The dominating height of mountainous terrain permits excellent long-range observation. However, rapidly changing weather with frequent periods of high winds, rain, snow, sleet, hail, and fog can limit visibility. The rugged nature of the terrain often produces dead space at midranges.

a. Low cloud cover at higher elevations may neutralize the effectiveness of OPs established on peaks or mountaintops. High wind speeds and sound often mask the noises of troop movement. Several OPs may need to be established laterally, in depth, and at varying altitudes to provide visual coverage of the battle area.

b. Conversely, the nature of the terrain can be used to provide concealment from observation. This concealment can be obtained in the dead space. Mountainous regions are subject to intense shadowing effects when the sun is low in relatively clear skies. The contrast from lighted to shaded areas causes visual acuity in the shaded regions to be considerably reduced. These shadowed areas can provide increased concealment when combined with other camouflage and should be considered in maneuver plans.

1-10. FIELDS OF FIRE

Fields of fire, like observation, are excellent at long ranges. However, dead space is a problem at short ranges. When forces cannot be positioned to cover dead space with direct fires, mines and obstacles or indirect fire must be used. Range determination is deceptive in mountainous terrain. Soldiers must routinely train in range estimation in mountainous regions to maintain their proficiency.

Section II. MOUNTAIN WEATHER

Most people subconsciously "forecast" the weather. If they look outside and see dark clouds they may decide to take rain gear. If an unexpected wind strikes, people glance to the sky for other bad signs. A conscious effort to follow weather changes will ultimately lead to a more accurate forecast. An analysis of mountain weather and how it is affected by mountain terrain shows that such weather is prone to patterns and is usually severe, but patterns are less obvious in mountainous terrain than in other areas. Conditions greatly change with altitude, latitude, and exposure to atmospheric winds and air masses. Mountain weather can be extremely erratic. It varies from stormy winds to calm, and from extreme cold to warmth within a short time or with a minor shift in locality. The severity and variance of the weather causes it to have a major impact on military operations.

1-11. CONSIDERATIONS FOR PLANNING

Mountain weather can be either a dangerous obstacle to operations or a valuable aid, depending on how well it is understood and to what extent advantage is taken of its peculiar characteristics.

a. Weather often determines the success or failure of a mission since it is highly changeable. Military operations plans must be flexible, especially in planning airmobile and airborne operations. The weather must be anticipated to allow enough time for planning so that the leaders of subordinate units can use their initiative in turning an important weather factor in their favor. The clouds that often cover the tops of mountains and the fogs that cover valleys are an excellent means of concealing movements that normally are made during darkness or in smoke. Limited visibility can be used as a combat multiplier.

b. The safety or danger of almost all high mountain regions, especially in winter, depends upon a change of a few degrees of temperature above or below the freezing point. Ease and speed of travel depend mainly on the weather. Terrain that can be crossed swiftly and safely one day may become impassable or highly dangerous the next due to snowfall, rainfall, or a rise in temperature. The reverse can happen just as quickly. The prevalence of avalanches depends on terrain, snow conditions, and weather factors.

c. Some mountains, such as those found in desert regions, are dry and barren with temperatures ranging from extreme heat in the summer to extreme cold in the winter. In tropical regions, lush jungles with heavy seasonal rains and little temperature variation often cover mountains. High rocky crags with glaciated peaks can be found in mountain ranges at most latitudes along the western portion of the Americas and Asia.

d. Severe weather may decrease morale and increase basic survival problems. These problems can be minimized when men have been trained to accept the weather by being self-sufficient. Mountain soldiers properly equipped and trained can use the weather to their advantage in combat operations.

1-12. MOUNTAIN AIR

High mountain air is dry and may be drier in the winter. Cold air has a reduced capacity to hold water vapor. Because of this increased dryness, equipment does not rust as quickly and organic material decomposes slowly. The dry air also requires soldiers to increase consumption of water. The reduced water vapor in the air causes an increase in evaporation of moisture from the skin and in loss of water through transpiration in the respiratory system. Due to the cold, most soldiers do not naturally consume the quantity of fluids they would at higher temperatures and must be encouraged to consciously increase their fluid intake.

a. Pressure is low in mountainous areas due to the altitude. The barometer usually drops 2.5 centimeters for every 300 meters gained in elevation (3 percent).

b. The air at higher altitudes is thinner as atmospheric pressure drops with the increasing altitude. The altitude has a natural filtering effect on the sun's rays. Rays are absorbed or reflected in part by the molecular content of the atmosphere. This effect is greater at lower altitudes. At higher altitudes, the thinner, drier air has a reduced molecular content and, consequently, a reduced filtering effect on the sun's rays. The intensity of both visible and ultraviolet rays is greater with increased altitude. These conditions increase the chance of sunburn, especially when combined with a snow cover that reflects the rays upward.

1-13. WEATHER CHARACTERISTICS

The earth is surrounded by an atmosphere that is divided into several layers. The world's weather systems are in the lower of these layers known as the "troposphere." This layer reaches as high as 40,000 feet. Weather is a result of an atmosphere, oceans, land masses, unequal heating and cooling from the sun, and the earth's rotation. The weather found in any one place depends on many things such as the air temperature, humidity (moisture content), air pressure (barometric pressure), how it is being moved, and if it is being lifted or not.

a. Air pressure is the "weight" of the atmosphere at any given place. The higher the pressure, the better the weather will be. With lower air pressure, the weather will more than likely be worse. In order to understand this, imagine that the air in the atmosphere acts like a liquid. Areas with a high level of this "liquid" exert more pressure on an area and are called high-pressure areas. Areas with a lower level are called low-pressure areas. The average air pressure at sea level is 29.92 inches of mercury (hg) or 1,013 millibars (mb). The higher in altitude, the lower the pressure.

(1) ***High Pressure.*** The characteristics of a high-pressure area are as follows:

- The airflow is clockwise and out.
- Otherwise known as an "anticyclone".
- Associated with clear skies.
- Generally the winds will be mild.
- Depicted as a blue "H" on weather maps.

(2) ***Low Pressure.*** The characteristics of a low-pressure area are as follows:

- The airflow is counterclockwise and in.
- Otherwise known as a "cyclone".
- Associated with bad weather.
- Depicted as a red "L" on weather maps.

b. Air from a high-pressure area is basically trying to flow out and equalize its pressure with the surrounding air. Low pressure, on the other hand, is building up vertically by pulling air in from outside itself, which causes atmospheric instability resulting in bad weather.

c. On a weather map, these differences in pressure are depicted as isobars. Isobars resemble contour lines and are measured in either millibars or inches of mercury. The areas of high pressure are called "ridges" and lows are called "troughs."

1-14. WIND

In high mountains, the ridges and passes are seldom calm; however, strong winds in protected valleys are rare. Normally, wind speed increases with altitude since the earth's frictional drag is strongest near the ground. This effect is intensified by mountainous terrain. Winds are accelerated when they converge through mountain passes and canyons. Because of these funneling effects, the wind may blast with great force on an exposed mountainside or summit. Usually, the local wind direction is controlled by topography.

a. The force exerted by wind quadruples each time the wind speed doubles; that is, wind blowing at 40 knots pushes four times harder than a wind blowing at 20 knots. With increasing wind strength, gusts become more important and may be 50 percent higher than the average wind speed. When wind strength increases to a hurricane force of 64 knots or more, soldiers should lay on the ground during gusts and continue moving during lulls. If a hurricane- force wind blows where there is sand or snow, dense clouds fill the air. The rocky debris or chunks of snow crust are hurled near the surface. During the winter season, or at high altitudes, commanders must be constantly aware of the wind-chill factor and associated cold-weather injuries (see Chapter 2).

b. Winds are formed due to the uneven heating of the air by the sun and rotation of the earth. Much of the world's weather depends on a system of winds that blow in a set direction.

c. Above hot surfaces, air expands and moves to colder areas where it cools and becomes denser, and sinks to the earth's surface. The results are a circulation of air from the poles along the surface of the earth to the equator, where it rises and moves to the poles again.

d. Heating and cooling together with the rotation of the earth causes surface winds. In the Northern Hemisphere, there are three prevailing winds:

(1) ***Polar Easterlies.*** These are winds from the polar region moving from the east. This is air that has cooled and settled at the poles.

(2) ***Prevailing Westerlies.*** These winds originate from approximately 30 degrees north latitude from the west. This is an area where prematurely cooled air, due to the earth's rotation, has settled to the surface.

(3) ***Northeast Tradewinds.*** These are winds that originate from approximately 30° north from the northeast.

e. The jet stream is a long meandering current of high-speed winds often exceeding 250 miles per hour near the transition zone between the troposphere and the stratosphere known as the tropopause. These winds blow from a generally westerly direction dipping down and picking up air masses from the tropical regions and going north and bringing down air masses from the polar regions.

f. The patterns of wind mentioned above move air. This air comes in parcels called "air masses." These air masses can vary from the size of a small town to as large as a country. These air masses are named from where they originate:

- Maritime—over water.
- Continental—over land
- Polar—north of 60° north latitude.
- Tropical—south of 60° north latitude.

Combining these parcels of air provides the names and description of the four types of air masses:

- Continental Polar—cold, dry air mass.
- Maritime Polar—cold, wet air mass.
- Maritime Tropical—warm, wet air mass.
- Continental Tropical—warm, dry air mass.

g. Two types of winds are peculiar to mountain environments, but do not necessarily affect the weather.

(1) ***Anabatic Wind (Valley Winds).*** These winds blow up mountain valleys to replace warm rising air and are usually light winds.

(2) ***Katabatic Wind (Mountain Wind).*** These winds blow down mountain valley slopes caused by the cooling of air and are occasionally strong winds.

1-15. HUMIDITY

Humidity is the amount of moisture in the air. All air holds water vapor even if it cannot be seen. Air can hold only so much water vapor; however, the warmer the air, the more moisture it can hold. When air can hold all that it can the air is "saturated" or has 100 percent relative humidity.

a. If air is cooled beyond its saturation point, the air will release its moisture in one form or another (clouds, fog, dew, rain, snow, and so on). The temperature at which this happens is called the "condensation point". The condensation point varies depending on the amount of water vapor contained in the air and the temperature of the air. If the air contains a great deal of water, condensation can occur at a temperature of 68 degrees Fahrenheit, but if the air is dry and does not hold much moisture, condensation may not form until the temperature drops to 32 degrees Fahrenheit or even below freezing.

b. The adiabatic lapse rate is the rate at which air cools as it rises or warms as it descends. This rate varies depending on the moisture content of the air. Saturated (moist) air will warm and cool approximately 3.2 degrees Fahrenheit per 1,000 feet of elevation gained or lost. Dry air will warm and cool approximately 5.5 degrees Fahrenheit per 1,000 feet of elevation gained or lost.

1-16. CLOUD FORMATION

Clouds are indicators of weather conditions. By reading cloud shapes and patterns, observers can forecast weather with little need for additional equipment such as a barometer, wind meter, and thermometer. Any time air is lifted or cooled beyond its saturation point (100 percent relative humidity), clouds are formed. The four ways air gets lifted and cooled beyond its saturation point are as follows.

a. **Convective Lifting.** This effect happens due to the sun's heat radiating off the Earth's surface causing air currents (thermals) to rise straight up and lift air to a point of saturation.

b. **Frontal Lifting.** A front is formed when two air masses of different moisture content and temperature collide. Since air masses will not mix, warmer air is forced aloft over the colder air mass. From there it is cooled and then reaches its saturation point. Frontal lifting creates the majority of precipitation.

c. **Cyclonic Lifting.** An area of low pressure pulls air into its center from all over in a counterclockwise direction. Once this air reaches the center of the low pressure, it has nowhere to go but up. Air continues to lift until it reaches the saturation point.

d. **Orographic Lifting.** This happens when an air mass is pushed up and over a mass of higher ground such as a mountain. Air is cooled due to the adiabatic lapse rate until the air's saturation point is reached.

1-17. TYPES OF CLOUDS

Clouds are one of the signposts to what is happening with the weather. Clouds can be described in many ways. They can be classified by height or appearance, or even by the amount of area covered vertically or horizontally. Clouds are classified into five categories: low-, mid-, and high-level clouds; vertically-developed clouds; and less common clouds.

a. **Low-Level Clouds.** Low-level clouds (0 to 6,500 feet) are either cumulus or stratus (Figures 1-1 and 1-2, page 1-16). Low-level clouds are mostly composed of water droplets since their bases lie below 6,500 feet. When temperatures are cold enough, these clouds may also contain ice particles and snow.

(1) The two types of precipitating low-level clouds are nimbostratus and stratocumulus (Figures 1-3 and 1-4, page 1-17).

(a) Nimbostratus clouds are dark, low-level clouds accompanied by light to moderately falling precipitation. The sun or moon is not visible through nimbostratus clouds, which distinguishes them from mid-level altostratus clouds. Because of the fog and falling precipitation commonly found beneath and around nimbostratus clouds, the cloud base is typically extremely diffuse and difficult to accurately determine.

(b) Stratocumulus clouds generally appear as a low, lumpy layer of clouds that is sometimes accompanied by weak precipitation. Stratocumulus vary in color from dark gray to light gray and may appear as rounded masses with breaks of clear sky in between. Because the individual elements of stratocumulus are larger than those of altocumulus, deciphering between the two cloud types is easier. With your arm extended toward the sky, altocumulus elements are about the size of a thumbnail while stratocumulus are about the size of a fist.

Figure 1-1. Cumulus clouds.

Figure 1-2. Stratus clouds.

Figure 1-3. Nimbostratus clouds.

Figure 1-4. Stratocumulus clouds.

(2) Low-level clouds may be identified by their height above nearby surrounding relief of known elevation. Most precipitation originates from low-level clouds because rain or snow usually evaporate before reaching the ground from higher clouds. Low-level clouds usually indicate impending precipitation, especially if the cloud is more than 3,000 feet thick. (Clouds that appear dark at their bases are more than 3,000 feet thick.)

b. **Mid-Level Clouds.** Mid-level clouds (between 6,500 to 20,000 feet) have a prefix of alto. Middle clouds appear less distinct than low clouds because of their height. Alto clouds with sharp edges are warmer because they are composed mainly of water droplets. Cold clouds, composed mainly of ice crystals and usually colder than -30 degrees F, have distinct edges that grade gradually into the surrounding sky. Middle clouds usually indicate fair weather, especially if they are rising over time. Lowering middle clouds indicate potential storms, though usually hours away. There are two types of mid-level clouds, altocumulus and altostratus clouds (Figures 1-5 and 1-6).

(1) Altocumulus clouds can appear as parallel bands or rounded masses. Typically a portion of an altocumulus cloud is shaded, a characteristic which makes them distinguishable from high-level cirrocumulus. Altocumulus clouds usually form in advance of a cold front. The presence of altocumulus clouds on a warm humid summer morning is commonly followed by thunderstorms later in the day. Altocumulus clouds that are scattered rather than even, in a blue sky, are called "fair weather" cumulus and suggest arrival of high pressure and clear skies.

(2) Altostratus clouds are often confused with cirrostratus. The one distinguishing feature is that a halo is not observed around the sun or moon. With altostratus, the sun or moon is only vaguely visible and appears as if it were shining through frosted glass.

Figure 1-5. Altocumulus.

Figure 1-6. Altostratus.

c. **High-Level Clouds.** High-level clouds (more than 20,000 feet above ground level) are usually frozen clouds, indicating air temperatures at that elevation below -30 degrees Fahrenheit, with a fibrous structure and blurred outlines. The sky is often covered with a thin veil of cirrus that partly obscures the sun or, at night, produces a ring of light around the moon. The arrival of cirrus indicates moisture aloft and the approach of a traveling storm system. Precipitation is often 24 to 36 hours away. As the storm approaches, the cirrus thickens and lowers, becoming altostratus and eventually stratus. Temperatures are warm, humidity rises, and winds become southerly or south easterly. The two types of high-level clouds are cirrus and cirrostratus (Figure 1-7 and Figure 1-8, page 1-20).

(1) Cirrus clouds are the most common of the high-level clouds. Typically found at altitudes greater than 20,000 feet, cirrus are composed of ice crystals that form when super-cooled water droplets freeze. Cirrus clouds generally occur in fair weather and point in the direction of air movement at their elevation. Cirrus can be observed in a variety of shapes and sizes. They can be nearly straight, shaped like a comma, or seemingly all tangled together. Extensive cirrus clouds are associated with an approaching warm front.

(2) Cirrostratus clouds are sheet-like, high-level clouds composed of ice crystals. They are relatively transparent and can cover the entire sky and be up to several thousand feet thick. The sun or moon can be seen through cirrostratus. Sometimes the only indication of cirrostratus clouds is a halo around the sun or moon. Cirrostratus clouds tend to thicken as a warm front approaches, signifying an increased production of ice crystals. As a result, the halo gradually disappears and the sun or moon becomes less visible.

Figure 1-7. Cirrus.

Figure 1-8. Cirrostratus.

d. **Vertical-Development Clouds.** Clouds with vertical development can grow to heights in excess of 39,000 feet, releasing incredible amounts of energy. The two types of clouds with vertical development are fair weather cumulus and cumulonimbus.

(1) Fair weather cumulus clouds have the appearance of floating cotton balls and have a lifetime of 5 to 40 minutes. Known for their flat bases and distinct outlines, fair weather cumulus exhibit only slight vertical growth, with the cloud tops designating the limit of the rising air. Given suitable conditions, however, these clouds can later develop into towering cumulonimbus clouds associated with powerful thunderstorms. Fair weather cumulus clouds are fueled by buoyant bubbles of air known as thermals that rise up from the earth's surface. As the air rises, the water vapor cools and condenses forming water droplets. Young fair weather cumulus clouds have sharply defined edges and bases while the edges of older clouds appear more ragged, an artifact of erosion. Evaporation along the cloud edges cools the surrounding air, making it heavier and producing sinking motion outside the cloud. This downward motion inhibits further convection and growth of additional thermals from down below, which is why fair weather cumulus typically have expanses of clear sky between them. Without a continued supply of rising air, the cloud begins to erode and eventually disappears.

(2) Cumulonimbus clouds are much larger and more vertically developed than fair weather cumulus (Figure 1-9). They can exist as individual towers or form a line of towers called a squall line. Fueled by vigorous convective updrafts, the tops of cumulonimbus clouds can reach 39,000 feet or higher. Lower levels of cumulonimbus clouds consist mostly of water droplets while at higher elevations, where the temperatures are well below freezing, ice crystals dominate the composition. Under favorable conditions, harmless fair weather cumulus clouds can quickly develop into large cumulonimbus associated with powerful thunderstorms known as super-cells. Super-cells are large thunderstorms with deep rotating updrafts and can have a lifetime of several hours. Super-cells produce frequent lightning, large hail, damaging winds, and

tornadoes. These storms tend to develop during the afternoon and early evening when the effects of heating from the sun are the strongest.

Figure 1-9. Cumulonimbus.

e. **Other Cloud Types.** These clouds are a collection of miscellaneous types that do not fit into the previous four groups. They are orographic clouds, lenticulars, and contrails.

(1) Orographic clouds develop in response to the forced lifting of air by the earth's topography. Air passing over a mountain oscillates up and down as it moves downstream. Initially, stable air encounters a mountain, is lifted upward, and cools. If the air cools to its saturation temperature during this process, the water vapor condenses and becomes visible as a cloud. Upon reaching the mountain top, the air is heavier than the environment and will sink down the other side, warming as it descends. Once the air returns to its original height, it has the same buoyancy as the surrounding air. However, the air does not stop immediately because it still has momentum carrying it downward. With continued descent, the air becomes warmer then the surrounding air and accelerates back upwards towards its original height. Another name for this type of cloud is the lenticular cloud.

(2) Lenticular clouds are cloud caps that often form above pinnacles and peaks, and usually indicate higher winds aloft (Figure 1-10, page 1-22). Cloud caps with a lens shape, similar to a "flying saucer," indicate extremely high winds (over 40 knots). Lenticulars should always be watched for changes. If they grow and descend, bad weather can be expected.

Figure 1-10. Lenticular.

(3) Contrails are clouds that are made by water vapor being inserted into the upper atmosphere by the exhaust of jet engines (Figure 1-11). Contrails evaporate rapidly in fair weather. If it takes longer than two hours for contrails to evaporate, then there is impending bad weather (usually about 24 hours prior to a front).

Figure 1-11. Contrails.

f. **Cloud Interpretation.** Serious errors can occur in interpreting the extent of cloud cover, especially when cloud cover must be reported to another location. Cloud cover always appears greater on or near the horizon, especially if the sky is covered with cumulus clouds, since the observer is looking more at the sides of the clouds rather than between them. Cloud cover estimates should be restricted to sky areas more than 40 degrees above the horizon—that is, to the local sky. Assess the sky by dividing the 360 degrees of sky around you into eighths. Record the coverage in eighths and the types of clouds observed.

1-18. FRONTS

Fronts occur when two air masses of different moisture and temperature contents meet. One of the indicators that a front is approaching is the progression of the clouds. The four types of fronts are warm, cold, occluded, and stationary.

a. **Warm Front.** A warm front occurs when warm air moves into and over a slower or stationary cold air mass. Because warm air is less dense, it will rise up and over the cooler air. The cloud types seen when a warm front approaches are cirrus, cirrostratus, nimbostratus (producing rain), and fog. Occasionally, cumulonimbus clouds will be seen during the summer months.

b. **Cold Front.** A cold front occurs when a cold air mass overtakes a slower or stationary warm air mass. Cold air, being more dense than warm air, will force the warm air up. Clouds observed will be cirrus, cumulus, and then cumulonimbus producing a short period of showers.

c. **Occluded Front.** Cold fronts generally move faster than warm fronts. The cold fronts eventually overtake warm fronts and the warm air becomes progressively lifted from the surface. The zone of division between cold air ahead and cold air behind is called a "cold occlusion." If the air behind the front is warmer than the air ahead, it is a warm occlusion. Most land areas experience more occlusions than other types of fronts. The cloud progression observed will be cirrus, cirrostratus, altostratus, and nimbostratus. Precipitation can be from light to heavy.

d. **Stationary Front.** A stationary front is a zone with no significant air movement. When a warm or cold front stops moving, it becomes a stationary front. Once this boundary begins forward motion, it once again becomes a warm or cold front. When crossing from one side of a stationary front to another, there is typically a noticeable temperature change and shift in wind direction. The weather is usually clear to partly cloudy along the stationary front.

1-19. TEMPERATURE

Normally, a temperature drop of 3 to 5 degrees Fahrenheit for every 1,000 feet gain in altitude is encountered in motionless air. For air moving up a mountain with condensation occurring (clouds, fog, and precipitation), the temperature of the air drops 3.2 degrees Fahrenheit with every 1,000 feet of elevation gain. For air moving up a mountain with no clouds forming, the temperature of the air drops 5.5 degrees Fahrenheit for every 1,000 feet of elevation gain.

a. An expedient to this often occurs on cold, clear, calm mornings. During a troop movement or climb started in a valley, higher temperatures may often be encountered as altitude is gained. This reversal of the normal cooling with elevation is called temperature inversion. Temperature inversions are caused when mountain air is cooled by ice, snow, and heat loss through thermal radiation. This cooler, denser air settles into the valleys and low areas. The inversion continues until the sun warms the surface of the earth or a moderate wind causes a mixing of the warm and cold layers. Temperature inversions are common in the mountainous regions of the arctic, subarctic, and mid-latitudes.

b. At high altitudes, solar heating is responsible for the greatest temperature contrasts. More sunshine and solar heat are received above the clouds than below. The important effect of altitude is that the sun's rays pass through less of the atmosphere and more direct heat is received than at lower levels, where solar radiation is absorbed and reflected by dust and water vapor. Differences of 40 to 50 degrees Fahrenheit may occur between surface temperatures in the shade and surface temperatures in the sun. This is particularly true for dark metallic objects. The difference in temperature felt on the skin between the sun and shade is normally 7 degrees Fahrenheit. Special care must be taken to avoid sunburn and snow blindness. Besides permitting rapid heating, the clear air at high altitudes also favors rapid cooling at night. Consequently, the temperature rises fast after sunrise and drops quickly after sunset. Much of the chilled air drains downward, due to convection currents, so that the differences between day and night temperatures are greater in valleys than on slopes.

c. Local weather patterns force air currents up and over mountaintops. Air is cooled on the windward side of the mountain as it gains altitude, but more slowly (3.2 degrees Fahrenheit per 1,000 feet) if clouds are forming due to heat release when water vapor becomes liquid. On the leeward side of the mountain, this heat gained from the condensation on the windward side is added to the normal heating that occurs as the air descends and air pressure increases. Therefore, air and winds on the leeward slope are considerably warmer than on the windward slope, which is referred to as Chinook winds. The heating and cooling of the air affects planning considerations primarily with regard to the clothing and equipment needed for an operation.

1-20. WEATHER FORECASTING

The use of a portable aneroid barometer, thermometer, wind meter, and hygrometer help in making local weather forecasts. Reports from other localities and from any weather service, including USAF, USN, or the National Weather Bureau, are also helpful. Weather reports should be used in conjunction with the locally observed current weather situation to forecast future weather patterns.

a. Weather at various elevations may be quite different because cloud height, temperature, and barometric pressure will all be different. There may be overcast and rain in a lower area, with mountains rising above the low overcast into warmer clear weather.

b. To be effective, a forecast must reach the small-unit leaders who are expected to utilize weather conditions for assigned missions. Several different methods can be used to create a forecast. The method a forecaster chooses depends upon the forecaster's experience, the amount of data available, the level of difficulty that the forecast situation presents, and the degree of accuracy needed to make the forecast. The five ways to forecast weather are:

(1) ***Persistence Method.*** "Today equals tomorrow" is the simplest way of producing a forecast. This method assumes that the conditions at the time of the forecast will not

change; for example, if today was hot and dry, the persistence method predicts that tomorrow will be the same.

(2) ***Trends Method.*** "Nowcasting" involves determining the speed and direction of fronts, high- and low-pressure centers, and clouds and precipitation. For example, if a cold front moves 300 miles during a 24-hour period, we can predict that it will travel 300 miles in another 24-hours.

(3) ***Climatology Method.*** This method averages weather statistics accumulated over many years. This only works well when the pattern is similar to the following years.

(4) ***Analog Method.*** This method examines a day's forecast and recalls a day in the past when the weather looked similar (an analogy). This method is difficult to use because finding a perfect analogy is difficult.

(5) ***Numerical Weather Prediction.*** This method uses computers to analyze all weather conditions and is the most accurate of the five methods.

1-21. RECORDING DATA

An accurate observation is essential in noting trends in weather patterns. Ideally, under changing conditions, trends will be noted in some weather parameters. However, this may not always be the case. A minor shift in the winds may signal an approaching storm.

a. **Wind Direction.** Assess wind direction as a magnetic direction from which the wind is blowing.

b. **Wind Speed.** Assess wind speed in knots.

(1) If an anemometer is available, assess speed to the nearest knot.

(2) If no anemometer is available, estimate the speed in knots. Judge the wind speed by the way objects, such as trees, bushes, tents, and so forth, are blowing.

c. **Visibility in Meters.** Observe the farthest visible major terrain or man-made feature and determine the distance using any available map.

d. **Present Weather.** Include any precipitation or obscuring weather. The following are examples of present weather:

- Rain—continuous and steady liquid precipitation that will last at least one hour.
- Rain showers—short-term and potentially heavy downpours that rarely last more than one hour.
- Snow—continuous and steady frozen precipitation that will last at least one hour.
- Snow showers—short-term and potentially heavy frozen downpours that rarely last more than one hour.
- Fog, haze—obstructs visibility of ground objects.
- Thunderstorms—a potentially dangerous storm. Thunderstorms will produce lightning, heavy downpours, colder temperatures, tornadoes (not too frequently), hail, and strong gusty winds at the surface and aloft. Winds commonly exceed 35 knots.

e. **Total Cloud Cover.** Assess total cloud cover in eighths. Divide the sky into eight different sections measuring from horizon to horizon. Count the sections with cloud cover, which gives the total cloud cover in eighths. (For example, if half of the sections are covered with clouds, total cloud cover is 4/8.)

f. **Ceiling Height.** Estimate where the cloud base intersects elevated terrain. Note if bases are above all terrain. If clouds are not touching terrain, then estimate to the best of your ability.

g. **Temperature.** Assess temperature with or without a thermometer.

(1) With a thermometer, assess temperature in degrees Celsius (use Fahrenheit only if Celsius conversion is not available). To convert Fahrenheit to Celsius: C = F minus 32 times .55. To convert Celsius to Fahrenheit: F = 1.8 times C plus 32.

Example: 41 degrees F – 32 x .55 = 5 degrees C.
5 degrees C x 1.8 + 32 = 41 degrees F.

(2) Without a thermometer, estimate temperature as above or below freezing (0°C), as well as an estimated temperature.

h. **Pressure Trend.** With a barometer or altimeter, assess the pressure trend.

(1) A high pressure moving in will cause altimeters to indicate lower elevation.

(2) A low pressure moving in will cause altimeters to indicate higher elevation.

i. **Observed Weather.** Note changes or trends in observed weather conditions.

(1) Deteriorating trends include:

- Marked wind direction shifts. A high pressure system wind flows clockwise. A low pressure system wind flows counterclockwise. The closer the isometric lines are, the greater the differential of pressure (greater wind speeds).
- Marked wind speed increases.
- Changes in obstructions to visibility.
- Increasing cloud coverage.
- Increase in precipitation. A steady drizzle is usually a long-lasting rain.
- Lowering cloud ceilings.
- Marked cooler temperature changes, which could indicate that a cold front is passing through.
- Marked increase in humidity.
- Decreasing barometric pressure, which indicates a lower pressure system is moving through the area.

(2) Improving trends include:

- Steady wind direction, which indicates no change in weather systems in the area.
- Decreasing wind speeds.
- Clearing of obstructions to visibility.
- Decreasing or ending precipitation.
- Decreasing cloud coverage.
- Increasing height of cloud ceilings.
- Temperature changes slowly warmer.
- Humidity decreases.
- Increasing barometric pressure, which indicates that a higher pressure system is moving through the area.

j. **Update.** Continue to evaluate observed conditions and update the forecast.

Section III. MOUNTAIN HAZARDS

Hazards can be termed natural (caused by natural occurrence), man-made (caused by an individual, such as lack of preparation, carelessness, improper diet, equipment misuse), or as a combination (human trigger). There are two kinds of hazards while in the mountains—subjective and objective. Combinations of objective and subjective hazards are referred to as cumulative hazards.

1-22. SUBJECTIVE HAZARDS

Subjective hazards are created by humans; for example, choice of route, companions, overexertion, dehydration, climbing above one's ability, and poor judgment.

a. **Falling.** Falling can be caused by carelessness, over-fatigue, heavy equipment, bad weather, overestimating ability, a hold breaking away, or other reasons.

b. **Bivouac Site.** Bivouac sites must be protected from rockfall, wind, lightning, avalanche run-out zones, and flooding (especially in gullies). If the possibility of falling exists, rope in, the tent and all equipment may have to be tied down.

c. **Equipment.** Ropes are not total security; they can be cut on a sharp edge or break due to poor maintenance, age, or excessive use. You should always pack emergency and bivouac equipment even if the weather situation, tour, or a short climb is seemingly low of dangers.

1-23. OBJECTIVE HAZARDS

Objective hazards are caused by the mountain and weather and cannot be influenced by man; for example, storms, rockfalls, icefalls, lightning, and so on.

a. **Altitude.** At high altitudes (especially over 6,500 feet), endurance and concentration is reduced. Cut down on smoking and alcohol. Sleep well, acclimatize slowly, stay hydrated, and be aware of signs and symptoms of high-altitude illnesses. Storms can form quickly and lightning can be severe.

b. **Visibility.** Fog, rain, darkness, and or blowing snow can lead to disorientation. Take note of your exact position and plan your route to safety before visibility decreases. Cold combined with fog can cause a thin sheet of ice to form on rocks (verglas). Whiteout conditions can be extremely dangerous. If you must move under these conditions, it is best to rope up. Have the point man move to the end of the rope. The second man will use the first man as an aiming point with the compass. Use a route sketch and march table. If the tactical situation does not require it, plan route so as not to get caught by darkness.

c. **Gullies.** Rock, snow, and debris are channeled down gullies. If ice is in the gully, climbing at night may be better because the warming of the sun will loosen stones and cause rockfalls.

d. **Rockfall.** Blocks and scree at the base of a climb can indicate recurring rockfall. Light colored spots on the wall may indicate impact chips of falling rock. Spring melt or warming by the sun of the rock/ice/snow causes rockfall.

e. **Avalanches.** Avalanches are caused by the weight of the snow overloading the slope. (Refer to paragraph 1-25 for more detailed information on avalanches.)

f. **Hanging Glaciers and Seracs.** Avoid, if at all possible, hanging glaciers and seracs. They will fall without warning regardless of the time of day or time of year. One cubic meter of glacier ice weighs 910 kilograms (about 2,000 pounds). If you must cross these danger areas, do so quickly and keep an interval between each person.

g. **Crevasses.** Crevasses are formed when a glacier flows over a slope and makes a bend, or when a glacier separates from the rock walls that enclose it. A slope of only two to three degrees is enough to form a crevasse. As this slope increases from 25 to 30 degrees, hazardous icefalls can be formed. Likewise, as a glacier makes a bend, it is likely that crevasses will form at the outside of the bend. Therefore, the safest route on a glacier would be to the inside of bends, and away from steep slopes and icefalls. Extreme care must be taken when moving off of or onto the glacier because of the moat that is most likely to be present.

1-24. WEATHER HAZARDS

Weather conditions in the mountains may vary from one location to another as little as 10 kilometers apart. Approaching storms may be hard to spot if masked by local peaks. A clear, sunny day in July could turn into a snowstorm in less than an hour. Always pack some sort of emergency gear.

a. Winds are stronger and more variable in the mountains; as wind doubles in speed, the force quadruples.

b. Precipitation occurs more on the windward side than the leeward side of ranges. This causes more frequent and denser fog on the windward slope.

c. Above approximately 8,000 feet, snow can be expected any time of year in the temperate climates.

d. Air is dryer at higher altitudes, so equipment does not rust as quickly, but dehydration is of greater concern.

e. Lightning is frequent, violent, and normally attracted to high points and prominent features in mountain storms. Signs indicative of thunderstorms are tingling of the skin, hair standing on end, humming of metal objects, crackling, and a bluish light (St. Elmo's fire) on especially prominent metal objects (summit crosses and radio towers).

(1) Avoid peaks, ridges, rock walls, isolated trees, fixed wire installations, cracks that guide water, cracks filled with earth, shallow depressions, shallow overhangs, and rock needles. Seek shelter around dry, clean rock without cracks; in scree fields; or in deep indentations (depressions, caves). Keep at least half a body's length away from a cave wall and opening.

(2) Assume a one-point-of-contact body position. Squat on your haunches or sit on a rucksack or rope. Pull your knees to your chest and keep both feet together. If half way up the rock face, secure yourself with more than one point—lightning can burn through rope. If already rappelling, touch the wall with both feet together and hurry to the next anchor.

f. During and after rain, expect slippery rock and terrain in general and adjust movement accordingly. Expect flash floods in gullies or chimneys. A climber can be washed away or even drowned if caught in a gully during a rainstorm. Be especially alert for falling objects that the rain has loosened.

g. Dangers from impending high winds include frostbite (from increased wind-chill factor), windburn, being blown about (especially while rappelling), and debris being blown about. Wear protective clothing and plan the route to be finished before bad weather arrives.

h. For each 100-meter rise in altitude, the temperature drops approximately one degree Fahrenheit. This can cause hypothermia and frostbite even in summer, especially when combined with wind, rain, and snow. Always wear or pack appropriate clothing.

i. If it is snowing, gullies may contain avalanches or snow sloughs, which may bury the trail. Snowshoes or skis may be needed in autumn or even late spring. Unexpected snowstorms may occur in the summer with accumulations of 12 to 18 inches; however, the snow quickly melts.

j. Higher altitudes provide less filtering effects, which leads to greater ultraviolet (UV) radiation intensity. Cool winds at higher altitudes may mislead one into underestimating the sun's intensity, which can lead to sunburns and other heat injuries. Use sunscreen and wear hat and sunglasses, even if overcast. Drink plenty of fluids.

1-25. AVALANCHE HAZARDS

Avalanches occur when the weight of accumulated snow on a slope exceeds the cohesive forces that hold the snow in place. (Table 1-2, page 1-32, shows an avalanche hazard evaluation checklist.)

a. **Slope Stability.** Slope stability is the key factor in determining the avalanche danger.

(1) ***Slope Angle.*** Slopes as gentle as 15 degrees have avalanched. Most avalanches occur on slopes between 30 and 45 degrees. Slopes above 60 degrees often do not build up significant quantities of snow because they are too steep.

(2) ***Slope Profile.*** Dangerous slab avalanches are more likely to occur on convex slopes, but may occur on concave slopes.

(3) ***Slope Aspect.*** Snow on north facing slopes is more likely to slide in midwinter. South facing slopes are most dangerous in the spring and on sunny, warm days. Slopes on the windward side are generally more stable than leeward slopes.

(4) ***Ground Cover.*** Rough terrain is more stable than smooth terrain. On grassy slopes or scree, the snow pack has little to anchor to.

b. **Triggers.** Various factors trigger avalanches.

(1) ***Temperature.*** When the temperature is extremely low, settlement and adhesion occur slowly. Avalanches that occur during extreme cold weather usually occur during or immediately following a storm. At a temperature just below freezing, the snowpack stabilizes quickly. At temperatures above freezing, especially if temperatures rise quickly, the potential for avalanche is high. Storms with a rise in temperature can deposit dry snow early, which bonds poorly with the heavier snow deposited later. Most avalanches occur during the warmer midday.

(2) ***Precipitation.*** About 90 percent of avalanches occur during or within twenty-four hours after a snowstorm. The rate at which snow falls is important. High rates of snowfall (2.5 centimeters per hour or greater), especially when accompanied by wind, are usually responsible for major periods of avalanche activity. Rain falling on snow will increase its weight and weakens the snowpack.

(3) ***Wind.*** Sustained winds of 15 miles per hour and over transport snow and form wind slabs on the lee side of slopes.

(4) ***Weight.*** Most victims trigger the avalanches that kill them.

(5) ***Vibration.*** Passing helicopters, heavy equipment, explosions, and earth tremors have triggered avalanches.

c. **Snow Pits.** Snow pits can be used to determine slope stability.

(1) Dig the snow pit on the suspect slope or a slope with the same sun and wind conditions. Snow deposits may vary greatly within a few meters due to wind and sun variations. (On at least one occasion, a snow pit dug across the fall line triggered the suspect slope). Dig a 2-meter by 2-meter pit across the fall line, through all the snow, to the ground. Once the pit is complete, smooth the face with a shovel.

(2) Conduct a shovel shear test.

(a) A shovel shear test puts pressure on a representative sample of the snowpack. The core of this test is to isolate a column of the snowpack from three sides. The column should be of similar size to the blade of the shovel. Dig out the sides of the column without pressing against the column with the shovel (this affects the strength). To isolate the rear of the column, use a rope or string to saw from side to side to the base of the column.

(b) If the column remained standing while cutting the rear, place the shovel face down on the top of the column. Tap with varying degrees of strength on the shovel to see what force it takes to create movement on the bed of the column. The surface that eventually slides will be the layer to look at closer. This test provides a better understanding of the snowpack strength. For greater results you will need to do this test in many areas and formulate a scale for the varying methods of tapping the shovel.

(3) Conduct a Rutschblock test. To conduct the test, isolate a column slightly longer than the length of your snowshoes or skis (same method as for the shovel shear test). One person moves on their skis or snowshoes above the block without disturbing the block. Once above, the person carefully places one showshoe or ski onto the block with no body weight for the first stage of the test. The next stage is adding weight to the first leg. Next, place the other foot on the block. If the block is still holding up, squat once, then twice, and so on. The remaining stage is to jump up and land on the block.

d. **Types of Snow Avalanches.** There are two types of snow avalanches: loose snow (point) and slab.

(1) Loose snow avalanches start at one point on the snow cover and grow in the shape of an inverted "V." Although they happen most frequently during the winter snow season, they can occur at any time of the year in the mountains. They often fall as many small sluffs during or shortly after a storm. This process removes snow from steep upper slopes and either stabilizes lower slopes or loads them with additional snow.

(2) Wet loose snow avalanches occur in spring and summer in all mountain ranges. Large avalanches of this type, lubricated and weighed down by meltwater or rain can travel long distances and have tremendous destructive power. Coastal ranges that have high temperatures and frequent rain are the most common areas for this type of avalanche.

(3) Slab avalanches occur when cohesive snow begins to slide on a weak layer. The fracture line where the moving snow breaks away from the snowpack makes this type of avalanche easy to identify. Slab release is rapid. Although any avalanche can kill you, slab avalanches are generally considered more dangerous than loose snow avalanches.

(a) Most slab avalanches occur during or shortly after a storm when slopes are loaded with new snow at a critical rate. The old rule of never travel in avalanche terrain for a few days after a storm still holds true.

(b) As slabs become harder, their behavior becomes more unpredictable; they may allow several people to ski across before releasing. Many experts believe they are susceptible to rapid temperature changes. Packed snow expands and contracts with temperature changes. For normal density, settled snow, a drop in temperature of 10 degrees Celsius (18 degrees Fahrenheit) would cause a snow slope 300 meters wide to contract 2 centimeters. Early ski mountaineers in the Alps noticed that avalanches sometimes occurred when shadows struck a previously sun-warmed slope.

d. **Protective Measures.** Avoiding known or suspected avalanche areas is the easiest method of protection. Other measures include:

(1) ***Personal Safety.*** Remove your hands from ski pole wrist straps. Detach ski runaway cords. Prepare to discard equipment. Put your hood on. Close up your clothing to prepare for hypothermia. Deploy avalanche cord. Make avalanche probes and shovels accessible. Keep your pack on at all times—do not discard. Your pack can act as a flotation device, as well as protect your spine.

(2) ***Group Safety.*** Send one person across the suspect slope at a time with the rest of the group watching. All members of the group should move in the same track from safe zone to safe zone.

e. **Route Selection.** Selecting the correct route will help avoid avalanche prone areas, which is always the best choice. Always allow a wide margin of safety when making your decision.

(1) The safest routes are on ridge tops, slightly on the windward side; the next safest route is out in the valley, far from the bottom of slopes.

(2) Avoid cornices from above or below. Should you encounter a dangerous slope, either climb to the top of the slope or descend to the bottom—well out of the way of the run-out zone. If you must traverse, pick a line where you can traverse downhill as quickly as possible. When you must ascend a dangerous slope, climb to the side of the avalanche path, and not directly up the center.

(3) Take advantage of dense timber, ridges, or rocky outcrops as islands of safety. Use them for lunch and rest stops. Spend as little time as possible on open slopes.

(4) Since most avalanches occur within twenty-four hours of a storm and or at midday, avoid moving during these periods. Moving at night is tactically sound and may be safer.

f. **Stability Analysis.** Look for nature's billboards on slopes similar to the one you are on.

(1) ***Evidence of Avalanching.*** Look for recent avalanches and for signs of wind-loading and wind-slabs.

(2) ***Fracture Lines.*** Avoid any slopes showing cracks.

(3) ***Sounds.*** Beware of hollow sounds—a "whumping" noise. They may suggest a radical settling of the snowpack.

g. **Survival.** People trigger avalanches that bury people. If these people recognized the hazard and chose a different route, they would avoid the avalanche. The following steps should be followed if caught in an avalanche.

(1) Discard equipment. Equipment can injure or burden you; discarded equipment will indicate your position to rescuers.

(2) Swim or roll to stay on tope of the snow. FIGHT FOR YOUR LIFE. Work toward the edge of the avalanche. If you feel your feet touch the ground, give a hard push and try to "pop out" onto the surface.

(3) If your head goes under the snow, shut your mouth, hold your breath, and position your hands and arms to form an air pocket in front of your face. Many avalanche victims suffocate by having their mouths and noses plugged with snow.

(4) When you sense the slowing of the avalanche, you must try your hardest to reach the surface. Several victims have been found quickly because a hand or foot was sticking above the surface.

(5) When the snow comes to rest it sets up like cement and even if you are only partially buried, it may be impossible to dig yourself out. Don't shout unless you hear rescuers immediately above you; in snow, no one can hear you scream. Don't struggle to free yourself—you will only waste energy and oxygen.

(6) Try to relax. If you feel yourself about to pass out, do not fight it. The respiration of an unconscious person is more shallow, their pulse rate declines, and the body temperature is lowered, all of which reduce the amount of oxygen needed. (See Appendix C for information on search and rescue techniques.)

AVALANCHE HAZARD EVALUATION CHECKLIST

Critical Data PARAMETERS: KEY INFORMATION	Hazard Rating G	Y	R
TERRAIN: *Is the terrain capable of producing an avalanche?*			
-Slope angle (steep enough to slide? prime time?)	☐	☐	☐
-Slope aspect (leeward, shadowed, or extremely sunny?)	☐	☐	☐
-Slope configuration (anchoring? shape?)	☐	☐	☐
Overall Terrain Rating:	☐	☐	☐
SNOWPACK: *Could the snow fail?*			
-Slab Configuration (slab? depth and distribution?)	☐	☐	☐
-Bonding Ability (weak layer? tender spots?)	☐	☐	☐
-Sensitivity (how much force to fail? shear tests? clues?)	☐	☐	☐
Overall Snowpack Rating:	☐	☐	☐
Weather: *Is the weather contributing to instability?*			
-Precipitation (type, amount, intensity? added weight?)	☐	☐	☐
-Wind (snow transport? amount and rate of deposition?)	☐	☐	☐
-Temperature (storm trends? effects on snowpack?)	☐	☐	☐
Overall Weather Rating:	☐	☐	☐
Human: *What are your alternatives and their possible consequences?*			
-Attitude (toward life? risk? goals? assumptions?)	☐	☐	☐
-Technical Skill Level (traveling? evaluating aval. hazard?)	☐	☐	☐
-Strength/Equipment (strength? prepared for the worst?)	☐	☐	☐
Overall Human Rating:	☐	☐	☐

Decision/Action:
Overall Hazard Rating/GO or NO Go? **GO ☐ or NOGO ☐**

***HAZARD LEVEL SYMBOLS:**
R = Red light (stop/dangerous)
G = Green light (go/OK)
Y = Yellow light (caution/potentially dangerous).

Table 1-2. Avalanche hazard evaluation checklist.

CHAPTER 2

MOUNTAIN LIVING

Units deploying to high elevations must receive advanced training to survive in the harsh mountain environment. Normal activities (navigation, communications, and movement) require specialized techniques. Training should be conducted as realistically as possible, preferably under severe conditions so the soldier gains confidence. Extended training exercises test support facilities and expose the soldier to the isolation common to mountain operations. Training should reflect the harsh mountain environment and should consider the following:

- *Temperature and altitude extremes.*
- *Hygiene and sanitation.*
- *Limited living space (difficulty of bivouac).*
- *Clothing requirements.*

Section I. SURVIVAL

The soldier trained to fight and survive in a mountain environment will have increased confidence in himself. Training should include: psychological preparation, locating water, shelter considerations, fire building, health hazards, and techniques for obtaining food (see FM 21-76).

2-1. WATER SUPPLY

Mountain water should never be assumed safe for consumption. Training in water discipline should be emphasized to ensure soldiers drink water only from approved sources. Fluids lost through respiration, perspiration, and urination must be replaced if the soldier is to operate efficiently.

a. Maintaining fluid balance is a major problem in mountain operations. The sense of thirst may be dulled by high elevations despite the greater threat of dehydration. Hyperventilation and the cool, dry atmosphere bring about a three- to four-fold increase in water loss by evaporation through the lungs. Hard work and overheating increase the perspiration rate. The soldier must make an effort to drink liquids even when he does not feel thirsty. One quart of water, or the equivalent, should be drunk every four hours; more should be drunk if the unit is conducting rigorous physical activity.

b. Three to six quarts of water each day should be consumed. About 75 percent of the human body is liquid. All chemical activities in the body occur in water solution, which assists in removing toxic wastes and in maintaining an even body temperature. A loss of two quarts of body fluid (2.5 percent of body weight) decreases physical efficiency by 25 percent, and a loss of 12 quarts (15 percent of body weight) is usually fatal. Salt lost by sweating should be replaced in meals to avoid a deficiency and subsequent cramping. Consuming the usual military rations (three meals a day) provides sufficient sodium replacement. Salt tablets are not necessary and may contribute to dehydration.

c. Even when water is plentiful, thirst should be satisfied in increments. Quickly drinking a large volume of water may actually slow the soldier. If he is hot and the water is cold, severe cramping may result. A basic rule is to drink small amounts often. Pure water should always be kept in reserve for first aid use. Emphasis must be placed on the three rules of water discipline:

- Drink only treated water.
- Conserve water for drinking. Potable water in the mountains may be in short supply.
- Do not contaminate or pollute water sources.

d. Snow, mountain streams, springs, rain, and lakes provide good sources of water supply. Purification must be accomplished, however, no matter how clear the snow or water appears. Fruits, juices, and powdered beverages may supplement and encourage water intake (do not add these until the water has been treated since the purification tablets may not work). Soldiers cannot adjust permanently to a decreased water intake. If the water supply is insufficient, physical activity must be reduced. Any temporary deficiency should be replaced to maintain maximum performance.

e. All water that is to be consumed must be potable. Drinking water must be taken only from approved sources or purified to avoid disease or the possible use of polluted water. Melting snow into water requires an increased amount of fuel and should be planned accordingly. Nonpotable water must not be mistaken for drinking water. Water that is unfit to drink, but otherwise not dangerous, may be used for other purposes such as bathing. Soldiers must be trained to avoid wasting water. External cooling (pouring water over the head and chest) is a waste of water and an inefficient means of cooling. Drinking water often is the best way to maintain a cool and functioning body.

f. Water is scarce above the timberline. After setting up a perimeter (patrol base, assembly area, defense), a watering party should be employed. After sundown, high mountain areas freeze, and snow and ice may be available for melting to provide water. In areas where water trickles off rocks, a shallow reservoir may be dug to collect water (after the sediment settles). Water should be treated with purification tablets (iodine tablets or calcium hypochlorite), or by boiling at least one to two minutes. Filtering with commercial water purification pumps can also be conducted. Solar stills may be erected if time and sunlight conditions permit (see FM 21-76). Water should be protected from freezing by storing it next to a soldier or by placing it in a sleeping bag at night. Water should be collected at midday when the sun thaw available.

2-2. NUTRITION

Success in mountain operations depends on proper nutrition. Because higher altitudes affect eating habits, precautions must be taken. If possible, at least one hot meal each day should be eaten, which may require personnel to heat their individual rations.

a. The following elements are characteristic of nutritional acclimatization in mountain operations:

- Weight loss during the first two to three days at high elevation.
- A loss of appetite with symptoms of mountain sickness.
- Loss of weight usually stops with acclimatization.
- At progressively higher elevations (greater than 14,000 feet), the tolerance of fatty/high-protein foods rapidly decreases. A high carbohydrate diet may lessen the symptoms of acute mountain sickness and is digested better than fat at high altitudes.

b. Increased fatigue may cause soldiers to become disinterested in eating properly. Decreased consumption may result in malnutrition because of the unpleasant taste of cold rations. Leaders should ensure that fuel tablets and squad stoves are available, or that natural flammable materials are used if possible. Although there is no physiological need for hot food, it does increase morale and a sense of well being. Loss of weight in the first few days occurs because of dehydration, metabolic changes, and loss of appetite. Carbohydrate-containing beverages, such as fruit juices and sports drinks, are an effective means of increasing carbohydrates, energy, and liquid intake when the normal appetite response is blunted at altitude.

c. Three major food components are required to maintain a well-functioning body: proteins, fats, and carbohydrates. These food components provide energy, amino acids, vitamins, fiber, and minerals. All three components must be provided in the correct proportions to maintain a healthy body.

(1) ***Protein.*** Proteins consist of a large number of amino acid units that are linked together to form the protein. The amino acids, resulting from digestion of protein, are absorbed through the intestine into the blood, and are used to make or replace body proteins (muscle and body tissue). Sources of readily useable animal proteins include eggs, milk, cheese, poultry, fish, and meats. Other foods such as cereals, vegetables, and legumes also provide amino acids. These proteins are not as balanced in essential amino acid composition as meat, eggs, or milk proteins. The minimum daily protein requirement, regardless of physical activity, is 8 ounces for a 154-pound man. Since amino acids are either oxidized for energy or stored as fats, consuming excess protein is inefficient and may increase the water intake needed for urea nitrogen excretion. Protein requires water for digestion and may facilitate dehydration. Proteins provide the body about four kilocalories of energy per gram and require the most energy for the body to digest.

(2) ***Fats.*** Fats are the most concentrated form of food energy. Of the total daily caloric intake, 25 to 30 percent may be supplied as fats. Main sources of fats are meats, nuts, butter, eggs, milk, and cheese. Fats require more water and oxygen, and are harder to digest at higher altitudes. Fats are the body's natural stored source of energy. Fats provide the body around 9 kilocalories of energy per gram and require less energy for the body to digest than protein but more than carbohydrates.

(3) ***Carbohydrates.*** Carbohydrates are an important source of calories. In the form of glucose, carbohydrates are found in the most important energy-producing cycles in the body's cells. If carbohydrate intake exceeds energy needs, moderate amounts are stored in the muscles and liver. Larger amounts are converted into fats and stored in that form. Carbohydrates should compose up to 50 percent of the total daily caloric intake. Nutritionally, the most useful sources of carbohydrates are foods such as unrefined grains, vegetables, and fruit. Carbohydrates provide the body around four kilocalories of energy per gram and are the easiest to digest.

(4) ***Vitamins.*** Vitamins are classified into two groups on the basis of their ability to dissolve in fat or water. The fat-soluble vitamins include vitamins A, D, E, and K. The water-soluble vitamins include the B vitamins and vitamin C, which are found in cereals, vegetables, fruits, and meats. A well-balanced diet provides all of the required vitamins. Since most water-soluble vitamins are not stored, a proper diet is necessary to ensure adequate levels of these vitamins. If an improper and unbalanced diet is likely to occur

during a deployment, vitamin supplements should be considered, especially if this period is to exceed 10 days.

(5) ***Minerals.*** Mineral elements can be divided into two groups: those needed in the diet in amounts of 100 milligrams or more a day such as calcium, phosphorous, and magnesium; and trace elements needed in amounts of only a few milligrams a day such as iodine, iron, and zinc. Required minerals are contained in a balanced diet (meats, vegetables, fruits).

d. Eating a balanced diet provides the energy needed to conduct daily activities and to maintain the internal body processes. A balanced diet containing adequate amounts of vitamins and minerals ensures an efficient metabolism. Since climbing is a strenuous activity and demands high-energy use, a balanced diet is a necessity.

(1) The efficiency of the body to work above the basal metabolism varies from 20 to 40 percent, depending on the soldier. Over 50 percent of caloric intake is released as heat and is not available when the soldier works. (About 4,500 calories are expended for strenuous work and 3,500 calories for garrison activity.) Heat is a by-product of exertion. Exertion causes excessive bodily heat loss through perspiration and increased radiation. During inactivity in cold weather, the metabolism may not provide enough heat. The "internal thermostat" initiates and causes the muscles to shiver, thus releasing heat. Shivering also requires energy and burns up to 220 calories per hour (estimated for a 100-pound man).

(2) With an abrupt ascent to high altitudes, the soldier experiences physiological acclimatization. The circulatory system labors to provide the needed oxygen to the body. Large meals require the digestive system to work harder than usual to assimilate food. Large meals may be accompanied by indigestion, shortness of breath, cramps, and illness. Therefore, relatively light meals that are high in carbohydrates are best while acclimatizing at higher elevations. Personnel should eat moderately and rest before strenuous physical activity. Since fats and protein are harder to digest, less digestive disturbances may occur if meals are eaten before resting. A diet high in carbohydrates is not as dense in energy and may require eating more often. Carbohydrates, beginning in the morning and continuing through mid-afternoon, are important in maintaining energy levels.

(3) Extra food should be carried in case resupply operations fail. Food should be lightweight and easy to digest, and be eaten hot or cold. Meals-ready-to-eat (MREs) meet these criteria and provide all of the basic food groups. Commanders may consider supplementing MREs with breakfast bars, fruits, juices, candies, cereal bars, and chocolate. Bouillon cubes can replace water and salt as well as warming cold bodies and stimulating the appetite. Hot beverages of soup, juices, powdered milk, and cider should also be considered. Since coffee, tea, and hot chocolate are diuretics, the consumption of these beverages should not be relied upon for hydration.

(4) Warm meals should be provided when possible. When cooking, the heat source must be kept away from equipment and ammunition. At higher elevations, the cooking time may be doubled. To conserve fuel, stoves, fires, and fuel tablets should be protected from the wind. Extra fuel should be stored in tightly sealed, marked, metal containers. Use stoves and heat tabs for warming food and boiling water. Canteen cups and utensils should be cleaned after use. All food items and garbage are carried with the unit. If possible, garbage should be burned or deep buried. Caution must be taken to prevent animals from foraging through rucksacks, ahkios, and burial sites. As all missions are tactical, no trace of a unit should be detected.

(5) Certain drugs, medications, alcohol, and smoking have adverse effects on the circulation, perspiration, hydration, and judgment of soldiers. Therefore, they should be avoided when operating in extremely cold conditions or at high altitudes.

2-3. PERSONAL HYGIENE AND SANITATION

The principles of personal hygiene and sanitation that govern operations on low terrain also apply in the mountains. Commanders must conduct frequent inspections to ensure that personal habits of hygiene are not neglected. Standards must be maintained as a deterrent to disease, and as reinforcement to discipline and morale.

a. **Personal Hygiene.** This is especially important in the high mountains, mainly during periods of cold weather. In freezing weather, the soldier may neglect washing due to the cold temperatures and scarcity of water. This can result in skin infections and vermin infestation. If bathing is difficult for any extended period, the soldier should examine his skin and clean it often. Snow baths in lieu of a water bath are recommended. This helps reduce skin infections and aids the comfort of the soldier.

(1) Snow may be used instead of toilet paper. Soldiers should shave at rest periods in the shelter so that oils stripped in shaving will be replenished. A beard may mask the presence of frostbite or lice. Water-based creams and lotions should be avoided in cold environments since this will further dehydrate tissues and induce frostbite by freezing. The nonwater-based creams can be used for shaving in lieu of soap. Sunscreens and chap sticks should be used on lips, nose, and eyelids. Topical steroid ointments should be carried for rashes. The teeth must also be cleaned to avoid diseases of the teeth and gums. Underwear should be changed when possible, but this should not be considered a substitute for bathing. When operating in areas where resupply is not possible, each soldier should carry a complete change of clothing. If laundering of clothing is difficult, clothes should be shaken and air-dried. Sleeping bags must be regularly cleaned and aired.

(2) The principles of foot hygiene must be followed to protect the feet from cold injuries. The causes of such injuries are present throughout the year in high mountains. Boots should be laced tightly when climbing to provide needed support but not so tight as to constrict circulation. Socks should be worn with no wrinkles since this causes blisters on the feet. Feet should be washed daily, and kept as dry and clean as possible. If regular foot washing is impossible, socks should be changed often (at halts and rest periods or at least once a day) and feet massaged, dried, and sprinkled with foot powder. Talc or antifungal powder should be used when massaging; excess powder is brushed off to avoid clumping, which may cause blisters. Feet can be cleaned with snow, but must be quickly dried. Whenever changing socks, soldiers should closely examine their feet for wrinkles, cracks, blisters, and discoloration. Nails should be trimmed but not too short. Long nails wear out socks; short nails do not provide proper support for the ends of the toes. Medical attention should be sought for any possible problems.

(3) Feet should be sprayed two or three times a day with an aluminum chlorohydrate antiperspirant for a week and then once a day for the rest of the winter. If fissures or cracks occur in the feet, it is best to discontinue spraying until they are healed or to spray less often to control sweating. This process stops about 70 percent of the sweating in the feet.

(4) During periods of extreme cold, there is a tendency for the soldier to become constipated. This condition is brought about by the desire to avoid the inconvenience and

discomfort of defecating. Adequate water intake plus a low protein, high roughage diet can be helpful in preventing constipation.

b. **Sanitation.** In rocky or frozen ground, digging latrines is usually difficult. If latrines are constructed, they should be located downwind from the position and buried after use. In tactical situations, the soldier in a designated, downwind location away from water sources may dig "cat holes." Since waste freezes, it can be covered with snow and ice or pushed down a crevasse. In rocky areas above the timberline, waste may be covered with stones.

Section II. ACCLIMATIZATION AND CONDITIONING

Terrestrial altitude can be classified into five categories. Low altitude is sea level to 5,000 feet. Here, arterial blood is 96 percent saturated with oxygen in most people. Moderate altitude is from 5,000 to 8,000 feet. At these altitudes, arterial blood is greater than 92 percent saturated with oxygen, and effects of altitude are mild and temporary. High altitude extends from 8,000 to 14,000 feet, where arterial blood oxygen saturation ranges from 92 percent down to 80 percent. Altitude illness is common here. Very high altitude is the region from 14,000 to 18,000 feet, where altitude illness is the rule. Areas above 18,000 feet are considered extreme altitudes.

Soldiers deployed to high mountainous elevations require a period of acclimatization before undertaking extensive military operations. The expectation that freshly deployed, unacclimatized troops can go immediately into action is unrealistic, and could be disastrous if the opposing force is acclimatized. Even the physically fit soldier experiences physiological and psychological degradation when thrust into high elevations. Time must be allocated for acclimatization, conditioning, and training of soldiers. Training in mountains of low or medium elevation (5,000 to 8,000 feet) does not require special conditioning and acclimatization procedures. However, some soldiers will have some impairment of operating efficiency at these low altitudes. Above 8,000 feet (high elevation), most unacclimatized soldiers may display some altitude effects. Training should be conducted at progressively higher altitudes, starting at about 8,000 feet and ending at 14,000 feet. Attempts to acclimatize beyond 17,000 feet results in a degradation of the body greater than the benefits gained. The indigenous populations can out-perform even the most acclimatized and physically fit soldier who is brought to this altitude; therefore, employment of the local population may be advantageous.

2-4. SYMPTOMS AND ADJUSTMENTS

A person is said to be acclimatized to high elevations when he can effectively perform physically and mentally. The acclimatization process begins immediately upon arrival at the higher elevation. If the change in elevation is large and abrupt, some soldiers can suffer from acute mountain sickness (AMS), high-altitude pulmonary edema (HAPE), or high-altitude cerebral edema (HACE). Disappearance of the symptoms of acute mountain sickness (from four to seven days) does not indicate complete acclimatization. The process of adjustment continues for weeks or months. The altitude at which complete acclimatization is possible is not a set point but for most soldiers with proper ascent, nutrition and physical activity it is about 14,000 feet.

a. Immediately upon arrival at high elevations, only minimal physical work can be performed because of physiological changes. The incidence and severity of AMS symptoms vary with initial altitude, the rate of ascent, and the level of exertion and individual susceptibility. Ten to twenty percent of soldiers who ascend rapidly (in less than 24 hours) to altitudes up to 6,000 feet experience some mild symptoms. Rapid ascent to 10,000 feet causes mild symptoms in 75 percent of personnel. Rapid ascent to elevations of 12,000 to 14,000 feet will result in moderate symptoms in over 50 percent of the soldiers and 12 to 18 percent may have severe symptoms. Rapid ascent to 17,500 feet causes severe, incapacitating symptoms in almost all individuals. Vigorous activity during ascent or within the first 24 hours after ascent will increase both the incidence and severity of symptoms. Some of the behavioral effects that will be encountered in unacclimatized personnel include:

- Increased errors in performing simple mental tasks.
- Decreased ability for sustained concentration.
- Deterioration of memory.
- Decreased vigilance or lethargy.
- Increased irritability in some individuals.
- Impairment of night vision and some constriction in peripheral vision (up to 30 percent at 6,000 feet).
- Loss of appetite.
- Sleep disturbances.
- Irregular breathing.
- Slurred speech.
- Headache.

b. Judgment and self-evaluation are impaired the same as a person who is intoxicated. During the first few days at a high altitude, leaders have extreme difficulty in maintaining a coordinated, operational unit. The roughness of the terrain and the harshness and variability of the weather add to the problems of unacclimatized personnel. Although strong motivation may succeed in overcoming some of the physical handicaps imposed by the environment, the total impact still results in errors of judgment. When a soldier cannot walk a straight line and has a loss of balance, or he suffers from an incapacitating headache, he should be evacuated to a lower altitude (a descent of at least 1,000 feet for at least 24 hours).

2-5. PHYSICAL AND PSYCHOLOGICAL CONDITIONING

The commander must develop a conditioning/training program to bring his unit to a level where it can operate successfully in mountain conditions. Priorities of training must be established. As with all military operations, training is a major influence on the success of mountain operations.

a. U.S. forces do not routinely train in mountainous terrain. Therefore, extensive preparations are needed to ensure individual and unit effectiveness. Units must be physically and psychologically conditioned and adjusted before undertaking rigorous mountain operations. Units must be conditioned and trained as a team to cope with the terrain, environment, and enemy situation. Certain factors must be considered:

- What are the climatic and terrain conditions of the area of operations?
- How much time is available for conditioning and training?
- Will the unit conduct operations with other U.S. or Allied forces? Are there language barriers? What assistance will be required? Will training and conditioning be required for attached personnel?

- What additional personnel will accompany the unit? Will they be available for training and conditioning?
- What is the current level of physical fitness of the unit?
- What is the current level of individual expertise in mountaineering?
- What type of operations can be expected?
- What is the composition of the advance party? Will they be available to assist in training and acclimatization?
- What areas in the U.S. most closely resemble the area of operations?
- Are predeployment areas and ranges available?
- Does the unit have instructors qualified in mountain warfare?
- What type equipment will be required (to fit the season, mission, terrain)?
- Does the unit have enough of the required equipment? Do personnel know how to use the equipment? Will the equipment go with the advance party, with the unit, or follow after the unit's arrival?
- Does equipment require modification?
- Do weapons and equipment require special maintenance?

b. When the unit arrives in the area of operations, all personnel require a period of conditioning and acclimatization. The time schedule should allow for longer and more frequent periods of rest. The rigors of establishing an assembly area exhaust most unacclimatized personnel. Water, food, and rest must be considered as priorities, ensuring sufficient amounts while individual metabolisms and bodies become accustomed to functioning at higher elevations.

c. Since the acclimatization process cannot be shortened, and the absence of acclimatization hampers the successful execution of operations, deployment to higher elevations must consider the following:

(1) Above 8,000 feet, a unit should ascend at a rate of 1,000 to 2,000 feet per day. Units can leapfrog, taking an extended rest period.

(2) Units should not resort to the use of pharmaceutical pretreatment with carbonic anhydrase inhibitors such as acetazolamide (Diamox). These drugs have side effects that mimic the signs and symptoms of AMS. Inexperienced medics may have difficulty recognizing the differences between the side effects of the drug and a condition that could possibly be life threatening. Additionally, these drugs are diuretics, which results in higher hydration levels (at least 25 percent increase per man per day). These higher hydration levels create a larger logistical demand on the unit by requiring more water, time to acquire water, water purification supplies, and, if in a winter environment, fuels for melting snow and ice for water.

(3) Carbonic anhydrase inhibitors such as acetazolamide are effective in the treatment of mild and severe AMS. These drugs should accompany attached medical personnel because they can treat the soldier suffering the symptoms of AMS and, although rest may be required evacuation may not be needed.

(4) Do not move troops directly to high altitudes even if allowances can be made for inactivity for the first three to five days before mission commitment. Moving troops directly to high altitude can increase the probability of altitude sickness. Even if inactivity follows deployment, the incidence of altitude sickness is more likely than with a gradual ascent.

d. Training on high-altitude effects can prevent psychological preconceptions. Soldiers who have lived on flat terrain may have difficulty when learning to negotiate steep slopes or cliffs, developing a sense of insecurity and fear. They must be slowly introduced to the new terrain and encouraged to develop the confidence required to negotiate obstacles with assurance and ease. They must be taught the many climbing techniques and principles of mountain movement. They overcome their fear of heights by becoming familiar with the problem. The soldier cannot be forced to disregard this fear.

e. Regardless of previous training and the amount of flat cross-country movement practice, the untrained soldier finds mountain movement hard and tiring. A different group of muscles are used, which must be developed and hardened. A new technique of rhythmic movement must be learned. Such conditioning is attained through frequent marches and climbs, while carrying TOE and special equipment loads. This conditions the back and legs, which results in increased ability and endurance. At the same time, the men acquire confidence and ability to safely negotiate the terrain. The better the physical condition of the soldier, the better the chance of avoiding exhaustion. Proper physical conditioning ensures the soldier is an asset and not a liability. The body improves its capacity for exercise, the metabolism becomes more efficient, and blood and oxygen flow quickly and effectively.

f. A physical fitness training program that gradually increases in difficulty should include marches, climbing, and calisthenics. This increases the soldier's endurance. Through a sustained high level of muscular exertion, the soldier's capacity for exertion is increased. Physical conditioning should include long-distance running for aerobic conditioning; calisthenics and weight training to strengthen the heart, lungs, abdomen, legs, back, arms, and hands; a swimming program to increase lung efficiency; and road marches over mountainous terrain with all combat equipment. Upon deploying to high elevations, caution must be exercised by units that are in superior physical condition. The heart rate, metabolism, and lungs must become accustomed to the elevation and thinner air. A conditioning program must be set up on site and integrated in gradual stages where acclimatization, conditioning, and mountaineering skills are realized.

g. Conditioning should begin with basic climbing. It is equally important to instill the will to climb. Confidence goes hand in hand with physical conditioning and skill development. Repetitive practice, to the point of instinctive reaction, is key to learning and maintaining climbing proficiency and technical skills. There are no quick and easy methods to becoming acclimatized and conditioned. Training should gradually challenge the soldier over an extended period and reinforce learning skills.

Section III. MEDICAL CONSIDERATIONS

Improper acclimatization poses many problems for medical personnel. Facilities and supplies may be inadequate to treat all victims. After acclimatization, personnel can still become injured (sprains, strains, fractures, frostbite, hypothermia, and trench foot). Mountain sickness and other illnesses may also occur. Evacuation of the sick and wounded is compounded by the terrain and weather.

2-6. ILLNESS AND INJURY

Units operating in mountainous regions are exposed to varied types of injuries and illnesses not associated with other areas. Medical considerations are like those for other environments; however, there are some unique aspects of mountain operations to be considered if effective support is to be provided. Most injuries in the mountain environment are soft tissue injuries. These include sprains, strains, abrasions, contusions and fractures. As with any other injuries, the most life threatening are treated first with the emphasis on airway control, breathing management, and circulatory support. Skills in basic first aid are essential to the mountain leader and should be reinforced with regular sustainment training.

2-7. TREATMENT AND EVACUATION

In harsh mountain weather, the most important course of action is to provide injured soldiers with medical aid as soon as possible. Immediate first aid is given on site. Due to rough terrain, medical units can seldom reach unit aid stations by vehicle to evacuate casualties. Litter bearers are required to move casualties to the rear where they can be evacuated by ground or air to clearing stations. The victim is protected from the weather and shock during transportation. Rendezvous points are coordinated with medical units as far forward as possible. Training must be accomplished with all litter bearers on evacuation techniques and first aid. Lightly wounded personnel may need assistance to move over rough terrain.

2-8. SOLAR INJURIES

Solar injuries can happen in warm weather or in cold weather. These types of injuries can be just as incapacitating as most other injuries but usually are not fatal. The peak hours of ultraviolet (UV) radiation are between the hours of 1100 and 1500. Due to the long wavelengths of ultraviolet light, cloudy days can be more dangerous than sunny days. On sunny days the soldier takes more care due to the bright conditions. On cloudy days the soldier tends not to wear sunglasses or sunscreen.

a. **Sunburn.** Sunburn is the burning of exposed skin surfaces by ultraviolet radiation.

(1) Contributing factors include fair skin, improper use of para-amino benzoic acid (PABA)-based sunscreens, and exposure to intense ultraviolet rays for extended periods.

(2) Symptoms of sunburn are painful, burning, red or blistered skin with a slight swelling. The skin may be warm to the touch. In severe cases chills, fever, and headaches may occur.

(3) To treat sunburn, apply cool saline dressings to alleviate pain and swelling. Do not pop blisters. If blisters do break, wash thoroughly, bandage, and seek medical attention. A solution of vinegar (acetic) and water can be lightly applied with sterile gauze to alleviate burning. The tannic acid in used tea bags can also be applied to alleviate burning. Administer pain medication if needed.

(4) To prevent sunburn, skin should be covered with clothing or PABA-based sunscreens (at least sun protection factor [SPF] 15) should be applied liberally to exposed skin during the peak hours of UV exposure. The SPF means that you can stay exposed to the suns UV rays that many times longer than without it. (For example, an SPF of 15 means that skin can be exposed to UV rays 15 times longer than without sunscreen.) During sustained activity, the sunscreen should be regularly reapplied to maintain the SPF.

b. **Snowblindness.** Snowblindness is sunburn of the cornea of the eye caused by exposure to ultraviolet radiation.

(1) A contributing factor is the reflection of sunlight from all directions off the snow, ice, and water. Ultraviolet rays can cause vision problems even on cloudy days. They are less filtered at high altitudes than at low altitudes.

(2) Symptoms of snowblindness are painful, red, watery eyes; a gritty feeling; blurred vision; and a headache.

(3) To treat snowblindness, patch both eyes with cold compresses for 24 hours. Topical anesthetics such as Tetracaine Ophthalmic can be used to relieve pain. Avoid rubbing the eyes. If still painful, keep the victim's eyes patched and administer oral pain medication. Snowblindness will usually resolve in about 24 hours for mild to moderate cases. Victims are rarely in need of evacuation unless the case is unusually severe.

(4) To prevent snow blindness, use quality sunglasses even on cloudy days in snow-covered terrain. Proper sunglasses should provide 100 percent UVA and UVB protection and have hoods on the sides to prevent reflected light from entering the eye. (Currently, the U.S. Army does not have these types of "glacier" sunglasses in their inventory and they must be acquired from nonmilitary sources.) In an emergency, improvise slit glasses from materials such as cardboard or birch bark.

2-9. COLD-WEATHER INJURIES

Cold-weather injuries can occur during any season of the year. Death has resulted in temperatures as high as 10 degrees Celsius (50 degrees Fahrenheit). A loss of body heat combined with shock produces devastating results. However, most of these accidents can be prevented by proper planning to include: timely requisition and receipt of supplies and proper clothing; thorough training of personnel with respect to the hazards of cold weather; effective methods for the receipt, dissemination, and use of cold-weather data; periodic inspections of clothing, personnel, and equipment; and personnel receiving a balance of water, rest, and nutrition.

a. Soldiers must be prepared to survive, move, and fight in winter conditions. Intense cold affects the mind as well as the body. Simple tasks take longer to perform, and they take more effort than in a temperate climate. When weather conditions become extreme the problems of survival become more significant. Warmth and comfort become the top priorities. The effects of extreme cold and the probability of injury are magnified due to the lack of proper diet and sleep. The most important measure in the prevention of cold-weather injuries is the education of personnel and their leaders.

b. Cold injuries may be divided into two types: freezing and nonfreezing. The freezing type is known as frostbite. The nonfreezing type includes hypothermia, dehydration, and immersion foot. Cold injuries result from impaired circulation and the action of ice formation and cold upon the tissues of the body. Temperature alone is not a reliable guide as to whether a cold injury can occur. Low temperatures are needed for cold injuries to occur, but freezing temperatures are not. Wind speed can accelerate body heat loss under both wet and cold conditions. All commanders and subordinate leaders/instructors must be familiar with and carry GTA 5-8-12, which includes a wind chill equivalent temperature chart (Figure 2-1, page 2-12).

WIND CHILL FACTOR CHART												
COOLING POWER OF WIND EXPRESSED AS AN EQUIVALENT CHILL TEMPERATURE (UNDER CALM CONDITIONS)												
ESTIMATED WIND SPEED (IN MPH)	**ACTUAL THERMOMETER READING (F)**											
	50	40	30	20	10	0	-10	-20	-30	-40	-50	-60
	EQUIVALENT TEMPERATURES (F)											
Calm	50	40	30	20	10	0	-10	-20	-30	-40	-50	-60
5	48	37	27	16	6	-5	-15	-26	-36	-47	-57	-68
10	40	28	16	4	-9	-24	-33	-46	-58	-70	-83	-95
15	36	22	9	-5	-18	-32	-45	-58	-72	-85	-99	-112
20	32	18	4	-10	-25	-39	-53	-67	-82	-96	-110	-124
25	30	16	0	-15	-29	-44	-59	-74	-88	-104	-118	-133
30	28	13	-2	-18	-33	-48	-63	-79	-94	-109	-125	-140
35	27	11	-4	-21	-35	-51	-67	-82	-98	-113	-129	-145
40	26	10	-6	-21	-37	-53	-69	-85	-100	-116	-132	-148
Winds greater than 40 MPH have little additional effect.	**LITTLE DANGER** (For properly clothed person) Maximum danger of false sense of security.				**INCREASING DANGER** Danger from freezing of exposed flesh.			**GREAT DANGER**				
	Trench foot and immersion foot may occur at any point on this chart.											

Figure 2-1. Wind chill chart.

c. Many other factors in various combinations determine if cold injuries will occur.

(1) ***Previous Cold Injuries.*** If a soldier has had a cold injury before, he is at higher risk for subsequent cold injuries.

(2) ***Race.*** Blacks are more susceptible to cold-weather injuries than Caucasians.

(3) ***Geographic Origin.*** Personnel from warmer climates are more susceptible to cold injury than those from colder climates.

(4) ***Ambient Temperature.*** The temperature of the air (or water) surrounding the body is critical to heat regulation. For example, the body uses more heat to maintain the temperature of the skin when the temperature of the surrounding air is 37 degrees Fahrenheit than when it is 50 degrees Fahrenheit.

(5) ***Wind Chill Factor.*** The commander should know the wind chill factor. When the forecast gives a figure that falls within the increased danger zone or beyond, caution must be taken to minimize cold injury. The equivalent wind chill temperature is especially important when the ambient temperature is 0 degrees Celsius (32 degrees Fahrenheit) or less. Tissue can freeze if exposed for a prolonged period and if frequent warming is not practiced. The lower the wind chill, the faster tissue freezing can occur. Wind chill is the rate of cooling. Wind does not lower the ambient temperature. The ambient temperature alone determines freezing or nonfreezing injuries. Frostbite Wind chill may cause faster cooling due to increased convection, but not below the ambient temperature.

(6) ***Type of Mission.*** Combat action requiring prolonged immobility and long hours of exposure to low temperatures, or not having an opportunity to warm up increases the possibility of cold injuries.

(7) ***Terrain.*** Minimal cover and wet conditions increase the potential for cold injury.

(8) ***Clothing.*** Clothing for cold weather should be worn with the acronym **C.O.L.D.** in mind.

- **C**—Clothing should be clean since prolonged wear reduces its air-trapping abilities and clogs air spaces with dirt and body oils.
- **O**—Overheating. Avoid overheating. Appropriate measures should be taken when a change in weather or activity alters the amount of clothing needed to prevent overheating and, therefore, accumulation of perspiration.
- **L**—Loose and in layers (to trap air and to conserve body heat). The uniform should be worn completely and correctly to avoid injury to exposed body surfaces. The cold-weather uniform is complete when worn with gloves and inserts.
- **D**—Dry. Keep dry. Wet clothing loses insulation value.

(9) ***Moisture.*** Water conducts heat more rapidly than air (25 percent). When the skin or clothing becomes damp or wet, the risk of cold injury is greatly increased.

(10) ***Dehydration.*** The most overlooked factor causing cold injuries is dehydration. Individuals must retain their body fluids. In cold weather the human body needs special care, and the consumption of water is important to retain proper hydration.

(11) ***Age.*** Within the usual age range of combat personnel, age is not a significant factor.

(12) ***Fatigue.*** Mental weariness may cause apathy leading to neglect of duties vital to survival.

(13) ***Concomitant Injury.*** Injuries resulting in shock or blood loss reduce blood flow to extremities and may cause the injured individual to be susceptible to cold injury, which in turn can accelerate shock.

(14) ***Discipline, Training, and Experience.*** Well-trained and disciplined soldiers suffer less than others from the cold.

(15) ***Nutrition.*** Good nutrition is essential for providing the body with fuel to produce heat in cold weather. The number of calories consumed normally increases as the temperature becomes colder.

(16) ***Excess Activity.*** Excess activity (overheating) results in loss of large amounts of body heat by perspiration. This loss of body heat combined with the loss of insulation value provided by the clothing (due to perspiration dampening the clothing) can subject a soldier to cold injuries.

(17) ***Radical Changes in the Weather.*** Weather conditions in mountainous terrain are known to change considerably throughout the day. Weather can quickly change to extremely cold and wet conditions, especially in higher elevations.

d. Commanders should ensure that the following measures are taken.

(1) Soldiers' uniforms are kept as dry as possible and are protected from the elements.

(2) Soldiers are educated on proper use of clothing systems to avoid the effects of overheating and perspiration (layer dressing and ventilate).

(3) The buddy system is used to watch for early signs of cold-weather injuries.

(4) All soldiers waterproof their equipment.

(5) The rate of movement should be slow, deliberate, and careful. Soldiers should not move out at a force march pace and then be stationary after they have perspired heavily. Soldiers should not wear excessive cold-weather clothing while moving.

e. Medical procedures are needed when sickness and injuries occur. Leaders should—

- Assess the situation (tactical and environmental).
- Approach the victim safely (avoid rock or snow slide).
- Perform emergency first aid.
- Treat for shock (always assume that shock is present).
- Check for other injuries/cold injuries.
- Develop a course of action (decide on a means of evacuation).
- Execute the plan and monitor the victim's condition.

f. Body heat may be lost through radiation, conduction, convection, or evaporation.

(1) ***Radiation.*** The direct heat loss from the body to its surrounding atmosphere is called radiation heat loss. The head can radiate up to 80 percent of the total body heat output. On cold days, personnel must keep all extremities covered to retain heat. This accounts for the largest amount of heat lost from the body.

(2) ***Conduction.*** Conduction is the direct transfer of heat from one object in contact with another (being rained on or sitting in snow).

(3) ***Convection.*** Convection is the loss of heat due to moving air or water in contact with the skin. Wind chill is convection cooling. Clothing that ventilates, insulates, and protects must control the layer of warm air next to the skin.

(4) ***Evaporation.*** The evaporation of perspiration causes heat loss. Wet clothing can cause heat loss by conduction and evaporation. Dressing in layers allows soldiers to remove or add clothing as needed.

g. Some of the most common cold-weather injuries are described in the following paragraphs.

(1) ***Shock.*** Shock is the depressed state of vital organs due to the cardiovascular (heart) system not providing enough blood. Although shock is not a cold-weather injury, it is a symptom or a result of other injuries. Any illness or injury can produce shock, which increases the instance and severity of a cold-weather injury. Shock should be assumed in all injuries and treated accordingly. Even minor injuries can produce shock due to cold, pain, fear, and loss of blood.

(a) *Symptoms.* Initial symptoms of shock include apprehension, shortness of breath, sweating, cold skin, rapid and faint pulse, and excessive thirst. If the victim is not given adequate first aid immediately, his condition may digress into incoherence, slower heart beat, unconsciousness, and possibly death.

(b) *Treatment.* To treat shock, restore breathing and heart rate through artificial respiration or cardiopulmonary resuscitation. Treat the injury and control hemorrhaging. Make the victim as comfortable as possible and try to relieve the pain. Keep the victim warm but do not overheat him. Elevate the back and head, or feet. If the victim is conscious and has no abdominal injuries, administer water. The victim should receive proper medical attention as soon as possible.

(2) ***Dehydration.*** Dehydration is the loss of body fluids to the point that normal body functions are prevented or slowed. This is usually caused by overexertion and improper water intake. Dehydration precedes all cold-weather injuries and is a major symptom in acute mountain sickness. It contributes to poor performance in all physical activities—even more so than lack of food. Cold weather requirements for water are no different than in the desert. They may, in fact, exceed desert requirements because of the increased difficulty in moving with extra clothing and through the snow. At high altitudes, the air is dry. Combined

with a rapid rate of breathing, as much as two liters of liquid may be lost each day through respiration. A soldier needs about three to six quarts of water each day to prevent dehydration when living and performing physical labor in a cold or mountainous environment. Coffee and tea are diuretics and cause excessive urination and should be avoided. The adequacy of liquid intake can best be judged by the urine color and volume. Dark amber colored urine instead of light yellow or the absence of a need to urinate upon awakening from a night's sleep are indicators of dehydration. Thirst is not a good indicator of hydration.

(a) *Contributing Factors.* Factors that contribute to dehydration in cold weather are:

- The thirst mechanism does not function properly in cold weather.
- Water is often inconvenient to obtain and purify.
- The air in cold climates and at high altitudes lacks moisture.
- Cold causes frequent urination.

(b) *Symptoms.* Symptoms of dehydration include darkening urine, decreased amounts of urine being produced, dry mouth, tiredness, mental sluggishness, lack of appetite, headache, fainting, rapid heartbeat, dizziness, higher temperature, upset stomach, and unconsciousness. The symptoms of dehydration are similar to those of hypothermia. To distinguish between them, open the victim's clothes and feel the stomach. If the stomach is cold, the victim is probably hypothermic; if it is warm, he is probably dehydrated. However, this test is not conclusive since cold-weather dehydrating can also lead to total body cooling. The cold environment may act as a diuretic and impair the body's ability to conserve fluid (cold-induced diuresis and increased rate of urination).

(c) *Treatment.* Prevent dehydration by consuming three to six quarts of fluids each day (forced drinking in the absence of thirst is mandatory) and avoid caffeine and alcohol, which may chemically contribute to dehydration. Keep the victim warm and treat for shock. In advanced cases, administer fluids by mouth if the victim is conscious. Do not let him eat snow; eating snow uses body heat. Allow the victim to rest. If he fails to improve within one hour or is unconscious, evacuate him to a medical facility immediately.

(3) ***Hypothermia.*** Hypothermia is the lowering of the body core temperature at a rate faster than the body can produce heat. Hypothermia may be caused by exposure or by sudden wetting of the body such as falling into a lake or being sprayed with fuel or other liquid. Hypothermia can occur even on moderate days with temperatures of 40 to 50 degrees Fahrenheit with little precipitation if heat loss exceeds heat gain and the condition of the soldier is allowed to deteriorate. Hypothermia is classified as mild (core temperature above 90 degrees Fahrenheit or 32 degrees Celsius) or severe (core temperature below 90 degrees Fahrenheit or 32 degrees Celsius). An individual is considered to be "clinically hypothermic" when the core temperature is less than or equal to 95 degrees Fahrenheit.

(a) *Contributing Factors.* Factors that contribute to hypothermia are:

- Dehydration.
- Poor nutrition.
- Diarrhea.
- Decreased physical activity.
- Accidental immersion in water.
- Change in weather.
- High winds.
- Inadequate types or amounts of clothing.

(b) *Symptoms* The first symptom of hypothermia is when the body core (rectal) temperature falls to about 96 degrees Fahrenheit. Other symptoms include:

- Shivering, which may progress to an uncontrollable point making it hard for an individual to care for himself. Shivering begins after a drop in body temperature of one to two degrees. This is followed by clumsiness (stumbling or falling), slow reactions, mental confusion, and difficulty in speaking.
- Body temperature drop from 95 degrees Fahrenheit to 90 degrees Fahrenheit, which can cause sluggish thinking, irrational thought, apathy, and a false sense of warmth. The victim becomes cold and pale; cannot perform simple tasks; experiences amnesia and hallucinations; develops blueness of skin and decreased heart and respiratory rate with a weak pulse; pupils of the eyes dilate; speech becomes slurred; and visual disturbance occurs.
- Body temperature drop from 90 degrees Fahrenheit to 85 degrees Fahrenheit, which causes irrationality, incoherence, loss of contact with the environment, muscular rigidity, disorientation, and exhaustion. The soldier might stop shivering after his core temperature drops below 90 degrees Fahrenheit.
- Body temperature drop from 85 degrees Fahrenheit and below, which causes muscle rigidity, unconsciousness, comatose state, and faint vital signs. The pulse may be faint or impalpable, and breathing is too shallow to observe.

(c) *Prevention.* Prevent hypothermia by using the buddy system to watch each other for symptoms; consume adequate amounts of liquids daily; rest; and eat properly.

(d) *Avoidance.* Hypothermia can be avoided by dressing in layers, which permits easy additions or deletions to prevent overheating, becoming too cold, or getting wet or windblown. If the soldier is in a situation that precludes staying warm and dry, he should seek shelter. Sweets and physical activity help to produce body heat.

(e) *Treatment.* Treatment methods vary based on the severity of the hypothermia.

- Mild cases: If a soldier shows symptoms of hypothermia, prevent additional heat loss by getting the victim into a shelter; removing wet clothing and replacing it with dry, insulated clothing; insulating the victim from the ground; and sharing a sleeping bag (cover head) to transfer body heat. Make a diagnosis (rectal temperature). Rehydrate the victim with warm liquids, sweets, and food. If the tactical situation allows, build a fire. Above all else, keep the victim conscious until his vital signs are normal, and seek medical assistance. If possible, keep the victim physically active to produce body heat.
- Severe cases: If the victim is unconscious or appears dead without any obvious injury, prevent further heat loss. Rapid rewarming of an unconscious victim may create problems and should not be attempted. It is best to evacuate as soon as possible. At all times, the victim should be handled gently so as not to cause the cold blood from the extremities to rush to the heart. Do not allow the victim to perform ANY physical activity. Immediately transport the victim to the nearest medical facility. Field reheating is not effective and may be hazardous. Provide artificial respiration if breathing stops. If no pulse is detectable, be aware that in hypothermia there is often effective circulation for the victim's hypothermic state. In such a case, cardiac compression (such as CPR) may be fatal. The exception is acute hypothermia with near drowning.

- Breathing warm, moist air is the fastest way to warm the inside of the body. If breathing steam is not possible, place tubing under the rescuer's shirt so the victim will still breathe warm, moist air. This process can be done while on the move. In addition to breathing moist, warm air the victim must be gradually rewarmed using external heat sources. Padded hot water bottles or heated stones should be placed in the armpits.
- If conscious, the victim can be given warm, sweet drinks.
- The Hibler Pack is an improvised method of rewarming hypothermic victims in the field. This is used to heat the body core first so the vital organs are warmed and not the extremities. As the body warms up the warm blood will eventually warm all parts of the body. First lay out a blanket or sleeping bag and place a poncho or space blanket inside of it. The poncho or space blanket should go from the base of the skull to the base of the butt. This keeps the sleeping bag/blanket dry and acts like a vapor barrier. Lay the hypothermic patient inside the sleeping bag/blanket. Using a stove, warm water until it is hot to the touch (but not hot enough to burn the patient) and completely dampen any absorbable materials (such as T-shirt, towel, BDU top, and so on). Place the warm, wet items inside a plastic bag or directly in the armpits and chest of the patient. After the warm, wet item has been placed on the patient, wrap the patient tightly inside the poncho/space blanket and the blanket/sleeping bag. Continually check the temperature of the wet material and keep it warm.
- All bodily systems in hypothermia are brittle so treat the victim gently. As these attempts are being made, try to evacuate the victim. Severe complications may arise as the body temperature rises, which may result in cardiac arrest even though the victim seems to be doing well.

(4) ***Immersion or Trench Foot.*** This is damage to the circulatory and nervous systems of the feet that occurs from prolonged exposure to cold and wet at above freezing temperatures. This can happen wearing boots or not. A soldier may not feel uncomfortable until the injury has already begun.

(a) *Contributing Factors.* Factors that contribute to immersion or trench foot are:

- Stepping into water over the boot tops.
- Not changing socks often enough.
- Improper hygiene.
- Prolonged exposure (three to five days).

(b) *Symptoms.* Symptoms of immersion or trench foot include the sensation of tingling, numbness, and then pain. The toes are pale, and feel cold and stiff. The skin is wet and soggy with the color turning from red to bright red, progressing to pale and mottled, and then grayish blue. As symptoms progress and damage appears, the skin becomes red and then bluish or black. Swelling may occur. Because the early stages of trench foot are not painful, soldiers must be constantly aware to prevent it.

(c) *Treatment.* To prevent this condition, keep the feet dry and clean. Change socks often, drying the insides of boots, massaging the feet, and using foot powder. Drying the feet for 24 hours usually heals mild cases. Moderate cases usually heal within three to five days. The feet should be handled gently—NOT rubbed or massaged. They should be cleaned with soap and water, dried, elevated, and exposed to room temperature. The victim must stay off his feet and seek medical attention. Severe cases, when feet are not allowed to dry, are evacuated as a litter casualty.

(5) ***Blisters.*** When first noticed and before the formation of a blister, cover a hotspot with moleskin (over the area and beyond it). Use tincture benzoin to help the moleskin adhere to and toughen the skin. Once a blister has formed, cover it with a dressing large enough to fit over the blister, and then tape it. Never drain blisters unless they are surrounded by redness, or draining pus indicates infection. If this occurs, drain the blister from the side with a clean sterile needle. After cleaning with soap and water, gently press out the fluid leaving the skin intact. Make a doughnut of moleskin to go around the blister and apply to the skin. For toe blisters, wrap the entire toe with adhesive tape over the moleskin. (Toenails should be trimmed straight across the top, leaving a 90-degree angle on the sides. This provides an arch so that the corners do not irritate the skin.)

(6) ***Frostbite.*** Frostbite is the freezing or crystallization of living tissues due to heat being lost faster than it can be replaced by blood circulation, or from direct exposure to extreme cold or high winds. Exposure time can be minutes or instantaneous. The extremities are usually the first to be affected. Damp hands and feet may freeze quickly since moisture conducts heat away from the body and destroys the insulating value of clothing. Heat loss is compounded with intense cold and inactivity. With proper clothing and equipment, properly maintained and used, frostbite can be prevented. The extent of frostbite depends on temperature and duration of exposure. Frostbite is one of the major nonfatal cold-weather injuries encountered in military operations, but does not occur above an ambient temperature of 32 degrees Fahrenheit.

(a) *Categories of Frostbite.* Superficial (mild) frostbite involves only the skin (Figure 2-2). The layer immediately below usually appears white to grayish with the surface feeling hard, but the underlying tissue is soft. Deep (severe) frostbite extends beyond the first layer of skin and may include the bone (Figure 2-3). Discoloration continues from gray to black, and the texture becomes hard as the tissue freezes deeper. This condition requires immediate evacuation to a medical facility.

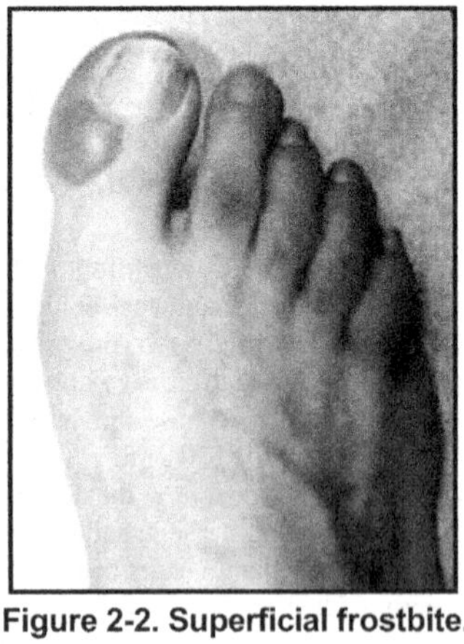

Figure 2-2. Superficial frostbite.

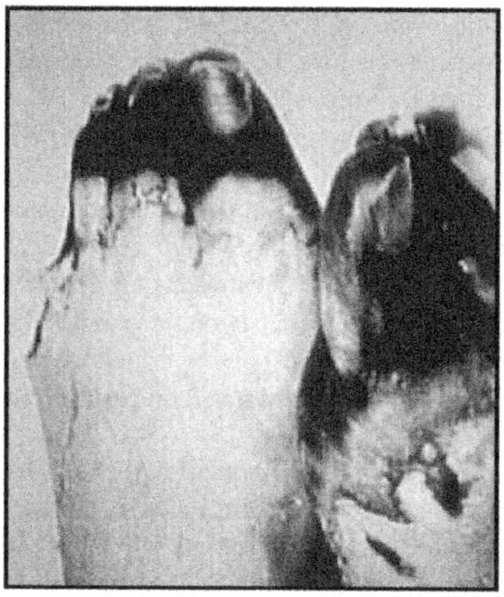

Figure 2-3. Deep frostbite.

(b) *Contributing Factors.* Factors that contribute to frostbite are:

- Dehydration.
- Below-freezing temperatures.
- Skin contact with super cooled metals or liquids.
- Use of caffeine, tobacco, or alcohol.
- Neglect.

(c) *Symptoms.* Symptoms of frostbite vary and may include a cold feeling, pain, burning, numbness, and, in the final stages, a false sense of warmth. The skin first turns red, then pale. It may be bluish in color and then may appear frosty or waxy white. The skin may feel hard, may not be movable over the joints and bony prominences, or may be frozen. Identification of deep versus superficial frostbite is difficult to determine and often requires three to seven days after rewarming for medical personnel to diagnose. Blisters, swelling, and pain may occur after thawing.

(d) *Treatment.* Using the buddy system is one of the primary ways to prevent frostbite. Buddies must watch each other for symptoms of frostbite and provide mutual aid if frostbite occurs. Frostbite should be identified early with prompt first-aid care applied to prevent further damage.

- Treat early signs of frostbite by rewarming with skin-to-skin contact or by sheltering the body part under the clothing next to the body. *Do this immediately.* If tissues have frozen, evacuate the victim before they thaw. If the feet are involved, evacuate the victim as a litter patient.
- Thawing of a frostbitten victim is a hospital procedure. If the victim has frostbite with frozen extremities, protect the frozen parts and evacuate as a litter patient.
- If frostbite is not recognized before it thaws, do not let the area refreeze since this causes more damage. The most often-affected body parts are the hands, fingers, toes, feet, ears, chin, and nose. If evacuation of the victim as a litter case

is not possible and the body part has not yet thawed, have the victim walk out on his own. Walking out on frozen feet is better than having them thaw and refreeze. Self-evacuation may be tactically necessary. Walking on frozen feet does less harm than walking on thawed feet.

- If reheating is inevitable, do not overheat the affected body parts near flame; the warming temperature should not be greater than normal body temperature. Do not rub the parts—the crystallized tissues may break internally and cause more damage. Do not pop blisters; cover them with a dry, sterile dressing. Keep the victim warm (apply loose, bulky bandages to separate toes and fingers.)
- Once a part is rewarmed it will become painful. Pain may be managed with narcotic analgesics.
- Once the foot is rewarmed it will swell and putting the boot back on will not be possible.

(7) ***Constipation.*** Constipation is the infrequent or difficult passage of stools.

(a) *Contributing Factors.* Factors that contribute to constipation are a lack of fluids, improper nutrition, and not defecating when needed.

(b) *Symptoms.* Symptoms include headache, cramping, lack of bowel movement, painful bowel movement, and loss of appetite.

(c) *Treatment.* Constipation is prevented by consuming adequate amounts and varieties of food, drinking from four to six liters of liquid each day, and defecating regularly. If allowed to progress beyond self-care stages, victims will need medical aid.

(8) ***Carbon Monoxide Poisoning.*** This is the replacement of oxygen in the blood with carbon monoxide.

(a) *Contributing Factor.* A contributing factor is inhaling fumes from burning fuel, such as fires, stoves, heaters, and running engines, without proper ventilation.

(b) *Symptoms.* Symptoms are similar to other common illnesses and include headaches, fatigue, excessive yawning, nausea, dizziness, drowsiness, confusion, and unconsciousness. Death may occur. The one visible symptom is bright red lips, mouth, and inside of the eyelids.

(c) *Treatment.* Remove the victim from the source of contamination; administer oxygen, if available; and evacuate to a medical facility. Severe complications may develop even in casualties who appear to have recovered. If the victim is unconscious, administer rescue breathing and CPR as needed.

2-10. HEAT INJURIES

Heat injuries, although associated with hot weather, can occur in cold-weather environments. Most heat injuries can be avoided by planning, periodic inspections of personnel clothing (ventilation) and equipment, a balance of water and food intake, and rest.

a. **Heat Cramps.** Heat cramps are caused by an accumulation of lactic acid in the muscles and a loss of salt through perspiration.

(1) ***Contributing Factor.*** Strenuous exertion causes the body to heat up and to produce heavy perspiration.

(2) ***Symptoms.*** Symptoms of heat cramps include pain and cramping in the arms, legs, back, and stomach. The victim sweats profusely and cannot quench his thirst.

(3) ***Treatment.*** Have the victim rest in a cool, shady area, breath deeply, and stretch the cramped muscle as soon as possible to obtain relief. Loosen the victim's clothing and have him drink cool water. Monitor his condition and seek medical attention if pain and cramps continue.

b. **Heat Exhaustion.** Heat exhaustion may occur when a soldier exerts himself in any environment and he overheats. The blood vessels in the skin become so dilated that the blood flow to the brain and other organs is reduced.

(1) ***Contributing Factors.*** Factors that contribute to heat exhaustion are strenuous activity in hot areas, unacclimatized troops, inappropriate diet, and not enough water or rest.

(2) ***Symptoms.*** Symptoms of heat exhaustion may be similar to fainting but may also include weakness; dizziness; confusion; headache; cold, clammy skin; and nausea. The victim may also have a rapid but weak pulse.

(3) ***Treatment.*** Move the victim to a cool, shady area and loosen his clothes and boots. Have the victim drink water and, if possible, immerse him in water to aid in cooling. Elevate the victim's legs to help restore proper circulation. Monitor his condition and seek medical attention if the symptoms persist.

c. **Heat Stroke.** Heat stroke is a life-threatening situation caused by overexposure to the sun. The body is so depleted of liquids that its internal cooling mechanisms fail to function.

(1) ***Contributing Factors.*** Factors that contribute to heat stroke are prolonged exposure to direct sunlight, overexertion, dehydration, and depletion of electrolytes.

(2) ***Symptoms.*** Symptoms of heat stroke include hot, dry skin; dizziness; confusion and incoherency; headache; nausea; seizures; breathing difficulty; a slow pulse; and loss of consciousness.

(3) ***Treatment.*** Cool the victim at once, and restore breathing and circulation. If the victim is conscious, administer water. If possible, submerge the victim in water to reduce his temperature, treat for shock, and prepare for immediate evacuation.

2-11. ACUTE MOUNTAIN SICKNESS

Acute mountain sickness is a temporary illness that may affect both the beginner and experienced climber. Soldiers are subject to this sickness in altitudes as low as 5,000 feet. Incidence and severity increases with altitude, and when quickly transported to high altitudes. Disability and ineffectiveness can occur in 50 to 80 percent of the troops who are rapidly brought to altitudes above 10,000 feet. At lower altitudes, or where ascent to altitudes is gradual, most personnel can complete assignments with moderate effectiveness and little discomfort.

a. Personnel arriving at moderate elevations (5,000 to 8,000 feet) usually feel well for the first few hours; a feeling of exhilaration or well-being is not unusual. There may be an initial awareness of breathlessness upon exertion and a need for frequent pauses to rest. Irregular breathing can occur, mainly during sleep; these changes may cause apprehension. Severe symptoms may begin 4 to 12 hours after arrival at higher altitudes with symptoms of nausea, sluggishness, fatigue, headache, dizziness, insomnia, depression, uncaring attitude, rapid and labored breathing, weakness, and loss of appetite.

b. A headache is the most noticeable symptom and may be severe. Even when a headache is not present, some loss of appetite and a decrease in tolerance for food occurs. Nausea, even without food intake, occurs and leads to less food intake. Vomiting may occur and contribute to dehydration. Despite fatigue, personnel are unable to sleep. The symptoms usually develop and increase to a peak by the second day. They gradually subside over the next several days so that the total course of AMS may extend from five to seven days. In some instances, the headache may become incapacitating and the soldier should be evacuated to a lower elevation.

c. Treatment for AMS includes the following:

- Oral pain medications such as ibuprofen or aspirin.
- Rest.
- Frequent consumption of liquids and light foods in small amounts.
- Movement to lower altitudes (at least 1,000 feet) to alleviate symptoms, which provides for a more gradual acclimatization.
- Realization of physical limitations and slow progression.
- Practice of deep-breathing exercises.
- Use of acetazolamide in the first 24 hours for mild to moderate cases.

d. AMS is nonfatal, although if left untreated or further ascent is attempted, development of high-altitude pulmonary edema (HAPE) and or high-altitude cerebral edema (HACE) can be seen. A severe persistence of symptoms may identify soldiers who acclimatize poorly and, thus, are more prone to other types of mountain sickness.

2-12. CHRONIC MOUNTAIN SICKNESS

Although not commonly seen in mountaineers, chronic mountain sickness (CMS) (or Monge's disease) can been seen in people who live at sufficiently high altitudes (usually at or above 10,000 feet) over a period of several years. CMS is a right-sided heart failure characterized by chronic pulmonary edema that is caused by years of strain on the right ventricle.

2-13. UNDERSTANDING HIGH-ALTITUDE ILLNESSES

As altitude increases, the overall atmospheric pressure decreases. Decreased pressure is the underlying source of altitude illnesses. Whether at sea level or 20,000 feet the surrounding atmosphere has the same percentage of oxygen. As pressure decreases the body has a much more difficult time passing oxygen from the lungs to the red blood cells and thus to the tissues of the body. This lower pressure means lower oxygen levels in the blood and increased carbon dioxide levels. Increased carbon dioxide levels in the blood cause a systemic vasodilatation, or expansion of blood vessels. This increased vascular size stretches the vessel walls causing leakage of the fluid portions of the blood into the interstitial spaces, which leads to cerebral edema or HACE. Unless treated, HACE will continue to progress due to the decreased atmospheric pressure of oxygen. Further ascent will hasten the progression of HACE and could possibly cause death.

While the body has an overall systemic vasodilatation, the lungs initially experience pulmonary vasoconstriction. This constricting of the vessels in the lungs causes increased workload on the right ventricle, the chamber of the heart that receives de-oxygenated blood from the right atrium and pushes it to the lungs to be re-oxygenated. As the right ventricle works harder to force blood to the lungs, its overall output is decreased thus decreasing the overall pulmonary perfusion. Decreased pulmonary perfusion causes decreased cellular respiration—the transfer of oxygen from the alveoli to the red blood cells. The body is now experiencing increased carbon dioxide levels due to the decreased oxygen levels, which now causes pulmonary vasodilatation. Just as in HACE, this expanding of the vascular structure causes leakage into interstitial space resulting in pulmonary edema or HAPE. As the edema or fluid in the lungs increases, the capability to pass oxygen to the red blood cells decreases thus creating a vicious cycle, which can quickly become fatal if left untreated.

2-14. HIGH-ALTITUDE PULMONARY EDEMA

HAPE is a swelling and filling of the lungs with fluid, caused by rapid ascent. It occurs at high altitudes and limits the oxygen supply to the body.

a. HAPE occurs under conditions of low oxygen pressure, is encountered at high elevations (over 8,000 feet), and can occur in healthy soldiers. HAPE may be considered a form of, or manifestation of, AMS since it occurs during the period of susceptibility to this disorder.

b. HAPE can cause death. Incidence and severity increase with altitude. Except for acclimatization to altitude, no known factors indicate resistance or immunity. Few cases have been reported after 10 days at high altitudes. When remaining at the same altitude, the incidence of HAPE is less frequent than that of AMS. No common indicator dictates how a soldier will react from one exposure to another. Contributing factors are:

- A history of HAPE.
- A rapid or abrupt transition to high altitudes.
- Strenuous physical exertion.
- Exposure to cold.
- Anxiety.

c. Symptoms of AMS can mask early pulmonary difficulties. Symptoms of HAPE include:

- Progressive dry coughing with frothy white or pink sputum (this is usually a later sign) and then coughing up of blood.
- Cyanosis—a blue color to the face, hands, and feet.
- An increased ill feeling, labored breathing, dizziness, fainting, repeated clearing of the throat, and development of a cough.
- Respiratory difficulty, which may be sudden, accompanied by choking and rapid deterioration.
- Progressive shortness of breath, rapid heartbeat (pulse 120 to 160), and coughing (out of contrast to others who arrived at the same time to that altitude).
- Crackling, cellophane-like noises (rales) in the lungs caused by fluid buildup (a stethoscope is usually needed to hear them).
- Unconsciousness, if left untreated. Bubbles form in the nose and mouth, and death results.

d. HAPE is prevented by good nutrition, hydration, and gradual ascent to altitude (no more than 1,000 to 2,000 feet per day to an area of sleep). A rest day, with no gain in altitude or heavy physical exertion, is planned for every 3,000 feet of altitude gained. If a soldier develops symptoms despite precautions, immediate descent is mandatory where he receives prompt treatment, rest, warmth, and oxygen. He is quickly evacuated to lower altitudes as a litter patient. A descent of 300 meters may help; manual descent is not delayed to await air evacuation. If untreated, HAPE may become irreversible and cause death. Cases that are recognized early and treated promptly may expect to recover with no aftereffects. Soldiers who have had previous attacks of HAPE are prone to second attacks.

e. Treatment of HAPE includes:

- Immediate descent (2,000 to 3,000 feet minimum) if possible; if not, then treatment in a monoplace hyperbaric chamber.
- Rest (litter evacuation)
- Supplemental oxygen if available.
- Morphine for the systemic vasodilatation and reduction of preload. This should be carefully considered due to the respiratory depressive properties of the drug.
- Furosemide (Lasix), which is a diuretic, given orally can also be effective.
- The use of mannitol should not be considered due to the fact that it crystallizes at low temperatures. Since almost all high-altitude environments are cold, using mannitol could be fatal.
- Nifidipine (Procardia), which inhibits calcium ion flux across cardiac and smooth muscle cells, decreasing contractility and oxygen demand. It may also dilate coronary arteries and arterioles.
- Diphenhydramine (Benadryl), which can help alleviate the histamine response that increases mucosal secretions.

2-15. HIGH-ALTITUDE CEREBRAL EDEMA

HACE is the accumulation of fluid in the brain, which results in swelling and a depression of brain function that may result in death. It is caused by a rapid ascent to altitude without progressive acclimatization. Prevention of HACE is the same as for HAPE. HAPE and HACE may occur in experienced, well-acclimated mountaineers without warning or obvious predisposing conditions. They can be fatal; when the first symptoms occur, immediate descent is mandatory.

a. Contributing factors include rapid ascent to heights over 8,000 feet and aggravation by overexertion.

b. Symptoms of HACE include mild personality changes, paralysis, stupor, convulsions, coma, inability to concentrate, headaches, vomiting, decrease in urination, and lack of coordination. The main symptom of HACE is a severe headache. A headache combined with any other physical or psychological disturbances should be assumed to be manifestations of HACE. Headaches may be accompanied by a loss of coordination, confusion, hallucinations, and unconsciousness. These may be combined with symptoms of HAPE. The victim is often mistakenly left alone since others may think he is only irritable or temperamental; no one should ever be ignored. The symptoms may rapidly progress to death. Prompt descent to a lower altitude is vital.

c. Preventive measures include good eating habits, maintaining hydration, and using a gradual ascent to altitude. Rest, warmth, and oxygen at lower elevations enhance recovery. Left untreated, HACE can cause death.

d. Treatment for HACE includes:

- Dexamethasone injection immediately followed by oral dexamethasone.
- Supplemental oxygen.
- Rapid descent and medical attention.
- Use of a hyberbaric chamber if descent is delayed.

2-16. HYDRATION IN HAPE AND HACE

HAPE and HACE cause increased proteins in the plasma, or the fluid portion of the blood, which in turn increases blood viscosity. Increased viscosity increases vascular pressure. Vascular leakage caused by stretching of the vessel walls is made worse because of this increased vascular pressure. From this, edema, both cerebral and pulmonary, occurs. Hydration simply decreases viscosity.

CHAPTER 3
MOUNTAINEERING EQUIPMENT

Commanders at every level must understand the complexity of operations in a mountainous environment where every aspect of combat operations becomes more difficult. Leaders must understand that each individual has a different metabolism and, therefore, cools down and heats up differently, which requires soldiers to dress-up and dress-down at different intervals. Provided all tactical concerns are met, the concept of uniformity is outdated and only reduces the unit's ability to fight and function at an optimum level. The extreme cold weather clothing system (ECWCS) is specifically designed to allow for rapid moisture transfer and optimum heat retention while protecting the individual from the elements. Every leader is responsible for ensuring that the ECWCS is worn in accordance with the manufacturers' recommendations. Commanders at all levels must also understand that skills learned at an Army mountaineering school are perishable and soldiers need constant practice to remain proficient. The properly trained mountain soldier of today can live better, move faster, and fight harder in an environment that is every bit as hostile as the enemy.

Section I. EQUIPMENT DESCRIPTION AND MAINTENANCE

With mountainous terrain encompassing a large portion of the world's land mass, the proper use of mountaineering equipment will enhance a unit's combat capability and provide a combat multiplier. The equipment described in this chapter is produced by many different manufacturers; however, each item is produced and tested to extremely high standards to ensure safety when being used correctly. The weak link in the safety chain is the user. Great care in performing preventative maintenance checks and services and proper training in the use of the equipment is paramount to ensuring safe operations. The manufacturers of each and every piece of equipment provide recommendations on how to use and care for its product. It is imperative to follow these instructions explicitly.

3-1. FOOTWEAR

Currently, CTA 50-900 provides adequate footwear for most operations in mountainous terrain. In temperate climates a combination of footwear is most appropriate to accomplish all tasks.

a. The hot weather boot provides an excellent all-round platform for movement and climbing techniques and should be the boot of choice when the weather permits. The intermediate cold weather boot provides an acceptable platform for operations when the weather is less than ideal. These two types of boots issued together will provide the unit with the footwear necessary to accomplish the majority of basic mountain missions.

b. Mountain operations are encumbered by extreme cold, and the extreme cold weather boot (with vapor barrier) provides an adequate platform for many basic mountain missions. However, plastic mountaineering boots should be incorporated into training as soon as possible. These boots provide a more versatile platform for any condition that would be encountered in the mountains, while keeping the foot dryer and warmer.

c. Level 2 and level 3 mountaineers will need mission-specific footwear that is not currently available in the military supply system. The two types of footwear they will need are climbing shoes and plastic mountaineering boots.

(1) Climbing shoes are made specifically for climbing vertical or near vertical rock faces. These shoes are made with a soft leather upper, a lace-up configuration, and a smooth "sticky rubber" sole (Figure 3-1). The smooth "sticky rubber" sole is the key to the climbing shoe, providing greater friction on the surface of the rock, allowing the climber access to more difficult terrain.

(2) The plastic mountaineering boot is a double boot system (Figure 3-1). The inner boot provides support, as well as insulation against the cold. The inner boot may or may not come with a breathable membrane. The outer boot is a molded plastic (usually with a lace-up configuration) with a lug sole. The welt of the boot is molded in such a way that crampons, ski bindings, and snowshoes are easily attached and detached.

Note: Maintenance of all types of footwear must closely follow the manufacturers' recommendations.

Figure 3-1. Climbing shoes and plastic mountaineering boots.

3-2. CLOTHING

Clothing is perhaps the most underestimated and misunderstood equipment in the military inventory. The clothing system refers to every piece of clothing placed against the skin, the insulation layers, and the outer most garments, which protect the soldier from the elements. When clothing is worn properly, the soldier is better able to accomplish his tasks. When worn improperly, he is, at best, uncomfortable and, at worst, develops hypothermia or frostbite.

a. **Socks.** Socks are one of the most under-appreciated part of the entire clothing system. Socks are extremely valuable in many respects, if worn correctly. As a system, socks provide cushioning for the foot, remove excess moisture, and provide insulation from cold temperatures. Improper wear and excess moisture are the biggest causes of hot spots and blisters. Regardless of climatic conditions, socks should always be worn in layers.

(1) The first layer should be a hydrophobic material that moves moisture from the foot surface to the outer sock.

(2) The outer sock should also be made of hydrophobic materials, but should be complimented with materials that provide cushioning and abrasion resistance.

(3) A third layer can be added depending upon the climatic conditions.

(a) In severe wet conditions, a waterproof type sock can be added to reduce the amount of water that would saturate the foot. This layer would be worn over the first two layers if conditions were extremely wet.

(b) In extremely cold conditions a vapor barrier sock can be worn either over both of the original pairs of socks or between the hydrophobic layer and the insulating layer. If the user is wearing VB boots, the vapor barrier sock is not recommended.

b. **Underwear.** Underwear should also be made of materials that move moisture from the body. Many civilian companies manufacture this type of underwear. The primary material in this product is polyester, which moves moisture from the body to the outer layers keeping the user drier and more comfortable in all climatic conditions. In colder environments, several pairs of long underwear of different thickness should be made available. A lightweight set coupled with a heavyweight set will provide a multitude of layering combinations.

c. **Insulating Layers.** Insulating layers are those layers that are worn over the underwear and under the outer layers of clothing. Insulating layers provide additional warmth when the weather turns bad. For the most part, today's insulating layers will provide for easy moisture movement as well as trap air to increase the insulating factor. The insulating layers that are presently available are referred to as pile or fleece. The ECWCS (Figure 3-2, page 3-4) also incorporates the field jacket and field pants liner as additional insulating layers. However, these two components do not move moisture as effectively as the pile or fleece.

d. **Outer Layers.** The ECWCS provides a jacket and pants made of a durable waterproof fabric. Both are constructed with a nylon shell with a laminated breathable membrane attached. This membrane allows the garment to release moisture to the environment while the nylon shell provides a degree of water resistance during rain and snow. The nylon also acts as a barrier to wind, which helps the garment retain the warm air trapped by the insulating layers. Leaders at all levels must understand the importance of wearing the ECWCS correctly.

Note: Cotton layers must not be included in any layer during operations in a cold environment.

Figure 3-2. Extreme cold weather clothing system.

e. **Gaiters.** Gaiters are used to protect the lower leg from snow and ice, as well as mud, twigs, and stones. The use of waterproof fabrics or other breathable materials laminated to the nylon makes the gaiter an integral component of the cold weather clothing system. Gaiters are not presently fielded in the standard ECWCS and, in most cases, will need to be locally purchased. Gaiters are available in three styles (Figure 3-3).

(1) The most common style of gaiter is the open-toed variety, which is a nylon shell that may or may not have a breathable material laminated to it. The open front allows the boot to slip easily into it and is closed with a combination of zipper, hook-pile tape, and snaps. It will have an adjustable neoprene strap that goes under the boot to keep it snug to the boot. The length should reach to just below the knee and will be kept snug with a drawstring and cord lock.

(2) The second type of gaiter is referred to as a full or randed gaiter. This gaiter completely covers the boot down to the welt. It can be laminated with a breathable material and can also be insulated if necessary. This gaiter is used with plastic mountaineering boots and should be glued in place and not removed.

(3) The third type of gaiter is specific to high-altitude mountaineering or extremely cold temperatures and is referred to as an overboot. It is worn completely over the boot and must be worn with crampons because it has no traction sole.

Figure 3-3. Three types of gaiters.

f. **Hand Wear.** During operations in mountainous terrain the use of hand wear is extremely important. Even during the best climatic conditions, temperatures in the mountains will dip below the freezing point. While mittens are always warmer than gloves, the finger dexterity needed to do most tasks makes gloves the primary cold weather hand wear (Figure 3-4, page 3-6).

(1) The principals that apply to clothing also apply to gloves and mittens. They should provide moisture transfer from the skin to the outer layers—the insulating layer must insulate the hand from the cold and move moisture to the outer layer. The outer layer must be weather resistant and breathable. Both gloves and mittens should be required for all soldiers during mountain operations, as well as replacement liners for both. This will provide enough flexibility to accomplish all tasks and keep the users' hands warm and dry.

(2) Just as the clothing system is worn in layers, gloves and mittens work best using the same principle. Retention cords that loop over the wrist work extremely well when the wearer needs to remove the outer layer to accomplish a task that requires fine finger dexterity. Leaving the glove or mitten dangling from the wrist ensures the wearer knows where it is at all times.

Figure 3-4. Hand wear.

g. **Headwear.** A large majority of heat loss (25 percent) occurs through the head and neck area. The most effective way to counter heat loss is to wear a hat. The best hat available to the individual soldier through the military supply system is the black watch cap. Natural fibers, predominately wool, are acceptable but can be bulky and difficult to fit under a helmet. As with clothes and hand wear, man-made fibers are preferred. For colder climates a neck gaiter can be added. The neck gaiter is a tube of man-made material that fits around the neck and can reach up over the ears and nose (Figure 3-5). For extreme cold, a balaclava can be added. This covers the head, neck, and face leaving only a slot for the eyes (Figure 3-5). Worn together the combination is warm and provides for moisture movement, keeping the wearer drier and warmer.

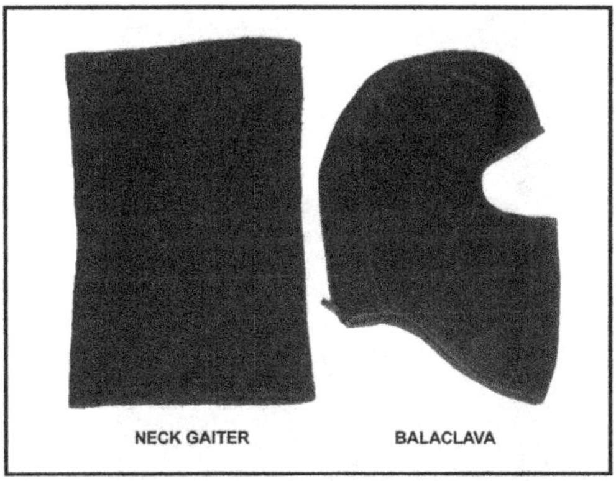

Figure 3-5. Neck gaiter and balaclava.

h. **Helmets**. The Kevlar ballistic helmet can be used for most basic mountaineering tasks. It must be fitted with parachute retention straps and the foam impact pad (Figure 3-6). The level 2 and 3 mountaineer will need a lighter weight helmet for specific climbing scenerios. Several civilian manufacturers produce an effective helmet. Whichever helmet is selected, it should be designed specifically for mountaineering and adjustable so the user can add a hat under it when needed.

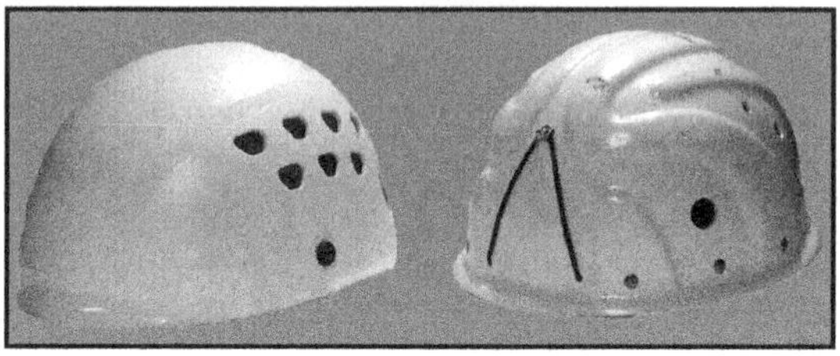

Figure 3-6. Helmets.

i. **Eyewear.** The military supply system does not currently provide adequate eyewear for mountaineering. Eyewear is divided into two catagories: glacier glasses and goggles (Figure 3-7). Glacier glasses are sunglasses that cover the entire eye socket. Many operations in the mountains occur above the tree line or on ice and snow surfaces where the harmful UV rays of the sun can bombard the eyes from every angle increasing the likelihood of snowblindness. Goggles for mountain operations should be antifogging. Double or triple lenses work best. UV rays penetrate clouds so the goggles should be UV protected. Both glacier glasses and goggles are required equipment in the mountains. The lack of either one can lead to severe eye injury or blindness.

Figure 3-7. Glacier glasses and goggles.

j. **Maintenance of Clothing.** Clothing and equipment manufacturers provide specific instructions for proper care. Following these instructions is necessary to ensure the equipment works as intended.

3-3. CLIMBING SOFTWARE

Climbing software refers to rope, cord, webbing, and harnesses. All mountaineering specific equipment, to include hardware (see paragraph 3-4), should only be used if it has the UIAA certificate of safety. UIAA is the organization that oversees the testing of mountaineering equipment. It is based in Paris, France, and comprises several commissions. The safety commission has established standards for mountaineering and climbing equipment that have become well recognized throughout the world. Their work continues as new equipment develops and is brought into common use. Community Europe (CE) recognizes UIAA testing standards and, as the broader-based testing facility for the combined European economy, meets or exceeds the UIAA standards for all climbing and mountaineering equipment produced in Europe. European norm (EN) and CE have been combined to make combined European norm (CEN). While the United States has no specific standards, American manufacturers have their equipment tested by UIAA to ensure safe operating tolerances.

a. **Ropes and Cord.** Ropes and cords are the most important pieces of mountaineering equipment and proper selection deserves careful thought. These items are your lifeline in the mountains, so selecting the right type and size is of the utmost importance. All ropes and cord used in mountaineering and climbing today are constructed with the same basic configuration. The construction technique is referred to as Kernmantle, which is, essentially, a core of nylon fibers protected by a woven sheath, similar to parachute or 550 cord (Figure 3-8).

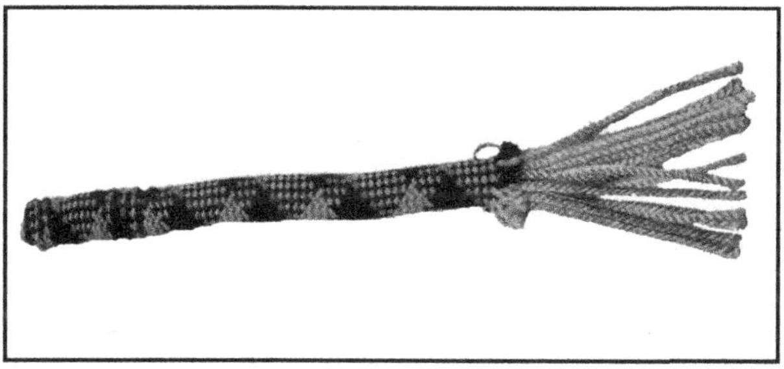

Figure 3-8. Kernmantle construction.

(1) Ropes come in two types: static and dynamic. This refers to their ability to stretch under tension. A static rope has very little stretch, perhaps as little as one to two percent, and is best used in rope installations. A dynamic rope is most useful for climbing and general mountaineering. Its ability to stretch up to 1/3 of its overall length makes it the right choice any time the user might take a fall. Dynamic and static ropes come in various

diameters and lengths. For most military applications, a standard 10.5- or 11-millimeter by 50-meter dynamic rope and 11-millimeter by 45-meter static rope will be sufficient.

(2) When choosing dynamic rope, factors affecting rope selection include intended use, impact force, abrasion resistance, and elongation. Regardless of the rope chosen, it should be UIAA certified.

(3) Cord or small diameter rope is indispensable to the mountaineer. Its many uses make it a valuable piece of equipment. All cord is static and constructed in the same manner as larger rope. If used for Prusik knots, the cord's diameter should be 5 to 7 millimeters when used on an 11-mm rope.

b. **Webbing and Slings.** Loops of tubular webbing or cord, called slings or runners, are the simplest pieces of equipment and some of the most useful. The uses for these simple pieces are endless, and they are a critical link between the climber, the rope, carabiners, and anchors. Runners are predominately made from either 9/16-inch or 1-inch tubular webbing and are either tied or sewn by a manufacturer (Figure 3-9). Runners can also be made from a high-performance fiber known as spectra, which is stronger, more durable, and less susceptible to ultraviolet deterioration. Runners should be retired regularly following the same considerations used to retire a rope. For most military applications, a combination of different lengths of runners is adequate.

(1) Tied runners have certain advantages over sewn runners—they are inexpensive to make, can be untied and threaded around natural anchors, and can be untied and retied to other pieces of webbing to create extra long runners.

(2) Sewn runners have their own advantages—they tend to be stronger, are usually lighter, and have less bulk than the tied version. They also eliminate a major concern with the homemade knotted runner—the possibility of the knot untying. Sewn runners come in four standard lengths: 2 inches, 4 inches, 12 inches, and 24 inches. They also come in three standard widths: 9/16 inch, 11/16 inch, and 1 inch.

Figure 3-9. Tied or sewn runners.

c. **Harnesses.** Years ago climbers secured themselves to the rope by wrapping the rope around their bodies and tying a bowline-on-a-coil. While this technique is still a viable way of attaching to a rope, the practice is no longer encouraged because of the increased possibility of injury from a fall. The bowline-on-a-coil is best left for low-angle climbing or an emergency situation where harness material is unavailable. Climbers today can select from a wide range of manufactured harnesses. Fitted properly, the harness should ride high on the hips and have snug leg loops to better distribute the force of a fall to the entire pelvis. This type of harness, referred to as a seat harness, provides a comfortable seat for rappelling (Figure 3-10).

(1) Any harness selected should have one very important feature—a double-passed buckle. This is a safety standard that requires the waist belt to be passed over and back through the main buckle a second time. At least 2 inches of the strap should remain after double-passing the buckle.

(2) Another desirable feature on a harness is adjustable leg loops, which allows a snug fit regardless of the number of layers of clothing worn. Adjustable leg loops allow the soldier to make a latrine call without removing the harness or untying the rope.

(3) Equipment loops are desirable for carrying pieces of climbing equipment. For safety purposes always follow the manufacturer's directions for tying-in.

(4) A field-expedient version of the seat harness can be constructed by using 22 feet of either 1-inch or 2-inch (preferred) tubular webbing (Figure 3-10). Two double-overhand knots form the leg loops, leaving 4 to 5 feet of webbing coming from one of the leg loops. The leg loops should just fit over the clothing. Wrap the remaining webbing around the waist ensuring the first wrap is routed through the 6- to 10-inch long strap between the double-overhand knots. Finish the waist wrap with a water knot tied as tightly as possible. With the remaining webbing, tie a square knot without safeties over the water knot ensuring a minimum of 4 inches remains from each strand of webbing.

(5) The full body harness incorporates a chest harness with a seat harness (Figure 3-10). This type of harness has a higher tie-in point and greatly reduces the chance of flipping backward during a fall. This is the only type of harness that is approved by the UIAA. While these harnesses are safer, they do present several disadvantages—they are more expensive, are more restrictive, and increase the difficulty of adding or removing clothing. Most mountaineers prefer to incorporate a separate chest harness with their seat harness when warranted.

(6) A separate chest harness can be purchased from a manufacturer, or a field-expedient version can be made from either two runners or a long piece of webbing. Either chest harness is then attached to the seat harness with a carabiner and a length of webbing or cord.

Figure 3-10. Seat harness, field-expedient harness, and full body harness.

3-4. CLIMBING HARDWARE

Climbing hardware refers to all the parts and pieces that allow the trained mountain soldier to accomplish many tasks in the mountains. The importance of this gear to the mountaineer is no less than that of the rifle to the infantryman.

a. **Carabiners.** One of the most versatile pieces of equipment available to the mountaineer is the carabiner. This simple piece of gear is the critical connection between the climber, his rope, and the protection attaching him to the mountain. Carabiners must be strong enough to hold hard falls, yet light enough for the climber to easily carry a quantity of them. Today's high tech metal alloys allow carabiners to meet both of these requirements. Steel is still widely used, but is not preferred for general mountaineering, given other options. Basic carabiner construction affords the user several different shapes. The oval, the D-shaped, and the pear-shaped carabiner are just some of the types currently available. Most models can be made with or without a locking mechanism for the gate opening (Figure 3-11, page 3-12). If the carabiner does have a locking mechanism, it is usually referred to as a locking carabiner. When using a carabiner, great care should be taken to avoid loading the carabiner on its minor axis and to avoid three-way loading (Figure 3-12, page 3-12).

Note: Great care should be used to ensure all carabiner gates are closed and locked during use.

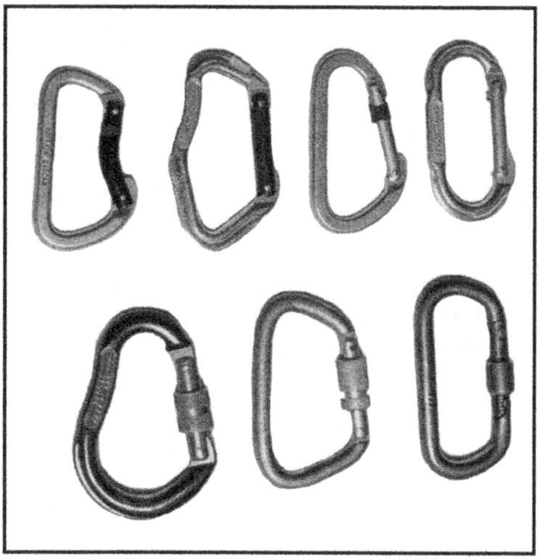

Figure 3-11. Nonlocking and locking carabiners.

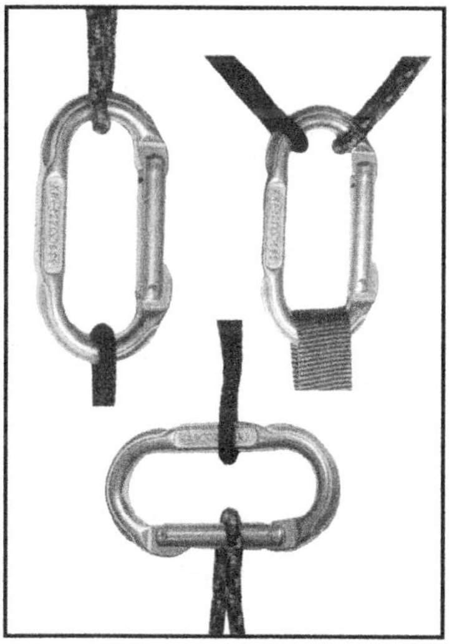

Figure 3-12. Major and minor axes and three-way loading.

(1) The major difference between the oval and the D-shaped carabiner is strength. Because of the design of the D-shaped carabiner, the load is angled onto the spine of the

carabiner thus keeping it off the gate. The down side is that racking any gear or protection on the D-shaped carabiner is difficult because the angle of the carabiner forces all the gear together making it impossible to separate quickly.

(2) The pear-shaped carabiner, specifically the locking version, is excellent for clipping a descender or belay device to the harness. They work well with the munter hitch belaying knot.

(3) Regardless of the type chosen, all carabiners should be UIAA tested. This testing is extensive and tests the carabiner in three ways—along its major axis, along its minor axis, and with the gate open.

b. **Pitons.** A piton is a metal pin that is hammered into a crack in the rock. They are described by their thickness, design, and length (Figure 3-13, page 3-14). Pitons provide a secure anchor for a rope attached by a carabiner. The many different kinds of pitons include: vertical, horizontal, wafer, and angle. They are made of malleable steel, hardened steel, or other alloys. The strength of the piton is determined by its placement rather than its rated tensile strength. The two most common types of pitons are: blades, which hold when wedged into tight-fitting cracks, and angles, which hold blade compression when wedged into a crack.

(1) ***Vertical Pitons***. On vertical pitons, the blade and eye are aligned. These pitons are used in flush, vertical cracks.

(2) ***Horizontal Pitons***. On horizontal pitons, the eye of the piton is at right angles to the blade. These pitons are used in flush, horizontal cracks and in offset or open-book type vertical or horizontal cracks. They are recommended for use in vertical cracks instead of vertical pitons because the torque on the eye tends to wedge the piton into place. This provides more holding power than the vertical piton under the same circumstances.

(3) ***Wafer Pitons***. These pitons are used in shallow, flush cracks. They have little holding power and their weakest points are in the rings provided for the carabiner.

(4) ***Knife Blade Pitons***. These are used in direct-aid climbing. They are small and fit into thin, shallow cracks. They have a tapered blade that is optimum for both strength and holding power.

(5) ***Realized Ultimate Reality Pitons***. Realized ultimate reality pitons (RURPs) are hatchet-shaped pitons about 1-inch square. They are designed to bite into thin, shallow cracks.

(6) ***Angle Pitons***. These are used in wide cracks that are flush or offset. Maximum strength is attained only when the legs of the piton are in contact with the opposite sides of the crack.

(7) ***Bong Pitons***. These are angle pitons that are more than 3.8 centimeters wide. Bongs are commonly made of steel or aluminum alloy and usually contain holes to reduce weight and accommodate carabiners. They have a high holding power and require less hammering than other pitons.

(8) ***Skyhook (Cliffhangers).*** These are small hooks that cling to tiny rock protrusions, ledges, or flakes. Skyhooks require constant tension and are used in a downward pull direction. The curved end will not straighten under body weight. The base is designed to prevent rotation and aid stability.

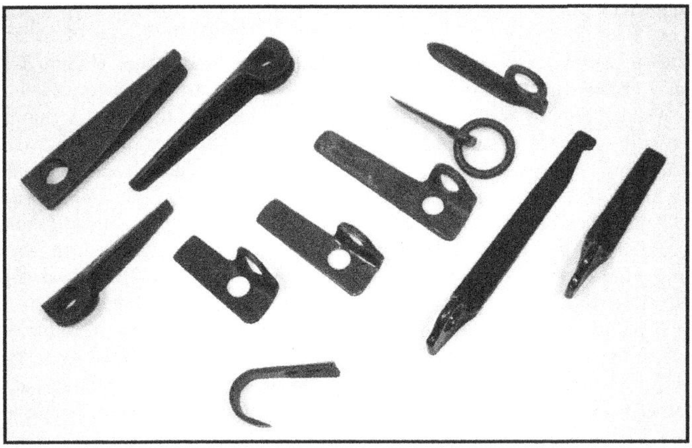

Figure 3-13. Various pitons.

c. **Piton Hammers.** A piton hammer has a flat metal head; a handle made of wood, metal, or fiberglass; and a blunt pick on the opposite side of the hammer (Figure 3-14). A safety lanyard of nylon cord, webbing, or leather is used to attach it to the climber The lanyard should be long enough to allow for full range of motion. Most hammers are approximately 25.5 centimeters long and weigh 12 to 25 ounces. The primary use for a piton hammer is to drive pitons, to be used as anchors, into the rock. The piton hammer can also be used to assist in removing pitons, and in cleaning cracks and rock surfaces to prepare for inserting the piton. The type selected should suit individual preference and the intended use.

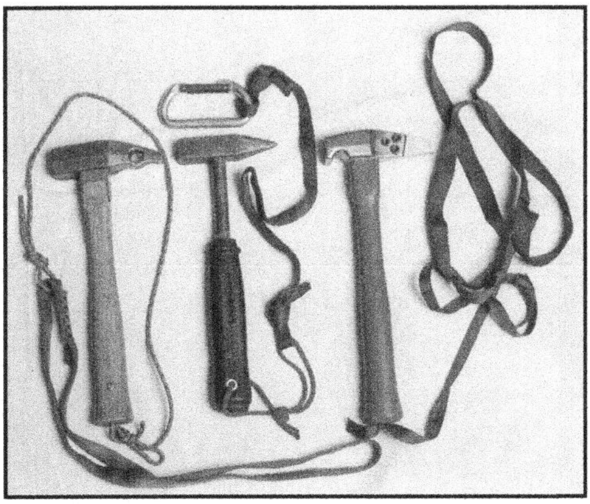

Figure 3-14. Piton hammer.

d. **Chocks.** "Chocks" is a generic term used to describe the various types of artificial protection other than bolts or pitons. Chocks are essentially a tapered metal wedge constructed in various sizes to fit different sized openings in the rock (Figure 3-15). The design of a chock will determine whether it fits into one of two categories—wedges or cams. A wedge holds by wedging into a constricting crack in the rock. A cam holds by slightly rotating in a crack, creating a camming action that lodges the chock in the crack or pocket. Some chocks are manufactured to perform either in the wedging mode or the camming mode. One of the chocks that falls into the category of both a wedge and cam is the hexagonal-shaped or "hex" chock. This type of chock is versatile and comes with either a cable loop or is tied with cord or webbing. All chocks come in different sizes to fit varying widths of cracks. Most chocks come with a wired loop that is stronger than cord and allows for easier placement. Bigger chocks can be threaded with cord or webbing if the user ties the chock himself. Care should be taken to place tubing in the chock before threading the cord. The cord used with chocks is designed to be stiffer and stronger than regular cord and is typically made of Kevlar. The advantage of using a chock rather than a piton is that a climber can carry many different sizes and use them repeatedly.

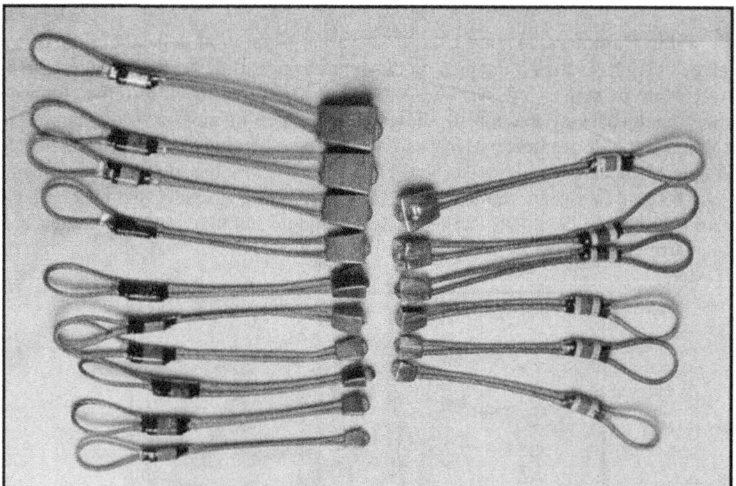

Figure 3-15. Chocks.

e. **Three-Point Camming Device.** The three-point camming device's unique design allows it to be used both as a camming piece and a wedging piece (Figure 3-16). Because of this design it is extremely versatile and, when used in the camming mode, will fit a wide range of cracks. The three-point camming device comes in several different sizes with the smaller sizes working in pockets that no other piece of gear would fit in.

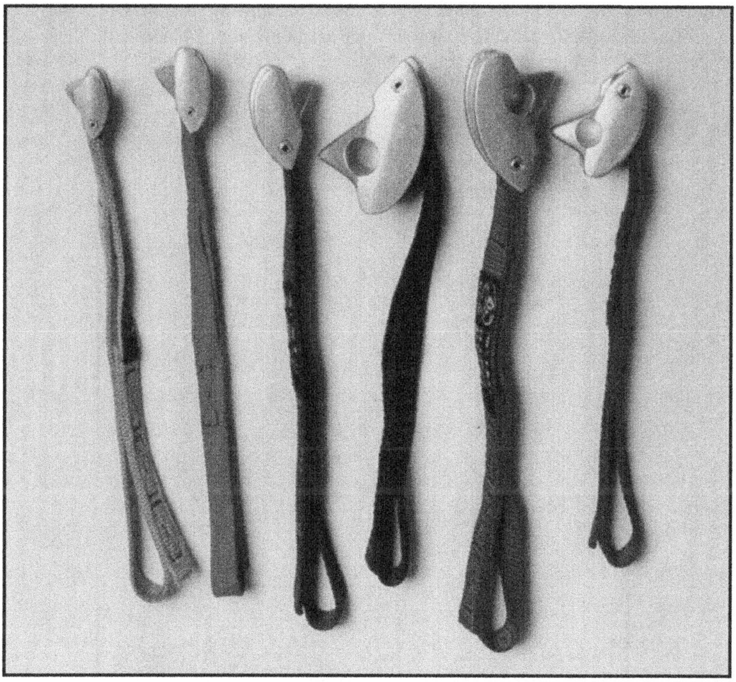

Figure 3-16. Three-point camming device.

f. **Spring-Loaded Camming Devices.** Spring-loaded camming devices (SLCDs) (Figure 3-17) provide convenient, reliable placement in cracks where standard chocks are not practical (parallel or flaring cracks or cracks under roofs). SLCDs have three or four cams rotating around a single or double axis with a rigid or semi-rigid point of attachment. These are placed quickly and easily, saving time and effort. SLCDs are available in many sizes to accommodate different size cracks. Each fits a wide range of crack widths due to the rotating cam heads. The shafts may be rigid metal or semi-rigid cable loops. The flexible cable reduces the risk of stem breakage over an edge in horizontal placements.

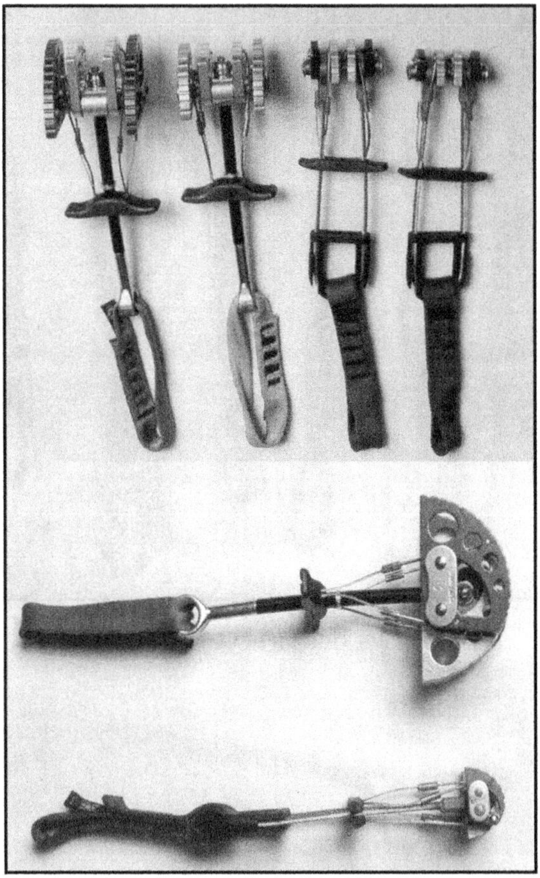

Figure 3-17. Spring-loaded camming devices.

g. **Chock Picks.** Chock picks are primarily used to extract chocks from rock when the they become severely wedged (Figure 3-18). They are also handy to clean cracks with. Made from thin metal, they can be purchased or homemade. When using a chock pick to extract a chock be sure no force is applied directly to the cable juncture. One end of the chock pick should have a hook to use on jammed SLCDs.

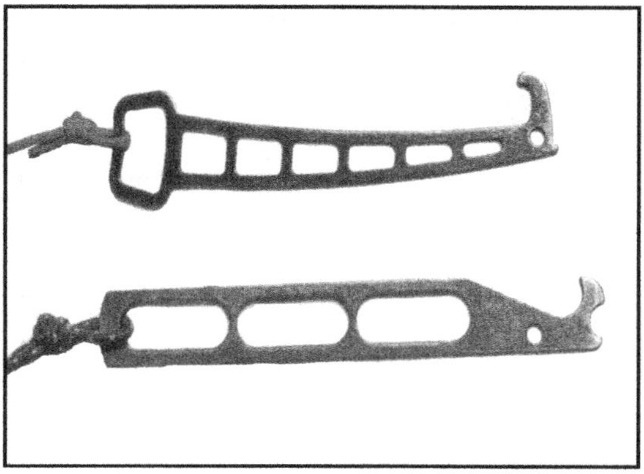

Figure18. Chock picks.

h. **Bolts.** Bolts are screw-like shafts made from metal that are drilled into rock to provide protection (Figure 3-19). The two types are contraction bolts and expansion bolts. Contraction bolts are squeezed together when driven into a rock. Expansion bolts press around a surrounding sleeve to form a snug fit into a rock. Bolts require drilling a hole into a rock, which is time-consuming, exhausting, and extremely noisy. Once emplaced, bolts are the most secure protection for a multidirectional pull. Bolts should be used only when chocks and pitons cannot be emplaced. A bolt is hammered only when it is the nail or self-driving type.

(1) A hanger (for carabiner attachment) and nut are placed on the bolt. The bolt is then inserted and driven into the hole. Because of this requirement, a hand drill must be carried in addition to a piton hammer. Hand drills (also called star drills) are available in different sizes, brands, and weights. A hand drill should have a lanyard to prevent loss.

(2) Self-driving bolts are quicker and easier to emplace. These require a hammer, bolt driver, and drilling anchor, which is driven into the rock. A bolt and carrier are then secured to the emplaced drilling anchor. All metal surfaces should be smooth and free of rust, corrosion, dirt, and moisture. Burrs, chips, and rough spots should be filed smooth and wire-brushed or rubbed clean with steel wool. Items that are cracked or warped indicate excessive wear and should be discarded.

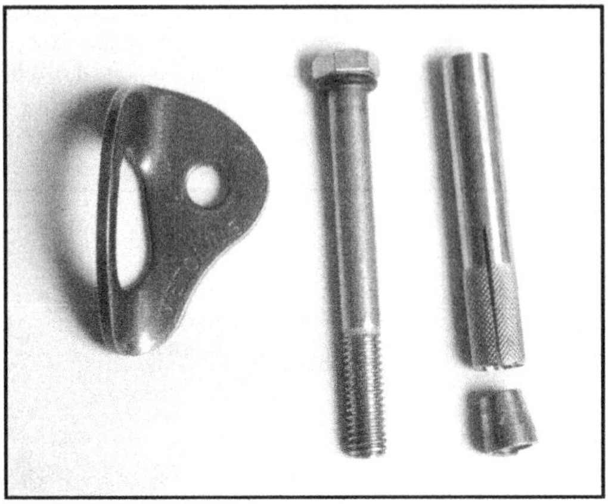

Figure 3-19. Bolts and hangers.

i. **Belay Devices.** Belay devices range from the least equipment intensive (the body belay) to high-tech metal alloy pieces of equipment. Regardless of the belay device choosen, the basic principal remains the same—friction around or through the belay device controls the ropes' movement. Belay devices are divided into three categories: the slot, the tuber, and the mechanical camming device (Figure 3-20).

(1) The slot is a piece of equipment that attaches to a locking carabiner in the harness; a bight of rope slides through the slot and into the carabiner for the belay. The most common slot type belay device is the Sticht plate.

(2) The tuber is used exactly like the slot but its shape is more like a cone or tube.

(3) The mechanical camming device is a manufactured piece of equipment that attaches to the harness with a locking carabiner. The rope is routed through this device so that when force is applied the rope is cammed into a highly frictioned position.

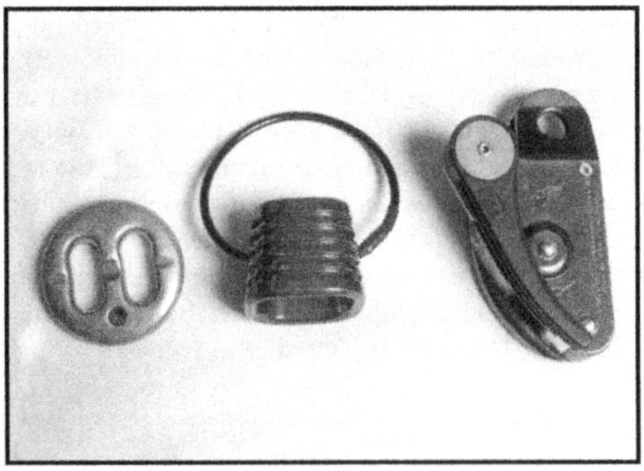

Figure 3-20. Slot, tuber, mechanical camming device.

j. **Descenders.** One piece of equipment used for generations as a descender is the carabiner. A figure-eight is another useful piece of equipment and can be used in conjunction with the carabiner for descending (Figure 3-21).

Note: All belay devices can also be used as descending devices.

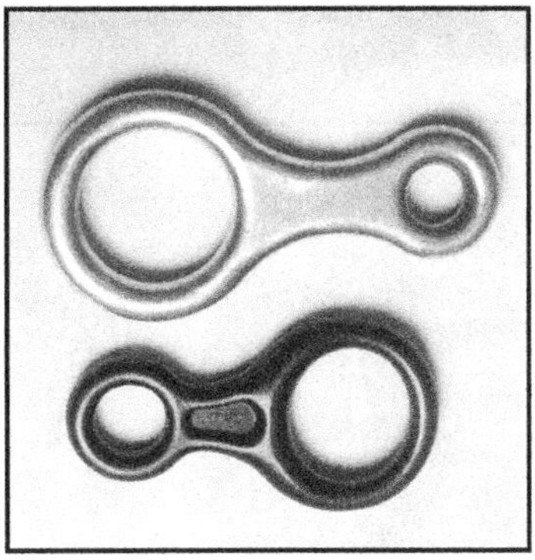

Figure 3-21. Figure-eights.

k. **Ascenders.** Ascenders may be used in other applications such as a personal safety or hauling line cam. All modern ascenders work on the principle of using a cam-like device to allow movement in one direction. Ascenders are primarily made of metal alloys and come in a variety of sizes (Figure 3-22). For difficult vertical terrain, two ascenders work best. For lower angle movement, one ascender is sufficient. Most manufacturers make ascenders as a right and left-handed pair.

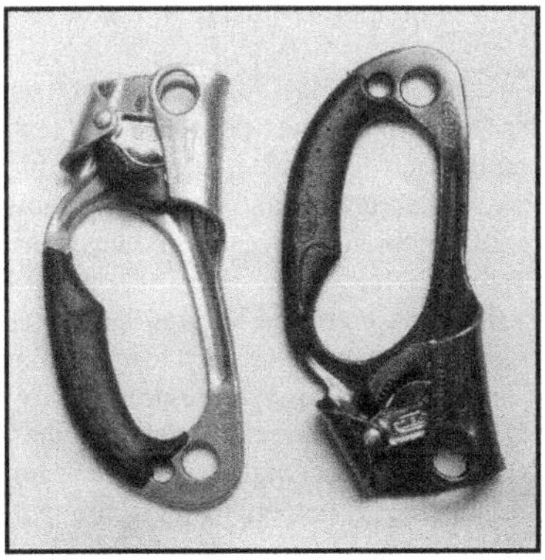

Figure 3-22. Ascenders.

l. **Pulleys.** Pulleys are used to change direction in rope systems and to create mechanical advantage in hauling systems. A pulley should be small, lightweight, and strong. They should accommodate the largest diameter of rope being used. Pulleys are made with several bearings, different-sized sheaves (wheel), and metal alloy sideplates (Figure 3-23). Plastic pulleys should always be avoided. The sideplate should rotate on the pulley axle to allow the pulley to be attached at any point along the rope. For best results, the sheave diameter must be at least four times larger than the rope's diameter to maintain high rope strength.

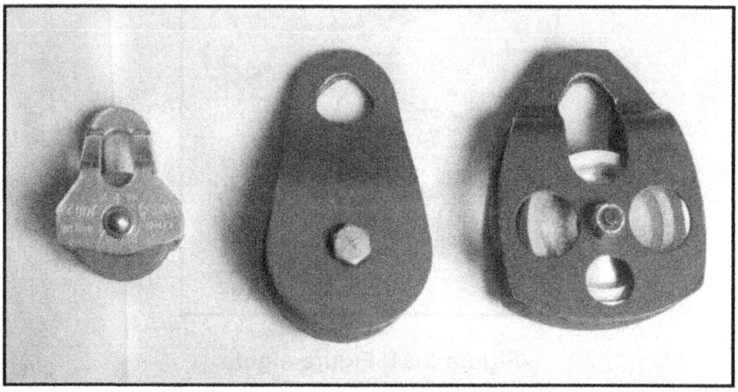

Figure 3-23. Pulley.

3-5. SNOW AND ICE CLIMBING HARDWARE

Snow and ice climbing hardware is the equipment that is particular to operations in some mountainous terrain. Specific training on this type of equipment is essential for safe use. Terrain that would otherwise be inaccessible—snowfields, glaciers, frozen waterfalls—can now be considered avenues of approach using the snow and ice climbing gear listed in this paragraph.

a. **Ice Ax.** The ice ax is one of the most important tools for the mountaineer operating on snow or ice. The climber must become proficient in its use and handling. The versatility of the ax lends itself to balance, step cutting, probing, self-arrest, belays, anchors, direct-aid climbing, and ascending and descending snow and ice covered routes.

(1) Several specific parts comprise an ice ax: the shaft, head (pick and adze), and spike (Figure 3-24, page 3-22).

(a) The shaft (handle) of the ax comes in varying lengths (the primary length of the standard mountaineering ax is 70 centimeters). It can be made of fiberglass, hollow aluminum, or wood; the first two are stronger, therefore safer for mountaineering.

(b) The head of the ax, which combines the pick and the adze, can have different configurations. The pick should be curved slightly and have teeth at least one-fourth of its length. The adze, used for chopping, is perpendicular to the shaft. It can be flat or curved along its length and straight or rounded from side to side. The head can be of one-piece construction or have replaceable picks and adzes. The head should have a hole directly above the shaft to allow for a leash to be attached.

(c) The spike at the bottom of the ax is made of the same material as the head and comes in a variety of shapes.

(2) As climbing becomes more technical, a shorter ax is much more appropriate, and adding a second tool is a must when the terrain becomes vertical. The shorter ax has all the attributes of the longer ax, but it is anywhere from 40 to 55 centimeters long and can have a straight or bent shaft depending on the preference of the user.

b. **Ice Hammer.** The ice hammer is as short or shorter than the technical ax (Figure 3-24). It is used for pounding protection into the ice or pitons into the rock. The only difference between the ice ax and the ice hammer is the ice hammer has a hammerhead instead of an adze. Most of the shorter ice tools have a hole in the shaft to which a leash is secured, which provides a more secure purchase in the ice.

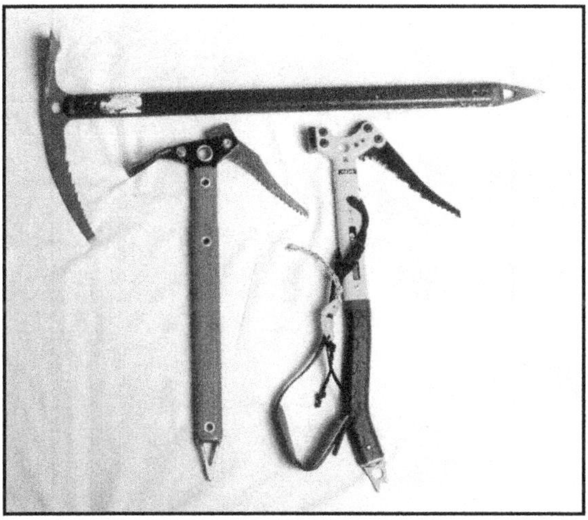

Figure 3-24. Ice ax and ice hammers.

c. **Crampons.** Crampons are used when the footing becomes treacherous. They have multiple spikes on the bottom and spikes protruding from the front (Figure 3-25). Two types of crampons are available: flexible and rigid. Regardless of the type of crampon chosen, fit is the most important factor associated with crampon wear. The crampon should fit snugly on the boot with a minimum of 1 inch of front point protruding. Straps should fit snugly around the foot and any long, loose ends should be trimmed. Both flexible and rigid crampons come in pairs, and any tools needed for adjustment will be provided by the manufacturer.

(1) The hinged or flexible crampon is best used when no technical ice climbing will be done. It is designed to be used with soft, flexible boots, but can be attached to plastic

mountaineering boots. The flexible crampon gets its name from the flexible hinge on the crampon itself. All flexible crampons are adjustable for length while some allow for width adjustment. Most flexible crampons will attach to the boot by means of a strap system. The flexible crampon can be worn with a variety of boot types.

(2) The rigid crampon, as its name implies, is rigid and does not flex. This type of crampon is designed for technical ice climbing, but can be used on less vertical terrain. The rigid crampon can only be worn with plastic mountaineering boots. Rigid crampons will have a toe and heel bail attachment with a strap that wraps around the ankle.

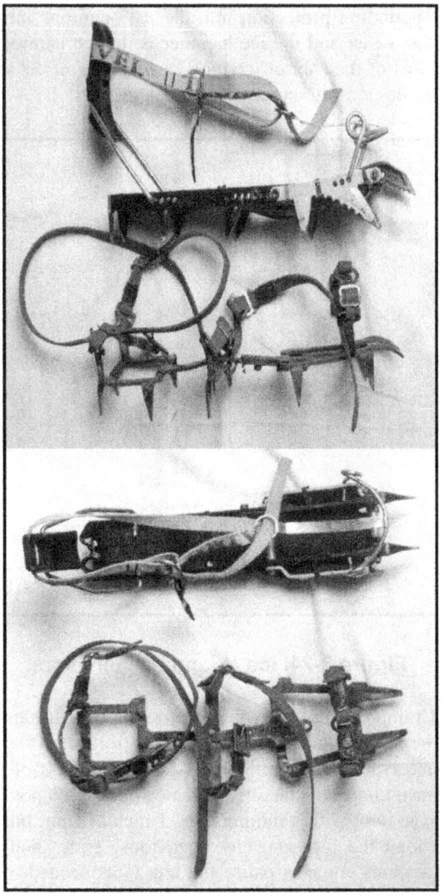

Figure 3-25. Crampons.

d. **Ice Screws.** Ice screws provide artificial protection for climbers and equipment for operations in icy terrain. They are screwed into ice formations. Ice screws are made of chrome-molybdenum steel and vary in lengths from 11 centimeters to 40 centimeters (Figure 3-26). The eye is permanently affixed to the top of the ice screw. The tip consists of milled or hand-ground teeth, which create sharp points to grab the ice when being emplaced. The ice screw has right-hand threads to penetrate the ice when turned clockwise.

(1) When selecting ice screws, choose a screw with a large thread count and large hollow opening. The close threads will allow for ease in turning and better strength. The large hollow opening will allow snow and ice to slide through when turning.

- Type I is 17 centimeters in length with a hollow inner tube.
- Type II is 22 centimeters in length with a hollow inner tube.
- Other variations are hollow alloy screws that have a tapered shank with external threads, which are driven into ice and removed by rotation.

(2) Ice screws should be inspected for cracks, bends, and other deformities that may impair strength or function. If any cracks or bends are noticed, the screw should be turned in. A file may be used to sharpen the ice screw points. Steel wool should be rubbed on rusted surfaces and a thin coat of oil applied when storing steel ice screws.

Note: Ice screws should always be kept clean and dry. The threads and teeth should be protected and kept sharp for ease of application.

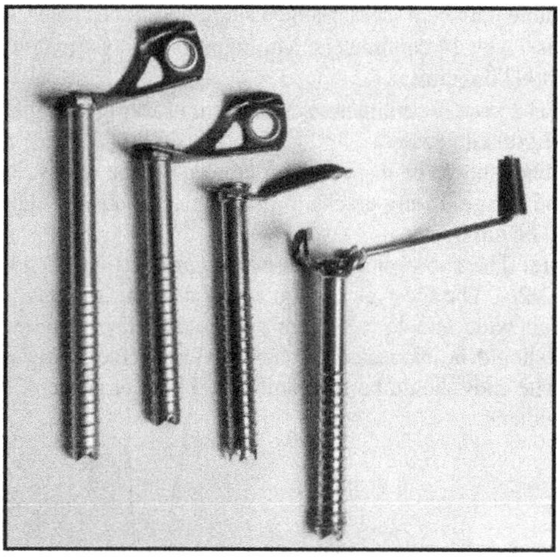

Figure 3-26. Ice screws.

e. **Ice Pitons.** Ice pitons are used to establish anchor points for climbers and equipment when conducting operations on ice. They are made of steel or steel alloys (chrome-molybdenum), and are available in various lengths and diameters (Figure 3-27). They are tubular with a hollow core and are hammered into ice with an ice hammer. The eye is permanently fixed to the top of the ice piton. The tip may be beveled to help grab the ice to facilitate insertion. Ice pitons are extremely strong when placed properly in hard ice. They can, however, pull out easily on warm days and require a considerable amount of effort to extract in cold temperatures.

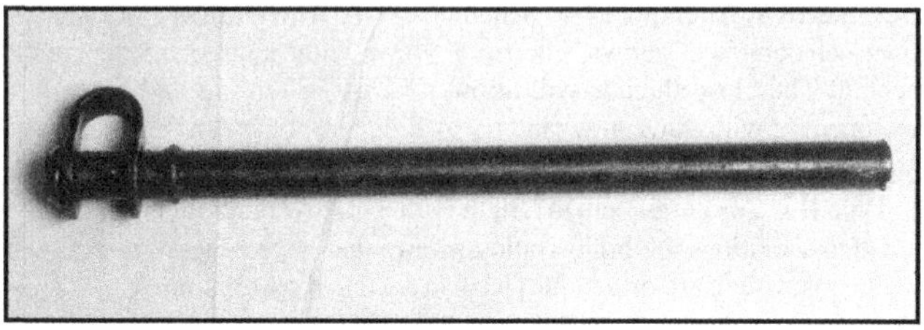

Figure 3-27. Ice piton.

f. **Wired Snow Anchors.** The wired snow anchor (or fluke) provides security for climbers and equipment in operations involving steep ascents by burying the snow anchor into deep snow (Figure 3-28, page 3-26). The fluted anchor portion of the snow anchor is made of aluminum. The wired portion is made of either galvanized steel or stainless steel. Fluke anchors are available in various sizes—their holding ability generally increases with size. They are available with bent faces, flanged sides, and fixed cables. Common types are:

- Type I is 22 by 14 centimeters. Minimum breaking strength of the swaged wire loop is 600 kilograms.
- Type II is 25 by 20 centimeters. Minimum breaking strength of the swaged wire loop is 1,000 kilograms.

The wired snow anchor should be inspected for cracks, broken wire strands, and slippage of the wire through the swage. If any cracks, broken wire strands, or slippage is noticed, the snow anchor should be turned in.

g. **Snow Picket.** The snow picket is used in constructing anchors in snow and ice (Figure 3-28, page 3-26). The snow picket is made of a strong aluminum alloy 3 millimeters thick by 4 centimeters wide, and 45 to 90 centimeters long. They can be angled or T–section stakes. The picket should be inspected for bends, chips, cracks, mushrooming ends, and other deformities. The ends should be filed smooth. If bent or cracked, the picket should be turned in for replacement.

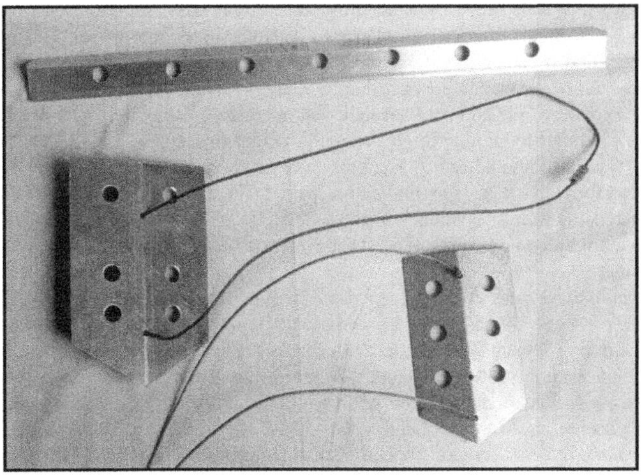

Figure 3-28. Snow anchors, flukes, and pickets.

3-6. SUSTAINABILITY EQUIPMENT

This paragraph describes all additional equipment not directly involved with climbing. This equipment is used for safety (avalanche equipment, wands), bivouacs, movement, and carrying gear. While not all of it will need to be carried on all missions, having the equipment available and knowing how to use it correctly will enhance the unit's capability in mountainous terrain.

a. **Snow Saw**. The snow saw is used to cut into ice and snow. It can be used in step cutting, in shelter construction, for removing frozen obstacles, and for cutting snow stability test pits. The special tooth design of the snow saw easily cuts into frozen snow and ice. The blade is a rigid aluminum alloy of high strength about 3 millimeters thick and 38 centimeters long with a pointed end to facilitate entry on the forward stroke. The handle is either wooden or plastic and is riveted to the blade for a length of about 50 centimeters. The blade should be inspected for rust, cracks, warping, burrs, and missing or dull teeth. A file can repair most defects, and steel wool can be rubbed on rusted areas. The handle should be inspected for cracks, bends, and stability. On folding models, the hinge and nuts should be secure. If the saw is beyond repair, it should not be used.

b. **Snow Shovel.** The snow shovel is used to cut and remove ice and snow. It can be used for avalanche rescue, shelter construction, step cutting, and removing obstacles. The snow shovel is made of a special, lightweight aluminum alloy. The handle should be telescopic, folding, or removable to be compact when not in use. The shovel should have a flat or rounded bottom and be of strong construction. The shovel should be inspected for cracks, bends, rust, and burrs. A file and steel wool can remove rust and put an edge on the blade of the shovel. The handle should be inspected for cracks, bends, and stability. If the shovel is beyond repair, it should be turned in.

c. **Wands.** Wands are used to identify routes, crevasses, snow-bridges, caches, and turns on snow and glaciers. Spacing of wands depends on the number of turns, number of hazards identified, weather conditions (and visibility), and number of teams in the climbing party. Carry too many wands is better than not having enough if they become lost. Wands are 1 to 1.25 meters long and made of lightweight bamboo or plastic shafts pointed on one end with a plastic or nylon flag (bright enough in color to see at a distance) attached to the other end. The shafts should be inspected for cracks, bends, and deformities. The flag should be inspected for tears, frays, security to the shaft, fading, and discoloration. If any defects are discovered, the wands should be replaced.

d. **Avalanche rescue equipment.** Avalanche rescue equipment (Figure 3-29) includes the following:

(1) ***Avalanche Probe.*** Although ski poles may be used as an emergency probe when searching for a victim in an avalanche, commercially manufactured probes are better for a thorough search. They are 9-millimeter thick shafts made of an aluminum alloy, which can be joined to probe up to 360 centimeters. The shafts must be strong enough to probe through avalanche debris. Some manufacturers of ski poles design poles that are telescopic and mate with other poles to create an avalanche probe.

(2) ***Avalanche Transceivers.*** These are small, compact radios used to identify avalanche burial sites. They transmit electromagnetic signals that are picked up by another transceiver on the receive mode.

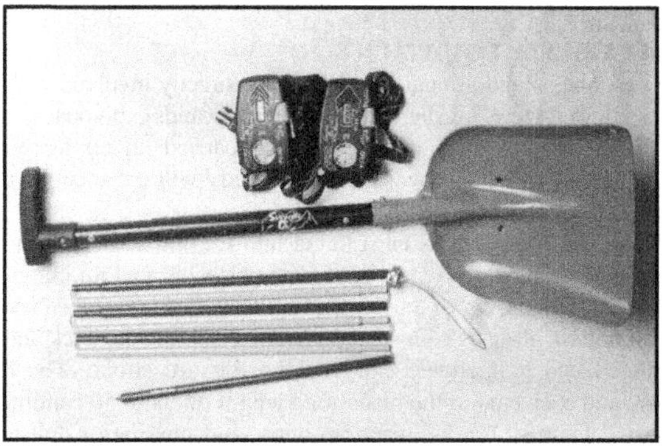

Figure 3-29. Avalanche rescue equipment.

e. **Packs.** Many types and brands of packs are used for mountaineering. The two most common types are internal and external framed packs.

(1) Internal framed packs have a rigid frame within the pack that help it maintain its shape and hug the back. This assists the climber in keeping their balance as they climb or ski. The weight in an internal framed pack is carried low on the body assisting with balance. The body-hugging nature of this type pack also makes it uncomfortable in warm weather.

(2) External framed packs suspend the load away from the back with a ladder-like frame. The frame helps transfer the weight to the hips and shoulders easier, but can be cumbersome when balance is needed for climbing and skiing.

(3) Packs come in many sizes and should be sized appropriately for the individual according to manufacturer's specifications. Packs often come with many unneeded features. A good rule of thumb is: The simpler the pack, the better it will be.

f. **Stoves.** When selecting a stove one must define its purpose—will the stove be used for heating, cooking or both? Stoves or heaters for large elements can be large and cumbersome. Stoves for smaller elements might just be used for cooking and making water, and are simple and lightweight. Stoves are a necessity in mountaineering for cooking and making water from snow and ice. When choosing a stove, factors that should be considered are weight, altitude and temperature where it will be used, fuel availability, and its reliability.

(1) There are many choices in stove design and in fuel types. White gas, kerosene, and butane are the common fuels used. All stoves require a means of pressurization to force the fuel to the burner. Stoves that burn white gas or kerosene have a hand pump to generate the pressurization and butane stoves have pressurized cartridges. All stoves need to vaporize the liquid fuel before it is burned. This can be accomplished by burning a small amount of fuel in the burner cup assembly, which will vaporize the fuel in the fuel line.

(2) Stoves should be tested and maintained prior to a mountaineering mission. They should be easy to clean and repair during an operation. The reliability of the stove has a huge impact on the success of the mission and the morale of personnel.

g. **Tents.** When selecting a tent, the mission must be defined to determine the number of people the tent will accommodate. The climate the tents will be used in is also of concern. A tent used for warmer temperatures will greatly differ from tents used in a colder, more harsh environment. Manufacturers of tents offer many designs of different sizes, weights, and materials.

(1) Mountaineering tents are made out of a breathable or weatherproof material. A single-wall tent allows for moisture inside the tent to escape through the tent's material. A double-wall tent has a second layer of material (referred to as a fly) that covers the tent. The fly protects against rain and snow and the space between the fly and tent helps moisture to escape from inside. Before using a new tent, the seams should be treated with seam sealer to prevent moisture from entering through the stitching.

(2) The frame of a tent is usually made of an aluminum or carbon fiber pole. The poles are connected with an elastic cord that allows them to extend, connect, and become long and rigid. When the tent poles are secured into the tent body, they create the shape of the tent.

(3) Tents are rated by a "relative strength factor," the speed of wind a tent can withstand before the frame deforms. Temperature and expected weather for the mission should be determined before choosing the tent.

h. **Skis.** Mountaineering skis are wide and short. They have a binding that pivots at the toe and allows for the heel to be free for uphill travel or locked for downhill. Synthetic skins with fibers on the bottom can be attached to the bottom of the ski and allow the ski to travel forward and prevent slipping backward. The skins aid in traveling uphill and slow down the rate of descents. Wax can be applied to the ski to aid in ascents instead of skins. Skis can decrease the time needed to reach an objective depending on the ability of the user. Skis can make crossing crevasses easier because of the load distribution, and they can become a makeshift stretcher for casualties. Ski techniques can be complicated and require thorough training for adequate proficiency.

i. **Snowshoes.** Snowshoes are the traditional aid to snow travel that attach to most footwear and have been updated into small, lightweight designs that are more efficient than older models. Snowshoes offer a large displacement area on top of soft snow preventing tiresome post-holing. Some snowshoes come equipped with a crampon like binding that helps in ascending steep snow and ice. Snowshoes are slower than skis, but are better suited for mixed terrain, especially if personnel are not experienced with the art of skiing. When carrying heavy packs, snowshoes can be easier to use than skis.

j. **Ski poles**. Ski poles were traditionally designed to assist in balance during skiing. They have become an important tool in mountaineering for aid in balance while hiking, snowshoeing, and carrying heavy packs. They can take some of the weight off of the lower body when carrying a heavy pack. Some ski poles are collapsible for ease of packing when not needed (Figure 3-30). The basket at the bottom prevents the pole from plunging deep into the snow and, on some models, can be detached so the pole becomes an avalanche or crevasse probe. Some ski poles come with a self-arrest grip, but should not be the only means of protection on technical terrain.

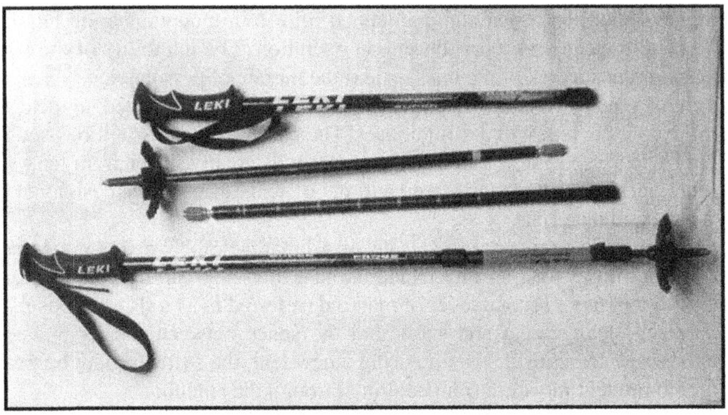

Figure 3-30. Collapsible ski poles.

k. **Sleds**. Sleds vary greatly in size, from the squad-size Ahkio, a component of the 10-man arctic tent system, to the one-person skow. Regardless of the size, sleds are an invaluable asset during mountainous operations when snow and ice is the primary surface on which to travel. Whichever sled is chosen, it must be attachable to the person or people that will be pulling it. Most sleds are constructed using fiberglass bottoms with or without exterior runners. Runners will aid the sleds ability to maintain a true track in the snow. The sled should also come with a cover of some sort—whether nylon or canvas, a cover is essential for keeping the components in the sled dry. Great care should be taken when packing the sled, especially when hauling fuel. Heavier items should be carried towards the rear of the sled and lighter items towards the front.

l. **Headlamps**. A headlamp is a small item that is not appreciated until it is needed. It is common to need a light source and the use of both hands during limited light conditions in mountaineering operations. A flashlight can provide light, but can be cumbersome when both hands are needed. Most headlamps attach to helmets by means of elastic bands.

(1) When choosing a headlamp, ensure it is waterproof and the battery apparatus is small. All components should be reliable in extreme weather conditions. When the light is being packed, care should be taken that the switch doesn't accidentally activate and use precious battery life.

(2) The battery source should compliment the resupply available. Most lights will accept alkaline, nickel-cadmium, or lithium batteries. Alkaline battery life diminishes quickly in cold temperatures, nickel-cadmium batteries last longer in cold but require a recharging unit, and lithium batteries have twice the voltage so modifications are required.

Section II. EQUIPMENT PACKING

Equipment brought on a mission is carried in the pack, worn on the body, or hauled in a sled (in winter). Obviously, the rucksack and sled (or Ahkio) can hold much more than a climber can carry. They would be used for major bivouac gear, food, water, first aid kits, climbing equipment, foul weather shells, stoves, fuel, ropes, and extra ammunition and demolition materials, if needed.

3-7. CHOICE OF EQUIPMENT

Mission requirements and unit SOP will influence the choice of gear carried but the following lists provide a sample of what should be considered during mission planning.

a. **Personal Gear.** Personal gear includes emergency survival kit containing signaling material, fire starting material, food procurement material, and water procurement material. Pocket items should include a knife, whistle, pressure bandage, notebook with pen or pencil, sunglasses, sunblock and lip protection, map, compass and or altimeter.

b. **Standard Gear.** Standard gear that can be individually worn or carried includes cushion sole socks; combat boots or mountain boots, if available; BDU and cap; LCE with canteens, magazine pouches, and first aid kit; individual weapon; a large rucksack containing waterproof coat and trousers, polypropylene top, sweater, or fleece top; helmet; poncho; and sleeping bag.

> **CAUTION**
> Cotton clothing, due to its poor insulating and moisture-wicking characteristics, is virtually useless in most mountain climates, the exception being hot, desert, or jungle mountain environments. Cotton clothing should be replaced with synthetic fabric clothing.

c. **Mountaineering Equipment and Specialized Gear.** This gear includes:

- Sling rope or climbing harness.
- Utility cord(s).
- Nonlocking carabiners.
- Locking carabiner(s).
- Rappelling gloves.
- Rappel/belay device.
- Ice ax.
- Crampons.
- Climbing rope, one per climbing team.
- Climbing rack, one per climbing team.

d. **Day Pack.** When the soldier plans to be away from the bivouac site for the day on a patrol or mountaineering mission, he carries a light day pack. This pack should contain the following items:

- Extra insulating layer: polypropylene, pile top, or sweater.
- Protective layer: waterproof jacket and pants, rain suit, or poncho.
- First aid kit.
- Flashlight or headlamp.
- Canteen.
- Cold weather hat or scarf.
- Rations for the time period away from the base camp.
- Survival kit.
- Sling rope or climbing harness.
- Carabiners.
- Gloves.
- Climbing rope, one per climbing team.
- Climbing rack, one per climbing team.

e. **Squad or Team Safety Pack.** When a squad-sized element leaves the bivouac site, squad safety gear should be carried in addition to individual day packs. This can either be loaded into one rucksack or cross-loaded among the squad members. In the event of an injury, casualty evacuation, or unplanned bivouac, these items may make the difference between success and failure of the mission.

- Sleeping bag.
- Sleeping mat.

- Squad stove.
- Fuel bottle.

f. **The Ten Essentials.** Regardless of what equipment is carried, the individual military mountaineer should always carry the "ten essentials" when moving through the mountains.

(1) ***Map.***

(2) ***Compass, Altimeter, and or GPS.***

(3) ***Sunglasses and Sunscreen.***

(a) In alpine or snow-covered sub-alpine terrain, sunglasses are a vital piece of equipment for preventing snow blindness. They should filter 95 to 100 percent of ultraviolet light. Side shields, which minimize the light entering from the side, should permit ventilation to help prevent lens fogging. At least one extra pair of sunglasses should be carried by each independent climbing team.

(b) Sunscreens should have an SPF factor of 15 or higher. For lip protection, a total UV blocking lip balm that resists sweating, washing, and licking is best. This lip protection should be carried in the chest pocket or around the neck to allow frequent reapplication.

(4) ***Extra Food.*** One day's worth extra of food should be carried in case of delay caused by bad weather, injury, or navigational error.

(5) ***Extra Clothing.*** The clothing used during the active part of a climb, and considered to be the basic climbing outfit, includes socks, boots, underwear, pants, blouse, sweater or fleece jacket, hat, gloves or mittens, and foul weather gear (waterproof, breathable outerwear or waterproof rain suit).

(a) Extra clothing includes additional layers needed to make it through the long, inactive hours of an unplanned bivouac. Keep in mind the season when selecting this gear.

- Extra underwear to switch out with sweat-soaked underwear.
- Extra hats or balaclavas.
- Extra pair of heavy socks.
- Extra pair of insulated mittens or gloves.
- In winter or severe mountain conditions, extra insulation for the upper body and the legs.

(b) To back up foul weather gear, bring a poncho or extra-large plastic trash bag. A reflective emergency space blanket can be used for hypothermia first aid and emergency shelter. Insulated foam pads prevent heat loss while sitting or lying on snow. Finally, a bivouac sack can help by protecting insulating layers from the weather, cutting the wind, and trapping essential body heat inside the sack.

(6) ***Headlamp and or Flashlight.*** Headlamps provide the climber a hands-free capability, which is important while climbing, working around the camp, and employing weapons systems. Miniature flashlights can be used, but commercially available headlamps are best. Red lens covers can be fabricated for tactical conditions. Spare batteries and spare bulbs should also be carried.

(7) ***First-aid Kit.*** Decentralized operations, the mountain environment—steep, slick terrain and loose rock combined with heavy packs, sharp tools, and fatigue—requires each climber to carry his own first-aid kit. Common mountaineering injuries that can be expected are punctures and abrasions with severe bleeding, a broken bone, serious sprain,

and blisters. Therefore, the kit should contain at least enough material to stabilize these conditions. Pressure dressings, gauze pads, elastic compression wrap, small adhesive bandages, butterfly bandages, moleskin, adhesive tape, scissors, cleanser, latex gloves and splint material (if above tree line) should all be part of the kit.

(8) ***Fire Starter.*** Fire starting material is key to igniting wet wood for emergency campfires. Candles, heat tabs, and canned heat all work. These can also be used for quick warming of water or soup in a canteen cup. In alpine zones above tree line with no available firewood, a stove works as an emergency heat source.

(9) ***Matches and Lighter.*** Lighters are handy for starting fires, but they should be backed up by matches stored in a waterproof container with a strip of sandpaper.

(10) ***Knife.*** A multipurpose pocket tool should be secured with cord to the belt, harness, or pack.

g. **Other Essential Gear.** Other essential gear may be carried depending on mission and environmental considerations.

(1) ***Water and Water Containers.*** These include wide-mouth water bottles for water collection; camel-back type water holders for hands-free hydration; and a small length of plastic tubing for water procurement at snow-melt seeps and rainwater puddles on bare rock.

(2) ***Ice Ax.*** The ice ax is essential for travel on snowfields and glaciers as well as snow-covered terrain in spring and early summer. It helps for movement on steep scree and on brush and heather covered slopes, as well as for stream crossings.

(3) ***Repair Kit.*** A repair kit should include:

- Stove tools and spare parts.
- Duct tape.
- Patches.
- Safety pins.
- Heavy-duty thread.
- Awl and or needles.
- Cord and or wire.
- Small pliers (if not carrying a multipurpose tool).
- Other repair items as needed.

(4) ***Insect Repellent.***

(5) ***Signaling Devices.***

(6) ***Snow Shovel.***

3-8. TIPS ON PACKING

When loading the internal frame pack the following points should be considered.

a. In most cases, speed and endurance are enhanced if the load is carried more by the hips (using the waist belt) and less by the shoulders and back. This is preferred for movement over trails or less difficult terrain. By packing the lighter, more compressible items (sleeping bag, clothing) in the bottom of the rucksack and the heavier gear (stove, food, water, rope, climbing hardware, extra ammunition) on top, nearer the shoulder blades, the load is held high and close to the back, thus placing the most weight on the hips.

b. In rougher terrain it pays to modify the pack plan. Heavy articles of gear are placed lower in the pack and close to the back, placing more weight on the shoulders and back. This lowers the climber's center of gravity and helps him to better keep his balance.

c. Equipment that may be needed during movement should be arranged for quick access using either external pockets or placing immediately underneath the top flap of the pack. As much as possible, this placement should be standardized across the team so that necessary items can be quickly reached without unnecessary unpacking of the pack in emergencies.

d. The pack and its contents should be soundly waterproofed. Clothing and sleeping bag are separately sealed and then placed in the larger wet weather bag that lines the rucksack. Zip-lock plastic bags can be used for small items, which are then organized into color-coded stuffsacks. A few extra-large plastic garbage bags should be carried for a variety of uses—spare waterproofing, emergency bivouac shelter, and water procurement, among others.

e. The ice ax, if not carried in hand, should be stowed on the outside of the pack with the spike up and the adze facing forward or to the outside, and be securely fastened. Mountaineering packs have ice ax loops and buckle fastening systems for this. If not, the ice ax is placed behind one of the side pockets, as stated above, and then tied in place.

f. Crampons should be secured to the outside rear of the pack with the points covered.

CHAPTER 4

ROPE MANAGEMENT AND KNOTS

The rope is a vital piece of equipment to the mountaineer. When climbing, rappelling, or building various installations, the mountaineer must know how to properly utilize and maintain this piece of equipment. If the rope is not managed or maintained properly, serious injury may occur. This chapter discusses common rope terminology, management techniques, care and maintenance procedures, and knots.

SECTION I. PREPARATION, CARE AND MAINTENANCE, INSPECTION, TERMINOLOGY

The service life of a rope depends on the frequency of use, applications (rappelling, climbing, rope installations), speed of descent, surface abrasion, terrain, climate, and quality of maintenance. Any rope may fail under extreme conditions (shock load, sharp edges, misuse).

4-1. PREPARATION

The mountaineer must select the proper rope for the task to be accomplished according to type, diameter, length, and tensile strength. It is important to prepare all ropes before departing on a mission. Avoid rope preparation in the field.

a. **Packaging.** New rope comes from the manufacturer in different configurations—boxed on a spool in various lengths, or coiled and bound in some manner. Precut ropes are usually packaged in a protective cover such as plastic or burlap. Do not remove the protective cover until the rope is ready for use.

b. **Securing the Ends of the Rope:** If still on a spool, the rope must be cut to the desired length. All ropes will fray at the ends unless they are bound or seared. Both static and dynamic rope ends are secured in the same manner. The ends must be heated to the melting point so as to attach the inner core strands to the outer sheath. By fusing the two together, the sheath cannot slide backward or forward. Ensure that this is only done to the ends of the rope. If the rope is exposed to extreme temperatures, the sheath could be weakened, along with the inner core, reducing overall tensile strength. The ends may also be dipped in enamel or lacquer for further protection.

4-2. CARE AND MAINTENANCE

The rope is a climber's lifeline. It must be cared for and used properly. These general guidelines should be used when handling ropes.

a. Do not step on or drag ropes on the ground unnecessarily. Small particles of dirt will be ground between the inner strands and will slowly cut them.

b. While in use, do not allow the rope to come into contact with sharp edges. Nylon rope is easily cut, particularly when under tension. If the rope must be used over a sharp edge, pad the edge for protection.

c. Always keep the rope as dry as possible. Should the rope become wet, hang it in large loops off the ground and allow it to dry. Never dry a rope with high heat or in direct sunlight.

d. Never leave a rope knotted or tightly stretched for longer than necessary. Over time it will reduce the strength and life of the rope.

e. Never allow one rope to continuously rub over or against another. Allowing rope-on-rope contact with nylon rope is extremely dangerous because the heat produced by the friction will cause the nylon to melt.

f. Inspect the rope before each use for frayed or cut spots, mildew or rot, or defects in construction (new rope).

g. The ends of the rope should be whipped or melted to prevent unraveling.

h. Do not splice ropes for use in mountaineering.

i. Do not mark ropes with paints or allow them to come in contact with oils or petroleum products. Some of these will weaken or deteriorate nylon.

j. Never use a mountaineering rope for any purpose except mountaineering.

k. Each rope should have a corresponding rope log (DA Form 5752-R, Rope History and Usage), which is also a safety record. It should annotate use, terrain, weather, application, number of falls, dates, and so on, and should be annotated each time the rope is used (Figure 4-1). DA Form 5752-R is authorized for local reproduction on 8 1/2- by 11-inch paper.

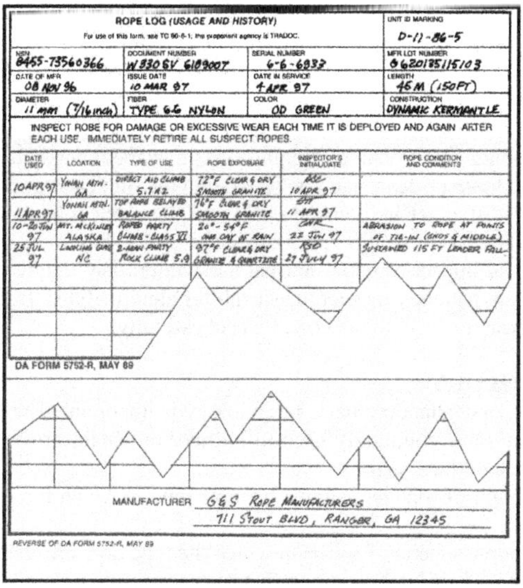

ROPE LOG (USAGE AND HISTORY)

For use of this form, see TC 90-6-1; the proponent agency is TRADOC.

NSN 8455-73560366	DOCUMENT NUMBER W330SV 6109007	SERIAL NUMBER 6-6-6933	MFR LOT NUMBER 0620135115103
DATE OF MFR 08 NOV 96	ISSUE DATE 10 MAR 97	DATE IN SERVICE 4 APR 97	LENGTH 46M (150FT)
DIAMETER 11 mm (7/16 inch)	FIBER TYPE 6.6 NYLON	COLOR OD GREEN	CONSTRUCTION DYNAMIC KERMANTLE

UNIT ID MARKING D-11-86-5

INSPECT ROBE FOR DAMAGE OR EXCESSIVE WEAR EACH TIME IT IS DEPLOYED AND AGAIN ARTER EACH USE. IMMEDIATELY RETIRE ALL SUSPECT ROPES.

DATE USED	LOCATION	TYPE OF USE	ROPE EXPOSURE	INSPECTOR'S INITIAL/DATE	ROPE CONDITION AND COMMENTS
10APR97	YONAH MTN-GA	DIRECT AID CLIMB 5.7A2	72°F CLEAR & DRY SMOOTH GRANITE	ABC 10APR 97	
11APR97	YONAH MTN. GA	TOP ROPE BELAYED BALANCE CLIMB	76°F CLEAR & DRY SMOOTH GRANITE	BH 11 APR 97	
10-20 JUN 97	MT. MCKINLEY ALASKA	ROPED PARTY CLIMB - CLASS VI	20°-54°F ONE DAY OF RAIN	CNR 22 JUN 97	ABRASION TO ROPE AT POINTS OF TIE-IN (ENDS & MIDDLE)
25 JUL 97	LOOKING GLASS NC	2-MAN PARTY ROCK CLIMB 5.9	97°F CLEAR & DRY GRANITE & QUARTZITE	RSD 27 JULY 97	SUSTAINED 115 FT LEADER FALL

DA FORM 5752-R, MAY 89

MANUFACTURER G&S ROPE MANUFACTURERS
711 STOUT BLVD, RANGER, GA 12345

REVERSE OF DA FORM 5752-R, MAY 89

Figure 4-1. Example of completed DA Form 5752-R.

l. Never subject the rope to high heat or flame. This will significantly weaken it.

m. All ropes should be washed periodically to remove dirt and grit, and rinsed thoroughly. Commercial rope washers are made from short pieces of modified pipe that connect to any faucet. Pinholes within the pipe force water to circulate around and scrub the rope as you slowly feed it through the washer. Another method is to machine wash, on a gentle cycle, in cold water with a nylon safe soap, never bleach or harsh cleansers. Ensure that only front loading washing machine are used to wash ropes.

n. Ultraviolet radiation (sunlight) tends to deteriorate nylon over long periods of time. This becomes important if rope installations are left in place over a number of months.

o. When not in use, ropes should be loosely coiled and hung on wooden pegs rather than nails or other metal objects. Storage areas should be relatively cool with low humidity levels to prevent mildew or rotting. Rope may also be loosely stacked and placed in a rope bag and stored on a shelf. Avoid storage in direct sunlight, as the ultraviolet radiation will deteriorate the nylon over long periods

4-3. INSPECTION

Ropes should be inspected before and after each use, especially when working around loose rock or sharp edges.

a. Although the core of the kernmantle rope cannot be seen, it is possible to damage the core without damaging the sheath. Check a kernmantle rope by carefully inspecting the sheath before and after use while the rope is being coiled. When coiling, be aware of how the rope feels as it runs through the hands. Immediately note and tie off any lumps or depressions felt.

b. Damage to the core of a kernmantle rope usually consists of filaments or yarn breakage that results in a slight retraction. If enough strands rupture, a localized reduction in the diameter of the rope results in a depression that can be felt or even seen.

c. Check any other suspected areas further by putting them under tension (the weight of one person standing on a Prusik tensioning system is about maximum). This procedure will emphasize the lump or depression by separating the broken strands and enlarging the dip. If a noticeable difference in diameter is obvious, retire the rope immediately.

d. Many dynamic kernmantle ropes are quite soft. They may retain an indention occasionally after an impact or under normal use without any trauma to the core. When damage is suspected, patiently inspect the sheath for abnormalities. Damage to the sheath does not always mean damage to the core. Inspect carefully.

4-4. TERMINOLOGY

When using ropes, understanding basic terminology is important. The terms explained in this section are the most commonly used in military mountaineering. (Figure 4-2, page 4-4, illustrates some of these terms.)

a. **Bight**. A bight of rope is a simple bend of rope in which the rope does not cross itself.

b. **Loop.** A loop is a bend of a rope in which the rope does cross itself.

c. **Half Hitch**. A half hitch is a loop that runs around an object in such a manner as to lock or secure itself.

d. **Turn.** A turn wraps around an object, providing 360-degree contact.

e. **Round Turn**. A round turn wraps around an object one and one-half times. A round turn is used to distribute the load over a small diameter anchor (3 inches or less). It may also be used around larger diameter anchors to reduce the tension on the knot, or provide added friction.

f. **Running End**. A running end is the loose or working end of the rope.

g. **Standing Part**. The standing part is the static, stationary, or nonworking end of the rope.

h. **Lay.** The lay is the direction of twist used in construction of the rope.

i. **Pigtail.** The pigtail (tail) is the portion of the running end of the rope between the safety knot and the end of the rope.

j. **Dress.** Dress is the proper arrangement of all the knot parts, removing unnecessary kinks, twists, and slack so that all rope parts of the knot make contact.

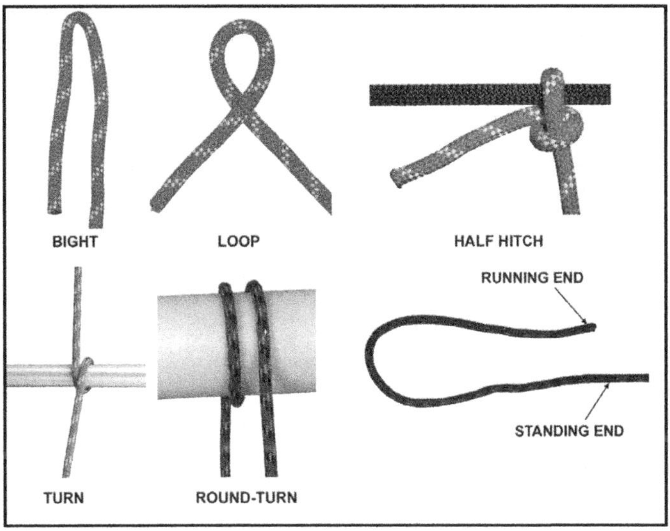

Figure 4-2. Examples of roping terminology.

Section II. COILING, CARRYING, THROWING

The ease and speed of rope deployment and recovery greatly depends upon technique and practice.

4-5. COILING AND CARRYING THE ROPE

Use the butterfly or mountain coil to coil and carry the rope. Each is easy to accomplish and results in a minimum amount of kinks, twists, and knots later during deployment.

a. **Mountain Coil.** To start a mountain coil, grasp the rope approximately 1 meter from the end with one hand. Run the other hand along the rope until both arms are outstretched. Grasping the rope firmly, bring the hands together forming a loop, which is laid in the hand closest to the end of the rope. This is repeated, forming uniform loops that run in a clockwise direction, until the rope is completely coiled. The rope may be given a 1/4 twist as each loop is formed to overcome any tendency for the rope to twist or form figure-eights.

(1) In finishing the mountain coil, form a bight approximately 30 centimeters long with the starting end of the rope and lay it along the top of the coil. Uncoil the last loop and, using this length of the rope, begin making wraps around the coil and the bight, wrapping toward the closed end of the bight and making the first wrap bind across itself so as to lock it into place. Make six to eight wraps to adequately secure the coil, and then route the end of the rope through the closed end of the bight. Pull the running end of the bight tight, securing the coil.

(2) The mountain coil may be carried either in the pack (by forming a figure eight), doubling it and placing it under the flap, or by placing it over the shoulder and under the opposite arm, slung across the chest. (Figure 4-3 shows how to coil a mountain coil.)

Figure 4-3. Mountain coil.

b. **Butterfly Coil.** The butterfly coil is the quickest and easiest technique for coiling (Figure 4-4).

Figure 4-4. Butterfly coil.

(1) ***Coiling.*** To start the double butterfly, grasp both ends of the rope and begin back feeding. Find the center of the rope forming a bight. With the bight in the left hand, grasp both ropes and slide the right hand out until there is approximately one arms length of rope. Place the doubled rope over the head, draping it around the neck and on top of the shoulders. Ensure that it hangs no lower than the waist. With the rest of the doubled rope in front of you, make doubled bights placing them over the head in the same manner as the first bight. Coil alternating from side to side (left to right, right to left) while maintaining equal-length bights. Continue coiling until approximately two arm-lengths of rope remain. Remove the coils from the neck and shoulders carefully, and hold the center in one hand. Wrap the two ends around the coils a minimum of three doubled wraps, ensuring that the first wrap locks back on itself.

(2) ***Tie-off and Carrying.*** Take a doubled bight from the loose ends of rope and pass it through the apex of the coils. Pull the loose ends through the doubled bight and dress it down. Place an overhand knot in the loose ends, dressing it down to the apex of the bight securing coils. Ensure that the loose ends do not exceed the length of the coils. In this configuration the coiled rope is secure enough for hand carrying or carrying in a rucksack, or for storage. (Figure 4-5 shows a butterfly coil tie-off.)

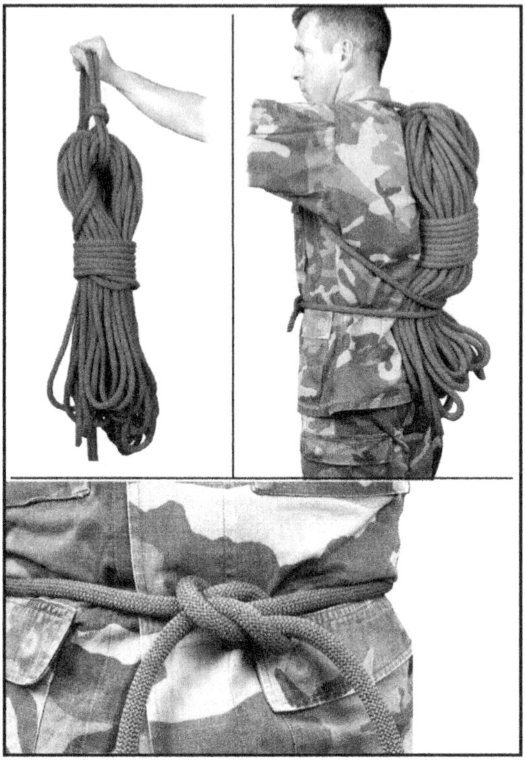

Figure 4-5. Butterfly coil tie-off.

c. **Coiling Smaller Diameter Rope.** Ropes of smaller diameters may be coiled using the butterfly or mountain coil depending on the length of the rope. Pieces 25 feet and shorter (also known as cordage, sling rope, utility cord) may be coiled so that they can be hung from the harness. Bring the two ends of the rope together, ensuring no kinks are in the rope. Place the ends of the rope in the left hand with the two ends facing the body. Coil the doubled rope in a clockwise direction forming 6- to 8-inch coils (coils may be larger depending on the length of rope) until an approximate 12-inch bight is left. Wrap that bight around the coil, ensuring that the first wrap locks on itself. Make three or more wraps. Feed the bight up through the bights formed at the top of the coil. Dress it down tightly. Now the piece of rope may be hung from a carabiner on the harness.

e. **Uncoiling, Back-feeding, and Stacking.** When the rope is needed for use, it must be uncoiled and stacked on the ground properly to avoid kinks and snarls.

(1) Untie the tie-off and lay the coil on the ground. Back-feed the rope to minimize kinks and snarls. (This is also useful when the rope is to be moved a short distance and

coiling is not desired.) Take one end of the rope in the left hand and run the right hand along the rope until both arms are outstretched. Next, lay the end of the rope in the left hand on the ground. With the left hand, re-grasp the rope next to the right hand and continue laying the rope on the ground.

(2) The rope should be laid or stacked in a neat pile on the ground to prevent it from becoming tangled and knotted when throwing the rope, feeding it to a lead climber, and so on. This technique can also be started using the right hand.

4-6. THROWING THE ROPE

Before throwing the rope, it must be properly managed to prevent it from tangling during deployment. The rope should first be anchored to prevent complete loss of the rope over the edge when it is thrown. Several techniques can be used when throwing a rope. Personal preference and situational and environmental conditions should be taken into consideration when determining which technique is best.

a. Back feed and neatly stack the rope into coils beginning with the anchored end of the rope working toward the running end. Once stacked, make six to eight smaller coils in the left hand. Pick up the rest of the larger coils in the right hand. The arm should be generally straight when throwing. The rope may be thrown underhanded or overhanded depending on obstacles around the edge of the site. Make a few preliminary swings to ensure a smooth throw. Throw the large coils in the right hand first. Throw up and out. A slight twist of the wrist, so that the palm of the hand faces up as the rope is thrown, allows the coils to separate easily without tangling. A smooth follow through is essential. When a slight tug on the left hand is felt, toss the six to eight smaller coils out. This will prevent the ends of the rope from becoming entangled with the rest of the coils as they deploy. As soon as the rope leaves the hand, the thrower should sound off with a warning of "ROPE" to alert anyone below the site.

b. Another technique may also be used when throwing rope. Anchor, back feed, and stack the rope properly as described above. Take the end of the rope and make six to eight helmet-size coils in the right hand (more may be needed depending on the length of the rope). Assume a "quarterback" simulated stance. Aiming just above the horizon, vigorously throw the rope overhanded, up and out toward the horizon. The rope must be stacked properly to ensure smooth deployment.

c. When windy weather conditions prevail, adjustments must be made. In a strong cross wind, the rope should be thrown angled into the wind so that it will land on the desired target. The stronger the wind, the harder the rope must be thrown to compensate.

SECTION III. KNOTS

All knots used by a mountaineer are divided into four classes: Class I—joining knots, Class II—anchor knots, Class III—middle rope knots, and Class IV—special knots. The variety of knots, bends, bights, and hitches is almost endless. These classes of knots are intended only as a general guide since some of the knots discussed may be appropriate in more than one class. The skill of knot tying can perish if not used and practiced. With experience and practice, knot tying becomes instinctive and helps the mountaineer in many situations.

4-7. SQUARE KNOT

The square knot is used to tie the ends of two ropes of equal diameter (Figure 4-6). It is a joining knot.

a. **Tying the Knot.**

STEP 1. Holding one working end in each hand, place the working end in the right hand over the one in the left hand.

STEP 2. Pull it under and back over the top of the rope in the left hand.

STEP 3. Place the working end in the left hand over the one in the right hand and repeat STEP 2.

STEP 4. Dress the knot down and secure it with an overhand knot on each side of the square knot.

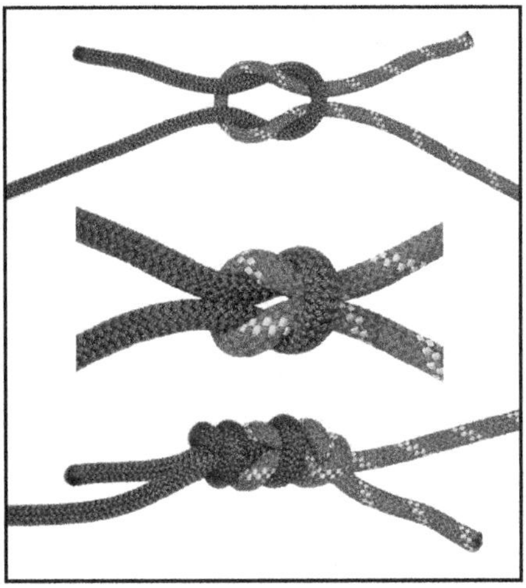

Figure 4-6. Square knot.

b. **Checkpoints**.

(1) There are two interlocking bights.

(2) The running end and standing part are on the same side of the bight formed by the other rope.

(3) The running ends are parallel to and on the same side of the standing ends with 4-inch minimum pig tails after the overhand safeties are tied.

4-8. FISHERMAN'S KNOT

The fisherman's knot is used to tie two ropes of the same or approximately the same diameter (Figure 4-7, page 4-10). It is a joining knot.

a. **Tying the Knot.**

STEP 1. Tie an overhand knot in one end of the rope.

STEP 2. Pass the working end of the other rope through the first overhand knot. Tie an overhand knot around the standing part of the first rope with the working end of the second rope.

STEP 3. Tightly dress down each overhand knot and tightly draw the knots together.

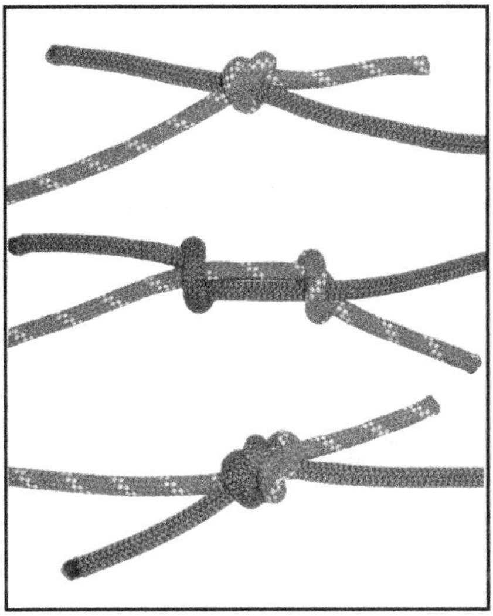

Figure 4-7. Fisherman's knot.

b. **Checkpoints.**

(1) The two separate overhand knots are tied tightly around the long, standing part of the opposing rope.

(2) The two overhand knots are drawn snug.

(3) Ends of rope exit knot opposite each other with 4-inch pigtails.

4-9. DOUBLE FISHERMAN'S KNOT

The double fisherman's knot (also called double English or grapevine) is used to tie two ropes of the same or approximately the same diameter (Figure 4-8). It is a joining knot.

a. **Tying the Knot.**

STEP 1. With the working end of one rope, tie two wraps around the standing part of another rope.

STEP 2. Insert the working end (STEP 1) back through the two wraps and draw it tight.

STEP 3. With the working end of the other rope, which contains the standing part (STEPS 1 and 2), tie two wraps around the standing part of the other rope (the working end in STEP 1). Insert the working end back through the two wraps and draw tight.

STEP 4. Pull on the opposing ends to bring the two knots together.

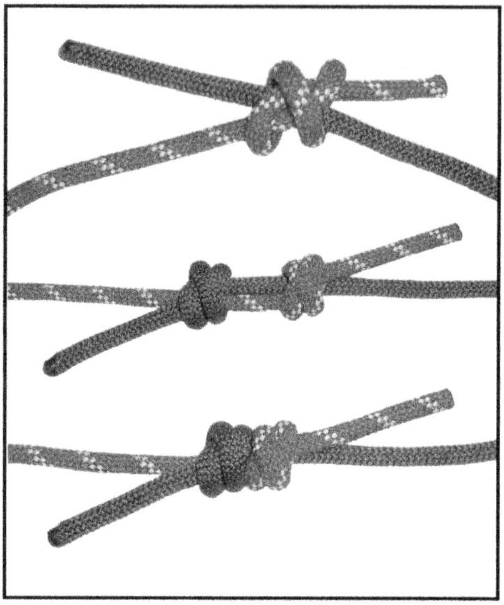

Figure 4-8. Double fisherman's knot.

b. **Checkpoints.**

(1) Two double overhand knots securing each other as the standing parts of the rope are pulled apart.

(2) Four rope parts on one side of the knot form two "x" patterns, four rope parts on the other side of the knot are parallel.

(3) Ends of rope exit knot opposite each other with 4-inch pigtails.

4-10. FIGURE-EIGHT BEND

The figure-eight bend is used to join the ends of two ropes of equal or unequal diameter within 5-mm difference (Figure 4-9, page 4-12).

a. **Tying the Knot.**

STEP 1. Grasp the top of a 2-foot bight.

STEP 2. With the other hand, grasp the running end (short end) and make a 360-degree turn around the standing end.

STEP 3. Place the running end through the loop just formed creating an in-line figure eight.

STEP 4. Route the running end of the other ripe back through the figure eight starting from the original rope's running end. Trace the original knot to the standing end.

STEP 5. Remove all unnecessary twists and crossovers. Dress the knot down.

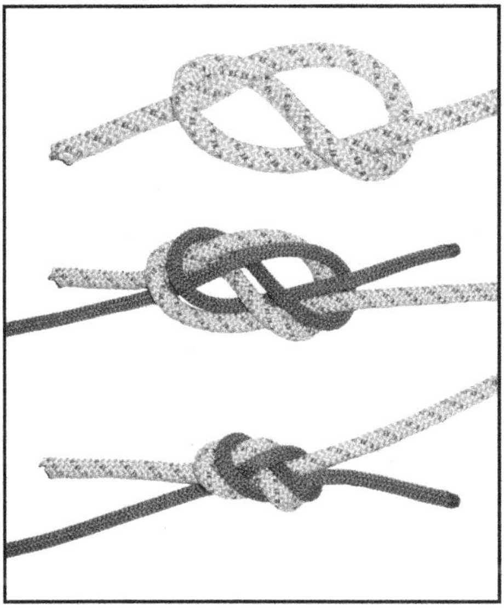

Figure 4-9. Figure-eight bend.

b. **Checkpoints.**

(1) There is a figure eight with two ropes running side by side.

(2) The running ends are on opposite sides of the knot.

(3) There is a minimum 4-inch pigtail.

4-11. WATER KNOT

The water knot is used to attach two webbing ends (Figure 4-10). It is also called a ring bend, overhand retrace, or tape knot. It is used in runners and harnesses and is a joining knot.

a. **Tying the Knot.**

STEP 1. Tie an overhand knot in one of the ends.

STEP 2. Feed the other end back through the knot, following the path of the first rope in reverse.

STEP 3. Draw tight and pull all of the slack out of the knot. The remaining tails must extend at least 4 inches beyond the knot in both directions.

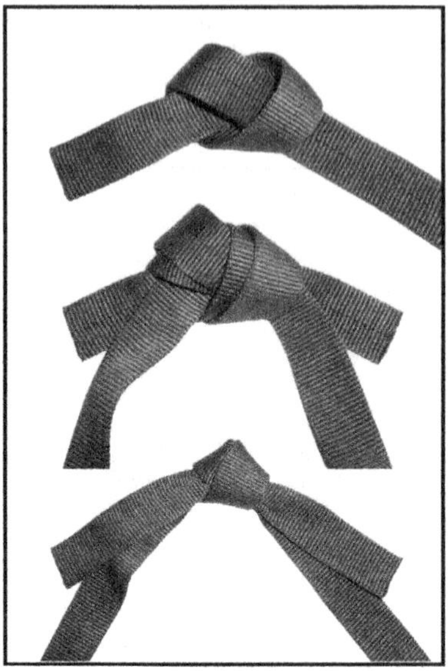

Figure 4-10. Water knot.

b. **Checkpoints.**

(1) There are two overhand knots, one retracing the other.

(2) There is no slack in the knot, and the working ends come out of the knot in opposite directions.

(3) There is a minimum 4-inch pigtail.

4-12. BOWLINE

The bowline is used to tie the end of a rope around an anchor. It may also be used to tie a single fixed loop in the end of a rope (Figure 4-11, page 4-14). It is an anchor knot.

a. **Tying the Knot.**

STEP 1. Bring the working end of the rope around the anchor, from right to left (as the climber faces the anchor).

STEP 2. Form an overhand loop in the standing part of the rope (on the climber's right) toward the anchor.

STEP 3. Reach through the loop and pull up a bight.

STEP 4. Place the working end of the rope (on the climber's left) through the bight, and bring it back onto itself. Now dress the knot down.

STEP 5. Form an overhand knot with the tail from the bight.

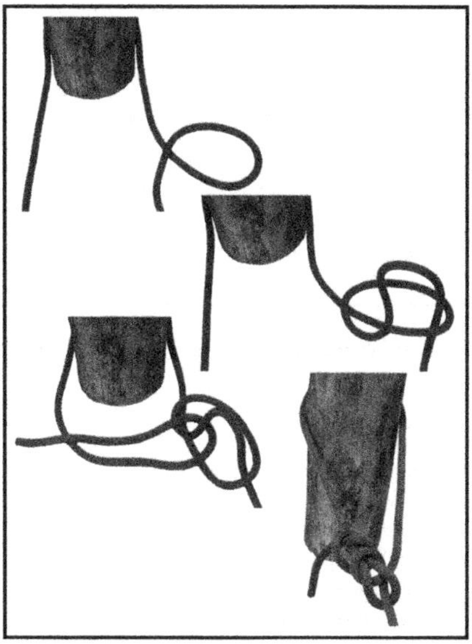

Figure 4-11. Bowline knot.

b. **Checkpoints.**

(1) The bight is locked into place by a loop.

(2) The short portion of the bight is on the inside and on the loop around the anchor (or inside the fixed loop).

(3) There is a minimum 4-inch pigtail after tying the overhand safety.

4-13. ROUND TURN AND TWO HALF HITCHES

This knot is used to tie the end of a rope to an anchor, and it must have constant tension (Figure 4-12). It is an anchor knot.

a. **Tying the Knot.**

STEP 1. Route the rope around the anchor from right to left and wrap down (must have two wraps in the rear of the anchor, and one in the front). Run the loop around the object to provide 360-degree contact, distributing the load over the anchor.

STEP 2. Bring the working end of the rope left to right and over the standing part, forming a half hitch (first half hitch).

STEP 3. Repeat STEP 2 (last half hitch has a 4 inch pigtail).

STEP 4. Dress the knot down.

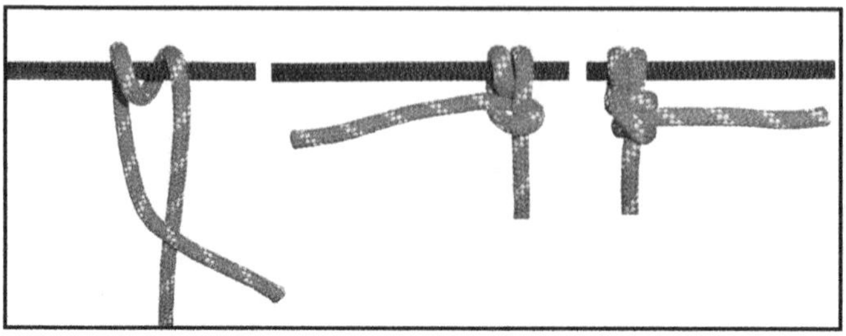

Figure 4-12. Round turn and two half hitches.

b. **Checkpoints.**

(1) A complete round turn should exist around the anchor with no crosses.

(2) Two half hitches should be held in place by a diagonal locking bar with no less than a 4-inch pigtail remaining.

4-14. FIGURE-EIGHT RETRACE (REROUTED FIGURE-EIGHT)

The figure-eight retrace knot produces the same result as a figure-eight loop. However, by tying the knot in a retrace, it can be used to fasten the rope to trees or to places where the loop cannot be used (Figure 4-13, page 4-16). It is also called a rerouted figure-eight and is an anchor knot.

a. **Tying the Knot.**

STEP 1. Use a length of rope long enough to go around the anchor, leaving enough rope to work with.

STEP 2. Tie a figure-eight knot in the standing part of the rope, leaving enough rope to go around the anchor. To tie a figure-eight knot form a loop in the rope, wrap the working end around the standing part, and route the working end through the loop. The finished knot is dressed loosely.

STEP 3. Take the working end around the anchor point.

STEP 4. With the working end, insert the rope back through the loop of the knot in reverse.

STEP 5. Keep the original figure eight as the outside rope and retrace the knot around the wrap and back to the long-standing part.

STEP 6. Remove all unnecessary twists and crossovers; dress the knot down.

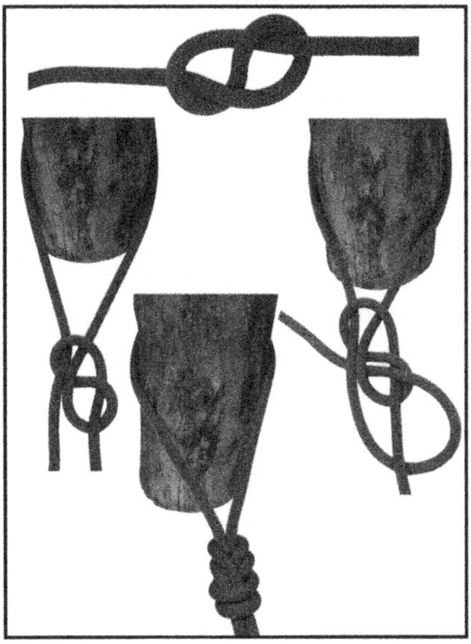

Figure 4-13. Figure-eight retrace.

b. **Checkpoints**

(1) A figure eight with a doubled rope running side by side, forming a fixed loop around a fixed object or harness.

(2) There is a minimum 4-inch pigtail.

4-15. CLOVE HITCH

The clove hitch is an anchor knot that can be used in the middle of the rope as well as at the end (Figure 4-14). The knot must have constant tension on it once tied to prevent slipping. It can be used as either an anchor or middle of the rope knot, depending on how it is tied.

a. **Tying the Knot.**

(1) ***Middle of the Rope.***

STEP 1. Hold rope in both hands, palms down with hands together. Slide the left hand to the left from 20 to 25 centimeters.

STEP 2. Form a loop away from and back toward the right.

STEP 3. Slide the right hand from 20 to 25 centimeters to the right. Form a loop inward and back to the left hand.

STEP 4. Place the left loop on top of the right loop. Place both loops over the anchor and pull both ends of the rope in opposite directions. The knot is tied.

(2) ***End of the Rope.***

Note: For instructional purposes, assume that the anchor is horizontal.

STEP 1. Place 76 centimeters of rope over the top of the anchor. Hold the standing end in the left hand. With the right hand, reach under the horizontal anchor, grasp the working end, and bring it inward.

STEP 2. Place the working end of the rope over the standing end (to form a loop). Hold the loop in the left hand. Place the working end over the anchor from 20 to 25 centimeters to the left of the loop.

STEP 3. With the right hand, reach down to the left hand side of the loop under the anchor. Grasp the working end of the rope. Bring the working end up and outward.

STEP 4. Dress down the knot.

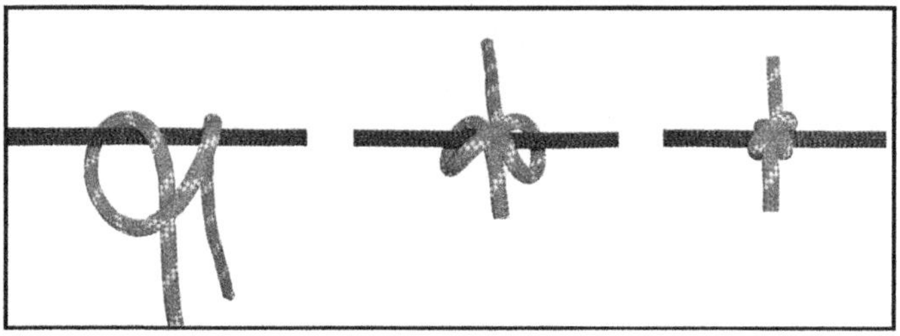

Figure 4-14. Clove hitch.

b. **Checkpoints.**

(1) The knot has two round turns around the anchor with a diagonal locking bar.

(2) The locking bar is facing 90 degrees from the direction of pull.

(3) The ends exit 180 degrees from each other.

(4) The knot has more than a 4-inch pigtail remaining.

4-16. WIREMAN'S KNOT

The wireman's knot forms a single, fixed loop in the middle of the rope (Figure 4-15, page 4-18). It is a middle rope knot.

a. **Tying the Knot.**

STEP 1. When tying this knot, face the anchor that the tie-off system will be tied to. Take up the slack from the anchor, and wrap two turns around the left hand (palm up) from left to right.

STEP 2. A loop of 30 centimeters is taken up in the second round turn to create the fixed loop of the knot.

STEP 3. Name the wraps from the palm to the fingertips: heel, palm, and fingertip.

STEP 4. Secure the palm wrap with the right thumb and forefinger, and place it over the heel wrap.

STEP 5. Secure the heel wrap and place it over the fingertip wrap.

STEP 6. Secure the fingertip wrap and place it over the palm wrap.

STEP 7. Secure the palm wrap and pull up to form a fixed loop.

STEP 8. Dress the knot down by pulling on the fixed loop and the two working ends.

STEP 9. Pull the working ends apart to finish the knot.

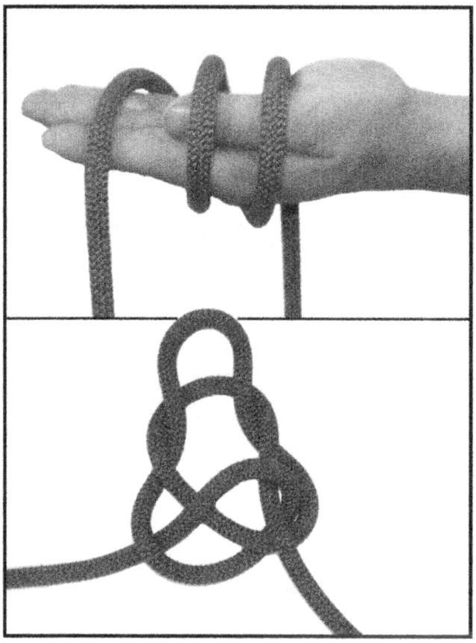

Figure 4-15. Wireman's knot.

b. **Checkpoints.**

(1) The completed knot should have four separate bights locking down on themselves with the fixed loop exiting from the top of the knot and laying toward the near side anchor point.

(2) Both ends should exit opposite each other without any bends.

4-17. DIRECTIONAL FIGURE-EIGHT

The directional figure-eight knot forms a single, fixed loop in the middle of the rope that lays back along the standing part of the rope (Figure 4-16). It is a middle rope knot.

a. **Tying the Knot.**

STEP 1. Face the far side anchor so that when the knot is tied, it lays inward.

STEP 2. Lay the rope from the far side anchor over the left palm. Make one wrap around the palm.

STEP 3. With the wrap thus formed, tie a figure-eight knot around the standing part that leads to the far side anchor.

STEP 4. When dressing the knot down, the tail and the bight must be together.

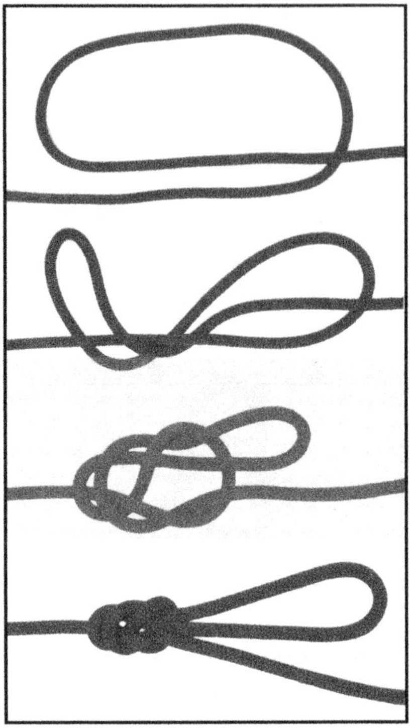

Figure 4-16. Directional figure-eight.

b. **Checkpoints.**

(1) The loop should be large enough to accept a carabiner but no larger than a helmet-size loop.

(2) The tail and bight must be together.

(3) The figure eight is tied tightly.

(4) The bight in the knot faces back toward the near side.

4-18. BOWLINE-ON-A-BIGHT (TWO-LOOP BOWLINE)

The bowline-on-a-bight is used to form two fixed loops in the middle of a rope (Figure 4-17, page 4-20). It is a middle rope knot.

a. **Tying the Knot.**

STEP 1. Form a bight in the rope about twice as long as the finished loops will be.

STEP 2. Tie an overhand knot on a bight.

STEP 3. Hold the overhand knot in the left hand so that the bight is running down and outward.

STEP 4. Grasp the bight with the right hand; fold it back over the overhand knot so that the overhand knot goes through the bight.

STEP 5. From the end (apex) of the bight, follow the bight back to where it forms the cross in the overhand knot. Grasp the two ropes that run down and outward and pull up, forming two loops.

STEP 6. Pull the two ropes out of the overhand knot and dress the knot down.

STEP 7. A final dress is required: grasp the ends of the two fixed loops and pull, spreading them apart to ensure the loops do not slip.

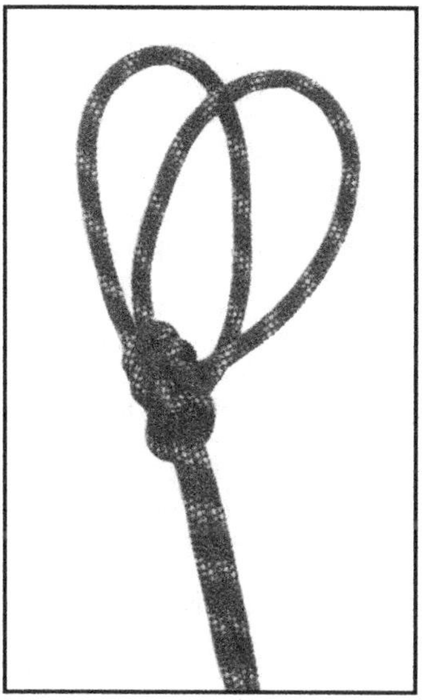

Figure 4-17. Bowline-on-a-bight.

b. **Checkpoints.**

(1) There are two fixed loops that will not slip.

(2) There are no twists in the knot.

(3) A double loop is held in place by a bight.

4-19. TWO-LOOP FIGURE-EIGHT

The two-loop figure-eight is used to form two fixed loops in the middle of a rope (Figure 4-18.) It is a middle rope knot.

a. **Tying the Knot.**

STEP 1. Using a doubled rope, form an 18-inch bight in the left hand with the running end facing to the left.

STEP 2. Grasp the bight with the right hand and make a 360-degree turn around the standing end in a counterclockwise direction.

STEP 3. With the working end, form another bight and place that bight through the loop just formed in the left hand.

STEP 4. Hold the bight with the left hand, and place the original bight (moving toward the left hand) over the knot.

STEP 5. Dress the knot down.

Figure 4-18. Two-loop figure-eight.

b. **Checkpoints.**

(1) There is a double figure-eight knot with two loops that share a common locking bar.

(2) The two loops must be adjustable by means of a common locking bar.

(3) The common locking bar is on the bottom of the double figure-eight knot.

4-20. FIGURE-EIGHT LOOP (FIGURE-EIGHT-ON-A-BIGHT)

The figure-eight loop, also called the figure-eight-on-a-bight, is used to form a fixed loop in a rope (Figure 4-19). It is a middle of the rope knot.

a. **Tying the Knot.**

STEP 1. Form a bight in the rope about as large as the diameter of the desired loop.

STEP 2. With the bight as the working end, form a loop in rope (standing part).

STEP 3. Wrap the working end around the standing part 360 degrees and feed the working end through the loop. Dress the knot tightly.

Figure 4-19. Figure-eight loop.

b. **Checkpoints.**

(1) The loop is the desired size.

(2) The ropes in the loop are parallel and do not cross over each other.

(3) The knot is tightly dressed.

4-21. PRUSIK KNOT

The Prusik knot is used to put a moveable rope on a fixed rope such as a Prusik ascent or a tightening system. This knot can be tied as a middle or end of the rope Prusik. It is a specialty knot.

a. **Tying the Knot.**

(1) ***Middle-of-the-Rope Prusik.*** The middle-of-the-rope Prusik knot can be tied with a short rope to a long rope as follows (Figure 4-20.):

STEP 1. Double the short rope, forming a bight, with the working ends even. Lay it over the long rope so that the closed end of the bight is 12 inches below the long rope and the remaining part of the rope (working ends) is the closest to the climber; spread the working end apart.

STEP 2. Reach down through the 12-inch bight. Pull up both of the working ends and lay them over the long rope. Repeat this process making sure that the working ends pass in the middle of the first two wraps. Now there are four wraps and a locking bar working across them on the long rope.

STEP 3. Dress the wraps and locking bar down to ensure they are tight and not twisted. Tying an overhand knot with both ropes will prevent the knot from slipping during periods of variable tension.

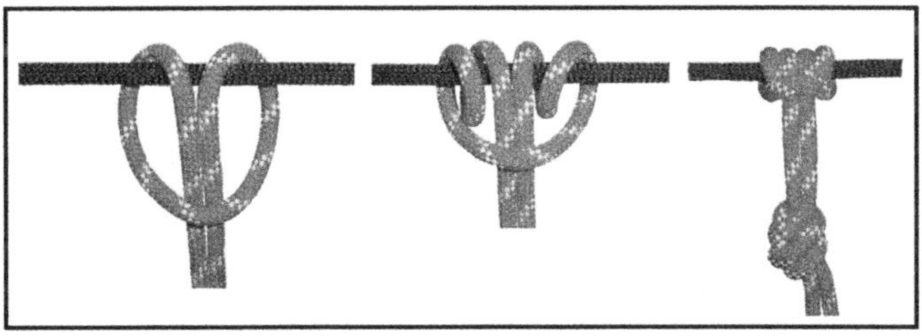

Figure 4-20. Middle-of-the-rope Prusik.

(2) ***End-of-the-Rope Prusik*** (Figure 4-21).

STEP 1. Using an arm's length of rope, and place it over the long rope.

STEP 2. Form a complete round turn in the rope.

STEP 3. Cross over the standing part of the short rope with the working end of the short rope.

STEP 4. Lay the working end under the long rope.

STEP 5. Form a complete round turn in the rope, working back toward the middle of the knot.

STEP 6. There are four wraps and a locking bar running across them on the long rope. Dress the wraps and locking bar down. Ensure they are tight, parallel, and not twisted.

STEP 7. Finish the knot with a bowline to ensure that the Prusik knot will not slip out during periods of varying tension.

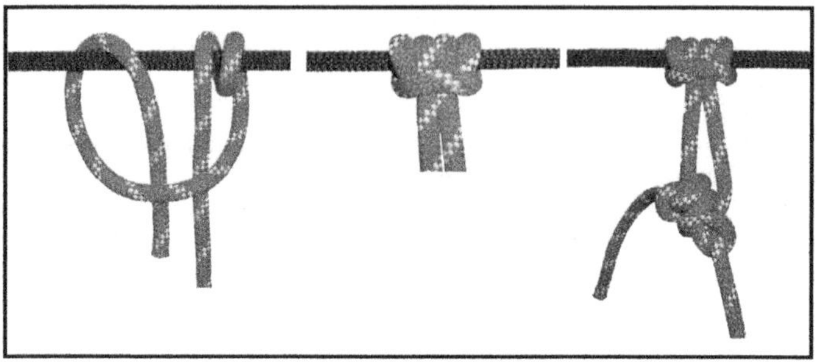

Figure 4-21. End-of-the-rope Prusik knot.

b. **Checkpoints.**

(1) Four wraps with a locking bar.

(2) The locking bar faces the climber.

(3) The knot is tight and dressed down with no ropes twisted or crossed.

(4) Other than a finger Prusik, the knot should contain an overhand or bowline to prevent slipping.

4-22. BACHMAN KNOT

The Bachman knot provides a means of using a makeshift mechanized ascender (Figure 4-22, page 4-24). It is a specialty knot.

a. **Tying the Knot.**

STEP 1. Find the middle of a utility rope and insert it into a carabiner.

STEP 2. Place the carabiner and utility rope next to a long climbing rope.

STEP 3. With the two ropes parallel from the carabiner, make two or more wraps around the climbing rope and through the inside portion of the carabiner.

Note: The rope can be tied into an etrier (stirrup) and used as a Prusik-friction principle ascender.

b. **Checkpoints.**

(1) The bight of the climbing rope is at the top of the carabiner.

(2) The two ropes run parallel without twisting or crossing.

(3) Two or more wraps are made around the long climbing rope and through the inside portion of the carabiner.

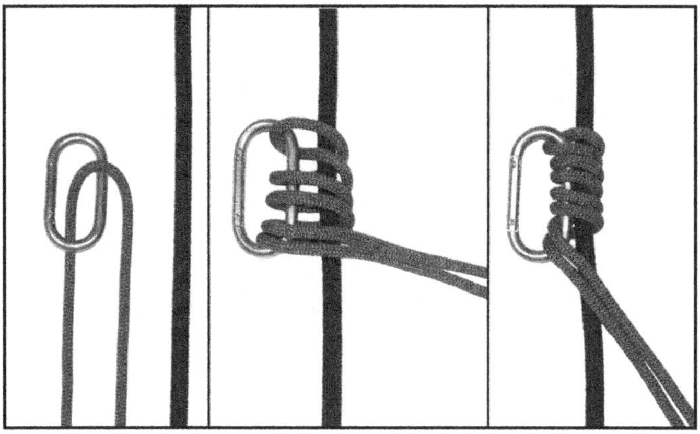

Figure 4-22. Bachman knot.

4-23. BOWLINE-ON-A-COIL

The bowline-on-a-coil is an expedient tie-in used by climbers when a climbing harness is not available (Figure 4-23). It is a specialty knot.

a. **Tying the Knot.**

STEP 1. With the running end, place 3 feet of rope over your right shoulder. The running end is to the back of the body.

STEP 2. Starting at the bottom of your rib cage, wrap the standing part of the rope around your body and down in a clockwise direction four to eight times.

STEP 3. With the standing portion of the rope in your left hand, make a clockwise loop toward the body. The standing portion is on the bottom.

STEP 4. Ensuring the loop does not come uncrossed, bring it up and under the coils between the rope and your body.

STEP 5. Using the standing part, bring a bight up through the loop. Grasp the running end of the rope with the right hand. Pass it through the bight from right to left and back on itself.

STEP 6. Holding the bight loosely, dress the knot down by pulling on the standing end.

STEP 7. Safety the bowline with an overhand around the top, single coil. Then, tie an overhand around all coils, leaving a minimum 4-inch pigtail.

b. **Checkpoints.**

(1) A minimum of four wraps, not crossed, with a bight held in place by a loop.

(2) The loop must be underneath all wraps.

(3) A minimum 4-inch pigtail after the second overhand safety is tied.

(4) Must be centered on the mid-line of the body.

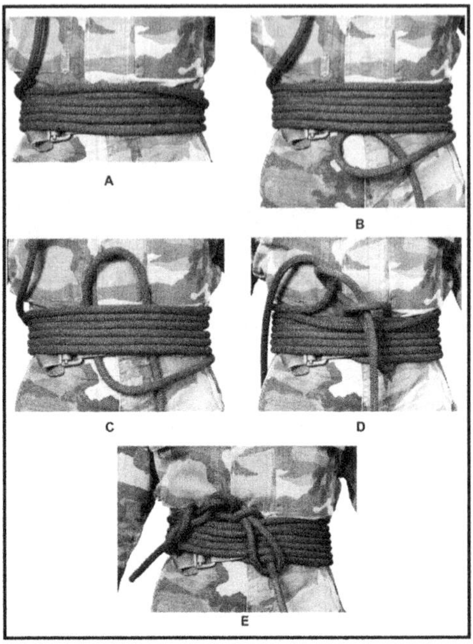

Figure 4-23. Bowline-on-a-coil.

4-24. THREE-LOOP BOWLINE

The three-loop bowline is used to form three fixed loops in the middle of a rope (Figure 4-24, page 4-26). It is used in a self-equalizing anchor system. It is a specialty knot.

a. **Tying the Knot.**

STEP 1. Form an approximate 24-inch bight.

STEP 2. With the right thumb facing toward the body, form a doubled loop in the standing part by turning the wrist clockwise. Lay the loops to the right.

STEP 3. With the right hand, reach down through the loops and pull up a doubled bight from the standing part of the rope.

STEP 4. Place the running end (bight) of the rope (on the left) through the doubled bight from left to right and bring it back on itself. Hold the running end loosely and dress the knot down by pulling on the standing parts.

STEP 5. Safety it off with a doubled overhand knot.

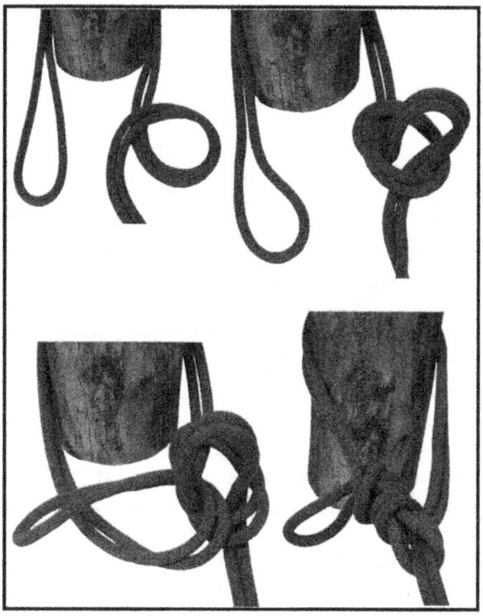

Figure 4-24. Three-loop bowline.

b. **Checkpoints.**

(1) There are two bights held in place by two loops.

(2) The bights form locking bars around the standing parts.

(3) The running end (bight) must be on the inside of the fixed loops.

(4) There is a minimum 4-inch pigtail after the double overhand safety knot is tied.

4-25. FIGURE-EIGHT SLIP KNOT

The figure eight slip knot forms an adjustable bight in a rope (Figure 4-25). It is a specialty knot.

a. **Tying the Knot.**

STEP 1. Form a 12-inch bight in the end of the rope.

STEP 2. Hold the center of the bight in the right hand. Hold the two parallel ropes from the bight in the left hand about 12 inches up the rope.

STEP 3. With the center of the bight in the right hand, twist two complete turns clockwise.

STEP 4. Reach through the bight and grasp the long, standing end of the rope. Pull another bight (from the long standing end) back through the original bight.

STEP 5. Pull down on the short working end of the rope and dress the knot down.

STEP 6. If the knot is to be used in a transport tightening system, take the working end of the rope and form a half hitch around the loop of the figure eight knot.

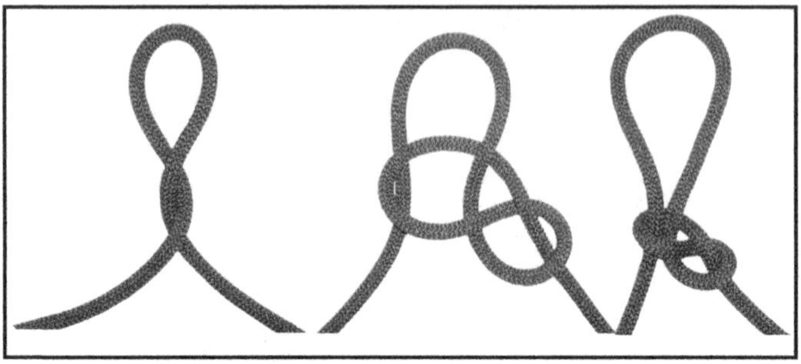

Figure 4-25. Figure-eight slip knot.

b. **Checkpoints.**

(1) The knot is in the shape of a figure eight.

(2) Both ropes of the bight pass through the same loop of the figure eight.

(3) The sliding portion of the rope is the long working end of the rope.

4-26. TRANSPORT KNOT (OVERHAND SLIP KNOT/MULE KNOT)

The transport knot is used to secure the transport tightening system (Figure 4-26, page 4-28). It is simply an overhand slip knot.

a. **Tying the Knot.**

STEP 1. Pass the running end of the rope around the anchor point passing it back under the standing portion (leading to the far side anchor) forming a loop.

STEP 2. Form a bight with the running end of the rope. Pass over the standing portion and down through the loop and dress it down toward the anchor point.

STEP 3. Secure the knot by tying a half hitch around the standing portion with the bight.

Figure 4-26. Transport knot.

b. **Check Points.**

(1) There is a single overhand slip knot.

(2) The knot is secured using a half hitch on a bight.
(3) The bight is a minimum of 12 inches long.

4-27. KLEIMHIEST KNOT

The Kleimhiest knot provides a moveable, easily adjustable, high-tension knot capable of holding extremely heavy loads while being pulled tight (Figure 4-27). It is a special-purpose knot.

a. **Tying the Knot.**

STEP 1. Using a utility rope or webbing offset the ends by 12 inches. With the ends offset, find the center of the rope and form a bight. Lay the bight over a horizontal rope.

STEP 2. Wrap the tails of the utility rope around the horizontal rope back toward the direction of pull. Wrap at least four complete turns.

STEP 3. With the remaining tails of the utility rope, pass them through the bight (see STEP 1).

STEP 4. Join the two ends of the tail with a joining knot.

STEP 5. Dress the knot down tightly so that all wraps are touching.

Note: Spectra should not be used for the Kleimhiest knot. It has a low melting point and tends to slip .

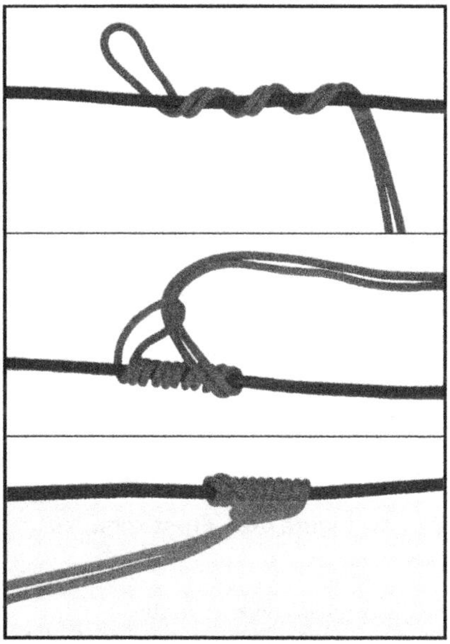

Figure 4-27. Kleimhiest knot.

b. **Checkpoints.**

(1) The bight is opposite the direction of pull.

(2) All wraps are tight and touching.

(3) The ends of the utility rope are properly secured with a joining knot.

4-28. FROST KNOT

The frost knot is used when working with webbing (Figure 4-28, page 4-30). It is used to create the top loop of an etrier. It is a special-purpose knot.

a. **Tying the Knot.**

STEP 1. Lap one end (a bight) of webbing over the other about 10 to 12 inches.

STEP 2. Tie an overhand knot with the newly formed triple-strand webbing; dress tightly.

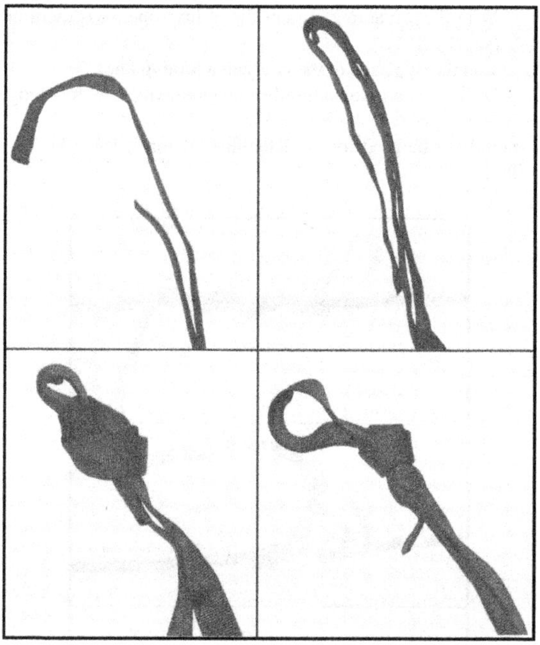

Figure 4-28. Frost knot.

b. **Checkpoints.**

(1) The tails of the webbing run in opposite directions.

(2) Three strands of webbing are formed into a tight overhand knot.

(3) There is a bight and tail exiting the top of the overhand knot.

4-29. GIRTH HITCH

The girth hitch is used to attach a runner to an anchor or piece of equipment (Figure 4-29). It is a special-purpose knot.

a. **Tying the Knot.**

STEP 1: Form a bight.
STEP 2: Bring the runner back through the bight.
STEP 3: Cinch the knot tightly.

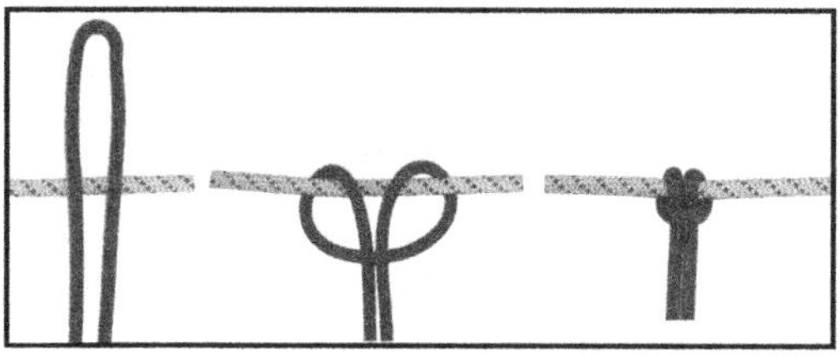

Figure 4-29. Girth hitch.

b. **Checkpoint.**

(1) Two wraps exist with a locking bar running across the wraps.
(2) The knot is dressed tightly.

4-30. MUNTER HITCH

The munter hitch, when used in conjunction with a pear-shaped locking carabiner, is used to form a mechanical belay (Figure 4-30).

a. **Tying the Knot.**

STEP 1. Hold the rope in both hands, palms down about 12 inches apart.
STEP 2. With the right hand, form a loop away from the body toward the left hand. Hold the loop with the left hand.
STEP 3. With the right hand, place the rope that comes from the bottom of the loop over the top of the loop.
STEP 4. Place the bight that has just been formed around the rope into the pear shaped carabiner. Lock the locking mechanism.

b. **Check Points.**

(1) A bight passes through the carabiner, with the closed end around the standing or running part of the rope.
(2) The carabiner is locked.

Figure 4-30. Munter hitch.

4-31. RAPPEL SEAT

The rappel seat is an improvised seat rappel harness made of rope (Figure 4-31, page 4-32). It usually requires a sling rope 14 feet or longer.

a. **Tying the Knot.**

STEP 1. Find the middle of the sling rope and make a bight.

STEP 2. Decide which hand will be used as the brake hand and place the bight on the opposite hip.

STEP 3. Reach around behind and grab a single strand of rope. Bring it around the waist to the front and tie two overhands on the other strand of rope, thus creating a loop around the waist.

STEP 4. Pass the two ends between the legs, ensuring they do not cross.

STEP 5. Pass the two ends up under the loop around the waist, bisecting the pocket flaps on the trousers. Pull up on the ropes, tightening the seat.

STEP 6. From rear to front, pass the two ends through the leg loops creating a half hitch on both hips.

STEP 7. Bring the longer of the two ends across the front to the nonbrake hand hip and secure the two ends with a square knot safetied with overhand knots. Tuck any excess rope in the pocket below the square knot.

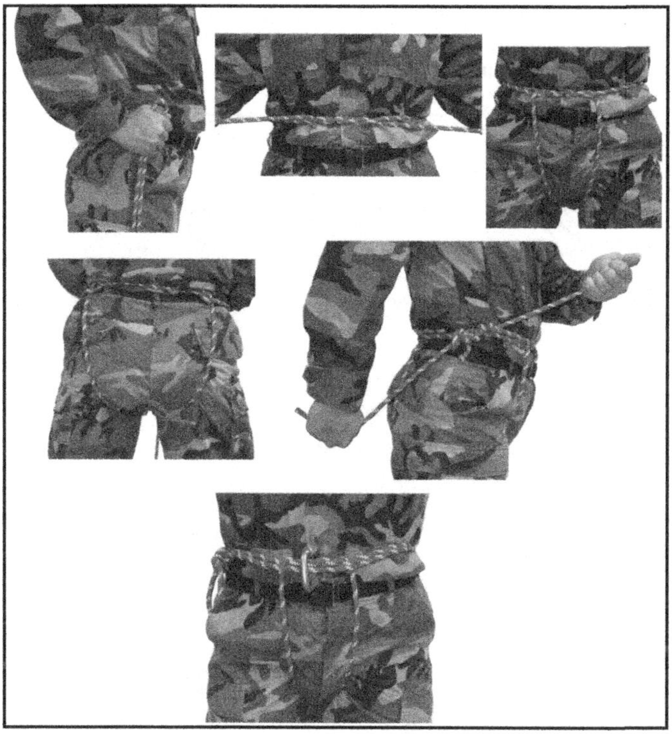

Figure 4-31. Rappel seat.

b. **Check Points.**

(1) There are two overhand knots in the front.

(2) The ropes are not crossed between the legs.

(3) A half hitch is formed on each hip.

(4) Seat is secured with a square knot with overhand safeties on the non-brake hand side.

(5) There is a minimum 4-inch pigtail after the overhand safeties are tied.

4-32. GUARDE KNOT

The guarde knot (ratchet knot, alpine clutch) is a special purpose knot primarily used for hauling systems or rescue (Figure 4-32). The knot works in only one direction and cannot be reversed while under load.

a. **Tying the Knot.**

STEP 1. Place a bight of rope into the two anchored carabiners (works best with two like carabiners, preferably ovals).

STEP 2. Take a loop of rope from the non-load side and place it down into the opposite cararabiner so that the rope comes out between the two carabiners.

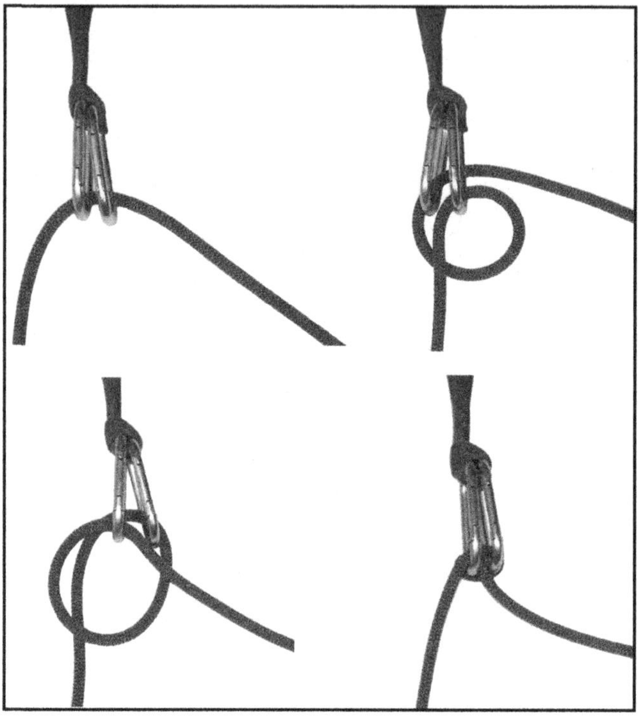

Figure 4-32. Guarde knot.

b. **Check Points.**

(1) When properly dressed, rope can only be pulled in one direction.

(2) The knot will not fail when placed under load.

CHAPTER 5
ANCHORS

This chapter discusses different types of anchors and their application in rope systems and climbing. Proper selection and placement of anchors is a critical skill that requires a great deal of practice. Failure of any system is most likely to occur at the anchor point. If the anchor is not strong enough to support the intended load, it will fail. Failure is usually the result of poor terrain features selected for the anchor point, or the equipment used in rigging the anchor was placed improperly or in insufficient amounts.

When selecting or constructing anchors, always try to make sure the anchor is "bombproof." A bombproof anchor is stronger than any possible load that could be placed on it. An anchor that has more strength than the climbing rope is considered bombproof.

Section I. NATURAL ANCHORS

Natural anchors should be considered for use first. They are usually strong and often simple to construct with minimal use of equipment. Trees, boulders, and other terrain irregularities are already in place and simply require a method of attaching the rope. However, natural anchors should be carefully studied and evaluated for stability and strength before use. Sometimes the climbing rope is tied directly to the anchor, but under most circumstances a sling is attached to the anchor and then the climbing rope is attached to the sling with a carabiner(s). (See paragraph 5-7 for slinging techniques.)

5-1. TREES

Trees are probably the most widely used of all natural anchors depending on the terrain and geographical region (Figure 5-1). However, trees must be carefully checked for suitability.

a. In rocky terrain, trees usually have a shallow root system. This can be checked by pushing or tugging on the tree to see how well it is rooted. Anchoring as low as possible to prevent excess leverage on the tree may be necessary.

b. Use padding on soft, sap producing trees to keep sap off ropes and slings.

Figure 5-1. Trees used as anchors.

5-2. BOULDERS

Boulders and rock nubbins make ideal anchors (Figure 5-2). The rock can be firmly tapped with a piton hammer to ensure it is solid. Sedimentary and other loose rock formations are not stable. Talus and scree fields are an indicator that the rock in the area is not solid. All areas around the rock formation that could cut the rope or sling should be padded.

Figure 5-2. Boulders used as anchors.

5-3. CHOCKSTONES

A chockstone is a rock that is wedged in a crack because the crack narrows downward (Figure 5-3). Chockstones should be checked for strength, security, and crumbling and should always be tested before use. All chockstones must be solid and strong enough to support the load. They must have maximum surface contact and be well tapered with the surrounding rock to remain in position.

a. Chockstones are often directional—they are secure when pulled in one direction but may pop out if pulled in another direction.

b. A creative climber can often make his own chockstone by wedging a rock into position, tying a rope to it, and clipping on a carabiner.

c. Slings should not be wedged between the chockstone and the rock wall since a fall could cut the webbing runner.

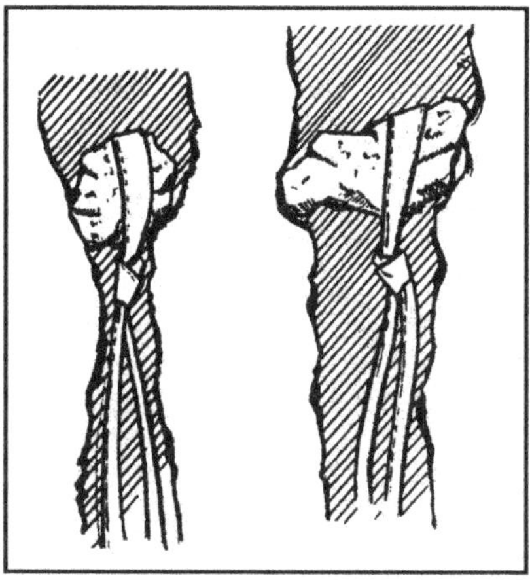

Figure 5-3. Chockstones.

5-4. ROCK PROJECTIONS

Rock projections (sometimes called nubbins) often provide suitable protection (Figure 5-4). These include blocks, flakes, horns, and spikes. If rock projections are used, their firmness is important. They should be checked for cracks or weathering that may impair their firmness. If any of these signs exist, the projection should be avoided.

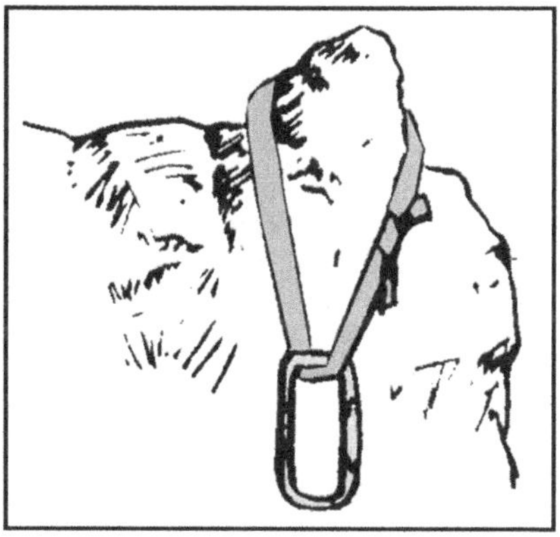

Figure 5-4. Rock projections.

5-5. TUNNELS AND ARCHES

Tunnels and arches are holes formed in solid rock and provide one of the more secure anchor points because they can be pulled in any direction. A sling is threaded through the opening hole and secured with a joining knot or girth hitch. The load-bearing hole must be strong and free of sharp edges (pad if necessary).

5-6. BUSHES AND SHRUBS

If no other suitable anchor is available, the roots of bushes can be used by routing a rope around the bases of several bushes (Figure 5-5). As with trees, the anchoring rope is placed as low as possible to reduce leverage on the anchor. All vegetation should be healthy and well rooted to the ground.

Figure 5-5. Bushes and shrubs.

5-7. SLINGING TECHNIQUES

Three methods are used to attach a sling to a natural anchor—drape, wrap, and girth. Whichever method is used, the knot is set off to the side where it will not interfere with normal carabiner movement. The carabiner gate should face away from the ground and open away from the anchor for easy insertion of the rope. When a locking carabiner cannot be used, two carabiners are used with gates opposed. Correctly opposed gates should open on opposite sides and form an "X" when opened (Figure 5-6).

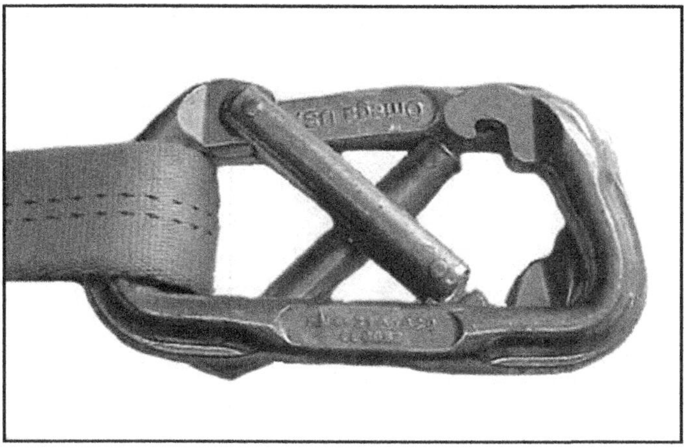

Figure 5-6. Correctly opposed carabiners.

a. **Drape.** Drape the sling over the anchor (Figure 5-7). Untying the sling and routing it around the anchor and then retying is still considered a drape.

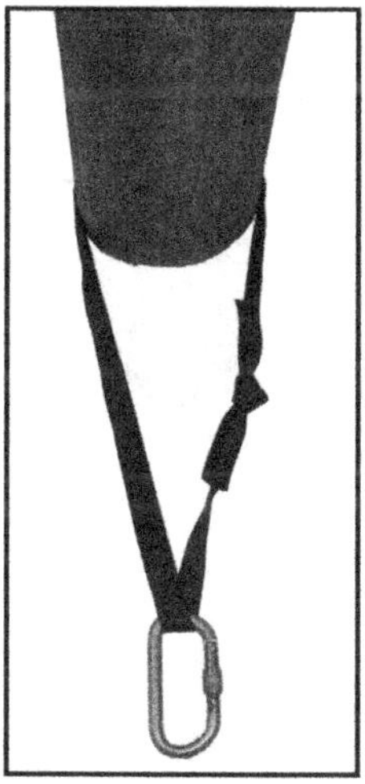

Figure 5-7. Drape.

b. **Wrap.** Wrap the sling around the anchor and connect the two ends together with a carabiner(s) or knot (Figure 5-8).

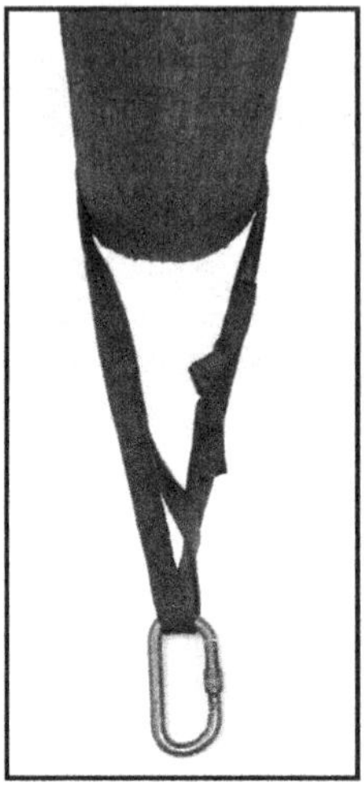

Figure 5-8. Wrap.

c. **Girth.** Tie the sling around the anchor with a girth hitch (Figure 5-9). Although a girth hitch reduces the strength of the sling, it allows the sling to remain in position and not slide on the anchor.

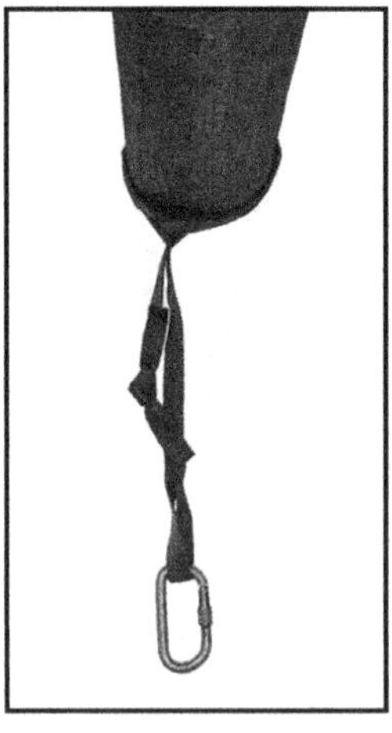

Figure 5-9. Girth.

Section II. ANCHORING WITH THE ROPE

The climbing or installation rope can be tied directly to the anchor using several different techniques. This requires less equipment, but also sacrifices some rope length to tie the anchor. The rope can be tied to the anchor using an appropriate anchor knot such as a bowline or a rerouted figure eight. Round turns can be used to help keep the rope in position on the anchor. A tensionless anchor can be used in high-load installations where tension on the attachment point and knot is undesirable.

5-8. ROPE ANCHOR

When tying the climbing or installation rope around an anchor, the knot should be placed approximately the same distance away from the anchor as the diameter of the anchor (Figure 5-10). The knot shouldn't be placed up against the anchor because this can stress and distort the knot under tension.

Figure 5-10. Rope tied to anchor with anchor knot.

5-9. TENSIONLESS ANCHOR

The tensionless anchor is used to anchor the rope on high-load installations such as bridging and traversing (Figure 5-11, page 5-8). The wraps of the rope around the anchor absorb the tension of the installation and keep the tension off the knot and carabiner. The anchor is usually tied with a minimum of four wraps, more if necessary, to absorb the tension. A smooth anchor may require several wraps, whereas a rough barked tree might only require a few. The rope is wrapped from top to bottom. A fixed loop is placed into the end of the rope and attached loosely back onto the rope with a carabiner.

Figure 5-11. Tensionless anchor.

Section III. ARTIFICIAL ANCHORS

Using artificial anchors becomes necessary when natural anchors are unavailable. The art of choosing and placing good anchors requires a great deal of practice and experience. Artificial anchors are available in many different types such as pitons, chocks, hexcentrics, and SLCDs. Anchor strength varies greatly; the type used depends on the terrain, equipment, and the load to be placed on it.

5-10. DEADMAN

A "deadman" anchor is any solid object buried in the ground and used as an anchor.

a. An object that has a large surface area and some length to it works best. (A hefty timber, such as a railroad tie, would be ideal.) Large boulders can be used, as well as a bundle of smaller tree limbs or poles. As with natural anchors, ensure timbers and tree limbs are not dead or rotting and that boulders are solid. Equipment, such as skis, ice axes, snowshoes, and ruck sacks, can also be used if necessary.

b. In extremely hard, rocky terrain (where digging a trench would be impractical, if not impossible) a variation of the deadman anchor can be constructed by building above the ground. The sling is attached to the anchor, which is set into the ground as deeply as possible. Boulders are then stacked on top of it until the anchor is strong enough for the load. Though normally not as strong as when buried, this method can work well for light-load installations as in anchoring a hand line for a stream crossing.

Note: Artificial anchors, such as pitons and bolts, are not widely accepted for use in all areas because of the scars they leave on the rock and the environment. Often they are left in place and become unnatural, unsightly fixtures in the natural environment. For training planning, local laws and courtesies should be taken into consideration for each area of operation.

5-11. PITONS

Pitons have been in use for over 100 years. Although still available, pitons are not used as often as other types of artificial anchors due primarily to their impact on the environment. Most climbers prefer to use chocks, SLCDs and other artificial anchors rather than pitons because they do not scar the rock and are easier to remove. Eye protection should always be worn when driving a piton into rock.

Note: The proper use and placement of pitons, as with any artificial anchor, should be studied, practiced, and tested while both feet are firmly on the ground and there is no danger of a fall.

a. **Advantages.** Some advantages in using pitons are:

- Depending on type and placement, pitons can support multiple directions of pull.
- Pitons are less complex than other types of artificial anchors.
- Pitons work well in thin cracks where other types of artificial anchors do not.

b. **Disadvantages.** Some disadvantages in using pitons are:

- During military operations, the distinct sound created when hammering pitons is a tactical disadvantage.
- Due to the expansion force of emplacing a piton, the rock could spread apart or break causing an unsafe condition.
- Pitons are more difficult to remove than other types of artificial anchors.
- Pitons leave noticeable scars on the rock.
- Pitons are easily dropped if not tied off when being used.

c. **Piton Placement.** The proper positioning or placement of pitons is critical. (Figure 5-12, page 5-10, shows examples of piton placement.) Usually a properly sized piton for a rock crack will fit one half to two thirds into the crack before being driven with the piton hammer. This helps ensure the depth of the crack is adequate for the size piton selected. As pitons are driven into the rock the pitch or sound that is made will change with each hammer blow, becoming higher pitched as the piton is driven in.

(1) Test the rock for soundness by tapping with the hammer. Driving pitons in soft or rotten rock is not recommended. When this type of rock must be used, clear the loose rock, dirt, and debris from the crack before driving the piton completely in.

(2) While it is being driven, attach the piton to a sling with a carabiner (an old carabiner should be used, if available) so that if the piton is knocked out of the crack, it will not be lost. The greater the resistance overcome while driving the piton, the firmer the anchor will be. The holding power depends on the climber placing the piton in a sound crack, and on the type of rock. The piton should not spread the rock, thereby loosening the emplacement.

Note: Pitons that have rings as attachment points might not display much change in sound as they are driven in as long as the ring moves freely.

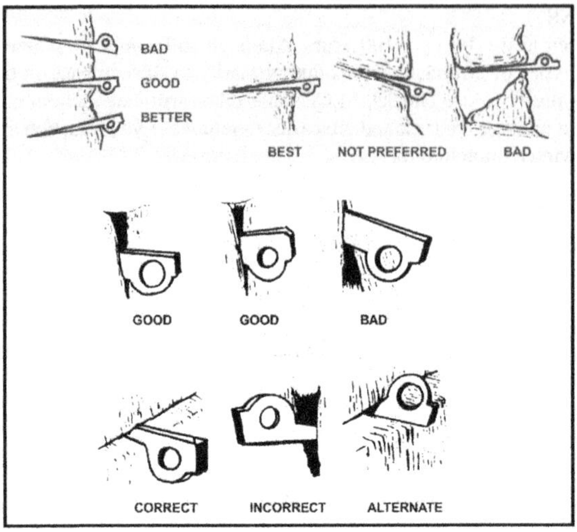

Figure 5-12. Examples of piton placements.

(3) Military mountaineers should practice emplacing pitons using either hand. Sometimes a piton cannot be driven completely into a crack, because the piton is too long. Therefore, it should be tied off using a hero-loop (an endless piece of webbing) (Figure 5-13). Attach this loop to the piton using a girth hitch at the point where the piton enters the rock so that the girth hitch is snug against the rock. Clip a carabiner into the loop.

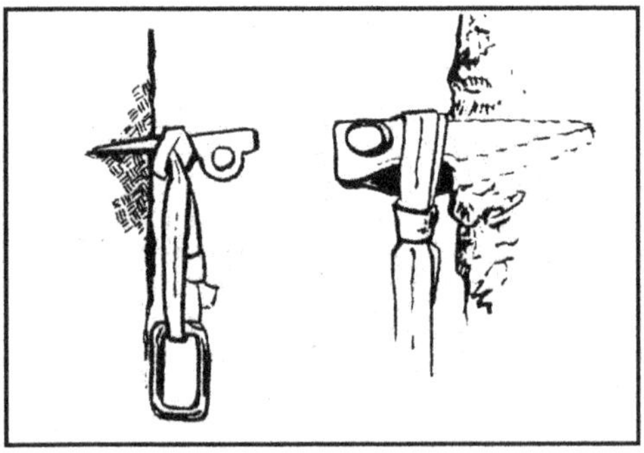

Figure 5-13. Hero-loop.

d. **Testing.** To test pitons pull up about 1 meter of slack in the climbing rope or use a sling. Insert this rope into a carabiner attached to the piton, then grasp the rope at least 1/2 meter from the carabiner. Jerk vigorously upward, downward, to each side, and then outward while observing the piton for movement. Repeat these actions as many times as necessary. Tap the piton to determine if the pitch has changed. If the pitch has changed greatly, drive the piton in as far as possible. If the sound regains its original pitch, the emplacement is probably safe. If the piton shows any sign of moving or if, upon driving it, there is any question of its soundness, drive it into another place. Try to be in a secure position before testing. This procedure is intended for use in testing an omni-directional anchor (one that withstands a pull in any direction). When a directional anchor (pull in one direction) is used, as in most free and direct-aid climbing situations, and when using chocks, concentrate the test in the direction that force will be applied to the anchor.

e. **Removing Pitons.** Attach a carabiner and sling to the piton before removal to eliminate the chance of dropping and losing it. Tap the piton firmly along the axis of the crack in which it is located. Alternate tapping from both sides while applying steady pressure. Pulling out on the attached carabiner eventually removes the piton (Figure 5-14).

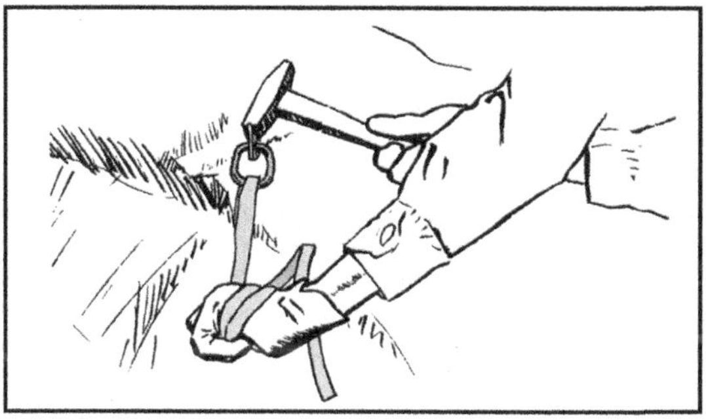

Figure 5-14. Piton removal.

f. **Reusing Pitons.** Soft iron pitons that have been used, removed, and straightened may be reused, but they must be checked for strength. In training areas, pitons already in place should not be trusted since weather loosens them in time. Also, they may have been driven poorly the first time. Before use, test them as described above and drive them again until certain of their soundness.

5-12. CHOCKS

Chock craft has been in use for many decades. A natural chockstone, having fallen and wedged in a crack, provides an excellent anchor point. Sometimes these chockstones are in unstable positions, but can be made into excellent anchors with little adjustment. Chock craft is an art that requires time and technique to master—simple in theory, but complex in practice. Imagination and resourcefulness are key principles to chock craft. The skilled climber must understand the application of mechanical advantage, vectors, and other forces that affect the belay chain in a fall.

a. **Advantages.** The advantages of using chocks are:

- Tactically quiet installation and recovery.
- Usually easy to retrieve and, unless severely damaged, are reusable.
- Light to carry.
- Easy to insert and remove.
- Minimal rock scarring as opposed to pitons.
- Sometimes can be placed where pitons cannot (expanding rock flakes where pitons would further weaken the rock).

b. **Disadvantages.** The disadvantages of using chocks are:

- May not fit in thin cracks, which may accept pitons.
- Often provide only one direction of pull.
- Practice and experience necessary to become proficient in proper placement.

c. **Placement.** The principles of placing chocks are to find a crack with a constriction at some point, place a chock of appropriate size above and behind the constriction, and set the chock by jerking down on the chock loop (Figure 5-15). Maximum surface contact with a tight fit is critical. Chocks are usually good for a single direction of pull.

(1) Avoid cracks that have crumbly (soft) or deteriorating rock, if possible. Some cracks may have loose rock, grass, and dirt, which should be removed before placing the chock. Look for a constriction point in the crack, then select a chock to fit it.

(2) When selecting a chock, choose one that has as much surface area as possible in contact with the rock. A chock resting on one small crystal or point of rock is likely to be unsafe. A chock that sticks partly out of the crack is avoided. Avoid poor protection. Ensure that the chock has a wire or runner long enough; extra ropes, cord, or webbing may be needed to extend the length of the runner.

(3) End weighting of the placement helps to keep the protection in position. A carabiner often provides enough weight

(4) Parallel-sided cracks without constrictions are a problem. Chocks designed to be used in this situation rely on camming principles to remain emplaced. Weighting the emplacement with extra hardware is often necessary to keep the chocks from dropping out.

(a) Emplace the wedge-shaped chock above and behind the constriction; seat it with a sharp downward tug.

(b) Place a camming chock with its narrow side into the crack, then rotate it to the attitude it will assume under load; seat it with a sharp downward tug.

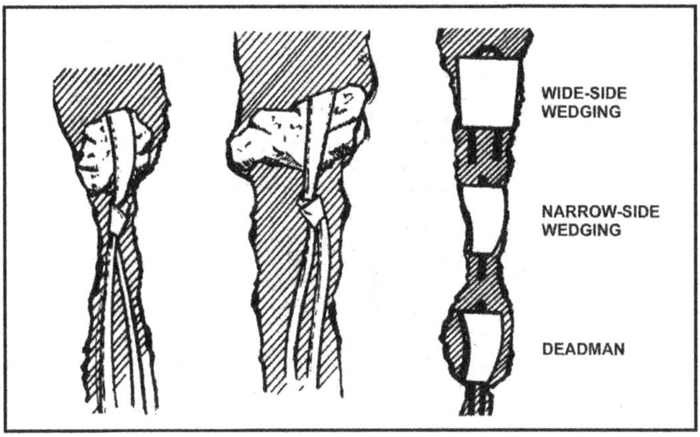

Figure 5-15. Chock placements.

d. **Testing.** After seating a chock, test it to ensure it remains in place. A chock that falls out when the climber moves past it is unsafe and offers no protection. To test it, firmly pull the chock in every anticipated direction of pull. Some chock placements fail in one or more directions; therefore, use pairs of chocks in opposition.

5-13. SPRING-LOADED CAMMING DEVICE

The SLCD offers quick and easy placement of artificial protection. It is well suited in awkward positions and difficult placements, since it can be emplaced with one hand. It can usually be placed quickly and retrieved easily (Figure 5-16, page 5-14).

a. To emplace an SLCD hold the device in either hand like a syringe, pull the retractor bar back, place the device into a crack, and release the retractor bar. The SLCD holds well in parallel-sided hand- and fist-sized cracks. Smaller variations are available for finger-sized cracks.

b. Careful study of the crack should be made before selecting the device for emplacement. It should be placed so that it is aligned in the direction of force applied to it. It should not be placed any deeper than is needed for secure placement, since it may be impossible to reach the extractor bar for removal. An SLCD should be extended with a runner and placed so that the direction of pull is parallel to the shaft; otherwise, it may rotate and pull out. The versions that have a semi-rigid wire cable shaft allow for greater flexibility and usage, without the danger of the shaft snapping off in a fall.

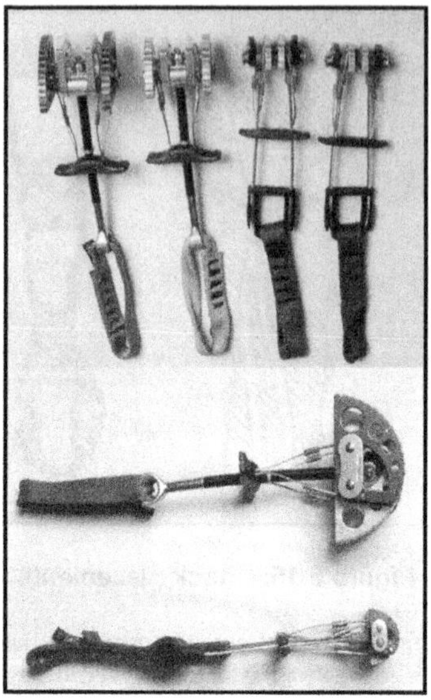

Figure 5-16. SLCD placements.

5-14. BOLTS

Bolts are often used in fixed-rope installations and in aid climbing where cracks are not available.

a. Bolts provide one of the most secure means of establishing protection. The rock should be inspected for evidence of crumbling, flaking, or cracking, and should be tested with a hammer. Emplacing a bolt with a hammer and a hand drill is a time-consuming and difficult process that requires drilling a hole in the rock deeper than the length of the bolt. This normally takes more than 20 minutes for one hole. Electric or even gas-powered drills can be used to greatly shorten drilling time. However, their size and weight can make them difficult to carry on the climbing route.

b. A hanger (carrier) and nut are placed on the bolt, and the bolt is inserted and then driven into the hole. A climber should never hammer on a bolt to test or "improve" it, since this permanently weakens it. Bolts should be used with carriers, carabiners, and runners.

c. When using bolts, the climber uses a piton hammer and hand drill with a masonry bit for drilling holes. Some versions are available in which the sleeve is hammered and turned into the rock (self-drilling), which bores the hole. Split bolts and expanding sleeves are common bolts used to secure hangers and carriers (Figure 5-17). Surgical tubing is useful in blowing dust out of the holes. Nail type bolts are emplaced by driving the nail with a hammer to expand the sleeve against the wall of the drilled hole. Safety glasses should always be worn when emplacing bolts.

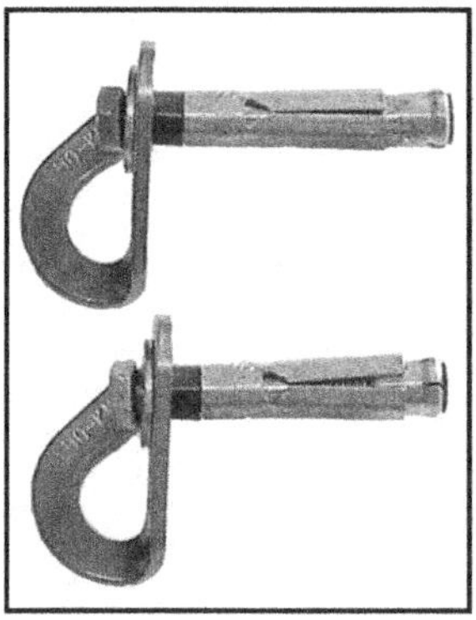

Figure 5-17. Bolt with expanding sleeve.

5-15. EQUALIZING ANCHORS

Equalizing anchors are made up of more than one anchor point joined together so that the intended load is shared equally. This not only provides greater anchor strength, but also adds redundancy or backup because of the multiple points.

a. **Self-equalizing Anchor.** A self-equalizing anchor will maintain an equal load on each individual point as the direction of pull changes (Figure 5-18). This is sometimes used in rappelling when the route must change left or right in the middle of the rappel. A self-equalizing anchor should only be used when necessary because if any one of the individual points fail, the anchor will extend and shock-load the remaining points or even cause complete anchor failure.

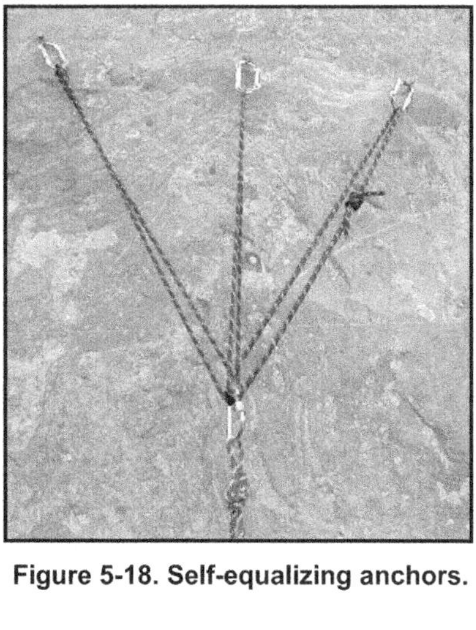

Figure 5-18. Self-equalizing anchors.

b. **Pre-equalized Anchor.** A pre-equalized anchor distributes the load equally to each individual point (Figure 5-19). It is aimed in the direction of the load. A pre-equalized anchor prevents extension and shock-loading of the anchor if an individual point fails. An anchor is pre-equalized by tying an overhand or figure-eight knot in the webbing or sling.

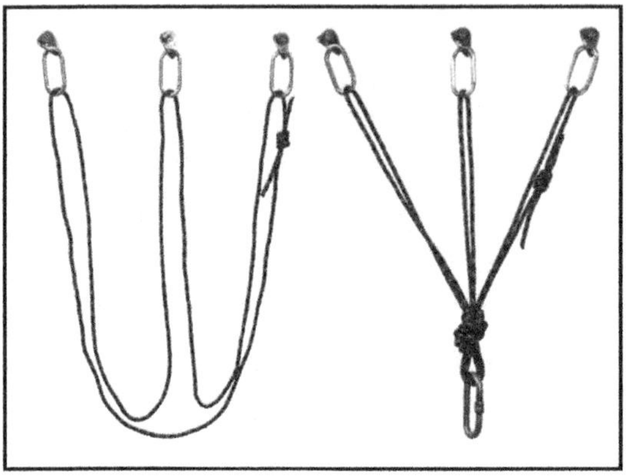

Figure 5-19. Pre-equalized anchor.

Note: When using webbing or slings, the angles of the webbing or slings directly affect the load placed on an anchor. An angle greater than 90 degrees can result in anchor failure (Figure 5-20).

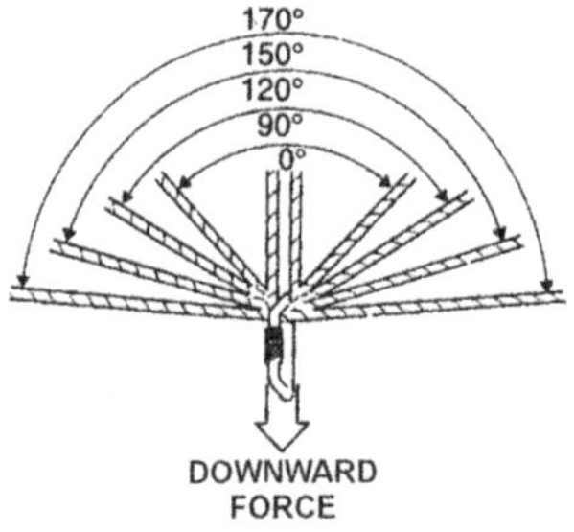

TENSION IN MULTIPLE ANCHOR RIGGING	
ANGLE	RESULTING LEG TENSION*
170°	1150%
150°	200%
120°	100%
90°	70%
0°	50%
*On each leg relative to downward force.	

Figure 5-20. Effects of angles on an anchor.

CHAPTER 6
CLIMBING

A steep rock face is a terrain feature that can be avoided most of the time through prior planning and good route selection. Rock climbing can be time consuming, especially for a larger unit with a heavy combat load. It can leave the climbing party totally exposed to weather, terrain hazards, and the enemy for the length of the climb.

Sometimes steep rock cannot be avoided. Climbing relatively short sections of steep rock (one or two pitches) may prove quicker and safer than using alternate routes. A steep rock route would normally be considered an unlikely avenue of approach and, therefore, might be weakly defended or not defended at all.

All personnel in a unit preparing for deployment to mountainous terrain should be trained in the basics of climbing. Forward observers, reconnaissance personnel, and security teams are a few examples of small units who may require rock climbing skills to gain their vantage points in mountainous terrain. Select personnel demonstrating the highest degree of skill and experience should be trained in roped climbing techniques. These personnel will have the job of picking and "fixing" the route for the rest of the unit.

Rock climbing has evolved into a specialized "sport" with a wide range of varying techniques and styles. This chapter focuses on the basics most applicable to military operations.

Section I. CLIMBING FUNDAMENTALS

A variety of refined techniques are used to climb different types of rock formations. The foundation for all of these styles is the art of climbing. Climbing technique stresses climbing with the weight centered over the feet, using the hands primarily for balance. It can be thought of as a combination of the balanced movement required to walk a tightrope and the technique used to ascend a ladder. No mountaineering equipment is required; however, the climbing technique is also used in roped climbing.

6-1. ROUTE SELECTION

The experienced climber has learned to climb with the "eyes." Even before getting on the rock, the climber studies all possible routes, or "lines," to the top looking for cracks, ledges, nubbins, and other irregularities in the rock that will be used for footholds and handholds, taking note of any larger ledges or benches for resting places. When picking the line, he mentally climbs the route, rehearsing the step-by-step sequence of movements that will be required to do the climb, ensuring himself that the route has an adequate number of holds and the difficulty of the climb will be well within the limit of his ability.

6-2. TERRAIN SELECTION FOR TRAINING

Route selection for military climbing involves picking the easiest and quickest possible line for all personnel to follow. However, climbing skill and experience can only be developed by increasing the length and difficulty of routes as training progresses. In the training environment, beginning lessons in climbing should be performed CLOSE to the ground on lower-angled rock with plenty of holds for the hands and feet. Personnel not climbing can act as "spotters" for those climbing. In later lessons, a "top-rope" belay can be used for safety, allowing the individual to increase the length and difficulty of the climb under the protection of the climbing rope.

6-3. PREPARATION

In preparation for climbing, the boot soles should be dry and clean. A small stick can be used to clean out dirt and small rocks that might be caught between the lugs of the boot sole. If the soles are wet or damp, dry them off by stomping and rubbing the soles on clean, dry rock. All jewelry should be removed from the fingers. Watches and bracelets can interfere with hand placements and may become damaged if worn while climbing. Helmets should be worn to protect the head from injury if an object, such as a rock or climbing gear, falls from climbers above. Most climbing helmets are not designed to provide protection from impact to the head if the wearer falls, but will provide a minimal amount of protection if a climber comes in contact with the rock during climbing.

CAUTION

Rings can become caught on rock facial features and or lodged into cracks, which could cause injuries during a slip or fall.

6-4. SPOTTING

Spotting is a technique used to add a level of safety to climbing without a rope. A second man stands below and just outside of the climbers fall path and helps (spots) the climber to land safely if he should fall. Spotting is only applicable if the climber is not going above the spotters head on the rock. Beyond that height a roped climbing should be conducted. If an individual climbs beyond the effective range of the spotter(s), he has climbed TOO HIGH for his own safety. The duties of the spotter are to help prevent the falling climber from impacting the head and or spine, help the climber land feet first, and reduce the impact of a fall.

CAUTION

The spotter should not catch the climber against the rock because additional injuries could result. If the spotter pushes the falling climber into the rock, deep abrasions of the skin or knee may occur. Ankle joints could be twisted by the fall if the climber's foot remained high on the rock. The spotter might be required to fully support the weight of the climber causing injury to the spotter.

6-5. CLIMBING TECHNIQUE

Climbing involves linking together a series of movements based on foot and hand placement, weight shift, and movement. When this series of movements is combined correctly, a smooth climbing technique results. This technique reduces excess force on the limbs, helping to minimize fatigue. The basic principle is based on the five body parts described here.

a. **Five Body Parts.** The five body parts used for climbing are the right hand, left hand, right foot, left foot, and body (trunk). The basic principle to achieve smooth climbing is to move only one of the five body parts at a time. The trunk is not moved in conjunction with a foot or in conjunction with a hand, a hand is not moved in conjunction with a foot, and so on. Following this simple technique forces both legs to do all the lifting simultaneously.

b. **Stance or Body Position.** Body position is probably the single most important element to good technique. A relaxed, comfortable stance is essential. (Figure 6-1 shows a correct climbing stance, and Figure 6-2, page 6-4, shows an incorrect stance.) The body should be in a near vertical or erect stance with the weight centered over the feet. Leaning in towards the rock will cause the feet to push outward, away from the rock, resulting in a loss of friction between the boot sole and rock surface. The legs are straight and the heels are kept low to reduce fatigue. Bent legs and tense muscles tire quickly. If strained for too long, tense muscles may vibrate uncontrollably. This vibration, known as "Elvis-ing" or "sewing-machine leg" can be cured by straightening the leg, lowering the heel, or moving on to a more restful position. The hands are used to maintain balance. Keeping the hands between waist and shoulder level will reduce arm fatigue.

Figure 6-1. Correct climbing stance—balanced over both feet.

Figure 6-2. Incorrect stance—stretched out.

(1) Whenever possible, three points of contact are maintained with the rock. Proper positioning of the hips and shoulders is critical. When using two footholds and one handhold, the hips and shoulders should be centered over both feet. In most cases, as the climbing progresses, the body is resting on one foot with two handholds for balance. The hips and shoulders must be centered over the support foot to maintain balance, allowing the "free" foot to maneuver.

(2) The angle or steepness of the rock also determines how far away from the rock the hips and shoulders should be. On low-angle slopes, the hips are moved out away from the rock to keep the body in balance with the weight over the feet. The shoulders can be moved closer to the rock to reach handholds. On steep rock, the hips are pushed closer to the rock. The shoulders are moved away from the rock by arching the back. The body is still in balance over the feet and the eyes can see where the hands need to go. Sometimes, when footholds are small, the hips are moved back to increase friction between the foot and the rock. This is normally done on quick, intermediate holds. It should be avoided in the rest position as it places more weight on the arms and hands. When weight must be placed on handholds, the arms should be kept straight to reduce fatigue. Again, flexed muscles tire quickly.

c. **Climbing Sequence.** The steps defined below provide a complete sequence of events to move the entire body on the rock. These are the basic steps to follow for a smooth climbing technique. Performing these steps in this exact order will not always be necessary because the nature of the route will dictate the availability of hand and foot placements. The basic steps are weight, shift, and movement (movement being either the foot, hand, or body). (A typical climbing sequence is shown in Figure 6-3, pages 6-6 through 6-8.)

STEP ONE: Shift the weight from both feet to one foot. This will allow lifting of one foot with no effect on the stance.

STEP TWO:	Lift the unweighted foot and place it in a new location, within one to two feet of the starting position, with no effect on body position or balance (higher placement will result in a potentially higher lift for the legs to make, creating more stress, and is called a high step) The trunk does not move during foot movement.
STEP THREE:	Shift the weight onto both feet. (Repeat steps 1 through 3 for remaining foot.)
STEP FOUR:	Lift the body into a new stance with both legs.
STEP FIVE:	Move one hand to a new position between waist and head height. During this movement, the trunk should be completely balanced in position and the removed hand should have no effect on stability.
STEP SIX:	Move the remaining hand as in Step 5.

Now the entire body is in a new position and ready to start the process again. Following these steps will prevent lifting with the hands and arms, which are used to maintain stance and balance. If both legs are bent, leg extension can be performed as soon as one foot has been moved. Hand movements can be delayed until numerous foot movements have been made, which not only creates shorter lifts with the legs, but may allow a better choice for the next hand movements because the reach will have increased.

Figure 6-3. Typical climbing sequence.

Figure 6-3. Typical climbing sequence (continued).

RIGHT HAND MOVED

Figure 6-3. Typical climbing sequence (continued).

(1) Many climbers will move more than one body part at a time, usually resulting in lifting the body with one leg or one leg and both arms. This type of lifting is inefficient, requiring one leg to perform the work of two or using the arms to lift the body. Proper climbing technique is lifting the body with the legs, not the arms, because the legs are much stronger.

(2) When the angle of the rock increases, these movements become more critical. Holding or pulling the body into the rock with the arms and hands may be necessary as the angle increases (this is still not lifting with the arms). Many climbing routes have angles greater than ninety degrees (overhanging) and the arms are used to support partial body weight. The same technique applies even at those angles.

(3) The climber should avoid moving on the knees and elbows. Other than being uncomfortable, even painful, to rest on, these bony portions of the limbs offer little friction and "feel" on the rock.

6-6. SAFETY PRECAUTIONS

The following safety precautions should be observed when rock climbing.

a. While ascending a seldom or never traveled route, you may encounter precariously perched rocks. If the rock will endanger your second, it may be possible to remove it from the route and trundle it, tossing it down. This is extremely dangerous to climbers below and should not be attempted unless you are absolutely sure no men are below. If you are not sure that the flight path is clear, do not do it. Never dislodge loose rocks carelessly. Should a rock become loose accidentally, immediately shout the warning "ROCK" to alert climbers below. Upon hearing the warning, personnel should seek immediate cover behind any rock bulges or overhangs available, or flatten themselves against the rock to minimize exposure.

b. Should a climber fall, he should do his utmost to maintain control and not panic. If on a low-angle climb, he may be able to arrest his own fall by staying in contact with the rock, grasping for any possible hold available. He should shout the warning "FALLING" to alert personnel below.

CAUTION
Grasping at the rock in a fall can result in serious injuries to the upper body. If conducting a roped climb, let the rope provide protection.

c. When climbing close to the ground and without a rope, a spotter can be used for safety. The duties of the spotter are to ensure the falling climber does not impact the head or spine, and to reduce the impact of a fall.

d. Avoid climbing directly above or below other climbers (with the exception of spotters). When personnel must climb at the same time, following the same line, a fixed rope should be installed.

e. Avoid climbing with gloves on because of the decreased "feel" for the rock. The use of gloves in the training environment is especially discouraged, while their use in the mountains is often mandatory when it is cold. A thin polypropylene or wool glove is best for rock climbing, although heavier cotton or leather work gloves are often used for belaying.

f. Be extremely careful when climbing on wet or moss-covered rock; friction on holds is greatly reduced.

g. Avoid grasping small vegetation for handholds; the root systems can be shallow and will usually not support much weight.

6-7. MARGIN OF SAFETY

Besides observing the standard safety precautions, the climber can avoid catastrophe by climbing with a wide margin of safety. The margin of safety is a protective buffer the climber places between himself and potential climbing hazards. Both subjective (personnel-related) and objective (environmental) hazards must be considered when applying the margin of safety. The leader must apply the margin of safety taking into account the strengths and weaknesses of the entire team or unit.

a. When climbing, the climber increases his margin of safety by selecting routes that are well within the limit of his ability. When leading a group of climbers, he selects a route well within the ability of the weakest member.

b. When the rock is wet, or when climbing in other adverse weather conditions, the climber's ability is reduced and routes are selected accordingly. When the climbing becomes difficult or exposed, the climber knows to use the protection of the climbing rope and belays. A lead climber increases his margin of safety by placing protection along the route to limit the length of a potential fall.

Section II. USE OF HOLDS

The climber should check each hold before use. This may simply be a quick, visual inspection if he knows the rock to be solid. When in doubt, he should grab and tug on the hold to test it for soundness BEFORE depending on it. Sometimes, a hold that appears weak can actually be solid as long as minimal force is applied to it, or the force is applied in a direction that strengthens it. A loose nubbin might not be strong enough to support the climber's weight, but it may serve as an adequate handhold. Be especially careful when climbing on weathered, sedimentary-type rock.

6-8. CLIMBING WITH THE FEET

"Climb with the feet and use the hands for balance" is extremely important to remember. In the early learning stages of climbing, most individuals will rely heavily on the arms, forgetting to use the feet properly. It is true that solid handholds and a firm grip are needed in some combination techniques; however, even the most strenuous techniques require good footwork and a quick return to a balanced position over one or both feet. Failure to climb any route, easy or difficult, is usually the result of poor footwork.

a. The beginning climber will have a natural tendency to look up for handholds. Try to keep the hands low and train your eyes to look down for footholds. Even the smallest irregularity in the rock can support the climber once the foot is positioned properly and weight is committed to it.

b. The foot remains on the rock as a result of friction. Maximum friction is obtained from a correct stance over a properly positioned foot. The following describes a few ways the foot can be positioned on the rock to maximize friction.

(1) ***Maximum Sole Contact.*** The principle of using full sole contact, as in mountain walking, also applies in climbing. Maximum friction is obtained by placing as much of the boot sole on the rock as possible. Also, the leg muscles can relax the most when the entire foot is placed on the rock. (Figure 6-4 shows examples of maximum and minimum sole contact.)

(a) Smooth, low-angled rock (slab) and rock containing large "bucket" holds and ledges are typical formations where the entire boot sole should be used.

(b) On some large holds, like bucket holds that extend deep into the rock, the entire foot cannot be used. The climber may not be able to achieve a balanced position if the foot is stuck too far underneath a bulge in the rock. In this case, placing only part of the foot on the hold may allow the climber to achieve a balanced stance. The key is to use as much of the boot sole as possible. Remember to keep the heels low to reduce strain on the lower leg muscles.

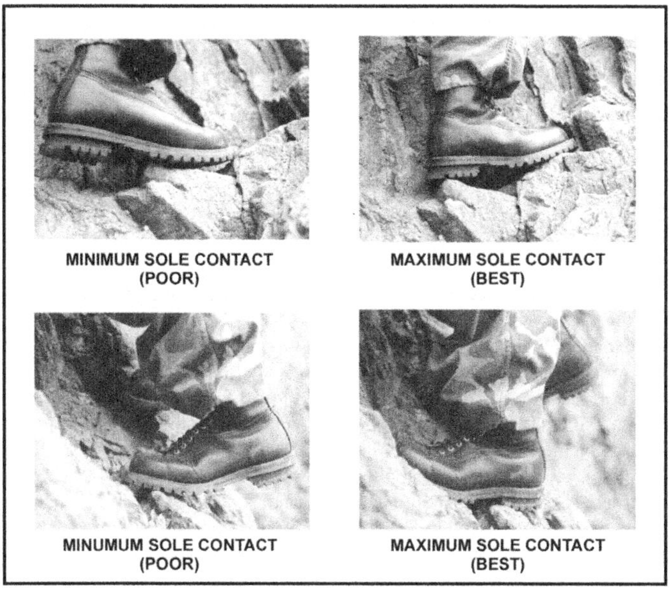

Figure 6-4. Examples of maximum and minimum sole contact.

(2) ***Edging.*** The edging technique is used where horizontal crack systems and other irregularities in the rock form small, well-defined ledges. The edge of the boot sole is placed on the ledge for the foothold. Usually, the inside edge of the boot or the edge area around the toes is used. Whenever possible, turn the foot sideways and use the entire inside edge of the boot. Again, more sole contact equals more friction and the legs can rest more when the heel is on the rock. (Figure 6-5, page 6-12, shows examples of the edging technique.)

(a) On smaller holds, edging with the front of the boot, or toe, may be used. Use of the toe is most tiring because the heel is off the rock and the toes support the climber's weight. Remember to keep the heel low to reduce fatigue. Curling and stiffening the toes in the boot increases support on the hold. A stronger position is usually obtained on small ledges by turning the foot at about a 45-degree angle, using the strength of the big toe and the ball of the foot.

(b) Effective edging on small ledges requires stiff-soled footwear. The stiffer the sole, the better the edging capability. Typical mountain boots worn by the US military have a relatively flexible lugged sole and, therefore, edging ability on smaller holds will be somewhat limited.

Figure 6-5. Examples of edging technique.

(3) ***Smearing.*** When footholds are too small to use a good edging technique, the ball of the foot can be "smeared" over the hold. The smearing technique requires the boot to adhere to the rock by deformation of the sole and by friction. Rock climbing shoes are specifically designed to maximize friction for smearing; some athletic shoes also work well. The Army mountain boot, with its softer sole, usually works better for smearing than for edging. Rounded, down-sloping ledges and low-angled slab rock often require good smearing technique. (Figure 6-6 shows examples of the smearing technique.)

(a) Effective smearing requires maximum friction between the foot and the rock. Cover as much of the hold as possible with the ball of the foot. Keeping the heel low will not only reduce muscle strain, but will increase the amount of surface contact between the foot and the rock.

(b) Sometimes flexing the ankles and knees slightly will place the climber's weight more directly over the ball of the foot and increase friction; however, this is more tiring and should only be used for quick, intermediate holds. The leg should be kept straight whenever possible.

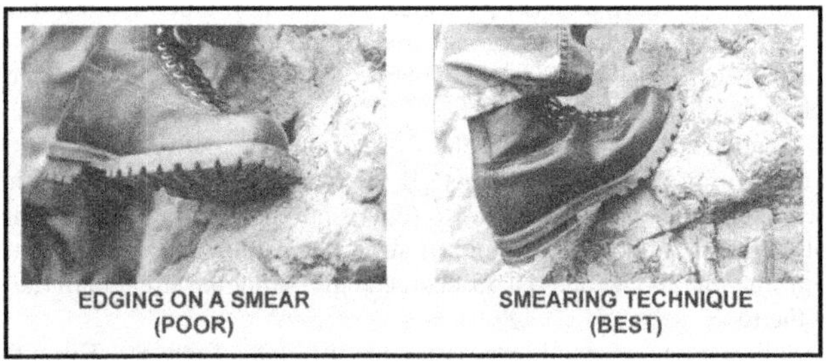

Figure 6-6. Examples of the smearing technique.

(4) ***Jamming.*** The jamming technique works on the same principal as chock placement. The foot is set into a crack in such a way that it "jams" into place, resisting a downward pull. The jamming technique is a specialized skill used to climb vertical or

near vertical cracks when no other holds are available on the rock face. The technique is not limited to just wedging the feet; fingers, hands, arms, even the entire leg or body are all used in the jamming technique, depending on the size of the crack. Jam holds are described in this text to broaden the range of climbing skills. Jamming holds can be used in a crack while other hand/foot holds are used on the face of the rock. Many cracks will have facial features, such as edges, pockets, and so on, inside and within reach. Always look or feel for easier to use features. (Figure 6-7 shows examples of jamming.)

(a) The foot can be jammed in a crack in different ways. It can be inserted above a constriction and set into the narrow portion, or it can be placed in the crack and turned, like a camming device, until it locks in place tight enough to support the climber's weight. Aside from these two basic ideas, the possibilities are endless. The toes, ball of the foot, or the entire foot can be used. Try to use as much of the foot as possible for maximum surface contact. Some positions are more tiring, and even more painful on the foot, than others. Practice jamming the foot in various ways to see what offers the most secure, restful position.

(b) Some foot jams may be difficult to remove once weight has been committed to them, especially if a stiffer sole boot is used. The foot is less likely to get stuck when it is twisted or "cammed" into position. When removing the boot from a crack, reverse the way it was placed to prevent further constriction.

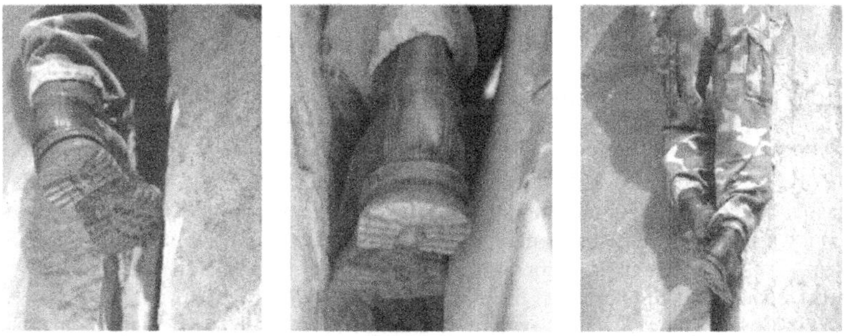

FOOT JAM

FOOT JAMMED INTO POSITION, HEEL IS PUSHED TO THE RIGHT

BOTH FEET JAMMED

Figure 6-7. Examples of jamming.

6-9. USING THE HANDS

The hands can be placed on the rock in many ways. Exactly how and where to position the hands and arms depends on what holds are available, and what configuration will best support the current stance as well as the movement to the next stance. Selecting handholds between waist and shoulder level helps in different ways. Circulation in the arms and hands is best when the arms are kept low. Secondly, the climber has less tendency to "hang" on his arms when the handholds are at shoulder level and below. Both of these contribute to a relaxed stance and reduce fatigue in the hands and arms.

a. As the individual climbs, he continually repositions his hands and arms to keep the body in balance, with the weight centered over the feet. On lower-angled rock, he may simply need to place the hands up against the rock and extend the arm to maintain balance; just like using an ice ax as a third point of contact in mountain walking. Sometimes, he will be able to push directly down on a large hold with the palm of the hand. More often though, he will need to "grip" the rock in some fashion and then push or pull against the hold to maintain balance.

b. As stated earlier, the beginner will undoubtedly place too much weight on the hands and arms. If we think of ourselves climbing a ladder, our body weight is on our legs. Our hands grip, and our arms pull on each rung only enough to maintain our balance and footing on the ladder. Ideally, this is the amount of grip and pull that should be used in climbing. Of course, as the size and availability of holds decreases, and the steepness of the rock approaches the vertical, the grip must be stronger and more weight might be placed on the arms and handholds for brief moments. The key is to move quickly from the smaller, intermediate holds to the larger holds where the weight can be placed back on the feet allowing the hands and arms to relax. The following describes some of the basic handholds and how the hand can be positioned to maximize grip on smaller holds.

(1) ***Push Holds.*** Push holds rely on the friction created when the hand is pushed against the rock. Most often a climber will use a push hold by applying "downward pressure" on a ledge or nubbin. This is fine, and works well; however, the climber should not limit his use of push holds to the application of down pressure. Pushing sideways, and on occasion, even upward on less obvious holds can prove quite secure. Push holds often work best when used in combination with other holds. Pushing in opposite directions and "push-pull" combinations are excellent techniques. (Figure 6-8 shows examples of push holds.)

(a) An effective push hold does not necessarily require the use of the entire hand. On smaller holds, the side of the palm, the fingers, or the thumb may be all that is needed to support the stance. Some holds may not feel secure when the hand is initially placed on them. The hold may improve or weaken during the movement. The key is to try and select a hold that will improve as the climber moves past it.

(b) Most push holds do not require much grip; however, friction might be increased by taking advantage of any rough surfaces or irregularities in the rock. Sometimes the strength of the hold can be increased by squeezing, or "pinching," the rock between the thumb and fingers (see paragraph on pinch holds).

Figure 6-8. Examples of push holds.

(2) ***Pull Holds.*** Pull holds, also called "cling holds," which are grasped and pulled upon, are probably the most widely used holds in climbing. Grip plays more of a role in a pull hold, and, therefore, it normally feels more secure to the climber than a push hold. Because of this increased feeling of security, pull holds are often overworked. These are the holds the climber has a tendency to hang from. Most pull holds do not require great strength, just good technique. Avoid the "death grip" syndrome by climbing with the feet. (Figure 6-9, page 6-16, shows examples of pull holds.)

(a) Like push holds, pressure on a pull hold can be applied straight down, sideways, or upward. Again, these are the holds the climber tends to stretch and reach for, creating an unbalanced stance. Remember to try and keep the hands between waist and shoulder level, making use of intermediate holds instead of reaching for those above the head.

(b) Pulling sideways on vertical cracks can be very secure. There is less tendency to hang from "side-clings" and the hands naturally remain lower. The thumb can often push against one side of the crack, in opposition to the pull by the fingers, creating a stronger hold. Both hands can also be placed in the same crack, with the hands pulling in opposite directions. The number of possible combinations is limited only by the imagination and experience of the climber.

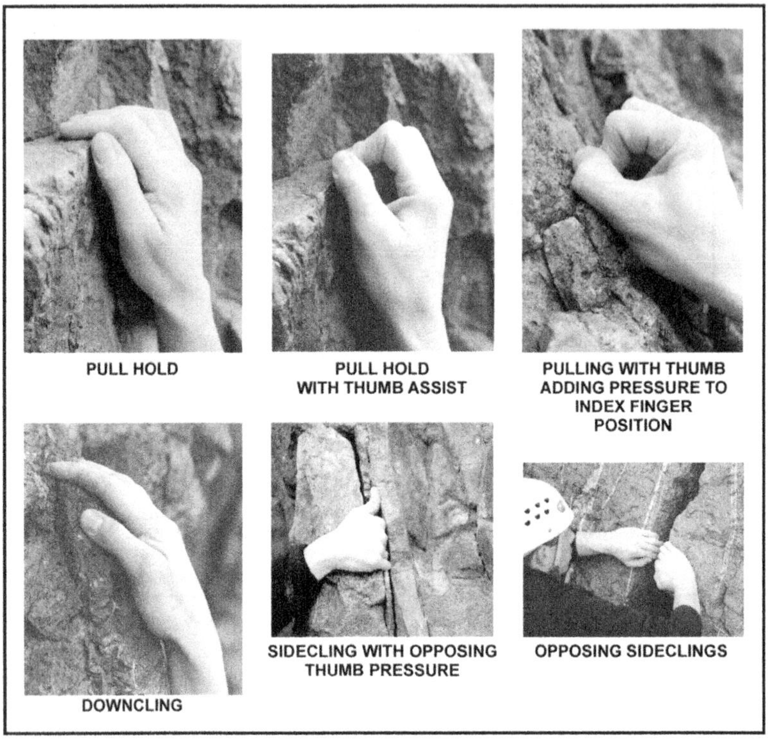

Figure 6-9. Examples of pull holds.

(c) Friction and strength of a pull hold can be increased by the way the hand grips the rock. Normally, the grip is stronger when the fingers are closed together; however, sometimes more friction is obtained by spreading the fingers apart and placing them between irregularities on the rock surface. On small holds, grip can often be improved by bending the fingers upward, forcing the palm of the hand to push against the rock. This helps to hold the finger tips in place and reduces muscle strain in the hand. Keeping the forearm up against the rock also allows the arm and hand muscles to relax more.

(d) Another technique that helps to strengthen a cling hold for a downward pull is to press the thumb against the side of the index finger, or place it on top of the index finger and press down. This hand configuration, known as a "ring grip," works well on smaller holds.

(3) ***Pinch Holds.*** Sometimes a small nubbin or protrusion in the rock can be "squeezed" between the thumb and fingers. This technique is called a pinch hold. Friction is applied by increasing the grip on the rock. Pinch holds are often overlooked by the

novice climber because they feel insecure at first and cannot be relied upon to support much body weight. If the climber has his weight over his feet properly, the pinch hold will work well in providing balance. The pinch hold can also be used as a gripping technique for push holds and pull holds. (Figure 6-10 shows examples of pinch holds.)

Figure 6-10. Examples of pinch holds.

(4) ***Jam Holds.*** Like foot jams, the fingers and hands can be wedged or cammed into a crack so they resist a downward or outward pull. Jamming with the fingers and hands can be painful and may cause minor cuts and abrasions to tender skin. Cotton tape can be used to protect the fingertips, knuckles, and the back of the hand; however, prolonged jamming technique requiring hand taping should be avoided. Tape also adds friction to the hand in jammed position. (Figure 6-11, page 6-18, shows examples of jam holds.)

(a) The hand can be placed in a crack a number of ways. Sometimes an open hand can be inserted and wedged into a narrower portion of the crack. Other times a clenched fist will provide the necessary grip. Friction can be created by applying cross pressure between the fingers and the back of the hand. Another technique for vertical cracks is to place the hand in the crack with the thumb pointed either up or down. The hand is then clenched as much as possible. When the arm is straightened, it will twist the hand and tend to cam it into place. This combination of clenching and camming usually produces the most friction, and the most secure hand jam in vertical cracks.

(b) In smaller cracks, only the fingers will fit. Use as many fingers as the crack will allow. The fingers can sometimes be stacked in some configuration to increase friction. The thumb is usually kept outside the crack in finger jams and pressed against the rock to increase friction or create cross pressure. In vertical cracks it is best to insert the fingers with the thumb pointing down to make use of the natural camming action of the fingers that occurs when the arm is twisted towards a normal position.

(c) Jamming technique for large cracks, or "off widths," requiring the use of arm, leg, and body jams, is another technique. To jam or cam an arm, leg, or body into an off width, the principle is the same as for fingers, hands, or feet—you are making the jammed appendage "fatter" by folding or twisting it inside the crack. For off widths, you may place your entire arm inside the crack with the arm folded and the palm pointing outward. The leg can be used, from the calf to the thigh, and flexed to fit the crack. Routes requiring this type of climbing should be avoided as the equipment normally used for protection might not be large enough to protect larger cracks and openings. However, sometimes a narrower section may be deeper in the crack allowing the use of "normal" size protection.

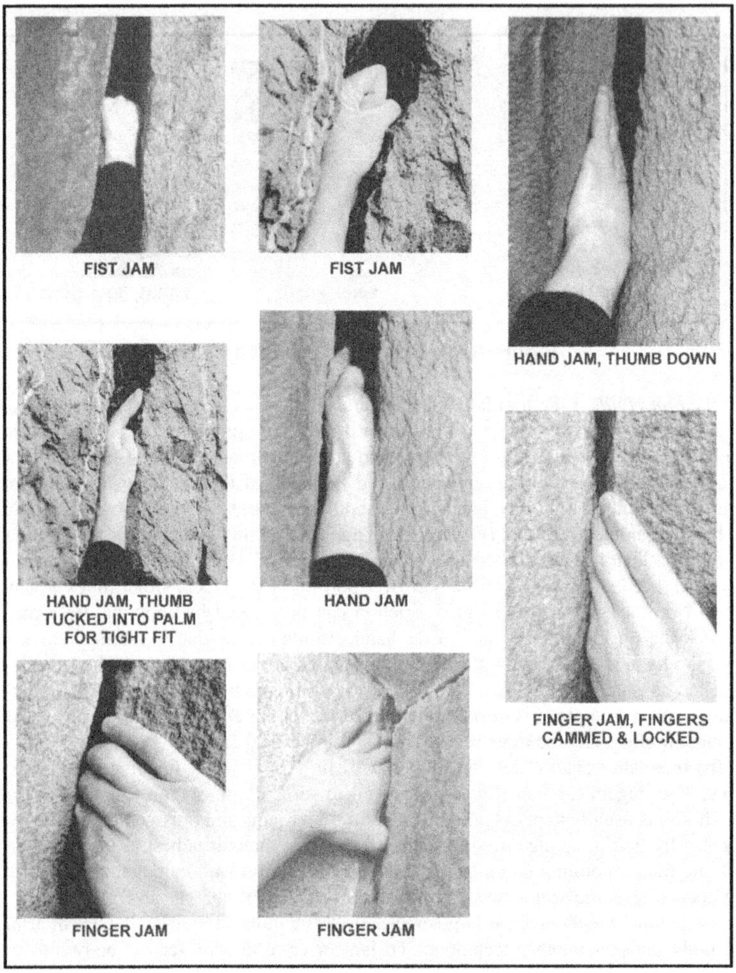

Figure 6-11. Examples of jam holds.

6-10. COMBINATION TECHNIQUES

The positions and holds previously discussed are the basics and the ones most common to climbing. From these fundamentals, numerous combination techniques are possible. As the climber gains experience, he will learn more ways to position the hands, feet, and body in relation to the holds available; however, he should always strive to climb with his weight on his feet from a balanced stance.

a. Sometimes, even on an easy route, the climber may come upon a section of the rock that defies the basic principles of climbing. Short of turning back, the only alternative is to figure out some combination technique that will work. Many of these type problems require the hands and feet to work in opposition to one another. Most will place more weight on the hands and arms than is desirable, and some will put the climber in an "out of balance" position. To make the move, the climber may have to "break the rules" momentarily. This is not a problem and is done quite frequently by experienced climbers. The key to using these type of combination techniques is to plan and execute them deliberately, without lunging or groping for holds, yet quickly, before the hands, arms, or other body parts tire. Still, most of these maneuvers require good technique more than great strength, though a certain degree of hand and arm strength certainly helps.

b. Combination possibilities are endless. The following is a brief description of some of the more common techniques.

(1) ***Change Step.*** The change step, or hop step, can be used when the climber needs to change position of the feet. It is commonly used when traversing to avoid crossing the feet, which might put the climber in an awkward position. To prevent an off balance situation, two solid handholds should be used. The climber simply places his weight on his handholds while he repositions the feet. He often does this with a quick "hop," replacing the lead foot with the trail foot on the same hold. Keeping the forearms against the rock during the maneuver takes some of the strain off the hands, while at the same time strengthening the grip on the holds.

(2) ***Mantling.*** Mantling is a technique that can be used when the distance between the holds increases and there are no immediate places to move the hands or feet. It does require a ledge (mantle) or projection in the rock that the climber can press straight down upon. (Figure 6-12, page 6-20, shows the mantling sequence.)

(a) When the ledge is above head height, mantling begins with pull holds, usually "hooking" both hands over the ledge. The climber pulls himself up until his head is above the hands, where the pull holds become push holds. He elevates himself until the arms are straight and he can lock the elbows to relax the muscles. Rotating the hands inward during the transition to push holds helps to place the palms more securely on the ledge. Once the arms are locked, a foot can be raised and placed on the ledge. The climber may have to remove one hand to make room for the foot. Mantling can be fairly strenuous; however, most individuals should be able to support their weight, momentarily, on one arm if they keep it straight and locked. With the foot on the ledge, weight can be taken off the arms and the climber can grasp the holds that were previously out of reach. Once balanced over the foot, he can stand up on the ledge and plan his next move.

(b) Pure mantling uses arm strength to raise the body; however, the climber can often smear the balls of the feet against the rock and "walk" the feet up during the maneuver to take some of the weight off the arms. Sometimes edges will be available for short steps in the process.

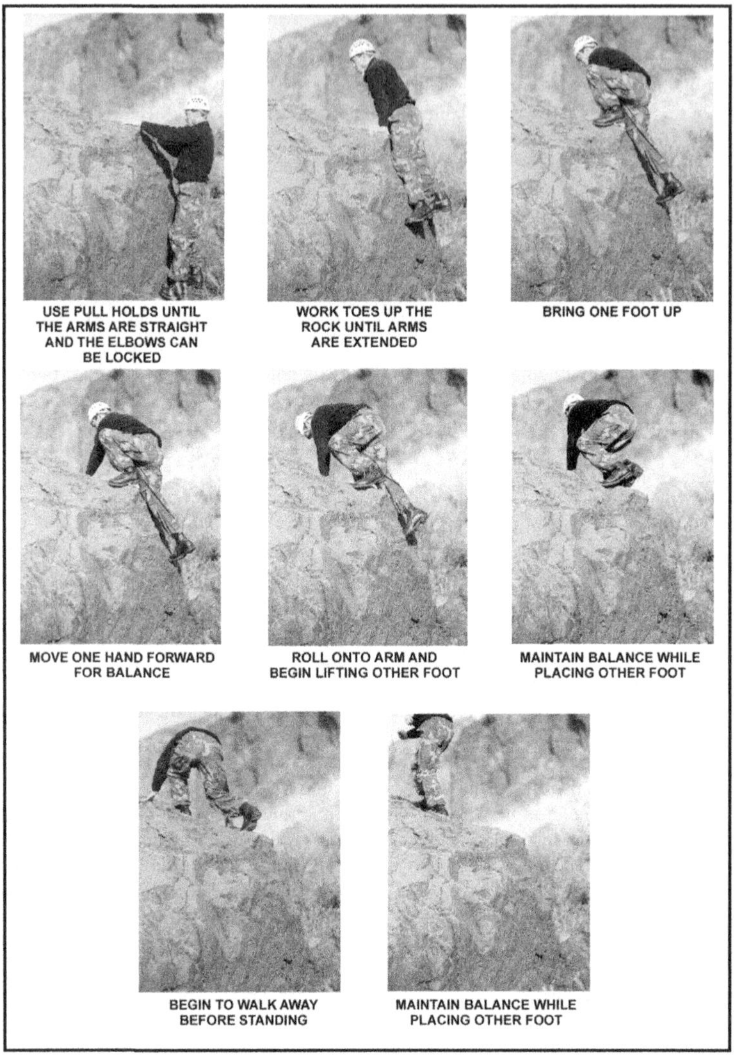

Figure 6-12. Mantling sequence.

(3) ***Undercling.*** An "undercling" is a classic example of handholds and footholds working in opposition (Figure 6-13). It is commonly used in places where the rock

projects outward, forming a bulge or small overhang. Underclings can be used in the tops of buckets, also. The hands are placed "palms-up" underneath the bulge, applying an upward pull. Increasing this upward pull creates a counterforce, or body tension, which applies more weight and friction to the footholds. The arms and legs should be kept as straight as possible to reduce fatigue. The climber can often lean back slightly in the undercling position, enabling him to see above the overhang better and search for the next hold.

Figure 6-13. Undercling.

(4) ***Lieback.*** The "lieback" is another good example of the hands working in opposition to the feet. The technique is often used in a vertical or diagonal crack separating two rock faces that come together at, more or less, a right angle (commonly referred to as a dihedral). The crack edge closest to the body is used for handholds while the feet are pressed against the other edge. The climber bends at the waist, putting the body into an L-shaped position. Leaning away from the crack on two pull holds, body tension creates friction between the feet and the hands. The feet must be kept relatively high to maintain weight, creating maximum friction between the sole and the rock surface. Either full sole contact or the smearing technique can be used, whichever seems to produce the most friction.

(a) The climber ascends a dihedral by alternately shuffling the hands and feet upward. The lieback technique can be extremely tiring, especially when the dihedral is near vertical. If the hands and arms tire out before completing the sequence, the climber will likely fall. The arms should be kept straight throughout the entire maneuver so the climber's weight is pulling against bones and ligaments, rather than muscle. The legs should be straightened whenever possible.

(b) Placing protection in a lieback is especially tiring. Look for edges or pockets for the feet in the crack or on the face for a better position to place protection from, or for a rest position. Often, a lieback can be avoided with closer examination of the available face features. The lieback can be used alternately with the jamming technique, or vice versa, for variation or to get past a section of a crack with difficult or nonexistent jam possibilities. The lieback can sometimes be used as a face maneuver (Figure 6-14).

Figure 6-14. Lieback on a face.

(5) ***Stemming.*** When the feet work in opposition from a relatively wide stance, the maneuver is known as stemming. The stemming technique can sometimes be used on faces, as well as in a dihedral in the absence of solid handholds for the lieback (Figure 6-15).

Figure 6-15. Stemming on a face.

(a) The classic example of stemming is when used in combination with two opposing push holds in wide, parallel cracks, known as chimneys. Chimneys are cracks in which the walls are at least 1 foot apart and just big enough to squeeze the body into. Friction is created by pushing outward with the hands and feet on each side of the crack. The climber ascends the chimney by alternately moving the hands and feet up the crack (Figure 6-16). Applying pressure with the back and bottom is usually necessary in wider chimneys. Usually, full sole contact of the shoes will provide the most friction, although smearing may work best in some instances. Chimneys that do not allow a full stemming position can be negotiated using the arms, legs, or body as an integral contact point. This technique will often feel more secure since there is more body to rock contact.

Figure 6-16. Chimney sequence.

Figure 6-16. Chimney sequence (continued).

(b) The climber can sometimes rest by placing both feet on the same side of the crack, forcing the body against the opposing wall. The feet must be kept relatively high up under the body so the force is directed sideways against the walls of the crack. The arms should be straightened with the elbows locked whenever possible to reduce muscle strain. The climber must ensure that the crack does not widen beyond the climbable width before committing to the maneuver. Remember to look for face features inside chimneys for more security in the climb.

(c) Routes requiring this type of climbing should be avoided as the equipment normally used for protection might not be large enough to protect chimneys. However, face features, or a much narrower crack in one or both corners, may sometimes be found deeper in the chimney allowing the use of normal size protection.

(6) ***Slab Technique.*** A slab is a relatively smooth, low-angled rock formation that requires a slightly modified climbing technique (Figure 6-17). Since slab rock normally contains few, if any holds, the technique requires maximum friction and perfect balance over the feet.

(a) On lower-angled slab, the climber can often stand erect and climb using full sole contact and other mountain walking techniques. On steeper slab, the climber will need to apply good smearing technique. Often, maximum friction cannot be attained on steeper slab from an erect stance. The climber will have to flex the ankles and knees so his weight is placed more directly over the balls of the feet. He may then have to bend at the waist to place the hands on the rock, while keeping the hips over his feet.

(b) The climber must pay attention to any changes in slope angle and adjust his body accordingly. Even the slightest change in the position of the hips over the feet can mean the difference between a good grip or a quick slip. The climber should also take advantage of any rough surfaces, or other irregularities in the rock he can place his hands or feet on, to increase friction.

Figure 6-17. Slab technique.

(7) ***Down Climbing.*** Descending steep rock is normally performed using a roped method; however, the climber may at some point be required to down climb a route. Even if climbing ropes and related equipment are on hand, down climbing easier terrain is often quicker than taking the time to rig a rappel point. Also, a climber might find himself confronted with difficulties part way up a route that exceed his climbing ability, or the abilities of others to follow. Whatever the case may be, down climbing is a skill well worth practicing.

CAUTIONS

1. Down climbing can inadvertently lead into an unforeseen dangerous position on a descent. When in doubt, use a roped descent.
2. Down climbing is accomplished at a difficulty level well below the ability of the climber. When in doubt, use a roped descent.

(a) On easier terrain, the climber can face outward, away from the rock, enabling him to see the route better and descend quickly. As the steepness and difficulty increase, he can often turn sideways, still having a good view of the descent route, but being better able to use the hands and feet on the holds available. On the steepest terrain, the climber will have to face the rock and down climb using good climbing techniques.

(b) Down climbing is usually more difficult than ascending a given route. Some holds will be less visible when down climbing, and slips are more likely to occur. The climber must often lean well away from the rock to look for holds and plan his movements. More weight is placed on the arms and handholds at times to accomplish this, as well as to help lower the climber to the next foothold. Hands should be moved to holds as low as waist level to give the climber more range of movement with each step. If the handholds are too high, he may have trouble reaching the next foothold. The climber must be careful not to overextend himself, forcing a release of his handholds before reaching the next foothold.

CAUTION

Do not drop from good handholds to a standing position. A bad landing could lead to injured ankles or a fall beyond the planned landing area.

(c) Descending slab formations can be especially tricky. The generally lower angle of slab rock may give the climber a false sense of security, and a tendency to move too quickly. Down climbing must be slow and deliberate, as in ascending, to maintain perfect balance and weight distribution over the feet. On lower-angle slab the climber may be able to stand more or less erect, facing outward or sideways, and descend using good flat foot technique. The climber should avoid the tendency to move faster, which can lead to uncontrollable speed.

(d) On steeper slab, the climber will normally face the rock and down climb, using the same smearing technique as for ascending. An alternate method for descending slab is to face away from the rock in a "crab" position (Figure 6-18). Weight is still concentrated over the feet, but may be shifted partly onto the hands to increase overall friction. The climber is able to maintain full sole contact with the rock and see the entire descent route. Allowing the buttocks to "drag behind" on the rock will decrease the actual weight on the footholds, reducing friction, and leading to the likelihood of a slip. Facing the rock, and down-climbing with good smearing technique, is usually best on steeper slab.

Figure 6-18. Descending slab in the crab position.

Section III. ROPED CLIMBING

When the angle, length, and difficulty of the proposed climbing route surpasses the ability of the climbers' safety margin (possibly on class 4 and usually on class 5 terrain), ropes must be used to proceed. Roped climbing is only safe if accomplished correctly. Reading this manual does not constitute skill with ropes—much training and practice is necessary. Many aspects of roped climbing take time to understand and learn. Ropes are normally not used in training until the basic principles of climbing are covered.

Note: A rope is completely useless for climbing unless the climber knows how to use it safely.

6-11. TYING-IN TO THE CLIMBING ROPE

Over the years, climbers have developed many different knots and procedures for tying-in to the climbing rope. Some of the older methods of tying directly into the rope require minimal equipment and are relatively easy to inspect; however, they offer little support to the climber, may induce further injuries, and may even lead to strangulation in a severe fall. A severe fall, where the climber might fall 20 feet or more and be left dangling on the end of the rope, is highly unlikely in most instances, especially for most personnel involved in military climbing. Tying directly into the rope is perfectly safe for many roped party climbs used in training on lower-angled rock. All climbers should know how to properly tie the rope around the waist in case a climbing harness is unavailable.

6-12. PRESEWN HARNESSES

Although improvised harnesses are made from readily available materials and take little space in the pack or pocket, presewn harnesses provide other aspects that should be considered. No assembly is required, which reduces preparation time for roped movement. All presewn harnesses provide a range of adjustability. These harnesses have a fixed buckle that, when used correctly, will not fail before the nylon materials connected to it. However, specialized equipment, such as a presewn harness, reduce the flexibility of gear. Presewn harness are bulky, also.

a. **Seat Harness.** Many presewn seat harnesses are available with many different qualities separating them, including cost.

(1) The most notable difference will be the amount and placement of padding. The more padding the higher the price and the more comfort. Gear loops sewn into the waist belt on the sides and in the back are a common feature and are usually strong enough to hold quite a few carabiners and or protection. The gear loops will vary in number from one model/manufacturer to another.

(2) Although most presewn seat harnesses have a permanently attached belay loop connecting the waist belt and the leg loops, the climbing rope should be run around the waist belt and leg loop connector. The presewn belay loop adds another link to the chain of possible failure points and only gives one point of security whereas running the rope through the waist belt and leg loop connector provides two points of contact.

(3) If more than two men will be on the rope, connect the middle position(s) to the rope with a carabiner routed the same as stated in the previous paragraph.

(4) Many manufactured seat harnesses will have a presewn loop of webbing on the rear. Although this loop is much stronger than the gear loops, it is not for a belay anchor. It is a quick attachment point to haul an additional rope.

b. **Chest Harness.** The chest harness will provide an additional connecting point for the rope, usually in the form of a carabiner loop to attach a carabiner and rope to. This type of additional connection will provide a comfortable hanging position on the rope, but otherwise provides no additional protection from injury during a fall (if the seat harness is fitted correctly).

(1) A chest harness will help the climber remain upright on the rope during rappelling or ascending a fixed rope, especially while wearing a heavy pack. (If rappelling or ascending long or multiple pitches, let the pack hang on a drop cord below the feet and attached to the harness tie-in point.)

(2) The presewn chest harnesses available commercially will invariably offer more comfort or performance features, such as padding, gear loops, or ease of adjustment, than an improvised chest harness.

c. **Full-Body Harness.** Full-body harnesses incorporate a chest and seat harness into one assembly. This is the safest harness to use as it relocates the tie-in point higher, at the chest, reducing the chance of an inverted position when hanging on the rope. This is especially helpful when moving on ropes with heavy packs. A full-body harness only affects the body position when hanging on the rope and will not prevent head injury in a fall.

CAUTION

This type of harness does not prevent the climber from falling head first. Body position during a fall is affected only by the forces that generated the fall, and this type of harness promotes an upright position only when hanging on the rope from the attachment point.

6-13. IMPROVISED HARNESSES

Without the use of a manufactured harness, many methods are still available for attaching oneself to a rope. Harnesses can be improvised using rope or webbing and knots.

a. **Swami Belt.** The swami belt is a simple, belt-only harness created by wrapping rope or webbing around the waistline and securing the ends. One-inch webbing will provide more comfort. Although an effective swami belt can be assembled with a minimum of one wrap, at least two wraps are recommended for comfort, usually with approximately ten feet of material. The ends are secured with an appropriate knot.

b. **Bowline-on-a-Coil.** Traditionally, the standard method for attaching oneself to the climbing rope was with a bowline-on-a-coil around the waist. The extra wraps distribute the force of a fall over a larger area of the torso than a single bowline would, and help prevent the rope from riding up over the rib cage and under the armpits. The knot must be tied snugly around the narrow part of the waist, just above the bony portions of the hips (pelvis). Avoid crossing the wraps by keeping them spread over the waist area. "Sucking in the gut" a bit when making the wraps will ensure a snug fit.

(1) The bowline-on-a-coil can be used to tie-in to the end of the rope (Figure 6-19). The end man should have a minimum of four wraps around the waist before completing the knot.

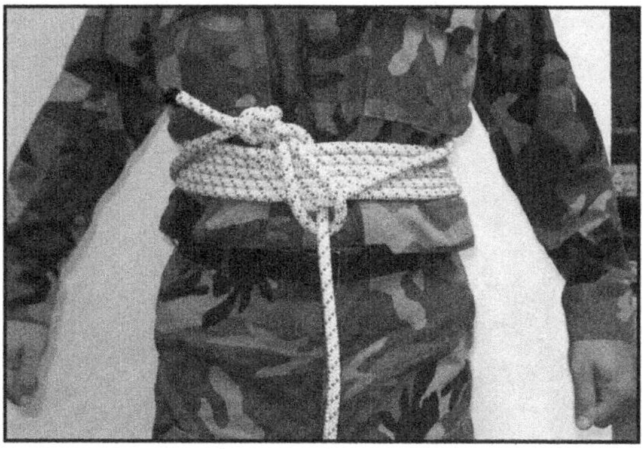

Figure 6-19. Tying-in with a bowline-on-a-coil.

(2) The bowline-on-a-coil is a safe and effective method for attaching to the rope when the terrain is low-angled, WITHOUT THE POSSIBILITY OF A SEVERE FALL. When the terrain becomes steeper, a fall will generate more force on the climber and this will be felt through the coils of this type of attachment. A hard fall will cause the coils to ride up against the ribs. In a severe fall, any tie-in around the waist only could place a "shock load" on the climber's lower back. Even in a relatively short fall, if the climber ends up suspended in mid-air and unable to regain footing on the rock, the rope around the waist can easily cut off circulation and breathing in a relatively short time.

(3) The climbing harness distributes the force of a fall over the entire pelvic region, like a parachute harness. Every climber should know how to tie some sort of improvised climbing harness from sling material. A safe, and comfortable, seat/chest combination harness can be tied from one-inch tubular nylon.

c. **Improvised Seat Harness.** A seat harness can be tied from a length of webbing approximately 25 feet long (Figure 6-20, page 6-30).

(1) Locate the center of the rope. Off to one side, tie two fixed loops approximately 6 inches apart (overhand loops). Adjust the size of the loops so they fit snugly around the thigh. The loops are tied into the sling "off center" so the remaining ends are different lengths. The short end should be approximately 4 feet long (4 to 5 feet for larger individuals).

(2) Slip the leg loops over the feet and up to the crotch, with the knots to the front. Make one complete wrap around the waist with the short end, wrapping to the outside, and hold it in place on the hip. Keep the webbing flat and free of twists when wrapping.

(3) Make two to three wraps around the waist with the long end in the opposite direction (wrapping to the outside), binding down on the short end to hold it in place. Grasping both ends, adjust the waist wraps to a snug fit. Connect the ends with the

appropriate knot between the front and one side so you will be able to see what you are doing.

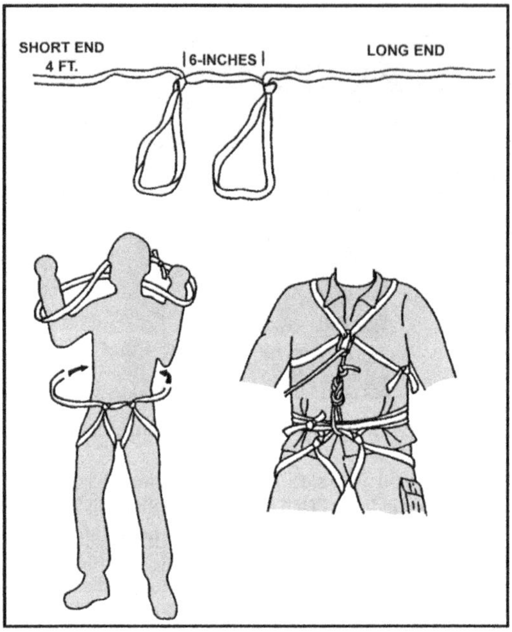

Figure 6-20. Improvised seat and chest harness.

d. **Improvised Chest Harness.** The chest harness can be tied from rope or webbing, but remember that with webbing, wider is better and will be more comfortable when you load this harness. Remember as you tie this harness that the remaining ends will need to be secured so choose the best length. Approximately 6 to 10 feet usually works.

(1) Tie the ends of the webbing together with the appropriate knot, making a sling 3 to 4 feet long.

(2) Put a single twist into the sling, forming two loops.

(3) Place an arm through each loop formed by the twist, just as you would put on a jacket, and drape the sling over the shoulders. The twist, or cross, in the sling should be in the middle of the back.

(4) Join the two loops at the chest with a carabiner. The water knot should be set off to either side for easy inspection (if a pack is to be worn, the knot will be uncomfortable if it gets between the body and the pack). The chest harness should fit just loose enough to allow necessary clothing and not to restrict breathing or circulation. Adjust the size of the sling if necessary.

e. **Improvised Full-Body Harness.** Full-body harnesses incorporate a chest and seat harness into one assembly.

(1) The full-body harness is the safest harness because it relocates the tie-in point higher, at the chest, reducing the chance of an inverted hanging position on the rope. This is especially helpful when moving on ropes with heavy packs. A full-body harness affects the body position only when hanging on the rope.

CAUTION

A full-body harness does not prevent falling head first; body position in a fall is caused by the forces that caused the fall.

(2) Although running the rope through the carabiner of the chest harness does, in effect, create a type of full-body harness, it is not a true full-body harness until the chest harness and the seat harness are connected as one piece. A true full-body harness can be improvised by connecting the chest harness to the seat harness, but not by just tying the rope into both—the two harnesses must be "fixed" as one harness. Fix them together with a short loop of webbing or rope so that the climbing rope can be connected directly to the chest harness and your weight is supported by the seat harness through the connecting material.

f. **Attaching the Rope to the Improvised Harness.** The attachment of the climbing rope to the harness is a CRITICAL LINK. The strength of the rope means nothing if it is attached poorly, or incorrectly, and comes off the harness in a fall. The climber ties the end of the climbing rope to the seat harness with an appropriate knot. If using a chest harness, the standing part of the rope is then clipped into the chest harness carabiner. The seat harness absorbs the main force of the fall, and the chest harness helps keep the body upright.

CAUTION

The knot must be tied around all the waist wraps and the 6-inch length of webbing between the leg loops.

(1) A middleman must create a fixed loop to tie in to. A rethreaded figure-eight loop tied on a doubled rope or the three loop bowline can be used. If using the three loop bowline, ensure the end, or third loop formed in the knot, is secured around the tie-in loops with an overhand knot. The standing part of the rope going to the lead climber is clipped into the chest harness carabiner.

Note: The climbing rope is not clipped into the chest harness when belaying.

(2) The choice of whether to tie-in with a bowline-on-a-coil or into a climbing harness depends entirely on the climber's judgment, and possibly the equipment

available. A good rule of thumb is: "Wear a climbing harness when the potential for severe falls exists and for all travel over snow-covered glaciers because of the crevasse fall hazard."

(3) Under certain conditions many climbers prefer to attach the rope to the seat harness with a locking carabiner, rather than tying the rope to it. This is a common practice for moderate snow and ice climbing, and especially for glacier travel where wet and frozen knots become difficult to untie.

CAUTION

Because the carabiner gate may be broken or opened by protruding rocks during a fall, tie the rope directly to the harness for maximum safety.

Section IV. BELAY TECHNIQUES

Tying-in to the climbing rope and moving as a member of a rope team increases the climber's margin of safety on difficult, exposed terrain. In some instances, such as when traveling over snow-covered glaciers, rope team members can often move at the same time, relying on the security of a tight rope and "team arrest" techniques to halt a fall by any one member. On steep terrain, however, simultaneous movement only helps to ensure that if one climber falls, he will jerk the other rope team members off the slope. For the climbing rope to be of any value on steep rock climbs, the rope team must incorporate "belays" into the movement.

Belaying is a method of managing the rope in such a way that, if one person falls, the fall can be halted or "arrested" by another rope team member (belayer). One person climbs at a time, while being belayed from above or below by another. The belayer manipulates the rope so that friction, or a "brake," can be applied to halt a fall. Belay techniques are also used to control the descent of personnel and equipment on fixed rope installations, and for additional safety on rappels and stream crossings.

Belaying is a skill that requires practice to develop proficiency. Setting up a belay may at first appear confusing to the beginner, but with practice, the procedure should become "second nature." If confronted with a peculiar problem during the setup of a belay, try to use common sense and apply the basic principles stressed throughout this text. Remember the following key points:

- Select the best possible terrain features for the position and use terrain to your advantage.
- Use a well braced, sitting position whenever possible.
- Aim and anchor the belay for all possible load directions.
- Follow the "minimum" rule for belay anchors—2 for a downward pull, 1 for an upward pull.
- Ensure anchor attachments are aligned, independent, and snug.
- Stack the rope properly.
- Choose a belay technique appropriate for the climbing.

- Use a guide carabiner for rope control in all body belays.
- Ensure anchor attachments, guide carabiner (if applicable), and rope running to the climber are all on the guidehand side.
- The brake hand remains on the rope when belaying.

CAUTION
Never remove the brake hand from the rope while belaying. If the brake hand is removed, there is no belay.

- Ensure you are satisfied with your position before giving the command "BELAY ON."
- The belay remains in place until the climber gives the command "OFF BELAY."

CAUTION
The belay remains in place the from the time the belayer commands *"BELAY ON"* until the climber commands *"OFF BELAY."*

6-14. PROCEDURE FOR MANAGING THE ROPE

A number of different belay techniques are used in modern climbing, ranging from the basic "body belays" to the various "mechanical belays," which incorporate some type of friction device.

a. Whether the rope is wrapped around the body, or run through a friction device, the rope management procedure is basically the same. The belayer must be able to perform three basic functions: manipulate the rope to give the climber slack during movement, take up rope to remove excess slack, and apply the brake to halt a fall.

b. The belayer must be able to perform all three functions while maintaining "total control" of the rope at all times. Total control means the brake hand is NEVER removed from the rope. When giving slack, the rope simply slides through the grasp of the brake hand, at times being fed to the climber with the other "feeling" or guide hand. Taking up rope, however, requires a certain technique to ensure the brake hand remains on the rope at all times. The following procedure describes how to take up excess rope and apply the brake in a basic body belay.

(1) Grasping the rope with both hands, place it behind the back and around the hips. The hand on the section of rope between the belayer and the climber would be the guide hand. The other hand is the brake hand.

(2) Take in rope with the brake hand until the arm is fully extended. The guide hand can also help to pull in the rope (Figure 6-21, step 1, page 6-34).

(3) Holding the rope in the brake hand, slide the guide hand out, extending the arm so the guide hand is father away from the body than the brake hand (Figure 6-21, step 2).

(4) Grasp both parts of the rope, to the front of the brake hand, with the guide hand (Figure 6-21, step 3).

(5) Slide the brake hand back towards the body (Figure 6-21, step 4).

(6) Repeat step 5 of Figure 6-21. The brake can be applied at any moment during the procedure. It is applied by wrapping the rope around the front of the hips while increasing grip with the brake hand (Figure 6-21, step 6).

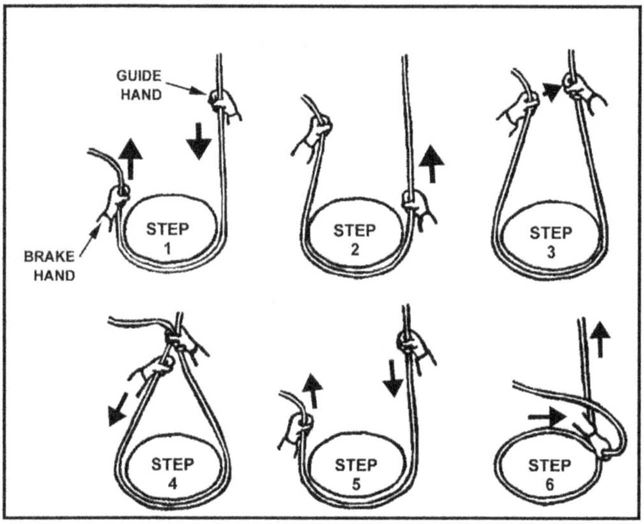

Figure 6-21. Managing the rope.

6-15. CHOOSING A BELAY TECHNIQUE

The climber may choose from a variety of belay techniques. A method that works well in one situation may not be the best choice in another. The choice between body belays and mechanical belays depends largely on equipment available, what the climber feels most comfortable with, and the amount of load, or fall force, the belay may have to absorb. The following describes a few of the more widely used techniques, and the ones most applicable to military mountaineering.

a. **Body Belay.** The basic body belay is the most widely used technique on moderate terrain. It uses friction between the rope and the clothed body as the rope is pressured across the clothing. It is the simplest belay, requiring no special equipment, and should be the first technique learned by all climbers. A body belay gives the belayer the greatest "feel" for the climber, letting him know when to give slack or take up rope. Rope management in a body belay is quick and easy, especially for beginners, and is effective in snow and ice climbing when ropes often become wet, stiff, and frozen. The body belay, in its various forms, will hold low to moderate impact falls well. It has been known to arrest some severe falls, although probably not without inflicting great pain on the belayer.

CAUTION

The belayer must ensure he is wearing adequate clothing to protect his body from rope burns when using a body belay. Heavy duty cotton or leather work gloves can also be worn to protect the hands.

(1) ***Sitting Body Belay.*** The sitting body belay is the preferred position and is usually the most secure (Figure 6-22). The belayer sits facing the direction where the force of a fall will likely come from, using terrain to his advantage, and attempts to brace both feet against the rock to support his position. It is best to sit in a slight depression, placing the buttocks lower than the feet, and straightening the legs for maximum support. When perfectly aligned, the rope running to the climber will pass between the belayer's feet, and both legs will equally absorb the force of a fall. Sometimes, the belayer may not be able to sit facing the direction he would like, or both feet cannot be braced well. The leg on the "guide hand" side should then point towards the load, bracing the foot on the rock when possible. The belayer can also "straddle" a large tree or rock nubbin for support, as long as the object is solid enough to sustain the possible load.

Figure 6-22. Sitting body belay.

(2) ***Standing Body Belay.*** The standing body belay is used on smaller ledges where there is no room for the belayer to sit (Figure 6-23). What appears at first to be a fairly unstable position can actually be quite secure when belay anchors are placed at or above shoulder height to support the stance when the force will be downward.

Figure 6-23. Standing body belay.

(a) For a body belay to work effectively, the belayer must ensure that the rope runs around the hips properly, and remains there under load when applying the brake. The rope should run around the narrow portion of the pelvic girdle, just below the bony high points of the hips. If the rope runs too high, the force of a fall could injure the belayer's midsection and lower rib cage. If the rope runs too low, the load may pull the rope below the buttocks, dumping the belayer out of position. It is also possible for a strong upward or downward pull to strip the rope away from the belayer, rendering the belay useless.

(b) To prevent any of these possibilities from happening, the belay rope is clipped into a carabiner attached to the guide hand side of the seat harness (or bowline-on-a-coil). This "guide carabiner" helps keep the rope in place around the hips and prevents loss of control in upward or downward loads (Figure 6-24).

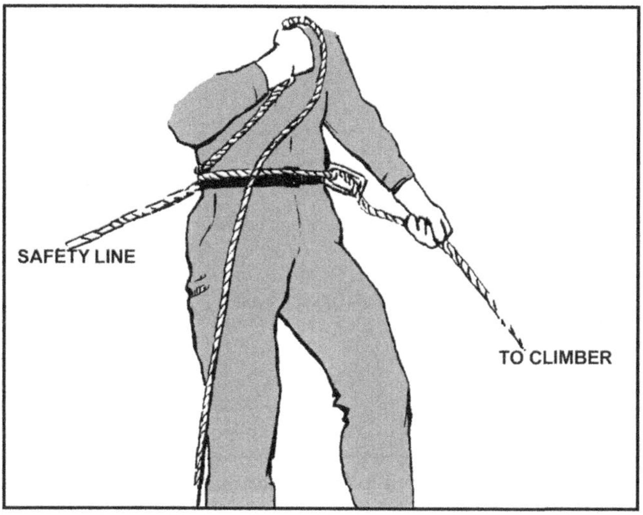

Figure 6-24. Guide carabiner for rope control in a body belay.

b. **Mechanical Belay.** A mechanical belay **must** be used whenever there is potential for the lead climber to take a severe fall. The holding power of a belay device is vastly superior to any body belay under high loads. However, rope management in a mechanical belay is more difficult to master and requires more practice. For the most part, the basic body belay should be totally adequate on a typical military route, as routes used during military operations should be the easiest to negotiate.

(1) ***Munter Hitch.*** The Munter hitch is an excellent mechanical belay technique and requires only a rope and a carabiner (Figure 6-25, page 6-38). The Munter is actually a two-way friction hitch. The Munter hitch will flip back and forth through the carabiner as the belayer switches from giving slack to taking up rope. The carabiner must be large enough, and of the proper design, to allow this function. The locking pear-shaped carabiner, or pearabiner, is designed for the Munter hitch.

(a) The Munter hitch works exceptionally well as a lowering belay off the anchor. As a climbing belay, the carabiner should be attached to the front of the belayer's seat harness. The hitch is tied by forming a loop and a bight in the rope, attaching both to the carabiner. It's fairly easy to place the bight on the carabiner backwards, which forms an obvious, useless hitch. Put some tension on the Munter to ensure it is formed correctly, as depicted in the following illustrations.

(b) The Munter hitch will automatically "lock-up" under load as the brake hand grips the rope. The brake is increased by pulling the slack rope away from the body, towards the load. The belayer must be aware that flipping the hitch DOES NOT change the function of the hands. The hand on the rope running to the climber, or load, is always the guide hand.

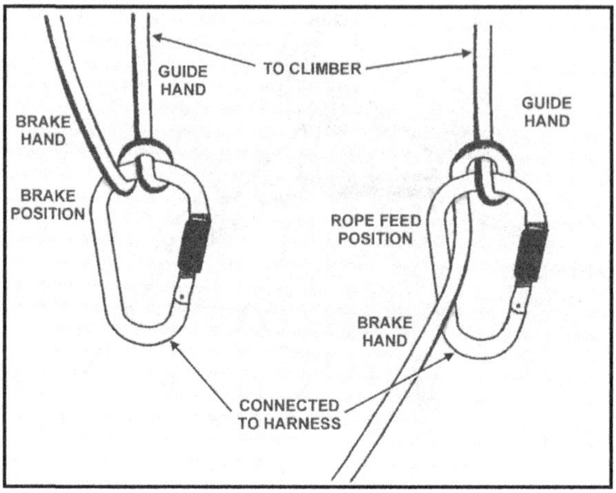

Figure 6-25. Munter hitch.

(2) ***Figure-Eight Device.*** The figure-eight device is a versatile piece of equipment and, though developed as a rappel device, has become widely accepted as an effective mechanical belay device (Figure 6-26). The advantage of any mechanical belay is friction required to halt a fall is applied on the rope through the device, rather than around the belayer's body. The device itself provides rope control for upward and downward pulls and excellent friction for halting severe falls. The main principle behind the figure-eight device in belay mode is the friction developing on the rope as it reaches and exceeds the 90-degree angle between the rope entering the device and leaving the device. As a belay device, the figure-eight works well for both belayed climbing and for lowering personnel and equipment on fixed-rope installations.

(a) As a climbing belay, a bight placed into the climbing rope is run through the "small eye" of the device and attached to a locking carabiner at the front of the belayer's seat harness. A short, small diameter safety rope is used to connect the "large eye" of the figure eight to the locking carabiner for control of the device. The guide hand is placed on the rope running to the climber. Rope management is performed as in a body belay. The brake is applied by pulling the slack rope in the brake hand towards the body, locking the rope between the device and the carabiner.

(b) As a lowering belay, the device is normally attached directly to the anchor with the rope routed as in rappelling.

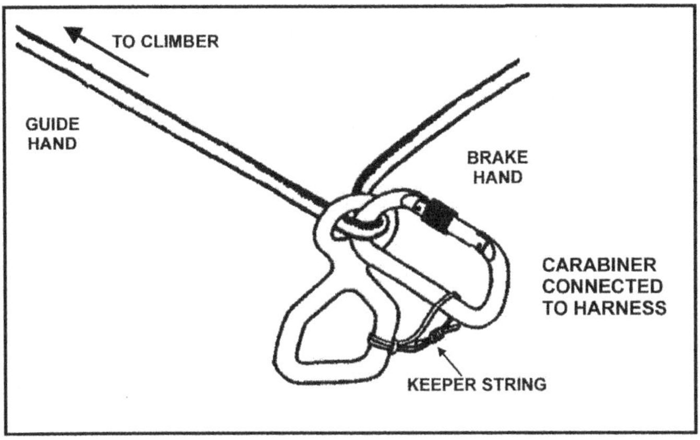

Figure 6-26. Figure-eight device.

Note: Some figure-eight descenders should not be used as belay devices due to their construction and design. Always refer to manufacturer's specifications and directions before use.

(3) ***Mechanical Camming Device.*** The mechanical camming device has an internal camming action that begins locking the rope in place as friction is increased. Unlike the other devices, the mechanical camming device can stop a falling climber without any input from the belayer. A few other devices perform similarly to this, but have no moving parts. Some limitations to these type devices are minimum and maximum rope diameters.

(4) ***Other Mechanical Belay Devices.*** There are many other commercially available mechanical belay devices. Most of these work with the same rope movement direction and the same braking principle. The air traffic controller (ATC), slotted plate, and other tube devices are made in many different shapes. These all work on the same principle as the figure-eight device—friction increases on the rope as it reaches and exceeds the 90-degree angle between the rope entering the device and leaving the device.

6-16. ESTABLISHING A BELAY

A belay can be established using either a direct or indirect connection. Each type has advantages and disadvantages. The choice will depend on the intended use of the belay.

a. **Direct Belay.** The direct belay removes any possible forces from the belayer and places this force completely on the anchor. Used often for rescue installations or to bring a second climber up to a new belay position in conjunction with the Munter hitch, the belay can be placed above the belayer's stance, creating a comfortable position and ease of applying the brake. Also, if the second falls or weights the rope, the belayer is not locked into a position. Direct belays provide no shock-absorbing properties from the belayer's attachment to the system as does the indirect belay; therefore, the belayer is apt to pay closer attention to the belaying process.

b. **Indirect Belay.** An indirect belay, the most commonly used, uses a belay device attached to the belayer's harness. This type of belay provides dynamic shock or weight absorption by the belayer if the climber falls or weights the rope, which reduces the direct force on the anchor and prevents a severe shock load to the anchor.

6-17. SETTING UP A BELAY

In rock climbing, climbers must sometimes make do with marginal protection placements along a route, but belay positions must be made as "bombproof" as possible. Additionally, the belayer must set up the belay in relation to where the fall force will come from and pay strict attention to proper rope management for the belay to be effective. All belay positions are established with the anchor connection to the front of the harness. If the belay is correctly established, the belayer will feel little or no force if the climber falls or has to rest on the rope. Regardless of the actual belay technique used, five basic steps are required to set up a sound belay.

a. **Select Position and Stance.** Once the climbing line is picked, the belayer selects his position. It's best if the position is off to the side of the actual line, putting the belayer out of the direct path of a potential fall or any rocks kicked loose by the climber. The position should allow the belayer to maintain a comfortable, relaxed stance, as he could be in the position for a fairly long time. Large ledges that allow a well braced, sitting stance are preferred. Look for belay positions close to bombproof natural anchors. The position must at least allow for solid artificial placements.

b. **Aim the Belay.** With the belay position selected, the belay must now be "aimed." The belayer determines where the rope leading to the climber will run and the direction the force of a fall will likely come from. When a lead climber begins placing protection, the fall force on the belayer will be in some upward direction, and in line with the first protection placement. If this placement fails under load, the force on the belay could be straight down again. The belayer must aim his belay for all possible load directions, adjusting his position or stance when necessary. The belay can be aimed through an anchor placement to immediately establish an upward pull; however, the belayer must always be prepared for the more severe downward fall force in the event intermediate protection placements fail.

c. **Anchor the Belay.** For a climbing belay to be considered bombproof, the belayer must be attached to a solid anchor capable of withstanding the highest possible fall force. A solid natural anchor would be ideal, but more often the belayer will have to place pitons or chocks. A single artificial placement should never be considered adequate for anchoring a belay (except at ground level). Multiple anchor points capable of supporting both upward and downward pulls should be placed. The rule of thumb is to place two anchors for a downward pull and one anchor for an upward pull as a MINIMUM. The following key points also apply to anchoring belays.

(1) Each anchor must be placed in line with the direction of pull it is intended to support.

(2) Each anchor attachment must be rigged "independently" so a failure of one will not shock load remaining placements or cause the belayer to be pulled out of position.

(3) The attachment between the anchor and the belayer must be snug to support the stance. Both belayer's stance and belay anchors should absorb the force of a fall.

(4) It is best for the anchors to be placed relatively close to the belayer with short attachments. If the climber has to be tied-off in an emergency, say after a severe fall, the belayer can attach a Prusik sling to the climbing rope, reach back, and connect the sling to one of the anchors. The load can be placed on the Prusik and the belayer can come out of the system to render help.

(5) The belayer can use either a portion of the climbing rope or slings of the appropriate length to connect himself to the anchors. It's best to use the climbing rope whenever possible, saving the slings for the climb. The rope is attached using either figure eight loops or clove hitches. Clove hitches have the advantage of being easily adjusted. If the belayer has to change his stance at some point, he can reach back with the guide hand and adjust the length of the attachment through the clove hitch as needed.

(6) The anchor attachments should also help prevent the force of a fall from "rotating" the belayer out of position. To accomplish this, the climbing rope must pass around the "guide-hand side" of the body to the anchors. Sling attachments are connected to the belayer's seat harness (or bowline-on-a-coil) on the guide-hand side.

(7) Arrangement of rope and sling attachments may vary according to the number and location of placements. Follow the guidelines set forth and remember the key points for belay anchors; "in line", "independent", and "snug". Figure 6-27 shows an example of a common arrangement, attaching the rope to the two "downward" anchors and a sling to the "upward" anchor. Note how the rope is connected from one of the anchors back to the belayer. This is not mandatory, but often helps "line-up" the second attachment.

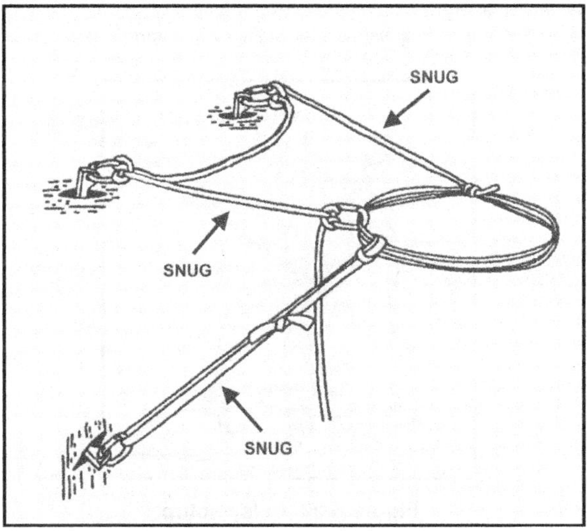

Figure 6-27. Anchoring a belay.

d. **Stack the Rope.** Once the belayer is anchored into position, he must stack the rope to ensure it is free of twists and tangles that might hinder rope management in the belay. The rope should be stacked on the ground, or on the ledge, where it will not get caught in cracks or nubbins as it is fed out to the climber.

(1) On small ledges, the rope can be stacked on top of the anchor attachments if there is no other place to lay it, but make sure to stack it carefully so it won't tangle with the anchored portion of the rope or other slings. The belayer must also ensure that the rope will not get tangled around his legs or other body parts as it "feeds" out.

(2) The rope should never be allowed to hang down over the ledge. If it gets caught in the rock below the position, the belayer may have to tie-off the climber and come out of the belay to free the rope; a time-consuming and unnecessary task. The final point to remember is the rope must be stacked "from the belayer's end" so the rope running to the climber comes off the "top" of the stacked pile.

e. **Attach the Belay.** The final step of the procedure is to attach the belay. With the rope properly stacked, the belayer takes the rope coming off the top of the pile, removes any slack between himself and the climber, and applies the actual belay technique. If using a body belay, ensure the rope is clipped into the guide carabiner.

(1) The belayer should make one quick, final inspection of his belay. If the belay is set up correctly, the anchor attachments, guide carabiner if applicable, and the rope running to the climber will all be on the "guide hand" side, which is normally closest to the rock (Figure 6-28). If the climber takes a fall, the force, if any, should not have any negative effect on the belayer's involvement in the system. The brake hand is out away from the slope where it won't be jammed between the body and the rock. The guide hand can be placed on the rock to help support the stance when applying the brake.

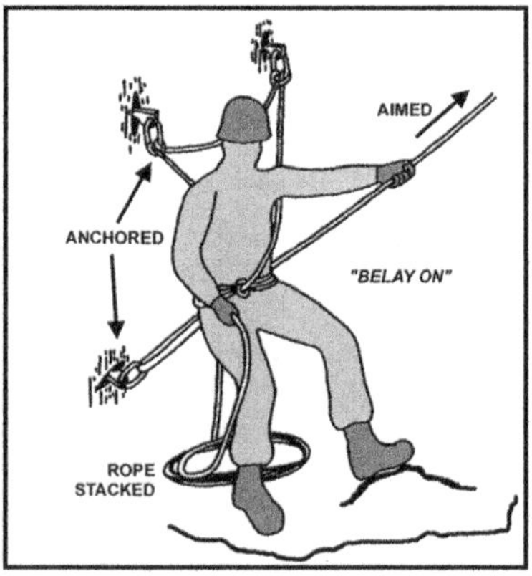

Figure 6-28. Belay setup.

(2) When the belayer is satisfied with his position, he gives the signal, "BELAY ON!". When belaying the "second", the same procedure is used to set up the belay. Unless the belay is aimed for an upward pull, the fall force is of course downward and the

belayer is usually facing away from the rock, the exception being a hanging belay on a vertical face. If the rope runs straight down to the climber and the anchors are directly behind the position, the belayer may choose to brake with the hand he feels most comfortable with. Anchor attachments, guide carabiner, and rope running to the climber through the guide hand must still be aligned on the same side to prevent the belayer from being rotated out of position, unless the belayer is using an improvised harness and the anchor attachment is at the rear.

6-18. TOP-ROPE BELAY

A "top-rope" is a belay setup used in training to protect a climber while climbing on longer, exposed routes. A solid, bombproof anchor is required at the top of the pitch. The belayer is positioned either on the ground with the rope running through the top anchor and back to the climber, or at the top at the anchor. The belayer takes in rope as the climber proceeds up the rock. If this is accomplished with the belayer at the bottom, the instructor can watch the belayer while he coaches the climber through the movements.

> **CAUTION**
> Do not use a body belay for top-rope climbing. The rope will burn the belayer if the climber has to be lowered.

Section V. CLIMBING COMMANDS

Communication is often difficult during a climb. As the distance between climber and belayer increases, it becomes harder to distinguish one word from another and the shortest sentence may be heard as nothing more than jumbled syllables. A series of standard voice commands were developed over the years to signal the essential rope management functions in a belayed climb. Each command is concise and sounds a bit different from another to reduce the risk of a misunderstanding between climber and belayer. They must be pronounced clearly and loudly so they can be heard and understood in the worst conditions.

6-19. VERBAL COMMANDS

Table 6-1, page 6-44, lists standard rope commands and their meanings in sequence as they would normally be used on a typical climb. (Note how the critical "BELAY" commands are reversed so they sound different and will not be confused.)

BELAYER	CLIMBER	MEANING/ACTION TAKEN
"BELAY ON"		The belay is on; you may climb when ready; the rope will be managed as needed.
	"CLIMBING" (as a courtesy)	I am ready to climb.
"CLIMB" (as a courtesy)		Proceed, and again, the rope will be managed as necessary.
"ROCK"	"ROCK"	**PROTECT YOURSELF FROM FALLING OBJECTS.** Signal will be echoed by all climbers in the area. If multipitch climbing, ensure climbers below hear.
	"TAKE ROPE"	Take in excess rope between us without pulling me off the route. Belayer takes in rope.
	"SLACK"	Release all braking/tension on the rope so I can have slack without pulling the rope. Belayer removes brake/tension.
	"TENSION"	Take all the slack, apply brake, and hold me. My weight will be on the rope. Belayer removes slack and applies brake.
	"FALLING"	I am falling. Belayer applies brake to arrest the fall.
"TWEN-TY-FIVE"		You have approximately 25 feet of rope left. Start looking for the next belay position. Climber selects a belay position.
"FIF-TEEN"		You have approximately 15 feet of rope left. Start looking for the next belay position. Climber selects a belay position within the next few feet.
"FIVE"	Set up the belay.	You have 5 feet of rope left. Set up the belay position. You have no more rope. Climber sets up the belay.
Removes the belay, remains anchored. Prepares to climb.	"OFF BELAY"	I have finished climbing and I am anchored. You may remove the belay. Belayer removes the belay and, remaining anchored, prepares to climb.

Table 6-1. Rope commands.

6-20. ROPE TUG COMMANDS

Sometimes the loudest scream cannot be heard when the climber and belayer are far apart. This is especially true in windy conditions, or when the climber is around a corner, above an overhang, or at the back of a ledge. It may be necessary to use a series of "tugs" on the rope in place of the standard voice commands. To avoid any possible confusion with interpretation of multiple rope tug commands, use only one.

a. While a lead climb is in progress, the most important command is "BELAY ON." This command is given only by the climber when the climber is anchored and is prepared for the second to begin climbing. With the issue of this command, the second knows the climber is anchored and the second prepares to climb.

b. For a rope tug command, the leader issues three distinct tugs on the rope AFTER anchoring and putting the second on belay. This is the signal for "BELAY ON" and signals the second to climb when ready. The new belayer keeps slack out of the rope.

Section VI. ROPED CLIMBING METHODS

In military mountaineering, the primary mission of a roped climbing team is to "fix" a route with some type of rope installation to assist movement of less trained personnel in the unit. This duty falls upon the most experienced climbers in the unit, usually working in two- or three-man groups or teams called assault climbing teams. Even if the climbing is for another purpose, roped climbing should be performed whenever the terrain becomes difficult and exposed.

6-21. TOP-ROPED CLIMBING

Top-roped climbing is used for training purposes only. This method of climbing is not used for movement due to the necessity of pre-placing anchors at the top of a climb. If you can easily access the top of a climb, you can easily avoid the climb itself.

a. For training, top-roped climbing is valuable because it allows climbers to attempt climbs above their skill level and or to hone present skills without the risk of a fall. Top-roped climbing may be used to increase the stamina of a climber training to climb longer routes as well as for a climber practicing protection placements.

b. The belayer is positioned either at the base of a climb with the rope running through the top anchor and back to the climber or at the top at the anchor. The belayer takes in rope as the climber moves up the rock, giving the climber the same protection as a belay from above. If this is accomplished with the belayer at the bottom, the instructor is able to keep an eye on the belayer while he coaches the climber through the movements.

6-22. LEAD CLIMBING

A lead climb consists of a belayer, a leader or climber, rope(s), and webbing or hardware used to establish anchors or protect the climb. As he climbs the route, the leader emplaces "intermediate" anchors, and the climbing rope is connected to these anchors with a carabiner. These "intermediate" anchors protect the climber against a fall-thus the term "protecting the climb."

Note: Intermediate anchors are commonly referred to as "protection," "pro," "pieces," "pieces of pro," "pro placements," and so on. For standardization within this publication, these specific anchors will be referred to as "protection;" anchors established for other purposes, such as rappel points, belays, or other rope installations, will be referred to as "anchors."

CAUTION
During all lead climbing, each climber in the team is either anchored or being belayed.

a. Lead climbing with two climbers is the preferred combination for movement on technically difficult terrain. Two climbers are at least twice as fast as three climbers, and are efficient for installing a "fixed rope," probably the most widely used rope installation in the mountains. A group of three climbers are typically used on moderate snow, ice, and snow-covered glaciers where the rope team can often move at the same time, stopping occasionally to set up belays on particularly difficult sections. A group or team of three climbers is sometimes used in rock climbing because of an odd number of personnel, a shortage of ropes (such as six climbers and only two ropes), or to protect and assist an individual who has little or no experience in climbing and belaying. Whichever technique is chosen, a standard roped climbing procedure is used for maximum speed and safety.

b. When the difficulty of the climbing is within the "leading ability" of both climbers, valuable time can be saved by "swinging leads." This is normally the most efficient method for climbing multipitch routes. The second finishes cleaning the first pitch and continues climbing, taking on the role of lead climber. Unless he requires equipment from the other rack or desires a break, he can climb past the belay and immediately begin leading. The belayer simply adjusts his position, re-aiming the belay once the new leader begins placing protection. Swinging leads, or "leap frogging," should be planned before starting the climb so the leader knows to anchor the upper belay for both upward and downward pulls during the setup.

c. The procedures for conducting a lead climb with a group of two are relatively simple. The most experienced individual is the "lead" climber or leader, and is responsible for selecting the route. The leader must ensure the route is well within his ability and the ability of the second. The lead climber carries most of the climbing equipment in order to place protection along the route and set up the next belay. The leader must also ensure that the second has the necessary equipment, such as a piton hammer, nut tool, etc., to remove any protection that the leader may place.

(1) The leader is responsible for emplacing protection frequently enough and in such a manner that, in the event that either the leader or the second should fall, the fall will be neither long enough nor hard enough to result in injury. The leader must also ensure that the rope is routed in a way that will allow it to run freely through the protection placements, thus minimizing friction, or "rope drag".

(2) The other member of the climbing team, the belayer (sometimes referred to as the "second"), is responsible for belaying the leader, removing the belay anchor, and retrieving the protection placed by the leader between belay positions (also called "cleaning the pitch").

(3) Before the climb starts, the second will normally set up the first belay while the leader is arranging his rack. When the belay is ready, the belayer signals, "BELAY ON", affirming that the belay is "on" and the rope will be managed as necessary. When the leader is ready, he double checks the belay. The leader can then signal, "CLIMBING", only as a courtesy, to let the belayer know he is ready to move. The belayer can reply with "CLIMB", again, only as a courtesy, reaffirming that the belay is "on" and the rope will be managed as necessary. The leader then begins climbing.

(4) While belaying, the second must pay close attention to the climber's every move, ensuring that the rope runs free and does not inhibit the climber's movements. If he cannot see the climber, he must "feel" the climber through the rope. Unless told otherwise by the climber, the belayer can slowly give slack on the rope as the climber proceeds on the route. The belayer should keep just enough slack in the rope so the climber does not have to pull it through the belay. If the climber wants a tighter rope, it can be called for. If the belayer notices too much slack developing in the rope, the excess

rope should be taken in quickly. It is the belayer's responsibility to manage the rope, whether by sight or feel, until the climber tells him otherwise.

(5) As the leader protects the climb, slack will sometimes be needed to place the rope through the carabiner (clipping), in a piece of protection above the tie-in point on the leaders harness. In this situation, the leader gives the command "SLACK" and the belayer gives slack, (if more slack is needed the command will be repeated). The leader is able to pull a bight of rope above the tie-in point and clip it into the carabiner in the protection above. When the leader has completed the connection, or the clip, the command "TAKE ROPE" is given by the leader and the belayer takes in the remaining slack.

(6) The leader continues on the route until either a designated belay location is reached or he is at the end of or near the end of the rope. At this position, the leader sets an anchor, connects to the anchor and signals "OFF BELAY". The belayer prepares to climb by removing all but at least one of his anchors and secures the remaining equipment. The belayer remains attached to at least one anchor until the command "BELAY ON" is given.

d. When the leader selects a particular route, he must also determine how much, and what types, of equipment might be required to safely negotiate the route. The selected equipment must be carried by the leader. The leader must carry enough equipment to safely protect the route, additional anchors for the next belay, and any other items to be carried individually such as rucksacks or individual weapons.

(1) The leader will assemble, or "rack," the necessary equipment onto his harness or onto slings around the head and shoulder. A typical leader "rack" consists of:

- Six to eight small wired stoppers on a carabiner.
- Four to six medium to large wired stoppers on a carabiner.
- Assorted hexentrics, each on a separate carabiner.
- SLCDs of required size, each on a separate carabiner.
- Five to ten standard length runners, with two carabiners on each.
- Two to three double length runners, with two carabiners on each.
- Extra carabiners.
- Nut tool.

Note: The route chosen will dictate, to some degree, the necessary equipment. Members of a climbing team may need to consolidate gear to climb a particular route.

(2) The belayer and the leader both should carry many duplicate items while climbing.

- Short Prusik sling.
- Long Prusik sling.
- Cordellette.
- 10 feet of 1-inch webbing.
- 20 feet of 1-inch webbing.
- Belay device (a combination belay/rappel device is multifunctional).
- Rappel device (a combination belay/rappel device is multifunctional).
- Large locking carabiner (pear shape carabiners are multifunctional).

- Extra carabiners.
- Nut tool (if stoppers are carried).

Note: If using an over the shoulder gear sling, place the items in order from smallest to the front and largest to the rear.

e. Leading a difficult pitch is the most hazardous task in roped climbing. The lead climber may be exposed to potentially long, hard falls and must exercise keen judgment in route selection, placement of protection, and routing of the climbing rope through the protection. The leader should try to keep the climbing line as direct as possible to the next belay to allow the rope to run smoothly through the protection with minimal friction. Protection should be placed whenever the leader feels he needs it, and BEFORE moving past a difficult section.

CAUTION

The climber must remember he will fall twice the distance from his last piece of protection before the rope can even begin to stop him.

(1) ***Placing Protection.*** Generally, protection is placed from one stable position to the next. The anchor should be placed as high as possible to reduce the potential fall distance between placements. If the climbing is difficult, protection should be placed more frequently. If the climbing becomes easier, protection can be placed farther apart, saving hardware for difficult sections. On some routes an extended diagonal or horizontal movement, known as a traverse, is required. As the leader begins this type of move, he must consider the second's safety as well as his own. The potential fall of the second will result in a pendulum swing if protection is not adequate to prevent this. The danger comes from any objects in the swinging path of the second.

CAUTION

Leader should place protection prior to, during, and upon completion of any traverse. This will minimize the potential swing, or pendulum, for both the leader and second if either should fall.

(2) ***Correct Clipping Technique.*** Once an anchor is placed, the climber "clips" the rope into the carabiner (Figure 6-29, page 6-50). As a carabiner hangs from the protection, the rope can be routed through the carabiner in two possible ways. One way will allow the rope to run smoothly as the climber moves past the placement; the other way will often create a dangerous situation in which the rope could become "unclipped" from the carabiner if the leader were to fall on this piece of protection. In addition, a

series of incorrectly clipped carabiners may contribute to rope drag. When placing protection, the leader must ensure the carabiner on the protection does not hang with the carabiner gate facing the rock; when placing protection in a crack ensure the carabiner gate is not facing into the crack.

- Grasp the rope with either hand with the thumb pointing down the rope towards the belayer
- Pull enough rope to reach the carabiner with a bight
- Note the direction the carabiner is hanging from the protection
- Place the bight into the carabiner so that, when released, the rope does not cause the carabiner to twist.

(a) If the route changes direction, clipping the carabiner will require a little more thought. Once leaving that piece of protection, the rope may force the carabiner to twist if not correctly clipped. If the clip is made correctly, a rotation of the clipped carabiner to ensure that the gate is not resting against the rock may be all that is necessary.

CAUTION

Ensure the carabiner gate is not resting against a protrusion or crack edge in the rock surface; the rock may cause the gate to open.

(b) Once the rope is clipped into the carabiner, the climber should check to see that it is routed correctly by pulling on the rope in the direction it will travel when the climber moves past that position.

(c) Another potential hazard peculiar to leading should be eliminated before the climber continues. The carabiner is attached to the anchor or runner with the gate facing away from the rock and opening down for easy insertion of the rope. However, in a leader fall, it is possible for the rope to run back over the carabiner as the climber falls below the placement. If the carabiner is left with the gate facing the direction of the route there is a chance that the rope will open the gate and unclip itself entirely from the placement. To prevent this possibility, the climber should ensure that after the clip has been made, the gate is facing away from the direction of the route. There are two ways to accomplish this: determine which direction the gate will face before the protection or runner is placed or once clipped, rotate the carabiner upwards 180 degrees. This problem is more apt to occur if bent gate carabiners are used. Straight gate ovals or "Ds" are less likely to have this problem and are stronger and are highly recommended. Bent gate carabiners are easier to clip the rope into and are used mostly on routes with bolts preplaced for protection. Bent gate carabiners are not recommended for many climbing situations.

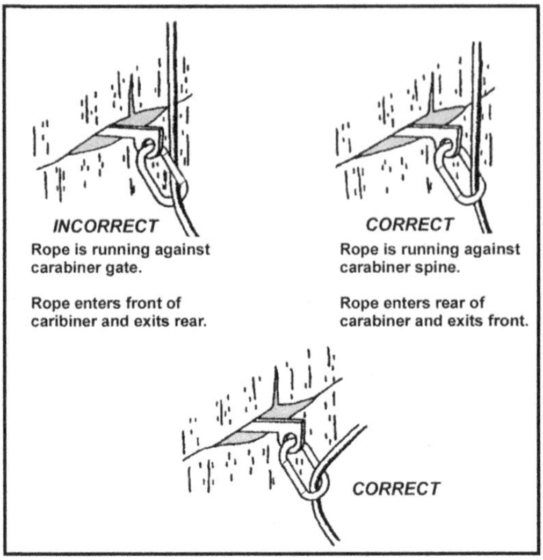

Figure 6-29. Clipping on to protection.

(3) ***Reducing Rope Drag; Using Runners***. No matter how direct the route, the climber will often encounter problems with "rope drag" through the protection positions. The friction created by rope drag will increase to some degree every time the rope passes through a carabiner, or anchor. It will increase dramatically if the rope begins to "zigzag" as it travels through the carabiners. To prevent this, the placements should be positioned so the rope creates a smooth, almost straight line as it passes through the carabiners (Figure 6-30). Minimal rope drag is an inconvenience; severe rope drag may actually pull the climber off balance, inducing a fall.

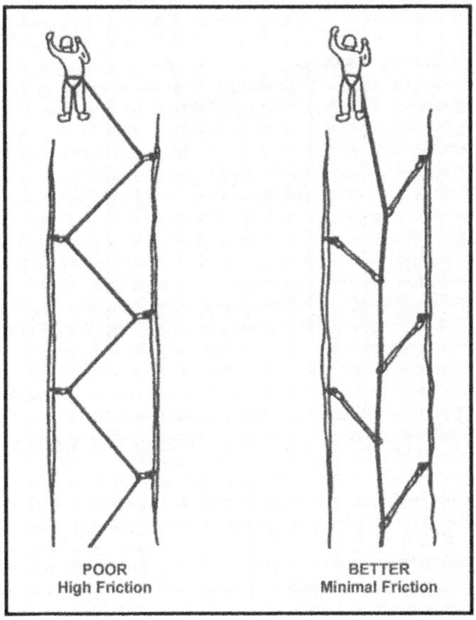

Figure 6-30. Use of slings on protection.

CAUTION

Rope drag can cause confusion when belaying the second or follower up to a new belay position. Rope drag can be mistaken for the climber, causing the belayer to not take in the necessary slack in the rope and possibly resulting in a serious fall.

(a) If it is not possible to place all the protection so the carabiners form a straight line as the rope moves through, you should "extend" the protection (Figure 6-31, page 6-52). Do this by attaching an appropriate length sling, or runner, to the protection to extend the rope connection in the necessary direction. The runner is attached to the protection's carabiner while the rope is clipped into a carabiner at the other end of the runner. Extending placements with runners will allow the climber to vary the route slightly while the rope continues to run in a relatively straight line.

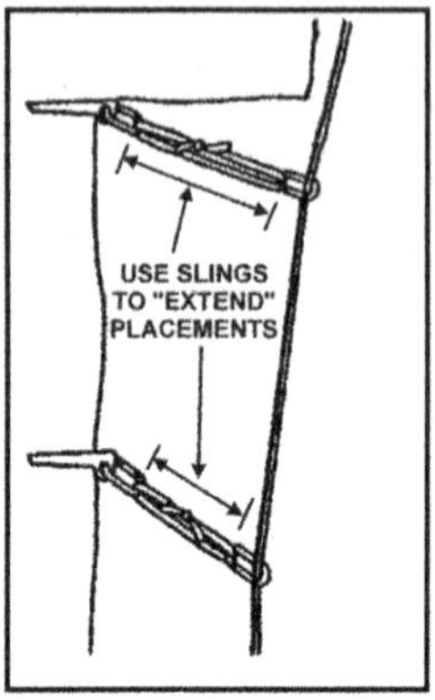

Figure 6-31. Use of slings to extend placement positions.

(b) Not only is rope drag a hindrance, it can cause undue movement of protection as the rope tightens between any "out of line" placements. Rope drag through chock placements can be dangerous. As the climber moves above the placements, an outward or upward pull from rope drag may cause correctly set chocks to pop out, even when used "actively". Most all chocks placed for leader protection should be extended with a runner, even if the line is direct to eliminate the possibility of movement.

(c) Wired chocks are especially prone to wiggling loose as the rope pulls on the stiff cable attachment. All wired chocks used for leader protection should be extended to reduce the chance of the rope pulling them out (Figure 6-32). Some of the larger chocks, such as roped Hexentrics and Tri-Cams, have longer slings pre-attached that will normally serve as an adequate runner for the placement. Chocks with smaller sling attachments must often be extended with a runner. Many of today's chocks are manufactured with pre-sewn webbing installed instead of cable.

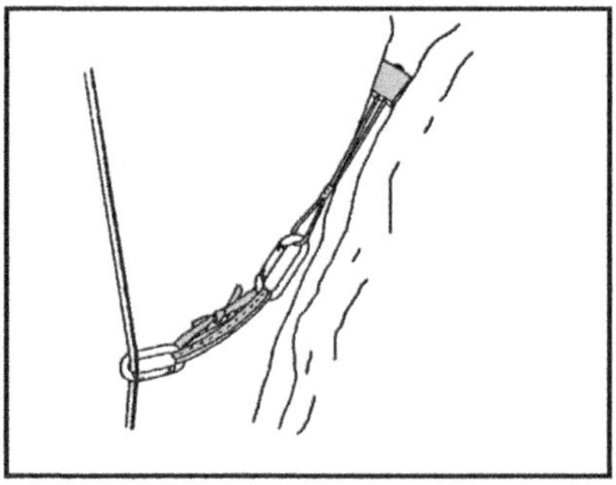

Figure 6-32. Use of sling on a wired stopper.

(d) When a correctly placed piton is used for protection, it will normally not be affected by rope drag. A correctly placed piton is generally a multi-directional anchor, therefore, rope drag through pitons will usually only affect the leader's movements but will continue to protect as expected.

(e) Rope drag will quite often move SLCDs out of position, or "walk" them deeper into the crack than initially placed, resulting in difficult removal or inability to remove them at all. Furthermore, most cases of SLCD movement result in the SLCD moving to a position that does not provide protection in the correct direction or no protection at all due to the lobes being at different angles from those at the original position.

Note: Any placement extended with a runner will increase the distance of a potential fall by the actual length of the sling. Try to use the shortest runners possible, ensuring they are long enough to function properly.

f. Belaying the follower is similar to belaying a top-roped climb in that the follower is not able to fall any farther than rope stretch will allow. This does not imply there is no danger in following. Sharp rocks, rock fall, and inadequately protected traverses can result in damage to equipment or injury to the second.

g. Following, or seconding, a leader has a variety of responsibilities. The second has to issue commands to the leader, as well as follow the leader's commands. Once the lead climber reaches a good belay position, he immediately establishes an anchor and connects to it. When this is completed he can signal "OFF BELAY" to the belayer. The second can now remove the leader's belay and prepare to climb. The second must remain attached to at least one of the original anchors while the leader is preparing the next belay position. The removed materials and hardware can be organized and secured on the second's rack in preparation to climb.

(1) When the leader has established the new belay position and is ready to belay the follower, the "new" belayer signals "BELAY ON." The second, now the climber, removes any remaining anchor hardware/materials and completes any final preparations. The belayer maintains tension on the rope, unless otherwise directed, while the final preparations are taking place, since removal of these remaining anchors can introduce slack into the rope. When the second is ready, he can, as a courtesy, signal "CLIMBING," and the leader can, again as a courtesy, reply with "CLIMB."

(2) Upon signaling "BELAY ON," the belayer must remove and keep all slack from the rope. (This is especially important as in many situations the belayer cannot see the follower. A long pitch induces weight and sometimes "drag" on the rope and the belayer above will have difficulty distinguishing these from a rope with no slack.)

h. When removing protection, the man cleaning the pitch should rack it properly to facilitate the exchange and or arrangement of equipment at the end of the pitch. When removing the protection, or "cleaning the pitch", SLCDs or chocks may be left attached to the rope to prevent loss if they are accidentally dropped during removal. If necessary, the hardware can remain on the rope until the second reaches a more secure stance. If removing a piton, the rope should be unclipped from the piton to avoid the possibility of damaging the rope with a hammer strike.

(1) The second may need to place full body weight on the rope to facilitate use of both hands for protection removal by giving the command "TENSION." The second

must also ensure that he does not climb faster than the rope is being taken in by the belayer. If too much slack develops, he should signal "TAKE ROPE" and wait until the excess is removed before continuing the climb. Once the second completes the pitch, he should immediately connect to the anchor. Once secured, he can signal "OFF BELAY." The leader removes the belay, while remaining attached to an anchor. The equipment is exchanged or organized in preparation for the next pitch or climb.

(2) When the difficulty of the climbing is within the "leading ability" of both climbers, valuable time can be saved by "swinging leads." This is normally the most efficient method for climbing multi-pitch routes. The second finishes cleaning the first pitch and continues climbing, taking on the role of lead climber. Unless he requires equipment from the belayer or desires a break, he can climb past the belay and immediately begin leading. The belayer simply adjusts his position, re-aiming the belay once the new leader begins placing protection. Swinging leads, or "leap frogging," should be planned before starting the climb so the leader knows to anchor the upper belay for both upward and downward pulls during the setup.

6-23. AID CLIMBING

When a route is too difficult to free climb and is unavoidable, if the correct equipment is available you might aid climb the route. Aid climbing consists of placing protection and putting full body weight on the piece. This allows you to hang solely on the protection you place, giving you the ability to ascend more difficult routes than you can free climb. Clean aid consists of using SLCDs and chocks, and is the simplest form of aid climbing.

a. **Equipment**. Aid climbing can be accomplished with various types of protection. Regardless of the type of protection used, the method of aid climbing is the same. In addition to the equipment for free climbing, other specialized equipment will be needed.

(1) ***Pitons***. Pitons are used the same as for free climbing. Most piton placements will require the use of both hands. Piton usage will usually leave a scar in the rock just by virtue of the hardness of the piton and the force required to set it with a hammer. Swinging a hammer to place pitons will lead to climber fatigue sooner than clean aid. Since pitons are multidirectional, the strength of a well-placed piton is more secure than most clean aid protection. Consider other forms of protection when noise could be hazardous to tactics.

(2) ***Bolts***. Bolts are used when no other protection will work. They are a more permanent form of protection and more time is needed to place them. Placing bolts creates more noise whether drilled by hand or by motorized drill. Bolts used in climbing are a multi-part expanding system pounded into predrilled holes and then tightened to the desired torque with a wrench or other tool. Bolts are used in many ways in climbing today. The most common use is with a hanger attached and placed for anchors in face climbing. However, bolts can be used for aid climbing, with or without the hanger.

(a) Placing bolts for aid climbing takes much more time than using pitons or clean aid. Bolting for aid climbing consists of consecutive bolts about 2 feet apart. Drilling a deep enough hole takes approximately thirty minutes with a hand drill and up to two minutes with a powered hammer drill. A lot of time and work is expended in a short distance no matter how the hole is drilled. (The weight of a powered hammer drill becomes an issue in itself.) Noise will also be a factor in both applications. A constant pounding with a hammer on the hand drill or the motorized pounding of the powered drill may alert the enemy to the position. The typical climbing bolt/hanger combination normally is left in the hole where it was placed.

(b) Other items that can be used instead of the bolt/hanger combination are the removable and reusable "spring-loaded removable bolts" such as rivets (hex head threaded bolts sized to fit tightly into the hole and pounded in with a hammer), split-shaft rivets, and some piton sizes that can be pounded into the holes. When using rivets or bolts without a hanger, place a loop of cable over the head and onto the shaft of the rivet or bolt and attach a carabiner to the other end of the loop (a stopper with the chock slid back will suffice). Rivet hangers are available that slide onto the rivet or bolt after it is placed and are easily removed for reuse. Easy removal means a slight loss of security while in use.

(3) ***SLCDs.*** SLCDs are used the same as for free climbing, although in aid climbing, full body weight is applied to the SLCD as soon as it is placed.

(4) ***Chocks.*** Chocks are used the same as for free climbing, although in aid climbing, weight is applied to the chock as soon as it is placed.

(5) ***Daisy Chains.*** Daisy chains are tied or presewn loops of webbing with small tied or presewn loops approximately every two inches. The small loops are just large enough for two or three carabiners. Two daisy chains should be girth-hitched to the tie-in point in the harness.

(6) ***Etriers (or Aiders).*** Etriers (aiders) are tied or presewn webbing loops with four to six tied or presewn internal loops, or steps, approximately every 12 inches. The internal loops are large enough to easily place one booted foot into. At least two etriers (aiders) should be connected by carabiner to the free ends of the daisy chains.

(7) ***Fifi Hook.*** A fifi hook is a small, smooth-surfaced hook strong enough for body weight. The fifi hook should be girth-hitched to the tie-in point in the harness and is used in the small loops of the daisy chain. A carabiner can be used in place of the fifi hook, although the fifi hook is simpler and adequate.

(8) ***Ascenders.*** Ascenders are mechanical devices that will move easily in one direction on the rope, but will lock in place if pushed or pulled the other direction. (Prusiks can be used but are more difficult than ascenders.)

b. **Technique.** The belay will be the same as in normal lead climbing and the rope will be routed through the protection the same way also. The big difference is the movement up the rock. With the daisy chains, aiders, and fifi hook attached to the rope tie-in point of the harness as stated above, and secured temporarily to a gear loop or gear sling, the climb continues as follows:

(1) The leader places the first piece of protection as high as can safely be reached and attaches the appropriate sling/carabiner

(2) Attach one daisy chain/aider group to the newly placed protection

(3) Clip the rope into the protection, (the same as for normal lead climbing)

(4) Insure the protection is sound by weighting it gradually; place both feet, one at a time, into the steps in the aider, secure your balance by grasping the top of the aider with your hands.

(5) When both feet are in the aider, move up the steps until your waist is no higher than the top of the aider.

(6) Place the fifi hook (or substituted carabiner) into the loop of the daisy chain closest to the daisy chain/aider carabiner, this effectively shortens the daisy chain;

maintain tension on the daisy chain as the hook can fall out of the daisy chain loop if it is unweighted.

Note: Moving the waist higher than the top of the aider is possible, but this creates a potential for a fall to occur even though you are on the aider and "hooked" close to the protection with the daisy chain. As the daisy chain tie-in point on the harness moves above the top of the aider, you are no longer supported from above by the daisy chain, you are now standing above your support. From this height, the fifi hook can easily fall out of the daisy chain loop if it is unweighted. If this happens, you could fall the full length of the daisy chain resulting in a static fall on the last piece of protection placed.

(7) Release one hand from the aider and place the next piece of protection, again, as high as you can comfortably reach; if using pitons or bolts you may need both hands free- "lean" backwards slowly, and rest your upper body on the daisy chain that you have "shortened" with the fifi hook

(8) Clip the rope into the protection

(9) Attach the other daisy chain/aider group to the next piece of protection

(10) Repeat entire process until climb is finished

c. **Seconding.** When the pitch is completed, the belayer will need to ascend the route. To ascend the route, use ascenders instead of Prusiks, ascenders are much faster and safer than Prusiks. Attach each ascender to a daisy chain/aider group with carabiners. To adjust the maximum reach/height of the ascenders on the rope, adjust the effective length of the daisy chains with a carabiner the same as with the fifi hook; the typical height will be enough to hold the attached ascender in the hand at nose level. When adjusted to the correct height, the arms need not support much body weight. If the ascender is too high, you will have difficulty reaching and maintaining a grip on the handle.

(1) Unlike lead climbing, there will be a continuous load on the rope during the cleaning of the route, this would normally increase the difficulty of removing protection. To make this easier, as you approach the protection on the ascenders, move the ascenders, one at a time, above the piece. When your weight is on the rope above the piece, you can easily unclip and remove the protection.

CAUTION

If both ascenders should fail while ascending the pitch, a serious fall could result. To prevent this possibility, *tie-in short* on the rope every 10-20 feet by tying a figure eight loop and clipping it into the harness with a separate locking carabiner as soon as the ascent is started. After ascending another 20 feet, repeat this procedure. Do not unclip the previous figure eight until the new knot is attached to another locking carabiner. Clear each knot as you unclip it.

Notes: 1. Ensure the loops formed by the short tie-ins do not catch on anything below as you ascend.

2. If the nature of the rock will cause the "hanging loop" of rope, formed by tying in at the end of the rope, to get caught as you move upward, do not tie into the end of the rope.

(2) Seconding an aid pitch can be done in a similar fashion as seconding free-climbed pitches. The second can be belayed from above as the second "climbs" the protection. However, the rope is unclipped from the protection before the aider/daisy chain is attached.

d. **Seconding Through a Traverse.** While leading an aid traverse, the climber is hanging on the protection placed in front of the current position. If the second were to clean the section by hanging on the rope while cleaning, the protection will be pulled in more than one direction, possibly resulting in the protection failing. To make this safer and easier, the second should hang on the protection just as the leader did. As the second moves to the beginning of the traverse, one ascender/daisy chain/aider group is removed from the rope and clipped to the protection with a carabiner, (keep the ascenders attached to the daisy chain/aider group for convenience when the traverse ends). The second will negotiate the traverse by leapfrogging the daisy chain/aider groups on the next protection just as the leader did. Cleaning is accomplished by removing the protection as it is passed when all weight is removed from it. This is in effect a self-belay. The second maintains a shorter safety tie-in on the rope than for vertical movement to reduce the possibility of a lengthy pendulum if the protection should pull before intended.

e. **Clean Aid Climbing**. Clean aid climbing consists of using protection placed without a hammer or drill involvement: chocks, SLCDs, hooks, and other protection placed easily by hand. This type of aid climbing will normally leave no trace of the climb when completed. When climbing the aiders on clean aid protection, ensure the protection does not "move" from it's original position.

(1) Hooks are any device that rests on the rock surface without a camming or gripping action. Hooks are just what the name implies, a curved piece of hard steel with a hole in one end for webbing attachment. The hook blade shape will vary from one model to another, some have curved or notched "blades" to better fit a certain crystal shape on a face placement. These types of devices due to their passive application, are only secure while weighted by the climber.

(2) Some featureless sections of rock can be negotiated with hook use, although bolts can be used. Hook usage is faster and quieter but the margin of safety is not there unless hooks are alternated with more active forms of protection. If the last twenty foot section of a route is negotiated with hooks, a forty foot fall could result.

6-24. THREE-MAN CLIMBING TEAM

Often times a movement on steep terrain will require a team of more than two climbers, which involves more difficulties. A four-man team (or more) more than doubles the difficulty found in three men climbing together. A four-man team should be broken down into two groups of two unless prevented by a severe lack of gear.

a. Given one rope, a three-man team is at a disadvantage on a steep, belayed climb. It takes at least twice as long to climb an average length pitch because of the third climber and the extra belaying required. The distance between belay positions will be halved if only one rope is used because one climber must tie in at the middle of the rope. Two ropes are recommended for a team of three climbers.

Note: Time and complications will increase when a three-man team uses only one rope. For example: a 100-foot climb with a 150-foot rope would normally require two belays for two climbers; a 100-foot climb with a 150-foot rope would require six belays for three climbers.

b. At times a three-man climb may be unavoidable and personnel should be familiar with the procedure. Although a team of three may choose from many different methods, only two are described below. If the climb is only one pitch, the methods will vary.

CAUTION

When climbing with a team of three, protected traverses will require additional time. The equipment used to protect the traverse must be left in place to protect both the second and third climbers.

(1) The first method can be used when the belay positions are not large enough for three men. If using one rope, two climbers tie in at each end and the other at the mid point. When using two ropes, the second will tie in at one end of both ropes, and the other two climbers will each tie in to the other ends. The most experienced individual is the leader, or number 1 climber. The second, or number 2 climber, is the stronger of the remaining two and will be the belayer for both number 1 and number 3. Number 3 will be the last to climb. Although the number 3 climber does no belaying in this method, each climber should be skilled in the belay techniques required. The sequence for this method (in one pitch increments) is as follows (repeated until the climb is complete):

(a) Number 1 ascends belayed by number 2. Number 2 belays the leader up the first pitch while number 3 is simply anchored to the rock for security (unless starting off at ground level) and manages the rope between himself and number 2. When the leader completes the pitch, he sets up the next belay and belays number 2 up.

(b) Number 2 ascends belayed by number 1, and cleans the route (except for traverses). Number 2 returns the hardware to the leader and belays him up the next pitch. When the leader completes this pitch, he again sets up a new belay. When number 2 receives "OFF BELAY" from the leader, he changes ropes and puts number 3 on belay. He should not have to change anchor attachments because the position was already aimed for a downward as well as an upward pull when he belayed the leader.

(c) Number 3 ascends belayed by number 2. When number 3 receives "BELAY ON," he removes his anchor and climbs to number 2's position. When the pitch is completed he secures himself to one of number 2's belay anchors. When number 1's belay is ready, he brings up number 2 while number 3 remains anchored for security. Number 2 again cleans the pitch and the procedure is continued until the climb is completed.

(d) In this method, number 3 performs no belay function. He climbs when told to do so by number 2. When number 3 is not climbing, he remains anchored to the rock for security. The standard rope commands are used; however, the number 2 climber may include the trailing climber's name or number in the commands to avoid confusion as to who should be climbing.

(d) Normally, only one climber would be climbing at a time; however, the number 3 climber could ascend a fixed rope to number 2's belay position using proper ascending technique, with no effect on the other two members of the team. This would save time for a team of three, since number 2 would not have to belay number 3 and could be either belaying number 1 to the next belay or climbing to number 1. If number 3 is to ascend a fixed rope to the next belay position, the rope will be loaded with number 3's weight, and positioned directly off the anchors established for the belay. The rope should be located so it does not contact any sharp edges. The rope to the ascending number 3 could be secured to a separate anchor, but this would require additional time and gear.

(2) The second method uses either two ropes or a doubled rope, and number 2 and number 3 climb simultaneously. This requires either a special belay device that accepts two ropes, such as the tuber type, or with two Munter hitches. The ropes must travel through the belay device(s) without affecting each other.

(a) As the leader climbs the pitch, he will trail a second rope or will be tied in with a figure eight in the middle of a doubled rope. The leader reaches the next belay position and establishes the anchor and then places both remaining climbers on belay. One remaining climber will start the ascent toward the leader and the other will start when a gap of at least 10 feet is created between the two climbers. The belayer will have to remain alert for differences in rope movement and the climbers will have to climb at the same speed. One of the "second" climbers also cleans the pitch.

(b) Having at least two experienced climbers in this team will also save time. The belayer will have additional requirements to meet as opposed to having just one second. The possible force on the anchor will be twice that of one second. The second that is not cleaning the pitch can climb off route, but staying on route will usually prevent a possible swing if stance is not maintained.

CHAPTER 7

ROPE INSTALLATIONS

Obstacles on the battlefield today are inevitable. They can limit the battlefield and, even worse, prevent a unit from accomplishing its mission. However, with highly skilled personnel trained on rope installations, leaders can be assured that even a unit with limited mountain skills and experience will be able to successfully move and operate in terrain that would otherwise have been impassable.

Section I. FIXED ROPE

A fixed rope is a rope anchored in place to assist soldiers in movement over difficult terrain. Its simplest form is a rope tied off on the top of steep terrain. As terrain becomes steeper or more difficult, fixed rope systems may require intermediate anchors along the route. Moving on a fixed rope requires minimal equipment. The use of harnesses, ascenders, and other technical gear makes fixed rope movement easier, faster, and safer, but adds to total mission weight.

7-1. INSTALLATION

To install a fixed rope, two experienced climbers rope up for a roped climb. The leader must have the necessary equipment to rig the anchor at the top of the pitch. Although leader protection is usually not needed on a typical slope, additional hardware can be brought along and placed at the leader's discretion. The second will establish a belay if protection is being placed. Otherwise, he will stack and manage the rope. He ensures the rope runs smoothly up the slope and does not get tangled as the climber ascends. Upon reaching the end of the pitch, the leader will establish the top anchor. Once the anchor is rigged, the leader will take up any remaining slack between himself and the second. He will anchor the installation rope and remain tied into the rope. The second unties from his end of the rope and begins to climb. If the leader placed protection, the second will clean the pitch on his way up.

7-2. UTILIZATION

All personnel using the fixed rope grasp the rope with the palm downward and use it for assistance as they ascend the slope (Figure 7-1, page 7-2). An individual can easily prevent a long fall by attaching himself to the rope with a sling using a friction knot (for example, Prusik, autoblock). The knot is slid along the rope as the individual ascends. If the climber slips and loses control of the rope, the friction knot will grab the rope and arrest the fall. The friction knot used in this manner is referred to as a self-belay (Figure 7-2, page 7-2).

Figure 7-1. Using a fixed rope.

Figure 7-2. Using a self-belay.

7-3. RETRIEVAL

If the fixed rope is to be used on the descent, it can be left in place and recovered after the last rappel. If not, the last climber will tie into the rope and be belayed from above. The climber now can easily free the rope if it gets caught on anything as it is taken up from the belayer.

7-4. FIXED ROPE WITH INTERMEDIATE ANCHORS

Whenever the route varies from the fall line of the slope, the fixed rope must be anchored at intermediate anchor points (Figure 7-3). Intermediate anchor points should also be used on any long routes that exceed the length of a single rope. The use of intermediate anchor points creates independent sections and allows for changes in direction from one section to the next. The independent sections allow for more personnel to move on the fixed rope. This type of fixed rope is commonly used along exposed ridges and narrow mountain passes.

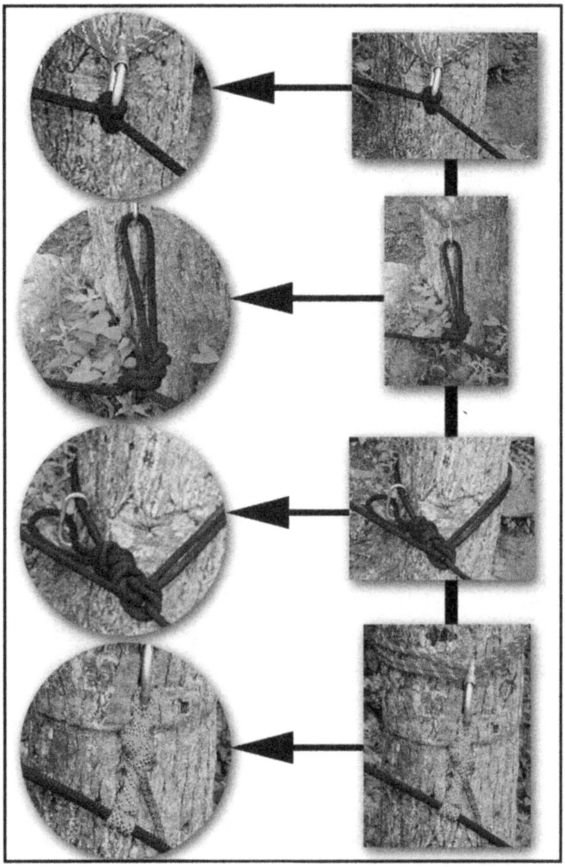

Figure 7-3. Fixed rope with intermediate anchors.

a. **Installation.** Two experienced climbers prepare for a roped climb. The leader will carry a typical rack with enough hardware to place an adequate number of intermediate anchor points. The second sets up a standard belay. The route they select must have the following characteristics:

- Most suitable location, ease of negotiation, avoids obstacles.
- Availability of anchors (natural and artificial).
- Area is safe from falling rock or ice.
- Tactical considerations are met.
- A rope routed between knee and chest height (waist high preferred).
- Rope crossovers should be avoided.

As the leader climbs the route, he will place the anchors and route the climbing rope as in a typical roped climb. The leader makes use of any available natural anchors.

Note: Sling attachments should be kept as short and snug as possible to ensure that a load on the fixed rope below the anchor is placed only on that anchor. This will prevent one section from affecting another section.

(1) The leader places an anchor at all points where a change of direction occurs. He also makes every attempt to route the rope so personnel will not have to cross back and forth over the rope between sections.

(2) When the leader reaches the end of the pitch, he temporarily anchors the rope. He should use a sling to anchor himself if there is any chance of slipping and falling. He then takes up any excess slack, and attaches the rope to the anchor.

Note: Enough slack must be left in the rope so the second can tie the knots necessary to fix the rope.

(3) The second unties from the rope and anchors it at the bottom. He attaches himself to the rope with a sling using a friction knot to create a self-belay. The self-belay will protect the second as he climbs and fixes the rope to the intermediate anchor points. When he reaches an anchor point, he unclips the climbing rope so he can advance the self-belay beyond the anchor point. He then takes the slack out of the section below the anchor point. He ensures that the fixed rope will be approximately knee to chest level as climbers negotiate the installation. He then attaches the rope using an anchor knot (for example, middle-of-the-rope clove hitch, double figure-eight). The second then moves to next anchor point and repeats the process.

(4) If a long runner is to be used at any anchor point, the second should adjust the section below it so the runner is oriented in the direction where the load or pull on the anchor will come from. This will help isolate the section.

(5) The sections are normally adjusted fairly snug between anchor points. A slack section may be necessary to move around obstacles in the route or large bulges in the terrain. If clove hitches are used, adjusting the clove hitches at each end of the section can leave any amount of slack.

(6) A middle-of-the-rope Prusik safetied with a figure eight may be used when utility ropes are available. These are used to adjust the rope height (either higher or lower).

(7) In addition to the fixed rope, the second could anchor etriers to be used as footholds.

(8) When the second reaches the end of the pitch, the rope is removed from the top anchor and the remaining slack is removed from the last section. The rope is reattached to the anchor. If additional fixed rope is required the procedure is repeated using another rope. The second will tie the ropes together before anchoring the next section, creating one continuous fixed rope.

b. **Utilization.** Personnel should be attached to the fixed rope during movement for safety reasons.

(1) If a self-belay is desired, a harness should be worn. A friction knot will be tied to the installation rope using a short sling. The sling will then be attached to the harness. Another short sling will be used as a safety line. One end of the sling will be attached to the harness and the other will have a carabiner inserted. This safety line is also attached to the fixed rope during movement. Once the climber reaches an anchor point, he removes his safety line and attaches it to the anchor or attaches it to the next section of rope. He will then untie the friction knot and tie another friction knot beyond the anchor point. The use of a mechanical ascender in the place of the friction knot could greatly speed up movement.

(2) There will be many situations where a self-belay may not be required. In these situations an individual may attach himself to the fixed rope using only a safety line. The individual will tie into the middle of a sling rope approximately 12 feet long. Fixed loops are tied into the running ends and a carabiner is attached into each of the fixed loops. The individual now has two points of attachments to the fixed rope. Upon reaching an anchor point, one safety line is removed and advanced beyond the anchor point onto the next section. Then the next safety is removed and placed on the next section. This way the individual is always secured to the fixed rope at all times.

(3) Personnel will move one at a time per section during the entire movement. Once an individual changes over to the next section he signals the next man to climb. When descending on the fixed rope, personnel can down climb using the installation for assistance. Another option would be to descend using a hasty rappel.

c. **Retrieval.** When the installation is retrieved, the next to last man on the system will untie the knots at the intermediate anchor points and reclips the rope as he ascends. He will be attached to rope using a self-belay. Once he reaches the top of the pitch, the rope should be running the same as when the leader initially placed it. The last man will untie the rope from the bottom anchor and tie into the rope. He will the clean the pitch as he climbs while being belayed from above.

Section II. RAPPELLING

When an individual or group must descend a vertical surface quickly, a rappel may be performed. Rappelling is a quick method of descent but it is extremely dangerous. These dangers include anchor failure, equipment failure, and individual error. Anchors in a mountainous environment should be selected carefully. Great care must be taken to load the anchor slowly and to ensure that no excessive stress is placed on the anchor. To ensure this, bounding rappels should be prohibited, and only walk down rappels used. Constant vigilance to every detail will guarantee a safe descent every time.

7-5. SELECTION OF A RAPPEL POINT

The selection of the rappel point depends on factors such as mission, cover, route, anchor points, and edge composition (loose or jagged rocks). There must be good anchors (primary and secondary). The anchor point should be above the rappeller's departure point. Suitable loading and off-loading platforms should be available.

7-6. INSTALLATION OF THE RAPPEL POINT

A rappel lane should have equal tension between all anchor points by establishing primary and secondary anchor points. The rappel rope should not extend if one anchor point fails. The following methods of establishing an anchor can be performed with a single or double rope. A double rope application should be used when possible for safety purposes.

a. If a rappel lane is less than half the rope length, the climber may apply one of the following techniques:

(1) Double the rope and tie a three-loop bowline around the primary anchor to include the primary anchor inside two loops and enough rope in the third loop to run to the secondary anchor (another three-loop bowline secured with an overhand knot).

(2) Bowline secured with an overhand knot (or any appropriate anchor knot).

(3) Double the rope and establish a self-equalizing anchor system with a three-loop bowline or any other appropriate anchor knot

b. If a rappel lane is greater than half the rope length, the climber may apply one of the following techniques:

(1) Use two ropes. With both ropes, tie a round turn anchor bowline around a primary anchor point. Take the remaining rope (the tail from the primary anchor bowline) and tie another round turn anchor bowline to a secondary anchor point. The secondary anchor point should be in a direct line behind the primary anchor point. The anchor can be either natural or artificial.

(2) Use two ropes. Establish a multi-point anchor system using a bowline on a bight or any other appropriate anchor knot.

c. Situations may arise where, due to the length of the rappel, the rappel rope cannot be tied to the anchor (if the rope is used to tie the knots, it will be too short to accomplish the rappel). The following techniques can be used:

(1) When using a natural anchor, tie a sling rope, piece of webbing, or another rope around the anchor using proper techniques for slinging natural anchors. The rappel rope will have a fixed loop tied in one end, which is attached to the anchor created.

(2) When using an artificial anchor, tie off a sling rope, piece of webbing, runner, or another rope to form a loop. Use this loop to create an equalizing or pre-equalized anchor, to which the rappel rope will be attached.

7-7. OPERATION OF THE RAPPEL POINT

Due to the inherent dangers of rappelling, special care must be taken to ensure a safe and successful descent.

a. **Communication.** Climbers at the top of a rappel point must be able to communicate with those at the bottom. During a tactical rappel, radios, hand signals, and rope signals are considered. For training situations use the commands shown in Table 7-1.

COMMAND	GIVEN BY	MEANING
LANE NUMBER ___, ON RAPPEL	Rappeller	I am ready to begin rappelling.
LANE NUMBER ___, ON BELAY	Belayer	I am on belay and you may begin your rappel.
LANE NUMBER ___, OFF RAPPEL	Rappeller	I have completed the rappel, cleared the rappel lane, and am off the rope.
LANE NUMBER ___, OFF BELAY	Belayer	I am off belay.

Table 7-1. Rappel commands.

Notes: 1. In a training environment, the lane number must be understood.

2. In a tactical situation, a series of tugs on the rope may be substituted for the oral commands to maintain noise discipline. The number of tugs used to indicate each command is IAW the unit SOP.

b. **Duties and Responsibilities.**

(1) Duties of the rappel point commander are as follows:

- Ensures that the anchors are sound and the knots are properly tied.
- Ensures that loose rock and debris are cleared from the loading platform.
- Allows only one man on the loading platform at a time and ensures that the rappel point is run orderly.
- Ensures that each man is properly prepared for the particular rappel: gloves on, sleeves down, helmet with chin strap fastened, gear prepared properly, and rappel seat and knots correct (if required). He also ensures that the rappeller is hooked up to the rope correctly and is aware of the proper braking position.
- Ensures that the proper signals or commands are used.
- Dispatches each man down the rope.
- Is the last man down the rope.

(2) Duties of the first rappeller down are as follows:

- Selects a smooth route, for the rope, that is clear of sharp rocks.
- Conducts a self-belay.
- Clears the route, placing loose rocks far enough back on ledges to be out of the way, which the rope may dislodge.
- Ensures the rope reaches the bottom or is at a place from which additional rappels can be made.
- Ensures that the rope will run freely around the rappel point when pulled from below.
- Clears the rappel lane by straightening all twists and tangles from the ropes.
- Belays subsequent rappellers down the rope or monitors subsequent belayers
- Takes charge of personnel as they arrive at the bottom (off-loading platform).

Note: A rappeller is always belayed from the bottom, except for the first man down. The first man belays himself down the rope using a self-belay attached to his rappel seat, which is hooked to the rappel rope with a friction knot. As the first man rappels down the rope, he "walks" the friction knot down with him.

(3) Each rappeller down clears the ropes, and shouts, "Off rappel," (if the tactical situation permits). After the rope is cleared and the rappeller is off rappel, he acts as the belayer for next rappeller.

(4) Soldiers wear gloves for all types of rappels to protect their hands from rope burns.

(5) Rappellers descend in a smooth, controlled manner.

(6) The body forms an L-shape with the feet shoulder-width apart, legs straight, and buttocks parallel to the ground. When carrying equipment or additional weight, a modified L-shape is used with the legs slightly lower than the buttocks to compensate for the additional weight. The rappeller's back is straight. He looks over the brake shoulder. The guide hand is extended on the rope with the elbow extended and locked. The rope slides freely through the guide hand. The guide hand is used to adjust equipment and assist balance during descent. The rappeller grasps the rope firmly with the brake hand and places it in the brake position. Releasing tension on the rope and moving the brake hand regulates the rate of descent. The rappeller never lets go of the ropes with his brake hand until the rappel is complete.

c. **Tying Off During the Rappel.** It may be necessary to stop during descent. This can be accomplished by passing the rope around the body and placing three or more wraps around the guide-hand-side leg, or by tying off using the appropriate knot for the rappel device.

7-8. RECOVERY OF THE RAPPEL POINT

After almost all personnel have descended, only two personnel will remain at the top of the rappel point. They will be responsible for establishing a retrievable rappel.

a. **Establishing the Retrievable Rappel.** To set up a retrievable rappel point, a climber must apply one of the following methods:

(1) Double the rope when the rappel is less than half the total length of the rope. Place the rope, with the bight formed by the midpoint, around the primary anchor. Join the tails of the rappel rope and throw the rope over the cliff. Tie a clove hitch around a carabiner, just below the anchor point, with the locking bar outside the carabiner away from the gate opening end and facing uphill. Snap the opposite standing portion into the carabiner. When the rappeller reaches the bottom, he pulls on that portion of the rope to which the carabiner is secured to allow the rope to slide around the anchor point.

(2) When the length of the rappel is greater than half the length of the rope used, join two ropes around the anchor point with an appropriate joining knot (except the square knot). Adjust the joining knot so that it is away from the anchor. Tie a clove hitch around a carabiner just below the anchor point with the locking bar outside the carabiner away from the gate opening end and facing uphill. Snap the opposite standing portion into the carabiner. Upon completion of the rappel, pull the rope to which the carabiner is secured to allow the rope to slide around the anchor point.

Notes: 1. When setting up a retrievable rappel, use only a primary point; care is taken in selecting the point.

2. Ensure the soldiers have a safety line when approaching the rappel point, with only the rappeller going near the edge.

b. **Retrieving the Rappel Rope.** The next to last rappeller will descend the lane, removing any twists, and routes the rope for easiest retrieval. Once he reaches the end of the rappel, he tests the rope for retrieval. If the rappel is retrievable, the last man will rappel down. Once he is off rappel, he pulls the lane down.

7-9. TYPES OF RAPPELS

During military mountaineering operations, many types of rappels may be used. The following paragraphs describe some these rappels.

a. **Hasty Rappel** (Figure 7-4). The hasty rappel is used only on moderate pitches. Its main advantage is that it is easier and faster than other methods. Gloves are worn to prevent rope burns.

(1) Facing slightly sideways to the anchor, the rappeller places the ropes horizontally across his back. The hand nearest to the anchor is his guide hand, and the other is the brake hand.

(2) To stop, the rappeller brings his brake hand across in front of his body locking the rope. At the same time, he turns to face up toward the anchor point.

Figure 7-4. Hasty rappel.

b. **Body Rappel** (Figure 7-5). The rappeller faces the anchor point and straddles the rope. He then pulls the rope from behind, and runs it around either hip, diagonally across the chest, and back over the opposite shoulder. From there, the rope runs to the brake hand, which is on the same side of the hip that the rope crosses (for example, the right hip to the left shoulder to the right hand). The rappeller leads with the brake hand down and faces slightly sideways. The foot corresponding to the brake hand precedes the guide hand at all times. The rappeller keeps the guide hand on the rope above him to guide himself--not to brake himself. He must lean out at a sharp angle to the rock. He keeps his legs spread well apart and relatively straight for lateral stability, and his back straight to reduce friction. The BDU collar is turned up to prevent rope burns on the neck. Gloves are worn, and other clothing may be used to pad the shoulders and buttocks. To brake, the rappeller leans back and faces directly toward the rock area so his feet are horizontal to the ground.

Figure 7-5. Body rappel.

Notes: 1. Hasty rappels and body rappels are not used on pitches that have overhangs; feet must maintain surface contact.

2. Hasty rappels and body rappels are not belayed from below.

c. **Seat-Hip Rappel** (Figure 7-6). The seat rappel differs from the body rappel in that the friction is absorbed by a carabiner that is inserted in a sling rope seat and fastened to the rappeller. This method provides a faster and more frictional descent than other methods. Gloves can be worn to prevent rope burns.

Figure 7-6. Seat-hip rappel.

(1) An alternate technique is to insert two carabiners opposite and opposed. Then insert a locking carabiner into the two carabiners with opening gate on brake hand side. Then run the rope through the single carabiner. This helps to keep the rappel rope away from the harness.

(2) To hook up for the seat-hip method, stand to one side of the rope. If using a right-hand brake, stand to the left of the rappel rope facing the anchor; if using a left-hand brake, stand to the right of the rappel rope. Place the rappel rope(s) into the locking carabiner; slack is taken between the locking carabiner and anchor point and wrapped around the shaft of the locking carabiner and placed into the gate so that a round turn is made around the shaft of the locking carabiner (Figure 7-7, page 7-12). Any remaining slack is pulled toward the uphill anchor point. If a single rope is used, repeat this process to place two round turns around the shaft of the locking carabiner. Face the anchor point and descend using the upper hand as the guide and the lower hand as the brake. This method has minimal friction, and is fast and safe. However, care is taken that the rope is hooked correctly into the carabiner to avoid the gate being opened by the rope. Loose clothing or equipment around the waist may be accidentally pulled into the locking carabiner and lock (stop) the rappel. For this reason, the rappeller must tuck in his shirt and keep his equipment out of the way during his descent.

Figure 7-7. Proper hookup using carabiner wrap.

d. **Figure-Eight Descender**. The figure-eight descender puts less kinks in the rope, and it can be used with one or two ropes (Figure 7-8).

(1) To use the figure-eight descender, pass a bight through the large eye and then over the small eye onto the neck. Place the small eye into a locking carabiner. To reduce the amount of friction on the figure-eight, place the original bight into the carabiner and not around the neck of the descender. (Less friction requires more braking force from the rappeller.)

(2) The guide hand goes on the rope that is running from the anchor. The brake hand goes on the slack rope. The brake is applied by moving the brake hand to the rear or downward.

Figure 7-8. Figure-eight descender.

d. **Other Devices**. Many different types of devices are similar in design and operation to the basic plate. These include slots or plates and tubers. Most of these devices can accommodate two ropes not greater than 7/16 of an inch in size. Follow manufacturer's directions for using these devices for rappelling.

e. **Extending the Rappel Device**. The rappel device can be extended using either a piece of webbing or cordage to move the device away from the body and the harness, preventing accidental damage (Figure 7-9, page 7-14). It also allows for easier self-belay.

f. **Self-Belay Techniques.** A friction knot can be used as a belay for a rappeller (Figure 7-9, page 7-14). The knot acts as the brake hand when the rappeller must work or negotiate an obstacle requiring the use of both hands. The knot acts as a belay if the rappeller loses control of the rope.

Figure 7-9. Extended hookup with self-belay.

Section III. ONE-ROPE BRIDGE

The one-rope bridge is constructed using a static rope. The rope is anchored with an anchor knot on the far side of the obstacle and is tied off at the near end with a tightening system. A one-rope bridge may be built many ways, depending upon the tactical situation and area to be crossed (crossing a gorge above the tree line may require constructing artificial anchors). However, they all share common elements to safely construct and use the bridge: two suitable anchors; good loading and unloading platforms; a rope about 1-meter (waist) high for loading and unloading; a tightening system; and a rope tight enough for ease of crossing. Which side the tightening system is utilized, or whether an anchor knot or retrievable bowline is used, depends on the technique.

7-10. SITE SELECTION

A suitable crossing site must have "bombproof" anchors on both the near side and far side. These anchors must be extremely strong due to the amount of tension that will be placed upon them. Natural anchors, such as large trees and solid rock formations, are always preferred. The site must also have suitable loading and off-loading platforms to facilitate safe personnel movement.

7-11. INSTALLATION USING TRANSPORT TIGHTENING SYSTEM

The transport tightening system provides a mechanical advantage without requiring additional equipment.

a. The rope must first be anchored on the far side of the obstacle. If crossing a stream, the swimmer must be belayed across. If crossing a ravine or gorge, crossing may involve rappelling and a roped climb. Once across, the swimmer/climber will temporarily anchor the installation rope.

b. One man on the near side ties a fixed-loop knot (for example, wireman's, figure-eight slip knot) approximately 3 feet from the near side anchor and places the carabiner into the loop of the knot. The opening gate must be up and away from the loop. If two carabiners are used, the gates will be opposing. At that time, soldiers route the remainder of the rope around the near side anchor point and hook the rope into the carabiner. This system is known as a transport-tightening system (Figure 7-10). The man on the far side pulls the knot out four to six feet from the near anchor.

c. Once the knot has been pulled out, the far side man anchors the rope using a tensionless anchor. The anchor should be waist high.

Figure 7-10. Transport tightening system.

d. A three-man pull team on the near side pulls the slack out of the installation rope. The knot should be close enough to the near side anchor to allow personnel to easily load the installation.

Note: No more than three personnel should be used to tighten the rope. Using more personnel can over-tighten the rope and bring the rope critically close to failure.

e. The rope the can be secured using one of three methods: transport knot (Figure 7-11), round turn around anchor and two half hitches on a bight (Figure 7-12), or a tensionless anchor knot (Figure 7-13).

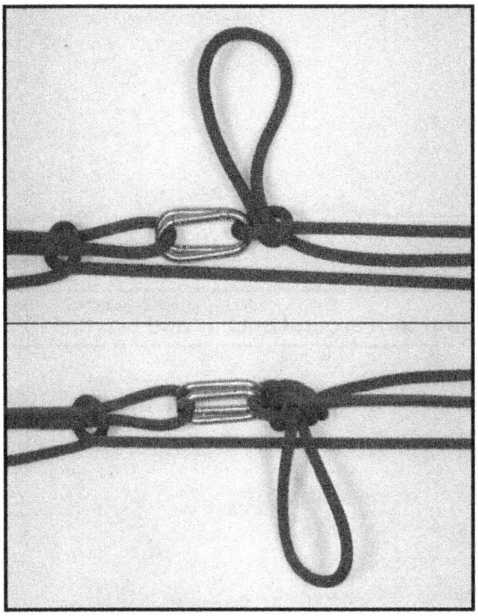

Figure 7-11. Transport knot.

Figure 7-12. Round turn around anchor and two half hitches on a bight.

Figure 7-13. Tensionless anchor knot.

Note: During training, a second static rope may be installed under less tension and alongside the tight rope to increase safety. An individual would clip into both ropes when crossing, thus having a backup in case of failure of the tighter rope.

7-12. INSTALLATION USING Z-PULLEY TIGHTENING SYSTEM

The Z-pulley tightening system (Figure 7-14) is another method for gaining a mechanical advantage.

a. The rope is brought across the obstacle the same way as discussed in paragraph 7-10.

b. Once across, the far side man anchors the rope.

c. One soldier ties a friction knot (autoblock, web wrap, Kleimheist) with a sling rope onto the bridging rope on the near side bank. Two steel carabiners are inserted with opposing gates into the friction knot.

d. The rope is routed around the near side anchor and through the carabiners, from inside to outside, and is run back to the near side anchor.

e. A second sling rope is tied to the bridge rope and then anchored to the near side anchor. This knot will be used as a progress capture device.

Figure 7-14. Z-pulley tightening system.

f. The three-man pull team on the near side then pulls on the rope, creating a pulley effect that tightens the system. As the rope is pulled tight, one man pushes the friction knot back toward the far side.

g. When the rope is tight, it is tied off with a tensionless anchor knot, transport knot, or round turn around anchor and two half hitches on a bight.

7-13. UTILIZATION

The rope bridge can be used to move personnel and equipment over obstacles. There are several methods of accomplishing this.

a. **Method of Crossing**. If dry crossing is impossible, soldiers will use the rope bridge as a hand line. Preferably, all soldiers will tie a safety line and attach it to the rope installation as they cross. If the soldier must cross with his rucksack, he may wear it over both shoulders, although the preferred method is to place another carabiner into the top of the rucksack frame, attach it to the bridge, and pull the rucksack across. Soldiers will always cross on the downstream side of the installation. If a dry crossing is possible soldiers will use one of three methods: commando crawl, monkey crawl, and Tyrolean traverse.

(1) ***Commando Crawl*** (Figure 7-15). The soldier lies on top of the rope with the upstream foot hooked on the rope and the knee bent close to the buttocks; the downstream leg hangs straight to maintain balance. He progresses by pulling with his hands and arms. To recover if he falls over, the soldier hooks one leg and the opposite arm over the rope, and then pushes down with the other hand to regain position.

Figure 7-15. Commando crawl.

Note: Only one man at a time is allowed on the bridge while conducting a commando crawl.

(2) ***Monkey Crawl*** (Figure 7-16, page 7-20). The soldier hangs below the rope suspended by his hands with both heels crossed over the rope. He pulls with his hands and arms, and pushes with his feet to make progress.

Figure 7-16. Monkey crawl.

(3) ***Rappel Seat Method*** (Figure 7-17). The soldier ties a rappel seat (or dons a seat harness) with the carabiner facing up and away from his body. He then faces the rope and clips into the rope bridge. He rotates under the rope and pulls with his hands and arms to make progress. The rappel seat method is the preferred method. If crossing with rucksacks, a carabiner is inserted into the frame and attached to the rope bridge. The soldier the places one or both legs through the shoulder carrying straps and pulls the rucksack across.

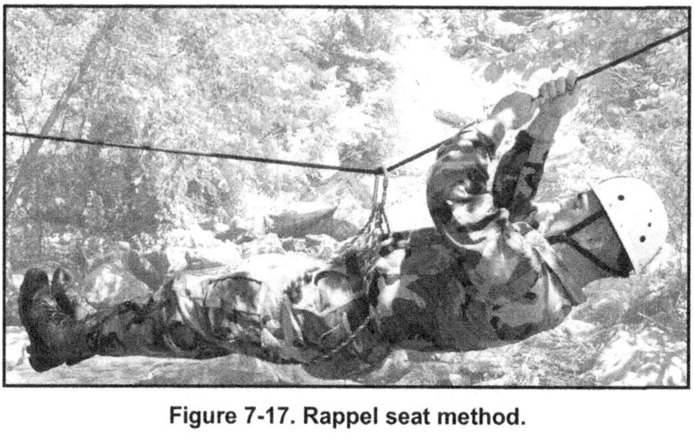

Figure 7-17. Rappel seat method.

b. **Rigging Special Equipment**. Any special equipment, such as crew-served weapons, ammunition, or supplies, must be rigged for movement across the rope bridge. A unit SOP may dictate the rigging of these items, but many expedient methods exist. The rigging should use various items that would be readily available to a deployed unit. Some of these items include tubular nylon webbing, cordage (various sizes), and carabiners.

(1) ***Machine Guns***. To rig machine guns, use a sling rope and tie a rerouted figure-eight around the spine of the front sight post. Then tie two evenly spaced fixed loops. Finally, anchor the sling rope to the buttstock of the machine gun. Additional tie downs may be necessary to prevent accidental disassembly of the weapon.

(2) ***ALICE Packs.*** ALICE packs can be joined together with a sling to facilitate moving more than one rucksack at one time.

7-14. HAULING LINE

A hauling line may be used to move rucksacks or casualties across the rope bridge (Figure 7-18).

a. **Construction**. An additional rope is brought across the rope bridge and anchored to the far side. The other end is anchored on the near side. All the slack is pulled to the near side, and a figure-eight slip knot is tied at the loading platform. A carabiner is inserted into the loop and clipped onto the rope bridge.

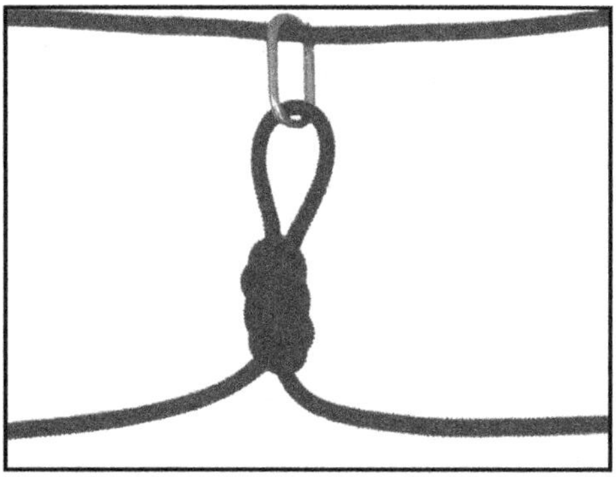

Figure 7-18. Hauling line.

b. **Moving Rucksacks**. Use carabiners to attach the rucksack frames to the rope bridge. Then clip the carabiner of the hauling line into the carabiner of the rucksack closest to the far side. Personnel on the far side pull the rucksacks across using the hauling line while personnel on the near side manages the slack at all times.

c. **Moving Litters**. The carabiner of the hauling line will remain on the rope bridge. On each side of this carabiner, using the hauling line tie a middle-of-the-rope clove hitch around both of the horizontal lift straps of the litter. Remove the slack between the carabiners. Then place the carabiners in each of the lift straps onto the rope bridge. The same technique used for the rucksacks is used to pull the litter across.

7-15. RETRIEVAL

Once all except two troops have crossed the rope bridge, the bridge team commander (BTC) chooses either the wet or dry method to dismantle the rope bridge.

a. If the BTC chooses the dry method, he should have anchored his tightening system with the transport knot.

(1) The BTC back-stacks all of the slack coming out of the transport knot, then ties a fixed loop and places a carabiner into the fixed loop.

(2) The next to last man to cross attaches the carabiner to his rappel seat or harness, and then moves across the bridge using the Tyrolean traverse method.

(3) The BTC then removes all knots from the system. The far side remains anchored. The rope should now only pass around the near side anchor.

(4) A three-man pull team, assembled on the far side, takes the end brought across by the next to last man and pulls the rope tight again and holds it.

(5) The BTC then attaches himself to the rope bridge and moves across.

(6) Once across, the BTC breaks down the far side anchor, removes the knots, and then pulls the rope across.

(b) If the BTC chooses a wet crossing, any method can be used to anchor the tightening system.

(1) All personnel cross except the BTC or the strongest swimmer.

(2) The BTC then removes all knots from the system.

(3) The BTC ties a fixed loop, inserts a carabiner, and attaches it to his rappel seat or harness. He then manages the rope as the slack is pulled to the far side.

(4) The BTC then moves across the obstacle while being belayed from the far side.

Section IV. SUSPENSION TRAVERSE

The suspension traverse is used to move personnel and equipment over rivers, ravines, chasms, and up or down a vertical rock face (Figure 7-19). The system may be established on a plane, varying from horizontal to near vertical.

Figure 7-19. Suspension traverse.

7-16. SITE SELECTION

The crossing site must have bombproof anchors at the near side and the far side, and suitable loading and off-loading platforms. If the anchors do not provide sufficient height to allow clearance, an A-frame must be used.

7-17. INSTALLATION

Installation of a suspension traverse can be time-consuming and equipment-intensive. All personnel must be well trained and well rehearsed in the procedures.

a. **A-frames**. Even in wooded mountainous terrain constructing an A-frame may be necessary due to the lack of height where the installation is needed. Site selection determines whether more height is needed; mission requirements determine site selection. The two main installations that use A-frames are the suspension traverse and vertical hauling line.

b. **Equipment**. Two sturdy poles are needed. The exact size of the poles depends on the type of load and location of the installation. The average size A-frame pole should be at least 3 inches in diameter and 9 to 12 feet long. Three to five 14-foot sling ropes are needed, depending on the size of the poles used for the A-frame.

c. **Construction**. Place two poles with the butt ends flush, and mark the apex on both poles.

(1) Ensure that proper height is attained and that the installation runs in a straight line between the two anchors. An A-frame placed out of proper alignment can cause the system to collapse. Try to find natural pockets in which to place the base of the A-frame poles.

(2) With a sling rope, tie a clove hitch around the left pole (standing at the base of the poles and facing the top) 3 inches above the apex marking, leaving about 18 inches of the sling rope free on top of the clove hitch. Place the locking bar on the outside edge of the pole. Make sure the rope end is pointing down as it is tied. (See Figure 7-20A, page 7-24.)

(3) Place the poles side by side and wrap the sling rope horizontally around both poles six to eight times, wrapping down from the clove hitch (Figure 7-20B, page 7-24). It may be necessary to join another sling rope to the first by using a square knot secured with overhand knots. Position this knot on the outside of one of the poles so as not to interfere with the vertical wraps. Make at least two additional wraps below the joining square knot. (See Figure 7-20C, page 7-24.)

(4) On the last horizontal wrap (ensure there are at least two wraps below the joining knot) to which the clove hitch is not tied, pass the rope between the poles below the wraps, and make four to six tight vertical wraps around the horizontal wraps (Figure 7-20D, page 7-24). Make the wraps as tight as possible. The vertical wraps must be as flat as possible next to each other. When starting the first vertical wrap, ensure it is in the same direction as the 18-inch tail on the top of the clove hitch. Insert a carabiner into the last two vertical wraps (Figure 7-20E, page 7-24).

(5) On the last vertical wrap, pass the rope between the poles above the horizontal wraps. Tie it off with a square knot in the section of rope coming from the clove hitch. Secure with overhand knots tied in the tails. (See Figure 7-20F, page 7-24.)

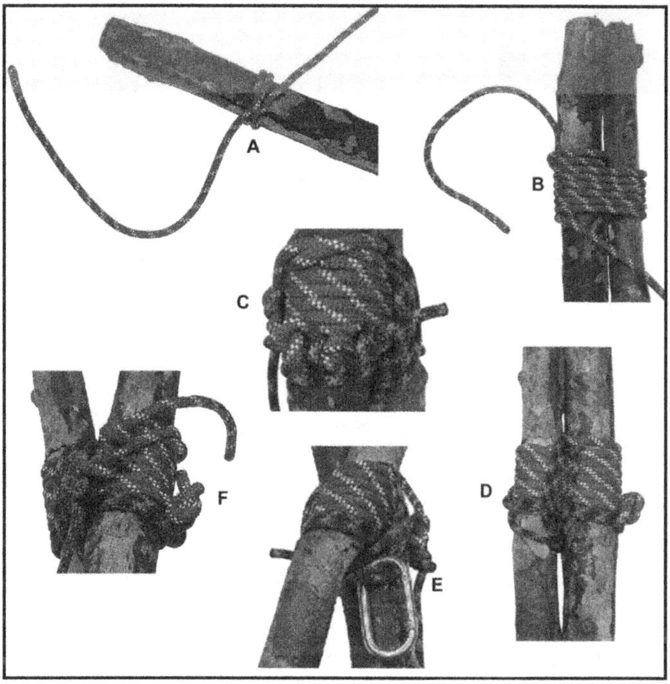

Figure 7-20. A-frame horizontal and vertical wraps.

(6) Use a spreader rope to prevent the A-frame from collapsing from pressure applied at the apex (Figure 7-21). If the ground is soft, dig the legs in about 6 inches. Tie a sling rope between the legs with a round turn with two half hitches around each leg. Remove all slack in the rope between the legs.

(7) If the ground is a hard surface, tie end-of-the-rope clove hitches with the locking portions facing to the rear, the direction of kick. Tie the tails off at a 45-degree angle with a round turn and two half hitches to a secondary anchor point. The spreader rope should be no more than 6 inches above ground level. The use of clove hitches and half hitches permits easy adjustment of the spreader rope. If more than one sling rope is needed, tie the two ropes together with a square knot and secure with half hitches or overhand knots.

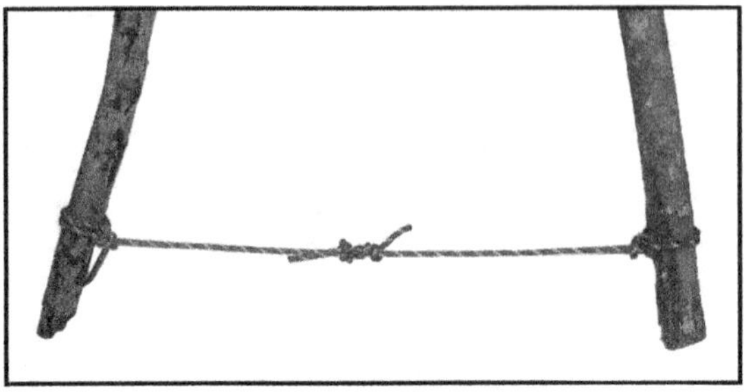

Figure 7-21. A-frame spreader.

d. **Installation Construction**. One man rappels down the pitch and secures two installation (traverse) ropes to the far anchor with an anchor knot. Place a transport tightening system in each installation rope at the near (upper) anchor. Run the installation ropes through or around the anchor in opposite directions and tie off. Anchor the traverse ropes as close together as possible so that the ropes do not cross.

(1) Place the A-frame (if needed) so that both traverse ropes run over the apex and the A-frame splits the angle formed between the near (upper) and far (lower) anchors, with the legs firmly emplaced or anchored with pitons. Ensure that the A-frame is in line with the anchors. Adjust the A-frame under the traverse ropes after tightening to firmly implant the A-frame.

(2) Tighten the installation ropes using either the transport tightening system (paragraph 7-11) or the z-pulley tightening system (paragraph 7-12).

(3) Anchor the A-frame to the traverse rope. Tie a clove hitch at the center of a sling rope. Place it over one of the poles above the apex and move down to the apex so that the locking bar of the clove hitch is to the inside of the A-frame. Secure each end of the sling rope to one of the tightened static lines with two Prusik knots-one forward and one to the rear of the A-frame on the same static line rope (Figure 7-22).

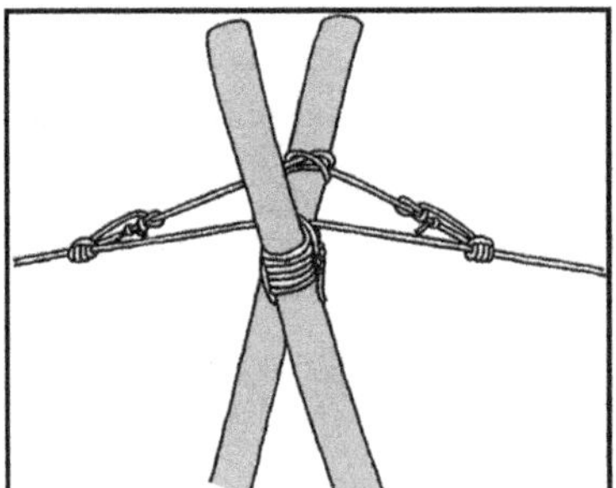

Figure 7-22. Anchoring the A-frame to the traverse rope.

Note: The A-frame should be positioned so that the angles created by the A-frame bisecting the installation rope are approximately equal on both sides. This creates downward pressure holding the A-frame in position, not forcing it in a lateral direction. It must also be placed in a straight line between the upper and lower anchor points.

(4) Use a carrying rope to attach loads to the traverse ropes (Figure 7-23). Join the ends of a 14-foot sling rope with a square knot and two overhand knots. Displace the knot one-third of the distance down the loop and tie an overhand knot both above and below the square knot. This forms two small loops and one large loop that is longer than the two small loops combined.

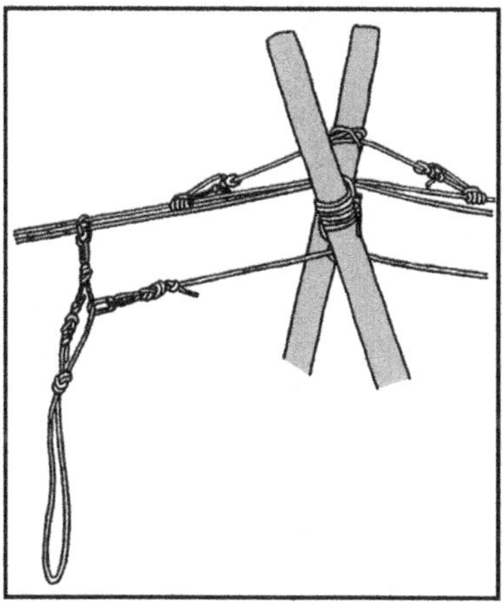

Figure 7-23. Carrying rope for use on a traverse.

(5) Attach the carrying rope to the traverse ropes with carabiners (or a pulley) that have the gates reversed and opening in opposite directions. Attach a belay rope to the center loop of the carrying rope using a fixed loop or locking carabiner on the side opposite the joining knot (Figure 7-23). When the suspension traverse is near horizontal, a second rope may be needed to pull the load across and should be attached to the carrying rope the same as the first.

(6) Insert second carabiner into the one placed into the wraps of the A-frame. This is where a belay rope will be attached

(7) With a sling rope, tie a six wrap middle-of-the-rope Prusik knot to both static ropes near the far side off-loading point. This acts as a stopper knot for the man descending, preventing him from hitting the lower anchor.

(8) Attach the load by running the long loop of the carrying rope through the load or through the soldier's harness and attaching the bottom loop to the traverse rope carabiner.

Descent must be belayed slowly and be controlled. Soldiers descending should hold onto the carrying rope and keep their feet high to avoid contact with the ground. Due to the constant tension maintained on the belayer, use a mechanical belay. If the belayer cannot view the entire descent route, use a relay man.

7-18. RETRIEVAL

The suspension traverse is not as readily retrievable as the one-rope bridge. Therefore, the installing unit should dismantle it after it is no longer needed.

Section V. VERTICAL HAULING LINE

The vertical hauling line is an installation used to move men and equipment up vertical or near-vertical slopes (Figure 7-24). It is often used with a fixed rope for personnel movement. The hauling line is used to move equipment, such as mortars or other crew-served weapons, rucksacks, or supplies

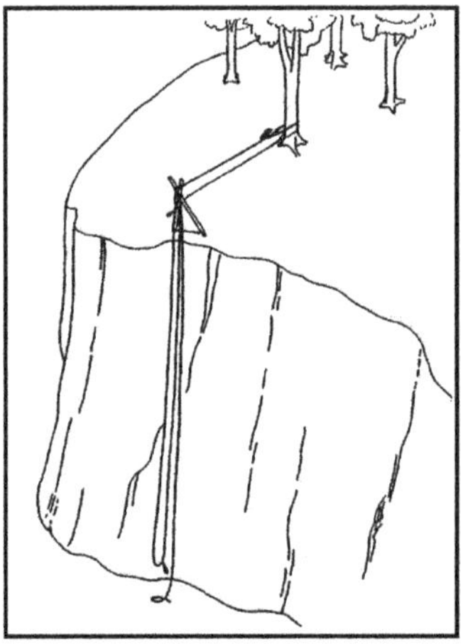

Figure 7-24. Vertical hauling line.

7-19. SITE SELECTION

The first and most important task is to determine where to construct the vertical hauling line. The site must have an appropriate top anchor that is secure enough to hold the system and load. Loading and unloading platforms should be easily accessible natural platforms that provide a safe working area. The ideal platform at the top allows construction of the vertical hauling line without the use of an A-frame. The site should also have sufficient clearance to allow for space between the slope and pulley rope for easy hauling of troops or equipment.

7-20. INSTALLATION

Construct an A-frame, if necessary, and anchor it. Double one installation rope, find the middle, and lay the middle of the installation rope over the apex of the A-frame; a 30-centimeter (12-inch) bight should hang below the apex.

a. To maintain the 12-inch bight, tie clove hitches above the A-frame lashing on each side of the apex with the installation rope, ensuring that the locking bars of the clove hitches are on the inside. Ensure that the portion of the rope that forms the bight comes out of the bottom of the clove hitch. (See Figure 7-25.)

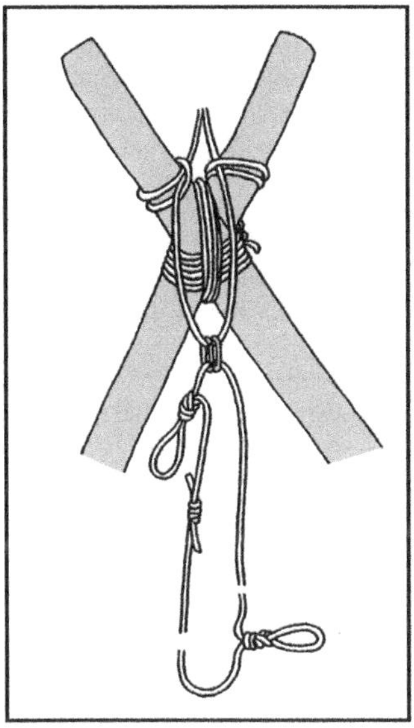

Figure 7-25. Attaching the anchor rope to the A-frame.

b. To anchor the A-frame, use a transport tightening system with the doubled rope, which is tied to the A-frame. Tie this off at an anchor point to the rear of the A-frame installation and adjust the angle of the A-frame so it leans out over the cliff edge. The angle should be 15 to 25 degrees unloaded. The A-frame should not lean outward more than 45 degrees once loaded since the legs can lose their position.

c. Tie the ends of another installation rope together with a joining knot to form the hauling line. Attach the rope to the system by two carabiners with gates up and opposed or one mountain rescue pulley with a locking steel carabiner in the 12-inch bight hanging from the apex of the A-frame. Tie fixed loops (wireman's, directional figure-eight, or single butterfly) on opposite sides of the endless rope at the loading and unloading platforms.

d. Attach equipment to the hauling line 12 inches above the joining knot by a carabiner in the fixed loop.

e. Additional fixed loops may be tied in the hauling line for more control over the object when moving large loads. Attach personnel to the hauling line by use of a rappel seat or seat harness.

Note: Mortar tubes and similar objects are attached to the line by two knots so that the tube stays parallel and as close to the hauling line as possible.

f. When personnel are moved using a vertical hauling line, make a knotted hand line; anchor it in line with, or to, the primary anchor (round turn with a bowline); and place it over the spreader on the legs of the A-frame. Space the overhand knots in the knotted hand line 12 inches apart, with about 20 feet of rope without knots at one end for the anchor. Throw the knotted hand line over the A-frame spreader rope and down the side of the cliff. Personnel ascending the vertical hauling line use this as a simple fixed rope.

g. Use as many men as needed to pull the load to the top by pulling on the rope opposite the load. If equipment and personnel are only being lowered, belay from the top using the hauling line. Station two climbers at the unloading platform to retrieve loads.

h. If only equipment is being hauled up, it is not necessary to use the knotted hand line rope, but it may be necessary to use a belay rope. To move materials or troops up on one side of the hauling line, pull the other side from below.

Note: Personnel using the hauling line for movement must apply all related principles of climbing. Always station two operators at the top of the vertical hauling line to aid men or to retrieve loads when they reach the top. They will always be safetied while working near the edge. When in use, the A-frame should lean slightly over the edge of the cliff to prevent excessive wear on the ropes that pass over sharp rocks. Reduce excessive friction on the system. Remove all obstacles and any loose objects that could be dislodged by personnel and equipment.

7-21. RETRIEVAL

The vertical hauling line is used along a main supply route. When it is no longer needed, the installing unit will return and dismantle the system.

Section VI. SIMPLE RAISING SYSTEMS

Moving heavy objects with limited manpower may be necessary in mountainous terrain. To reduce fatigue of those personnel moving the load, simple rigging techniques can be used to increase the mechanical advantage of the hauling system.

7-22. Z-PULLEY SYSTEM

The Z-pulley system is a simple, easily constructed hauling system (Figure 7-26, page 7-30).

a. **Considerations**. Anchors must be sturdy and able to support the weight of the load. Site selection is governed by different factors: tactical situation, weather, terrain, equipment, load weight, and availability of anchors.

b. **Theory.** Use carabiners as a substitute if pulleys are not available. The mechanical advantage obtained in theory is 3:1. The less friction involved the greater the mechanical advantage. Friction is caused by the rope running through carabiners, the load rubbing against the rock wall, and the rope condition.

c. **Construction**. Use the following procedures to construct a Z-pulley system.

(1) Establish an anchor (anchor pulley system [APS]). Place a carabiner on the runner at the anchor point, place a pulley into the carabiner, and run the hauling rope through the pulley.

(2) With a sling rope (preferably 7 millimeter), tie a middle-of-rope Prusik knot secured with a figure-eight knot on the load side of the pulley. This will be used as a progress capture device (PCD). A mechanical descender may be used in place of the Prusik knot. Take the tails exiting the figure-eight and tie a Munter hitch secured by a mule knot. Ensure the Munter hitch is loaded properly before tying the mule knot.

(3) At an angle away from the APS, establish a moveable pulley system (MPS) to create a "Z" in the hauling rope. Tie another Prusik knot on the load side of the hauling rope. Secure it with a figure-eight knot. Using the tails tie a double-double figure-eight knot. Insert a locking carabiner into the two loops formed, then place the working end into the carabiner. Mechanical ascenders should not be used as an MPS. Move the working end back on a parallel axis with the APS. Provide a pulling team on the working end with extra personnel to monitor the Prusik knots.

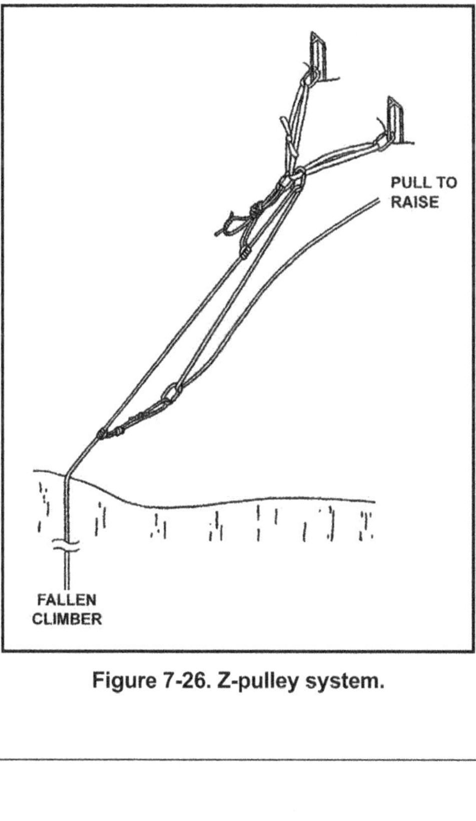

Figure 7-26. Z-pulley system.

d. **Other Factors**. If the two pulleys touch, the "Z" is lost along with the mechanical advantage. For greater efficiency, the main anchor should be well back from the edge and all ropes should pull parallel to the load.

Note: Avoid the possibility of overstressing the anchors. Be aware of reduced sensitivity to the load due to the mechanical advantage. Use belays and backup safeties. Protect the rope from edges and other abrasive parts of the rock.

7-23. U-PULLEY SYSTEM

The U-pulley system is another simple, easily-constructed hauling system (Figure 7-27, page 7-32).

a. **Considerations**. Anchors must be sturdy and able to support the weight of the load. Site selection is governed by different factors: tactical situation, weather, terrain, equipment, load weight, and availability of anchors.

b. **Theory.** Use carabiners as a substitute if pulleys are not available. The mechanical advantage obtained in theory is 2:1. The less friction involved the greater the mechanical advantage. Friction is caused by the rope running through carabiners, the load rubbing against the rock wall, and the rope condition.

c. **Construction.** Use the following procedures construct a U-pulley system.

(1) Anchor the hauling rope.

(2) Prepare the load or casualty for hauling. Place a locking carabiner the on to the harness or the rigged load.

(3) Lower a bight to the casualty or the load.

(4) Place the bight into the carabiner; or place the bight on to a pulley and then place pulley into the carabiner.

(5) Construct a second anchor. Attach a locking carabiner to the anchor.

(6) Tie a middle of the rope Prusik onto the haul rope exiting the pulley. Secure the Prusik with a double-double figure eight. This is the PCD. Place the fixed loops into the locking carabiner of the second anchor.

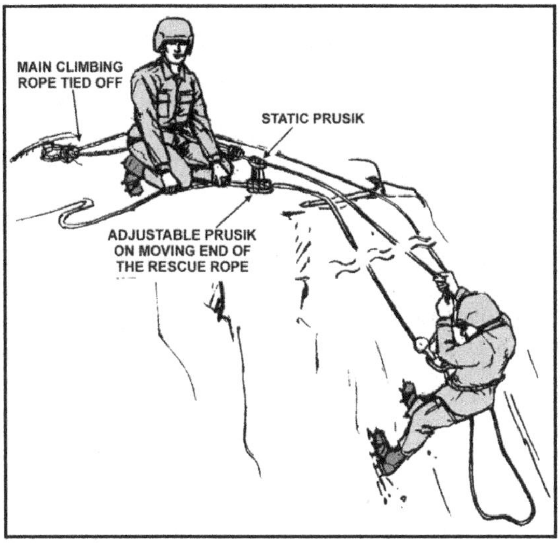

Figure 7-27. U-pulley system.

CHAPTER 8
MOUNTAIN WALKING TECHNIQUES

Mountain travel encompasses the full spectrum of techniques used to negotiate steep, rugged terrain. Mountain walking on rock and snow, technical rock and ice climbing, skiing or snow shoeing, rappelling, and stream crossing are the key travel skills a military mountaineer must possess.

8-1. BASIC PRINCIPLES

Up scree or talus, through boulder fields or steep wooded mountainsides, over snow or grass-covered slopes, the basic principles of mountain walking remain the same.

a. The soldier's weight is centered directly over the feet at all times. He places his foot flat on the ground to obtain as much (boot) sole-ground contact as possible. Then, he places his foot on the uphill side of grass tussocks, small talus and other level spots to avoid twisting the ankle and straining the Achilles tendon. He straightens the knee after each step to allow for rest between steps, and takes moderate steps at a steady pace. An angle of ascent or descent that is too steep is avoided, and any indentations in the slope are used to advantage.

b. In addition to proper technique, pace is adapted to conditions. The mountaineer sets a tempo, or number of steps per minute, according to the pace of the unit in which he is moving. (Physical differences mean that the tempos of two people moving at the same speed will not always be the same.) The soldier maintains tempo and compensates for changes of slope or terrain by adjusting the length of his stride. Tempo, pace, and rhythm are enhanced when an interval of three to five paces is kept between individuals. This interval helps lessen the "accordion" effect of people at the end of the file who must constantly stop and start.

c. The terrain, weather, and light conditions affect the rate of climb. The more adverse the conditions, the slower the pace. Moving too fast, even under ideal conditions, produces early fatigue, requires more rest halts, and results in loss of climbing time. A soldier can only move as fast as his lungs and legs will allow. The trained, conditioned and acclimatized soldier has greater endurance and moves more efficiently. Rest, good nutrition and hydration, conditioning, acclimatization, proper training, and the will to climb are key to successful mountain operations.

d. Breaks are kept to a minimum. When a moderate pace is set, the need for rest halts decreases, the chance of personnel overheating is lessened, and a unit can cover a given distance in a minimal time. If possible, rests should be taken on level ground avoiding steeper inclines.

(1) During the first half-hour of movement an adjustment halt should be taken. Soldiers will loosen or tighten bootlaces as needed, adjust packs and add or remove appropriate layers of clothing.

(2) Following the first halt, a well-conditioned party may take a short rest every 1 to 1.5 hours. If possible, soldiers lean against a tree, rock, or hillside to relieve the shoulders of pack weight, breathe deeply, hydrate, and snack on trail food. These halts are kept short enough to avoid muscles stiffening (one to two minutes).

(3) Later in the march longer halts may be necessary due to fatigue or mission requirements. At these halts soldiers should immediately put on additional clothing to avoid chilling—it is much easier to keep a warm body warm than to warm up a cold one.

(4) After a climb, a good rest is needed to revive tired muscles.

e. The rest step is used for steep slopes, snowfields, and higher elevations. It controls pace and limits fatigue by giving the lungs and legs a moment to recuperate between steps. Pace is kept slow and rhythmic.

(1) After each step forward, the soldier pauses briefly, relaxing the muscles of the forward leg while resting his entire bodyweight on the rear leg. The rear leg is kept straight with the knee locked so that bone, not muscle, supports the weight.

(2) Breathing is synchronized with the rest step. The number of breaths per step will change depending on the difficulty of the climb. Steeper slopes or higher elevations may require several breaths per step. When the air thins at altitude it is especially important to breathe deeply, using the "pressure breathing" technique. The soldier exhales strongly, enabling an easier, deeper inhale.

(3) This slow, steady, halting rest step is more efficient than spurts of speed, which are rapidly exhausting and require longer recovery.

f. Downhill walking uses less energy than uphill but is much harder on the body. Stepping down can hammer the full bodyweight onto the feet and legs. Blisters and blackened toenails, knee damage, and back pain may follow. To avoid these problems the soldier should start by tightening bootlaces to ensure a snug fit (also keep toenails trimmed). A ski pole, ice ax, or walking stick will help take some of the load and give additional stability. (Refer to Chapter 11 for techniques and use of the ice ax.) Keep a moderate pace and walk with knees flexed to absorb shock.

g. Side hill travel on any surface should be avoided whenever possible. Weighted down with a rucksack, the soldier is vulnerable to twisted ankles, back injury, and loss of balance. If side hill travel is necessary, try to switchback periodically, and use any lower angle flat areas such as rocks, animal trails, and the ground above grass or brush clumps to level off the route.

8-2. TECHNIQUES

Mountain walking techniques can be divided according to the general formation, surface, and ground cover such as walking on hard ground, on snow slopes and grassy slopes, through thick brush, and on scree and talus slopes.

a. **Hard Ground.** Hard ground is firmly compacted, rocky soil that does not give way under the weight of a soldier's step. It is most commonly found under mature forest canopy, in low brush or heather, and areas where animals have beaten out multiple trails.

(1) When ascending, employ the rest step to rest the leg muscles. Steep slopes can be traversed rather than climbed straight up. To turn at the end of each traverse, the soldier should step off in the new direction with the uphill foot. This prevents crossing the feet and possible loss of balance. While traversing, the full sole-to-ground principle is accomplished by rolling the ankle downhill on each step. For small stretches the herringbone step may be used—ascending straight up a slope with toes pointed out. A normal progression, as the slope steepens, would be from walking straight up, to a herringbone step, and then to a traverse on the steeper areas.

(2) Descending is best done by walking straight down the slope without traversing. The soldier keeps his back straight and bends at the knees to absorb the shock of each step. Body weight is kept directly over the feet and the full boot sole is placed on the ground with each step. Walking with a slight forward lean and with the feet in a normal position make the descent easier.

b. **Snow Slopes.** Snow-covered terrain can be encountered throughout the year above 1,500 meters in many mountainous areas. Talus and brush may be covered by hardened snowfields, streams made crossable with snowbridges. The techniques for ascending and descending moderate snow slopes are similar to walking on hard ground with some exceptions.

(1) ***Diagonal Traverse Technique.*** The diagonal traverse is the most efficient means to ascend snow. In conjunction with the ice ax it provides balance and safety for the soldier. This technique is a two-step sequence. The soldier performs a basic rest step, placing the leading (uphill) foot above and in front of the trailing (downhill) foot, and weighting the trail leg. This is the in-balance position. The ice ax, held in the uphill hand, is placed in the snow above and to the front. The soldier shifts his weight to the leading (uphill) leg and brings the unweighted trail (downhill) foot ahead of the uphill foot. He shifts weight to the forward (downhill) leg and then moves the uphill foot up and places it out ahead of the trail foot, returning to the in-balance position. At this point the ax is moved forward in preparation for the next step.

(2) ***Step Kicking.*** Step kicking is a basic technique used when crampons are not worn. It is best used on moderate slopes when the snow is soft enough to leave clear footprints. On softer snow the soldier swings his foot into the snow, allowing the leg's weight and momentum to carve the step. Fully laden soldiers will need to kick steps, which take half of the boot. The steps should be angled slightly into the slope for added security. Succeeding climbers will follow directly in the steps of the trailbreaker, each one improving the step as he ascends. Harder snow requires more effort to kick steps, and they will not be as secure. The soldier may need to slice the step with the side of his boot and use the diagonal technique to ascend.

(3) ***Descending Snow.*** If the snow is soft and the slope gentle, simply walk straight down. Harder snow or steeper slopes call for the plunge step, which must be done in a positive, aggressive manner. The soldier faces out, steps off, and plants his foot solidly, driving the heel into the snow while keeping his leg straight. He shifts his weight to the new foot plant and continues down with the other foot. On steeper terrain it may be necessary to squat on the weighted leg when setting the plunge step. The upper body should be kept erect or canted slightly forward.

(4) ***Tips on Snow Travel.*** The following are tips for travelling on snow.

(a) Often the best descent is on a different route than the ascent. When looking for a firmer travel surface, watch for dirty snow—this absorbs more heat and thus hardens faster than clean snow.

(b) In the Northern Hemisphere, slopes with southern and western exposures set up earlier in the season and quicker after storms, but are more prone to avalanches in the spring. These slopes generally provide firm surfaces while northern and eastern exposures remain unconsolidated.

(c) Travel late at night or early in the morning is best if daytime temperatures are above freezing and the sun heats the slopes. The night's cold hardens the snow surface.

(d) Avoid walking on snow next to logs, trees, and rocks as the subsurface snow has melted away creating hidden traps.

c. **Grassy Slopes.** Grassy slopes are usually composed of small tussocks of growth rather than one continuous field.

(1) When ascending, step on the upper side of each hummock or tussock, where the ground is more level.

(2) When descending a grassy slope, the traverse technique should be used because of the uneven nature of the ground. A climber can easily build up too much speed and fall if a direct descent is tried. The hop-skip step can be useful on this type of slope. In this technique, the lower leg takes all of the weight, and the upper leg is used only for balance. When traversing, the climber's uphill foot points in the direction of travel. The downhill foot points about 45 degrees off the direction of travel (downhill). This maintains maximum sole contact and prevents possible downhill ankle roll-out.

Note: Wet grass can be extremely slippery; the soldier must be aware of ground cover conditions.

d. **Thick Brush.** For the military mountaineer, brush is both a help and a hindrance. Brush-filled gullies can provide routes and rally points concealed from observation; on the other hand steep brushy terrain is hazardous to negotiate. Cliffs and steep ravines are hidden traps, and blow downs and thickets can obstruct travel as much as manmade obstacles. When brush must be negotiated take the most direct route across the obstacle; look for downed timber to use as raised paths through the obstacle; or create a tunnel through the obstacle by prying the brush apart, standing on lower branches and using upper limbs for support.

e. **Scree Slopes.** Slopes composed of the smallest rocks are called scree slopes. Scree varies in size from the smallest gravel to about the size of a man's fist.

(1) Ascending scree slopes is difficult and tiring and should be avoided, if possible. All principles of ascending hard ground and snow apply, but each step is carefully chosen so that the foot does not slide down when weighted. This is done by kicking in with the toe of the upper foot (similar to step-kicking in snow) so that a step is formed in the loose scree. After determining that the step is stable, weight is transferred to the upper leg, the soldier then steps up and repeats the process with the lower foot.

(2) The best method for descending scree slopes is to come straight down the slope using a short shuffling step with the knees bent, back straight, feet pointed downhill, and heels dug in. When several climbers descend a scree slope together, they should be as close together as possible (one behind the other at single arm interval) to prevent injury from dislodged rocks. Avoid running down scree as this can cause a loss of control. When the bottom of the slope (or run out zone) cannot be seen, use caution because drop-offs may be encountered.

(3) Scree slopes can be traversed using the ice ax as a third point of contact. Always keep the ax on the uphill side. When the herringbone or diagonal method is used to ascend scree, the ax can be used placing both hands on the top and driving the spike into the scree slope above the climber. The climber uses the ax for balance as he moves up to it, and then repeats the process.

f. **Talus Slopes.** Talus slopes are composed of rocks larger than a man's fist. When walking in talus, ascending or descending, climbers should always step on the uphill side of rocks and stay alert for movement underfoot. Disturbing unstable talus can cause rockslides. Climbers must stay in close columns while walking through talus so that dislodged rocks do not reach dangerous speeds before reaching lower soldiers. To prevent rock fall injuries, avoid traversing below other climbers. All other basics of mountain walking apply.

8-3. SAFETY CONSIDERATIONS

The mountain walking techniques presented here are designed to reduce the hazards of rock fall and loss of control leading to a fall. Carelessness can cause the failure of the best-planned missions.

a. Whenever a rock is kicked loose, the warning, "Rock!" is shouted immediately. Personnel near the bottom of the cliff immediately lean into the cliff to reduce their exposure, and do not look up. Personnel more than 3 meters away from the bottom of the cliff may look up to determine where the rock is heading and seek cover behind an obstacle. Lacking cover, personnel should anticipate which way the rock is falling and move out of its path to the left or right.

b. If a soldier slips or stumbles on sloping terrain (hard ground, grass, snow, or scree) he must immediately self-arrest, digging into the slope with hands, elbows, knees and toes. If he falls backwards and rolls over he must immediately try to turn over onto his stomach with his legs downhill and self-arrest with hands and toes.

c. When traveling through steep terrain, soldiers should be trained in the use of the ice ax for self-arrest. The ax can be used to arrest a fall on solid ground, grass and scree as well as snow. It may also be used as a third point of contact on difficult terrain. If not in use the ice ax is carried in or on the rucksack with its head down and secured.

8-4. NAVIGATION

Navigation is the process of determining one's present position, the location of a target objective, and selecting and following a route between these two points. Navigation consists of three distinct stages: orientation, navigation, and route finding.

- Orientation is simply figuring out exactly where one is. The use of the map, compass and identifiable terrain features, assisted by an altimeter and GPS, is the foundation of good navigation.
- Navigation includes the determination of the objective's location and the direction from the soldier's starting point to the objective. The same skills and equipment used in orientation are essential for good navigation.
- Route finding is picking the best line of travel that matches the equipment and capabilities of the team. Good route finding incorporates a comprehensive awareness of terrain, a solid base of mountaineering experience, good judgement and sound tactical instincts.

a. **Compasses.** The magnetic compass is the simplest and most widely used instrument for measuring directions and angles in the mountains. The lensatic compass is most commonly used in the military and can be employed in a variety of ways for either day or night navigation.

b. **Altimeters.** The altimeter is a vital piece of navigational equipment that can save valuable time in determining position through elevation.

(1) The standard altimeter is a modified barometer. A barometer is an instrument that measures the weight of a column of air above itself and displays the result on a scale marked in units of pressure, usually inches of mercury, millimeters of mercury, or millibars. Since air pressure drops uniformly as elevation is gained, it can be used to read altitude by means of the altimeter's scale, marked in feet or meters of elevation above sea level. By measuring air pressure, the altimeter/barometer gives the navigator new techniques for position finding, route planning, checking progress and terrain identification. It also gives the navigator valuable weather information specific to his immediate location.

(2) Changes in the weather are usually accompanied by air pressure changes, which are reflected in the altimeter. As the air pressure drops due to the approach of inclement weather for instance, the displayed elevation will rise by a corresponding amount. This means that a barometric pressure change of one inch of mercury equals roughly 1,000 feet of elevation. If the altimeter displays an elevation gain of 300 feet, a loss of barometric pressure of .3 inches has occurred, and bad weather should be expected.

(3) Altimeters come in two types: wrist-mounted digital altimeters and analog altimeters, usually attached to a cord.

(4) Because the altimeter is sensitive to changes in air pressure it must be recalibrated whenever a point of known elevation (summits, saddles, stream-trail intersections, survey monuments, and so forth) is reached. This is especially important when weather fronts are moving rapidly through the area.

(5) The altimeter may expand or contract because of changes in temperature. This can result in faulty elevation readings. Although some altimeters are temperature-compensated, rapid ascents or descents sometime overcome the adjustment, causing them to give poor readings.

(6) Keep the altimeter at a constant temperature. This is best accomplished by storing the altimeter (analog) in a pocket or on a cord around the neck, or on the wrist under the parka and hand gear (digital).

(7) Even though altimeters can be precise they are affected by both pressure and temperature changes and should be monitored carefully. The soldier should become familiar with the specific altimeter he employs and understand its capabilities and limitations.

c. **Global Positioning System.** The GPS is a space-based, global, all-weather, continuously available radio positioning navigation system. It is highly accurate in determining position location derived from a satellite constellation system. It can determine the latitude, longitude and elevation of the individual user. Location information is also displayed in military grid coordinates.

(1) The GPS provides precise steering information as well as position locations. The receiver can accept many checkpoints entered in any coordinate system by the user and convert them to the desired coordinate system. The user then calls up the desired checkpoint and the receiver will display direction and distance to the checkpoint. It can also compute travel time to the next checkpoint.

(2) Because the GPS does not need visible landmarks to operate, it can provide position (accurate up to 16 meters) in whiteouts or on featureless terrain. It also does not compound navigational errors as compass use can.

(3) During route planning, after choosing critical checkpoints, start point and objective, enter their coordinates as way points. The best use of the GPS is to verify these as they are reached, as a backup to terrain association and compass navigation.

(4) Since the 21-satellite constellation is not yet complete, coverage may be limited to specific hours of the day in certain areas of the world. The GPS navigational signals are similar to light rays, so anything that blocks light will reduce or block the effectiveness of the signals. The more unobstructed the view of the sky, the better the system performs. Although the GPS can be used in any terrain, it is performs best in more open areas such as the desert.

(5) Because the GPS requires horizon to horizon views for good satellite reception its use can be limited in the mountains. Canyons, deep valleys, saddles, and steep mountainsides are all problematic spots to use for shots. Ridgelines, spurs, summits, open valleys, or plateaus are better.

(6) When using GPS in regions with questionable surveying and mapping products, operational datum of the local maps must be reconciled with the datum used in navigational and targeting systems. Identify the spheroid and datum information on the pertinent map sheets and then check that the GPS receiver has the compatible datum loaded. If not then you must contact the S2 for updated datum or maps. Otherwise, the GPS will show different locations than those on the map.

(7) Extremely cold temperatures (-4 degrees F and below) and high elevations will adversely affect the operation of the GPS, due to the freezing of the batteries and the LCD screen. Battery life and overall performance can be improved by placing the GPS inside the parka or coat.

d. **Navigation Techniques.** The choice of movement technique often determines the route and navigational technique. For navigation, three techniques can be used: dead reckoning, terrain association, or altimeter navigation. The three are not mutually exclusive and are normally used together, with one chosen as the primary technique. The GPS can be used to supplement these techniques, but due to the problems associated with the restricted line of sight in the mountains, it should not be used as the main technique.

(1) ***Dead Reckoning.*** Because of the complex nature of mountainous terrain, dead reckoning is usually of limited value on most movements. The compass is generally employed more to support terrain association and to orient the map, than as a primary navigational aid. The main exception is during periods of limited visibility on featureless terrain. Heavy fog, snowy or whiteout conditions on a snowfield, glacier, large plateau or valley floor all would call for dead reckoning as a primary navigational technique.

(2) ***Terrain Association.*** The standard terrain association techniques all apply. Handrails, checkpoints, catching features, navigational corridors, boxing-in areas, and attack points are all used. When a small objective lies near or on an easily identifiable feature, that feature becomes an expanded objective. This simplifies the navigational problem by giving a large feature to navigate to first. The altimeter may finalize the search for the objective by identification through elevation. Rough compass headings are used to establish a general direction to the next checkpoint; used when the checkpoint headed toward is a linear feature, and not a precise point. The shape, orientation, size,

elevation, slope (SOSES) strategy is especially valuable in mountain terrain association and should be practiced extensively (FM 3-25.26).

(a) After extensive study of the map and all available sources of information it helps to create a mental image of the route. This will enable the navigator to make the terrain work in his favor. Avoid brush for speed and ease of movement; the military crest of spurs and ridgelines generally provides the best route while providing terrain masking effects. When clear cut, burned-over, or large avalanche slide areas are encountered, it may be necessary to box or contour around them as they may be full of slash or brushy second-growth small trees. Old-growth forest provides the easiest travel.

(b) The following situations will result in objects appearing closer than they actually are:

- When most of the object is visible and offers a clear outline.
- When you are looking across a partially cleared depression.
- When looking down a straight, open road or track.
- When looking over a smooth, uniform surface, such as snow, water, or desert.
- When the light is bright and the sun is shining from behind the observer.
- When the object is in sharp contrast to the background.
- When seen in the clear air of high altitude.
- When looking down from high ground to low ground.

(c) The following situations will result in objects appearing farther away than they actually are:

- When only part of the object is seen or it is small in relation to its surroundings.
- When you are looking across an exposed depression.
- When looking up from low ground to high ground.
- When your vision is narrowly confined.
- When the light is poor, such as dawn, dusk, or low visibility weather; or when the sun is in your eyes, but not behind the object being viewed.
- When the object blends into the background.

(2) ***Altimeter Navigation.*** Altimeters provide assistance to the navigator in several ways. They aid in orientation, in computing rates of ascent or descent, in resection, and in weather prediction.

(a) When moving along any linear feature such as a ridge, watercourse, or trail which is shown on the map, check the altimeter. The point where the indicated elevation contour crosses that feature is your location.

(b) The navigator frequently finds it necessary to determine his position through the use of resection. A modified resection can be performed by shooting an azimuth to a known, clearly visible summit or similar feature and then plotting the back azimuth on the map. By determining your present elevation and finding where that particular contour crosses the back azimuth you should locate your position. This can be difficult when in low ground, as mountain summits can rarely be clearly seen from valley floors. In addition, most mountaintops are so large that there is usually no specific point to shoot at. In this case, the soldier should take multiple azimuths to known features. If he is located on a good linear feature he will have a decent idea of where he is. The altimeter can be used to verify elevation and establish a notional linear feature a contour line. The point where the resecting back azimuths cross the contour line is the navigator's location.

(c) Using the altimeter to calculate rates of ascent can help in sound decision-making. Rates of travel, along with weather conditions, light conditions (time of day), and the physical condition of the team, are all key variables that can influence the success or failure of the mission.

(d) Altimeters can be used as barometers to assist in weather prediction.

e. **Approach Observations.** Watch the mountain during the approach march, studying it for climbing routes. Distant views can reveal large-scale patterns of ridges, cliffs, snowfields and glaciers. General angles of the large rock masses can be seen from afar.

(1) Closer viewing displays these patterns and angles on a smaller scale. Fault lines, gross bedding planes of rock, cliff bands, and crevasse zones become visible. Snowy or vegetated ledge systems appear. Weaknesses in the mountain walls, such as couloirs or gullies, may present themselves.

(2) Most of these features repeat themselves at increasingly finer levels, as they are generally derived from the overall structure of the particular mountain group. A basic knowledge of mountain geology, combined with the specific geological background of the operational area, pays off in more efficient travel.

f. **Natural Indicators of Direction in the Northern Hemisphere.** Southern slopes are sunnier and drier than northern slopes, with sparser or different types of vegetation. Northern slopes can be snowier and, because of more intense glaciation in past ages, are often steeper.

Note: Opposite rules apply in the Southern Hemisphere.

g. **Winter Route Selection.** The following must be considered when selecting a route in the winter.

(1) Conduct a thorough map reconnaissance considering the weather, individual ski abilities, avalanche danger, vegetation, water features, terrain relief, and the size of the unit.

(2) Weather conditions will affect the chosen route. During calm weather, your rate of movement will be significantly faster than during periods of inclement weather.

(3) Individual ski abilities will affect your rate of movement, constrain your choice of terrain, and impact on your route choices.

(4) Avalanche danger zones must be identified by map review and data gathered during route planning. During movement, snow pits, shovel tests, and ski shear tests must be conducted prior to crossing an avalanche danger zone. Bottom line: avoid avalanche danger areas. If you must cross one, cross above the starting zone or below the run-out zone.

(5) Vegetation can work for you or against you. Thickly forested areas usually have a deep snow pack. For weaker skiers, forested areas are full of potentially dangerous obstacles. On slopes with an angle of 30 to 45 degrees that are sparsely vegetated an avalanche danger is still present. If the weather turns bad, forested areas provide welcome relief from wind and blowing snow.

(6) Water features provide valuable navigation aids. Under deep snow pack small creeks and ponds may be hard to locate. Large frozen lakes and rivers can provide excellent means of increasing your rate of march.

(7) During ski movements, efficient use of terrain will greatly improve morale and reduce fatigue. While traveling in mountainous terrain, do not needlessly give up elevation gained. Maintain a steady climb rate and avoid over exertion. Avoid north, east, and south facing slopes when the avalanche danger is high. Avoid cornices and be aware of their probable and improbable fracture lines. Weather and tactical situation permitting, travel on the windward side of ridgelines. If weak skiers are in the group, stay away from restrictive terrain with sheer drop-offs. When touring use climbing skins to maintain control and lessen lost time per hour due to individuals falling.

(8) The following are additional hints for navigation in snowy conditions:

- Keep the compass warm.
- If no terrain features exist for steering marks, use your back azimuth and tracks to maintain course.
- Limit steering marks to shorter distances since visibility can change quickly.
- Never take azimuths near metallic objects. Hold the compass far enough from your weapon, ice ax, and so on to get accurate readings.
- Make frequent compass checks.
- Preset azimuths on your compass.
- Use a steady, unshifting wind to aid you in maintaining course.

h. **Problems.** The following conditions and characteristics of cold weather and mountainous regions make accurate navigation difficult.

(1) In winter, short hours of daylight, fog, snowfall, blizzards, whiteouts, and drifting snow, especially above tree line, drastically limit visibility. At times, an overcast sky and snow-covered terrain create a phenomenon called flat light, which makes recognition of irregularities in the terrain extremely difficult.

(2) Heavy snow may completely cover existing tracks, trails, outlines of small lakes and similar landmarks. Because the appearance of the terrain is quite different in winter from that in summer, particular attention must be paid to identifying landmarks, both on the ground and from aerial photographs.

(3) Magnetic disturbances, caused by large ore deposits, are frequently encountered and make magnetic compass readings difficult and sometimes unreliable.

(4) Handling maps, compasses, and other navigation instruments in low temperatures with bare hands is difficult. Removing hand wear may only be possible for short periods.

(5) Keeping count of pace is extremely difficult in winter and mountain environments. Thick vegetation and rough, steep slopes hamper attempts at accurate pace counts. The most reliable method is the use of a 50-meter long piece of field wire or rope.

8-5. ROUTE PLANNING

Proper route planning can make the difference between success and failure on long mountain movements. Careful map reconnaissance, knowledge of the enemy situation, terrain analysis of the operational area, and an accurate assessment of the unit's capabilities are all key parts of the planning process.

a. **Map Reconnaissance.** Topographic maps provide the primary source of information concerning the area of operations. A 1:25,000 map depicts greater detail than a 1:50,000 map and should be used whenever possible. Because examination of the micro-terrain is so important for mountain operations, even larger scale maps are desirable. Civilian 1:12,000 maps can be used if available. Aerial, oblique angle, photographs give details not always shown on maps (crags and overhangs). Sketch maps supplement other sources of information but should not be relied on for accuracy since they are seldom drawn to scale. Along with sketch maps, verbal descriptions, documented information gathered from units previously in the area, or published sources such as alpine journals or climbing guides may help. Forest service and logging and mining company maps provide additional information, often showing the most recent changes to logging trails and mining access roads. Standard military topographic maps are generally accurate graphic depictions of the operational area.

(1) When conducting a map reconnaissance, pay close attention to the marginal information. Mountain-specific terrain features may be directly addressed in the legend. In addition, such facilities as ski lifts, cable and tramways are often found. Check the datum descriptor (for foreign maps) to ensure compatibility with entered datum in GPS units. Along with the standard topographic map color scheme, there are some commonly seen applications for mountainous terrain. White with blue contours indicates glaciers or permanent snowfields. The outline of the snow or ice is shown by dashed blue lines while their contour lines are solid blue. High ice cliffs which are equal to or exceed the contour interval will be shown. Low ice cliffs and ice caves may be indicated if they provide local landmarks. Brown contour lines on white mean dry areas without significant forest cover. Areas above tree line, clear cuts, rock or avalanche slide paths and meadows are all possible. Study the surrounding terrain and the legend for other clues. An important point to remember is that thick brush in small gullies and streambeds may not be depicted by green, but should still be expected.

(2) Obstacles, such as rivers and gorges, will require technical equipment to cross if bridges are not present. Fords and river crossing sites should be identified. Due to the potential for hazardous weather conditions, potential bivouac sites are noted on the map. Ruins, barns, sheds and terrain-protected hollows are all possible bivouac sites. Danger areas in the mountains; isolated farms and hamlets, bridges, roads, trails, and large open areas above tree line, are factored in, and plans made to avoid them. Use of terrain-masking becomes essential because of the extended visibility offered by enemy observation points on the dominant high ground.

(3) Helicopter support, weather permitting, requires identification of tentative landing zones for insertions, extractions, resupply and medevac. The confined nature of mountain travel means that crucial passes become significant chokepoints and planners should designate overwatches/surveillance positions beforehand. Alternate routes should be chosen with weather imposed obstacles in mind: spring flood or afternoon snowmelt turns small streams into turbulent, impassable torrents. Avalanche danger prohibits travel on certain slopes or valley floors.

b. **Enemy Situation.** Route selection should only be done after reviewing all available information about the friendly and enemy situation.—Is the enemy force on his own ground? Are they accustomed to the terrain and the weather? Are they trained mountain troops with specialized equipment?—Only after answering these and other questions can an effective route plan be completed. If the enemy force is better prepared to maneuver in the mountains, they have a marked advantage, and route selection must be scrutinized.

c. **Analysis of the Operational Area.** Not all mountainous terrain is created equal and not all movement plans have the same expectation of success. Planners must undertake a thorough analysis of the general terrain to be crossed, including the geology, mountain structure and forms, and ground cover.

(1) Heavily glaciated granite mountains pose different problems than does river-carved terrain. The U-shaped valley bulldozed out by a glacier forces maneuver elements down to the valley floor or up to the ridge tops, while the water-cut V-shape of river valleys allows movement throughout the compartment.

(2) Routes through granite rock (long cracks, good friction; use of pitons, chocks and camming units) will call for different equipment and technique than that used for steep limestone (pockets, smooth rock; bolts, camming units).

(3) Operations above tree line in temperate climates or in the brushy zone of arid mountains means that material for suspension traverse A-frames must be packed. The thick brush and krummholtz mats of subalpine zones and temperate forested mountains can create obstacles that must be bypassed.

(4) Heavy spruce/fir tangles slow progress to a crawl, therefore planners should ensure routes do not blindly traverse these zones.

d. **Unit Assessment.** When assessing unit movement capabilities the key indices are training and conditioning levels. Soldiers who have received basic military mountaineer training, who know how to move through rough terrain, and who have been hardened with training hikes through the mountains, will perform better than troops without this background.

e. **Time-Distance Formulas.** Computing march rates in the mountains is extremely difficult, especially when there is snow cover. The following rates are listed as a guide (Table 8-1). Rates are given for movement over flat or gently rolling terrain for individuals carrying a rifle and loaded rucksack.

	UNBROKEN TRAIL	BROKEN TRAIL
On foot (no snow cover)	2 to 3 kph (cross-country)	3 to 4 kph (trail walking)
On foot (less than 1 foot of snow)	1.6 to 3.2 kph	2 to 3.2 kph
On foot (more than 1 foot of snow)	.4 to 1.2 kph	2 to 3.2 kph
Snowshoeing	1.6 to 3.2 kph	3.2 to 4 kph
Skiing	1. to 5.6 kph	4.8 to 5.6 kph
Skijoring	N/A	3 to 24 kph

Table 8-1. Time-distance formulas.

(1) March distances in mountainous terrain are often measured in time rather than distance units. In order to do this, first measure the map distance. This distance plus 1/3 is a good estimate of actual ground distance. Add one hour for each 1,000 feet of ascent or 2,000 feet of descent to the time required for marching a map distance.

(2) As Table 8-1 indicates, snow cover will significantly affect rates of march. Since snow can be expected in the mountains most months of the year, units should have some experience at basic snow travel.

(3) Individual loads also affect march rates. Combined soldier loads that exceed 50 pounds per man can be expected to slow movement significantly in mountainous terrain. Given the increased weight of extra ammunition for crew-served weapons, basic mountaineering gear, and clothing for mountain travel, it becomes obvious that soldiers will be carrying weights well in excess of that 50-pound limit. Units should conduct cross-country movements in the mountains with the expected rucksack and LCE weights in order to obtain accurate, realistic rates of march.

(4) In the harsh environment of the mountains, helicopter support cannot be relied on. The process of transporting extra equipment and sustainment supplies will result in vastly increased movement times. The heavier loads will exhaust soldiers mentally and physically. Tactical movements, such as patrolling or deliberate assaults, should take this into account.

8-6. ROUTE SELECTION

Many variables affect the selection of the proper route. The following guidelines apply to all situations.

a. **Select a Current Map.** Check the date of the map for an indication of the reliability of the map in depicting vegetation, clearings, roads, and trails accurately. The leader should use all the latest topographic data he can find.

b. **Gather Intelligence Information.** The most important consideration in every leader's mind when plotting a movement is "where is the enemy?" The latest intelligence reports are essential. Additionally, weather reports, snow condition reports, avalanche probability, aerial photos, and any past or recent history of operations in the area may be of help.

c. **Select a Route.** Identify the starting point and determine the movement objective. Plot start and end points. Carefully scrutinize the area in between and begin to select the route. Consider the following:

(1) ***Trafficability.*** This includes degree of slopes, vegetation, width of trails, snow depth, avalanche probability, and the likelihood of crevasses.

(2) ***Time-Distance Formula.*** Time allotted and distance to be covered must be considered.

(3) ***Required Equipment.*** Carry enough equipment to move along the route and to survive if an extended stay becomes necessary. Do not plan a route beyond the means of your equipment.

(4) ***Location of Enemy.*** Plan a route that allows maximum use of the masking effect of the terrain. Avoid danger areas or areas of recent enemy activity unless required by the mission. Use vegetation to mask your movement if possible (especially coniferous forests). Avoid silhouetting on ridgelines.

(5) ***Communications.*** Communications will be severely limited in the mountains. Dead spaces or communications holes are common. Use all available information and plan accordingly.

(6) ***Conditions/Capabilities of Unit.*** The unit must be able to negotiate the route chosen. Take into consideration their present health, as well as their training level when selecting your intended route.

(7) ***Checkpoints/Control Points.*** When plotting a route on the map, utilize prominent terrain features on either side of the route as checkpoints. Ensure that when you select

your checkpoints they are visually significant (elevation) and that they are easily identifiable. Avoid the use of manmade features as checkpoints due to their unreliability and lack of permanence. Select features that are unique to the area.

CHAPTER 9

MOUNTAIN STREAM CROSSINGS

Operations conducted in mountainous terrain may often require the crossing of swift flowing rivers or streams. Such crossings should not be taken lightly. The force of the flowing water may be extremely great and is most often underestimated. All rivers and streams are obstacles to movement. They should be treated as danger areas and avoided whenever possible. When rivers or streams must be crossed, there are a variety of techniques the small-unit leader may choose from, depending upon the type of stream, its width, speed of the current, and depth of the water.

There are limits on the safe use of these techniques. Not all mountain rivers or streams will be fordable with these techniques. If a water obstacle is too wide, swift, or deep, an alternate route should be used, or the crossing will require major bridging by engineers. It may require the use of rafts or boats. Reconnaissance of questionable crossing sites is essential. This chapter covers the techniques for crossing mountain streams that have a depth generally not exceeding waist deep.

9-1. RECONNAISSANCE

Reconnaissance of the route (map, photo, and or aerial) may not always reveal that a water obstacle exists. In a swamp, for example, unfordable sloughs may not show on the map, and they may be concealed from aerial observation by a canopy of vegetation. Whenever it is possible that a unit will be required to cross a water obstacle, its commander must plan some type of crossing capability.

a. Site selection is extremely important once you determine that you must make a crossing (Figure 9-1). Look for a high place from which you can get a good view of the obstacle and possible crossing sites. A distant view, perhaps from a ridge, is sometimes better than a hundred close views from a riverbank. Site selection must be made before the arrival of the main body.

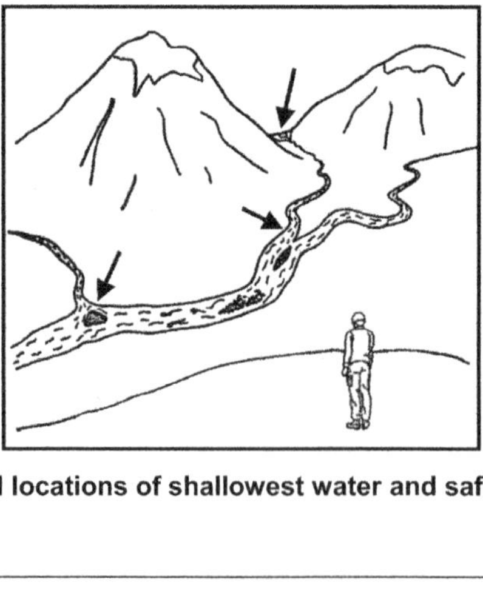

Figure 9-1. Normal locations of shallowest water and safest crossing sites.

b. A dry crossing on fallen timber or log jams is preferable to attempting a wet crossing. Depending upon the time of year, the closer you are to the source, or headwaters, the better your chances are of finding a natural snow or ice bridge for crossing. If a dry crossing is unavailable, the following considerations should be made:

(1) The time of day the crossing can be an important factor. Although early morning is generally best because the water level is normally lower during this period, recent weather is a big factor; there may have been heavy rain in the last eight hours. As glaciers, snow, or ice melt during the day, the rivers rise, reaching their maximum height between mid afternoon and late evening, depending on the distance from the source. Crossings, if made during the early morning, will also allow clothing to dry more quickly during the heat of the day.

(2) A crossing point should normally be chosen at the widest, and thus shallowest, point of the river or stream. Sharp bends in the river should be avoided since the water is likely to be deep and have a strong current on the outside of the bend. Crossings will be easiest on a smooth, firm bottom. Large rocks and boulders provide poor footing and cause a great deal of turbulence in the water.

(3) Many mountain streams, especially those which are fed by glacier run-off, contain sections with numerous channels. It is often easier to select a route through these braided sections rather than trying to cross one main channel. A drawback to crossing these braided channels, however, is the greater distance to the far bank may increase exposure time and often the sand and gravel bars between the channels will offer little cover or concealment, if any.

(4) The crossing site should have low enough banks on the near and far side to allow a man carrying equipment to enter and exit the stream with relative ease. If a handline or rope bridge is to be constructed, the crossing site should have suitable anchors on the near and far bank, along with safe loading and unloading areas. Natural anchors are not a necessity, however the time required to find a site with solid natural anchors will probably be less than the time required to construct artificial anchors. In some areas, above the tree line for example, artificial anchors may have to be constructed. Deadman anchors buried in the ground, or under a large pile of boulders work well.

(5) Log jams and other large obstructions present their own hazards. Logs floating downstream will generally get hung up in shallower sections creating the jam. Once a log jam is formed, however, the water forced to flow around it will erode the stream bottom. Eventually deep drop-offs or holes may develop, especially around the sides and off the downstream end of the log jam. A log jam that totally bridges a section of the stream may be the best way to cross. A wet crossing in the vicinity of a log jam should be performed a good distance below or above it. Some things to consider when crossing near log jams are:

- Cross well to the downstream side when possible.
- Keep a sharp lookout for floating timber that could knock you off your feet.
- If you must cross on the upstream side, stay well upstream from the log jam. If a person is swept off his feet and caught in the debris of the jam, he could easily drown. A handline will greatly increase safety here.

(6) When possible, select a crossing site that has enough natural protection on the near and far banks so that security teams may be placed out and enough cover and concealment is available for the size of the element making the crossing. When cover and

concealment is minimal, as in the higher alpine zones, the crossings must be conducted as efficiently as possible to minimize exposure to enemy observation.

9-2. PREPARATION OF TROOPS AND EQUIPMENT

Prepare men and equipment for a crossing as far in advance as feasible. Final preparation should be completed in a security perimeter on the near side just before crossing. Preparation includes the following.

a. Waterproof water-sensitive items. Wrap radios, binoculars, SOI, papers, maps and any extra clothing in waterproof bags (trash bags also work well), if available. These bags also provide additional buoyancy in case of a fall.

b. Trousers are unbloused and shirts are pulled out of the trousers. All pockets are buttoned. This allows water to escape through the clothing. Otherwise the clothing would fill up and retain water, which would weigh the body down. This is especially critical if an individual must swim to shore. Depending on the circumstances of the crossing (for example, tactical situation, temperature of the air and water), the crossing can be made in minimal clothing so that dry clothing is available after the crossing. Boots should be worn to protect feet from rocks; however, socks and inner soles should be removed. On the far side, the boots can be drained and dry socks replaced.

c. Load-carrying equipment harness and load-bearing vest (LBV) is unbuckled and worn loosely. It is extremely difficult to remove a buckled harness in the water in an emergency.

d. Helmets are normally removed and placed in the rucksack in slow moving streams with sandy or gravel bottoms. If you have to resort to swimming it is easier done without the helmet. However, when crossing swift flowing streams, especially those with large rocks and other debris, the risk of head injury if a person slips is high. In this case the helmet should be worn with the chinstrap fastened.

e. The rucksack should be worn well up on the shoulders and snug enough so it does not flop around and cause you to lose your balance. The waist strap MUST be unbuckled so you can get rid of the pack quickly if you are swept off your feet and have to resort to swimming. If a pack has a chest strap it must also be unbuckled. Secure everything well within your pack. It is easier to find one large pack than to find several smaller items.

f. Individual weapons should be attached to the pack or slung over the shoulder.

9-3. INDIVIDUAL CROSSINGS

Whenever possible, and when the degree of experience permits, streams should be forded individually for a speedier crossing. The average soldier should be able to cross most streams with mild to moderate currents and water depths of not much more than knee deep using proper techniques.

a. The individual should generally face upstream and slightly sideways, leaning slightly into the current to help maintain balance. At times, he may choose to face more sideways as this will reduce the surface area of the body against the current, thus reducing the current's overall force on the individual.

b. The feet should be shuffled along the bottom rather than lifted, with the downstream foot normally in the lead. He should take short, deliberate steps. Lunging steps and crossing the feet result in a momentary loss of balance and greatly increase the chance of a slip.

c. The individual should normally cross at a slight downstream angle so as not to fight the current. There is normally less chance of a slip when stepping off with the current as opposed to stepping off against the current.

d. The individual must constantly feel for obstacles, holes and drop-offs with the lead foot and adjust his route accordingly. If an obstacle is encountered, the feet should be placed on the upstream side of it where the turbulence is less severe and the water normally shallower.

e. To increase balance, and if available, a long ice ax, sturdy tree limb, or other staff can be used to give the individual a third point of contact (Figure 9-2). The staff should be used on the upstream side of the individual and slightly leaned upon for support. The staff should be moved first, then the feet shuffled forward to it. This allows two points of contact to be maintained with the streambed at all times. The individual still moves at a downstream angle with the downstream foot in the lead.

Figure 9-2. Individual crossing with staff.

9-4. TEAM CROSSING

When the water level begins to reach thigh deep, or anytime the current is too swift for personnel to safely perform an individual crossing, a team crossing may be used. For chain crossing, two or more individuals cross arms with each other and lock their hands in front of themselves (Figure 9-3). The line formed faces the far bank. The largest individual should be on the upstream end of the line to break the current for the group. The line formed will then move across the stream using the same principles as for individual crossings, but with the added support of each other. The line should cross parallel to the direction of the current. The team still moves at a slight downstream angle, stepping off with the downstream foot in the lead.

Figure 9-3. Chain method for stream crossing.

9-5. ROPE INSTALLATIONS

When the water level begins to reach waist deep or the current is too swift for even a team crossing, the chosen site must be closely examined. The stream at this point may be impassable. Many times though, a crossing site which may be unsafe for individual or team crossings can be made safe with the installation of a handline or rope bridge. Crossing on a handline will still require each individual to enter the water and get wet. If a one-rope bridge can be constructed, it may require only a couple of individuals to enter the water. Deciding whether to install a handline or a rope bridge will depend on the anchors available, height of the anchors above the water, and the distance from the near and far anchors. The maximum distance a one-rope bridge is capable of spanning is approximately 1/2 to 2/3 the length of the rope in use.

a. **Establishing the Far Anchor.** Whether a handline or rope bridge is to be installed, someone must cross the stream with one end of the rope and anchor it on the far side. This duty should be performed by the most capable and strongest swimmer in the party. The swimmer should be belayed across for his own safety. The belay position should be placed as far above the crossing as possible. In the event that the current is too strong for the individual, he will pendulum back to the near bank. Rescuers should be poised on the near bank at points where the individual will pendulum back, should he fail to reach the far bank. The initial crossing site should be free of obstacles that would snag the rope and prevent the pendulum back to the bank for an easy recovery.

(1) The individual may attach the belay rope to his seat harness or a swami belt with a carabiner. He should NEVER tie directly into the rope when being belayed for a stream crossing. If the swimmer should be swept away and become tangled, he must be able to release himself quickly from the rope and swim to shore as best he can. The individual may also choose to tie a fixed loop into the end of the belay rope and hang on to it, where he can immediately release it in an emergency.

(2) Anytime a crossing site must be used where the swimmer may encounter problems getting to the far bank, he should have on a life vest or other personal flotation

device (PFD). If the swimmer must release the rope at any time, he will have to rely on his own water survival skills and swimming ability to get to shore. A PFD will greatly increase his own personal safety. A PFD may also be used by the last man across, as he will release the rope from the anchor and be belayed across as the first man

b. **Installation of a Handline.** If it is possible to use a rope high enough above the water to enable soldiers to perform a dry crossing, then a rope bridge should be installed as such. If this is impossible, and the rope must be installed to assist in a wet crossing, then it should be installed as a handline (Figure 9-4).

(1) The far anchor should be downstream from the near anchor so that the rope will run at an angle downstream from the near anchor, approximately thirty to forty-five degrees, rather than straight across the stream. Here again, it is easier to move with the current as opposed to directly across or against it.

(2) The rope may be anchored immediately on the far bank, pulled tight, and anchored on the near bank, or it may be installed with a transport tightening system if a tighter rope is required.

(3) Crossing will always be performed on the downstream side of the handline, shuffling the feet with the downstream foot in the lead.

(4) A second climbing rope is used as a belay (Figure 9-5). One end of the belay rope will be on the near bank and the other end on the far bank. It should be sent across with the strong-swimmer. An appropriate knot is tied into the middle of the belay rope to form two fixed loops with each loop being approximately 6 inches long. One loop is connected to the handline with carabiner(s) and the individual crossing connects one loop to himself. The loops are short enough so the individual is always within arms reach of the handline should he slip and let go. The individuals are belayed from both the near and far banks. If a mishap should occur the individual can be retrieved from either shore, whichever appears easiest. The belay on the opposite shore can be released allowing the individual to pendulum to the bank. It is important that the belay rope NOT be anchored or tied to the belayer so that it may be quickly released if necessary.

Figure 9-4. Stream crossing using a handline.

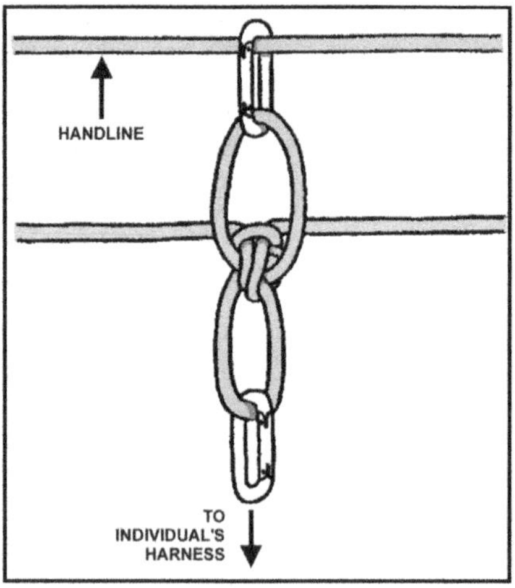

Figure 9-5. Belay rope for crossing using a handline.

(5) Under most circumstances, the handline should be crossed one person at a time. This keeps rope stretch and load on the anchors to a minimum.

(6) Rucksacks can be either carried on the back the same way as for individual crossings, or they can be attached to the handline and pulled along behind the individual.

(7) If a large amount of equipment must be moved across the stream, especially heavier weapons, such as mortars, recoilless rifles, and so on, then a site should be selected to install a rope bridge.

9-6. SAFETY

River and stream crossings present one of the most hazardous situations faced by the military mountaineer. The following safety procedures are minimum guidelines that should be followed when conducting a river or stream crossing.

a. All weak and nonswimmers should be identified before a crossing so that stronger swimmers may give assistance in crossing.

b. Not every river or stream can be crossed safely. It is always possible to cross at a different time or place, use a different technique, or choose another route.

c. The technique used is directly dependent upon water depth, speed of the current, stream bottom configuration, width of the stream, and individual experience.

d. The safest methods of crossing are always with the use of a handline or one-rope bridge.

e. If the installation of a handline or rope bridge becomes too difficult at a given crossing site, then that site should be considered too hazardous and another site selected.

f. A lookout should be posted 50 to 100 meters upstream to watch for any obstacles that may be carried downstream and interfere with the crossing.

g. When conducting individual crossings (those without a handline or rope bridge), lifeguards should be posted downstream with poles or ropes prepared to throw, for assistance or rescue.

h. When the unit knows a rope installation will be required for crossing, at least two life vests or other PFDs should be on hand to provide additional safety for the strong swimmer who must establish the far anchor, and the last man across who retrieves the system.

9-7. SWIMMING

There are times when you might be alone and have no choice but to swim across, or there may be a time that you find yourself suddenly plunged into a swift river or rapids. In either case, the following techniques could save your life.

a. Immediately jettison any equipment or clothing that restricts movement.

b. Do not try to fight the current. Maneuver towards shore in a position with the feet downstream, facing downstream, and fanning the hands alongside the body to add buoyancy and to fend off submerged rocks. Use the feet to protect the rest of the body and to fend off submerged rocks.

c. Keep the head above water to observe for obstacles and attempt to maneuver away from them.

d. Try to avoid backwater eddies and converging currents as they often contain dangerous swirls. Avoid bubbly water under falls as it has little buoyancy. Breath between the wave troughs.

e. If the shore is too difficult to reach, seek out the closest and safest spot, such as a sandbar, to get yourself out of the water as quickly as possible. Hypothermia will set in quickly in colder waters.

CHAPTER 10
MOVEMENT OVER SNOW AND ICE

Movement over snow- and ice-covered slopes presents its own unique problems. Movement on steeper slopes requires an ice ax, crampons, and the necessary training for this equipment. Personnel will also have to learn how to place solid anchors in snow and ice to protect themselves during these movements if roped. Snow-covered glaciers present crevasse fall hazards even when the slope is relatively flat, requiring personnel to learn unique glacier travel and crevasse rescue techniques.

All the principles of rock climbing, anchor placement, belays, and rope usage discussed throughout the previous chapters apply to snow and ice climbing as well. This chapter will focus on the additional skills and techniques required to move safely through snow-covered mountains and over glaciated terrain.

10-1. MOVEMENT OVER SNOW

The military mountaineer must be equally adept on both snow and ice due to route necessity and rapidly changing conditions. On steep slopes in deep snow, the climber may climb straight up facing the slope. The ice ax shaft, driven directly into the snow, provides a quick and effective self-belay in case of a slip—the deeper the shaft penetrates the snow, the better the anchor (Figure 10-1). It is usually best, however, to climb snow-covered slopes in a traversing fashion in order to conserve energy, unless there is significant avalanche danger.

Figure 10-1. Self-belay on snow.

a. The progression from walking on flat terrain to moving on steep terrain is the same as for moving over snow-free terrain. If the snow is packed the sole of the boot will generally hold by kicking steps, even on steep slopes. Where it is difficult to make an effective step with the boot, a cut made with the adze of the ice ax creates an effective step. In these situations crampons should be used for faster and easier movement.

b. When descending on snow, one can usually come straight downhill, even on steep terrain. Movement downhill should be slow and deliberate with the climber using an even pace. The heels should be kicked vigorously into the snow. The body may be kept erect with the aid of an ice ax, which may be jammed into the snow at each step for additional safety. Here again, crampons or step cutting may be necessary. A technique known as glissading may also be used as an easy method of descent and is covered in detail later in this chapter.

10-2. MOVEMENT OVER ICE

Ice is found in many areas of mountains when snow is present, and during the summer months also where perennial snowpack exists. Many times an ice area will be downslope of a snowfield and sometimes the ice pack itself will be lightly covered with snow. Even if using an ice ax and or crampons, movement will still be difficult without proper training.

10-3. USE OF ICE AX AND CRAMPONS

Movement over snow and ice is almost impossible without an ice ax and or crampons.

a. **Ice Ax.** When walking on snow or ice, the ice ax can be used as a third point of contact. When the terrain steepens, there are a number of ways to use the ice ax for snow or ice climbing. Some positions are more effective than others, depending on the intended result. You may find other ways to hold and use the ax, as long the security remains in effect.

(1) ***Cane Position.*** The ice ax can be used on gentle slopes as a walking stick or cane (Figure 10-2). The ax is held by the head with the spike down and the pick facing to the rear in preparation for self-arrest. When moving up or down gentle slopes the ice ax is placed in front as the third point of contact, and the climber moves toward it. When traversing, the ax is held on the uphill side, in preparation for a self-arrest.

Figure 10-2. Using the ice ax in the cane position.

(2) ***Cross Body Position or Port Arms Position.*** On steeper slopes the ax can be used in the port arms position, or cross body position (Figure 10-3). It is carried across the chest, upslope hand on the shaft, spike towards the slope. The head of the ax is held away from the slope with the pick to the rear in preparation for self-arrest. Ensure the leash is connected to the upslope hand, which allows the ax to be used in the hammer position on the upslope side of the climber. The spike, in this case, is used as an aid for maintaining balance.

Figure 10-3. Ice ax in the cross body or port arms position.

(3) ***Anchor Position***. As the slope continues to steepen, the ax may be used in the anchor position (Figure 10-4). The head is held in the upslope hand and the pick is driven into the slope. The spike is held in the downhill hand and pulled slightly away from the slope to increase the "bite" of the pick into the ice. If the climber is wearing a harness, the pick can be deeply inserted in the ice or hard snow and the ax leash could be connected to the tie-in point on the harness for an anchor (ensure the ax is placed for the intended direction of pull).

Figure 10-4. Ice ax in the anchor position.

(4) ***Push-Hold Position.*** Another variation on steep slopes is the push-hold position (Figure 10-5). The hand is placed on the shaft of the ax just below the head with the pick forward. The pick is driven into the slope at shoulder height. The hand is then placed on the top of the ax head for use as a handhold.

Figure 10-5. Ice ax in the push-hold position.

(5) ***Dagger Position.*** The dagger position is used on steep slopes to place a handhold above shoulder height (Figure 10-6). The hand grasps the head of the ax with the pick forward and the shaft hanging down. The ax is driven into the surface in a stabbing action. The hand is then placed on the ax head for use as a handhold.

Figure 10-6. Ice ax in the dagger position.

(6) ***Hammer Position.*** The hammer position will set the pick deepest in any snow or ice condition (Figure 10-7, page 10-6). The ax is used like a hammer with the pick being driven into the slope. On vertical or near-vertical sections, two axes used in the hammer position will often be required.

Figure 10-7. Ice ax in the hammer position.

b. **Crampons.** Walking in crampons is not complicated but it does present difficulties. When walking in crampons, the same principles are used as in mountain walking, except that when a leg is advanced it is swung in a slight arc around the fixed foot to avoid locking the crampons or catching them in clothing or flesh. The trousers should be bloused to prevent catching on crampons. All straps should be secured to prevent stepping on them and, potentially, causing a fall. The buckles should be located on the outside of each foot when the crampons are secured to prevent snagging. Remember, when the crampon snags on the pants or boots, a tear or cut usually results, and sometimes involves the skin on your leg and or a serious fall.

(1) Two methods of ascent are used on slopes: traversing and straight up.

(a) A traverse on ice or snow looks much like any mountain walking traverse, except that the ankles are rolled so that the crampons are placed flat on the surface (Figure 10-8). On snow the points penetrate easily; on ice the foot must be pressed or stamped firmly to obtain maximum penetration. At the turning points of a traverse, direction is changed with the uphill foot as in mountain walking.

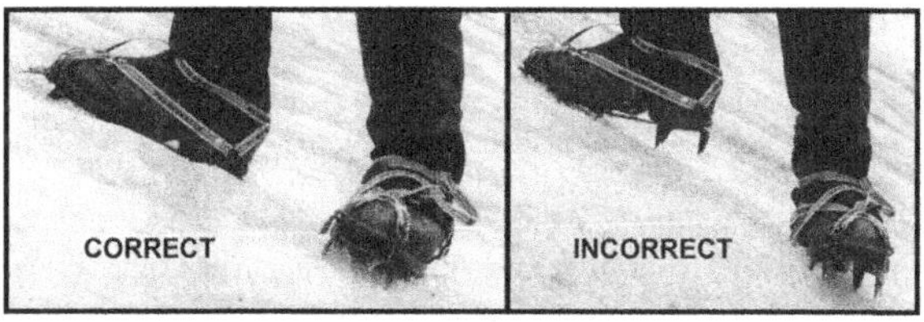

Figure 10-8. Correct and incorrect crampon technique.

(b) A straight up method is for relatively short pitches, since it is more tiring than a traverse. The climber faces directly up the slope and walks straight uphill. As the slope steepens, the herringbone step is used to maintain the flatfoot technique. For short steep pitches, the climber may also face downslope, squatting so the legs almost form a

90-degree angle at the knees, driving the spike of the ice ax into the slope at hip level, and then moving the feet up to the ax. By repeating these steps, the ax and crampon combination can be used to climb short, steep pitches without resorting to step cutting. This method can be tiring. The technique is similar to the crab position used for climbing on slab rock and can also be used for short descents.

(2) A technique known as "front-pointing" may be used for moving straight uphill (Figure 10-9). It is especially useful on steep terrain, in combination with the ice ax in the push-hold, dagger, or hammer position. Front-pointing is easiest with the use of more rigid mountain boots and rigid crampons. The technique is similar to doing calf raises on the tips of the toes and is much more tiring than flat-footing.

(a) The technique starts with the feet approximately shoulder width apart. When a step is taken the climber places the front points of the crampons into the ice with the toe of the boot pointing straight into the slope.

(b) When the front points have bitten into the ice the heel of the boot is lowered slightly so that the first set of vertical points can also bite. The body is kept erect, with the weight centered over the feet as in climbing on rock.

Figure 10-9. Front-pointing with crampons.

c. **Vertical Ice.** When a climb on ice reaches the 60- to 70-degree angle, two ice axes may be helpful, and will become necessary as the angle approaches 90 degrees. The same basic climbing techniques described in Chapter 6 should be applied. If leashes of the correct length and fit are attached to both axes, it may be possible to hang completely from the axes while moving the feet.

d. **Descending with Crampons and Ice Ax.** Whenever possible, descend straight down the fall line. As the slope steepens, gradually turn sideways; on steeper slopes, bend at the waist and knees as if sitting, keeping the feet flat to engage all vertical crampon points and keep the weight over the feet as in descending rock slab (Figures 10-10 and 10-11, page 10-8). On steep terrain, assume a cross body or port arms position with the ax, and traverse. The crab position or front-pointing may also be used for descending. Regardless of the technique used, always ensure the points of the crampons are inserted in the snow or ice and take short, deliberate steps to minimize the chance of tripping and falling down the slope.

Figure 10-10. Flat-footing in the crab position.

Figure 10-11. Use of ice ax in descent.

e. **Normal Progression.** The use of the ice ax and crampons follows a simple, logical progression. The techniques can be used in any combination, dictated by the terrain and skill of the individual. A typical progression could be as follows:

(1) ***Crampons.*** Use crampons in the following situations:

- Walking as on flat ground.

- Herringbone step straight up the slope.
- Traverse with feet flat.
- Backing up the slope (crab position).
- Front-pointing.

(2) ***Ice Ax.*** Use the ice ax in these situations:

- Cane position on flat ground.
- Cane position on uphill side as slope steepens.
- Port arms position with spike on uphill side.
- Anchor position with pick on uphill side.
- Push-hold position using front-pointing technique.
- Dagger position using front-pointing technique.
- Hammer position using front-pointing technique.

e. **Climbing Sequence.** Using most of these positions, a single ax can be "climbed" in steps to move upslope on low-angle to near vertical terrain (Figure 10-12). Begin by positioning the feet in a secure stance and placing the ax in the hammer position as high as possible. Slowly and carefully move the feet to higher positions alternately, and move the hand up the ax shaft. Repeat this until the hand is on top of the head of the ax. Remove the ax and place it at a higher position and begin again.

Figure 10-12. Climbing sequence.

f. **Step Cutting.** Step cutting is an extremely valuable technique that is a required skill for any military mountaineer (Figure 10-13). Using cut steps can save valuable time that would be spent in donning crampons for short stretches of ice and can, in some cases, save the weight of the crampons altogether. Steps may also have to be cut by the lead team to enable a unit without proper equipment to negotiate snow- or ice-covered terrain. As units continue to move up areas where steps have been cut they should continue to improve each step. In ascending, steps may be cut straight up the slope, although a traverse will normally be adopted. In descending, a traverse is also the preferred method. When changing direction, a step large enough for both feet and crampons must be made. Once the step is formed, the adze is best used to further shape and clean the step.

(1) ***Snow.*** On slopes of firm snow and soft ice, steps may be cut by swinging the ax in a near-vertical plane, using the inside corner of the adze for cutting. The step should be fashioned so that it slopes slightly inward and is big enough to admit the entire foot. Steps used for resting or for turning must be larger.

(2) ***Ice.*** Hard ice requires that the pick of the ax be used. Begin by directing a line of blows at right angles to the slope to make a fracture line along the base of the intended step. This technique will reduce the chance of an unwanted fracture in the ice breaking out the entire step. Next, chop above the fracture line to fashion the step. When using the pick it should be given an outward jerk as it is placed to prevent it from sticking in the ice.

(3) ***Step Cutting in a Traverse.*** When cutting steps in a traverse, the preferred cutting sequence is to cut one step at an arm's length from the highest step already cut, then cut one between those two. Cutting ahead one step then cutting an intermediate step keeps all of the steps relatively close to one another and maintains a suitable interval that all personnel can use.

(4) ***Handholds.*** If handholds are cut, they should be smaller than footholds, and angled more.

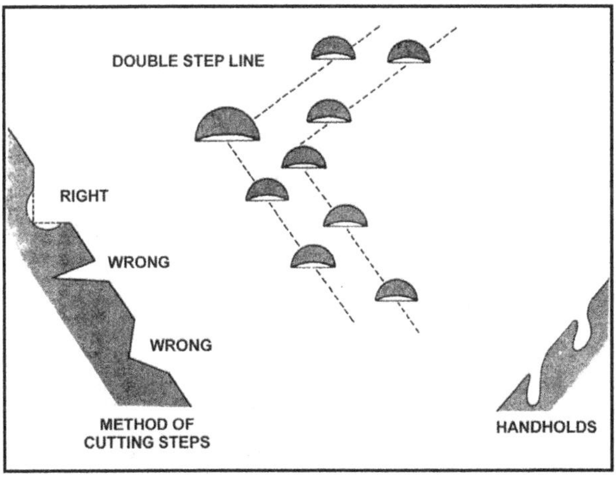

Figure 10-13. Step cutting and handhold cutting.

g. **Self-Arrest.** The large number of climbers injured or killed while climbing on snow and ice can be attributed to two major failings on the part of the climber: climbing unroped, and a lack of knowledge and experience in the techniques necessary to stop, or arrest, a fall (Figure 10-14). A climber should always carry an ice ax when climbing on steep snow or ice; if a fall occurs, he must retain possession and control of his ice ax if he is to successfully arrest the fall. During movement on steep ice, the ax pick will be in the ice solidly before the body is moved, which should prevent a fall of any significance (this is a self belay not a self-arrest).

CAUTION

Self-arrest requires the ax pick to gradually dig in to slow the descent. Self-arrest is difficult on steep ice because the ice ax pick instantly "bites" into the ice, possibly resulting in either arm or shoulder injury, or the ax is deflected immediately upon contact.

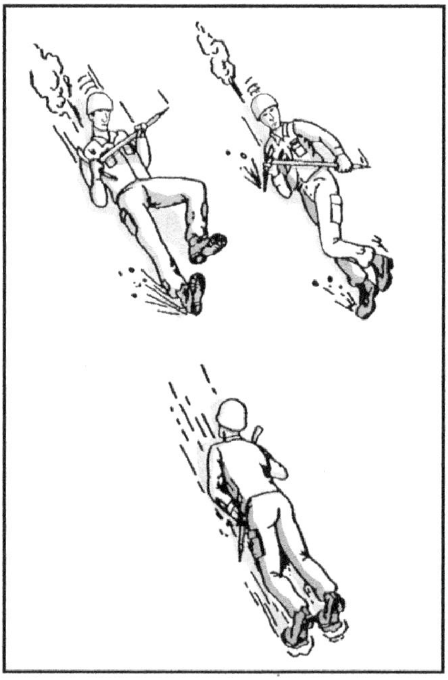

Figure 10-14. Self-arrest technique.

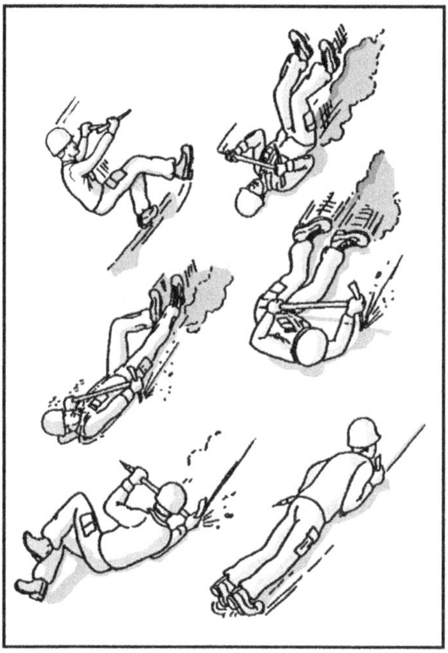

Figure 10-14. Self-arrest technique (continued).

(1) A climber who has fallen may roll or spin; if this happens, the climber must first gain control of his body, whether it is with his ice ax or simply by brute force. Once the roll or spin has been controlled, the climber will find himself in one of four positions.

- Head upslope, stomach on the slope, and feet pointed downslope.
- Head upslope, back to the slope, and feet pointed downslope.
- Head downslope, stomach on the slope, and feet pointed upslope.
- Head downslope, back to the slope, and feet pointed upslope.

(2) To place the body in position to arrest from the four basic fall positions the following must be accomplished.

(a) In the first position, the body is in proper relation to the slope for an arrest.

(b) In the second position, the body must first be rotated from face up to face down on the slope. This is accomplished by rolling the body toward the head of the ax.

(c) In the third position, the pick of the ice ax is placed upslope and used as a pivot to bring the body into proper position.

(d) In the fourth position, the head of the ax must be driven into the snow to the climber's side. This will cause the body to rotate into a head up, stomach down position.

(3) The final position when the arrest of the fall is completed should be with the head upslope, stomach on the slope, with the feet pointed downslope. If crampons are not

worn, the toe of the boots may be dug into the slope to help arrest the fall. The ax is held diagonally across the chest, with the head of the ax by one shoulder and the spike near the opposite hip. One hand grasps the head of the ax, with the pick pointed into the slope, while the other hand is on the shaft near the spike, lifting up on it to prevent the spike from digging into the slope.

Note: If crampons are worn, the feet must be raised to prevent the crampons from digging into the snow or ice too quickly. This could cause the climber to tumble and also, could severely injure his ankles and legs.

(4) When a fall occurs, the climber should immediately grasp the ax with both hands and hold it firmly as described above. Once sufficient control of the body is attained, the climber drives the pick of the ice ax into the slope, increasing the pressure until the fall is arrested. Raising the spike end of the shaft increases the biting action of the pick. It is critical that control of the ice ax be maintained at all times.

10-4. GLISSADING

Glissading is the intentional, controlled, rapid descent, or slide of a mountaineer down a steep slope covered with snow (Figure 10-15, page 10-14). Glissading is similar to skiing, except skis are not used. The same balance and control are necessary, but instead of skis the soles of the feet or the buttocks are used. The only piece of equipment required is the standard ice ax, which serves as the rudder, brake, and guide for the glissade. The two basic methods of glissading are:

a. **Squatting Glissade.** The squatting glissade is accomplished by placing the body in a semi-crouched position with both knees bent and the body weight directly over the feet. The ice ax is grasped with one hand on the head, pick, and adze outboard (away from the body), and the other hand on the shaft. The hand on the shaft grips it firmly in a position that allows control as well as the application of downward pressure on the spike of the ax.

b. **Sitting Glissade.** Using this method the glissader sits on the snow with the legs flat, and the heels and feet raised and pointed downslope. The ice ax is firmly grasped in the same manner as the squatting glissade, with the exception that the hand on the shaft must be locked against the hip for control. The sitting glissade is slower but easier to control than the squatting glissade.

c. **Safety.** A glissade should never be attempted on a slope where the bottom cannot be seen, since drop-offs may exist out of view. Also, a sitting glissade should not be used if the snow cover is thin, as painful injury could result.

Figure 10-15. Glissading techniques.

10-5. SNOW AND ICE ANCHORS

Ice and snow anchors consist of snow pickets, flukes, deadman-type anchors, ice screws, and ice pitons. Deadman anchors can be constructed from snowshoes, skis, backpacks, sleds, or any large items.

a. **Ice Pitons.** The ice piton is used to establish anchor points. The ice piton is not seen in modern ice climbing but may still be available to the military. The standard ice piton is made of tubular steel and is 10 inches in length. Ice pitons installed in pairs are a bombproof anchor; however, ice pitons have no threads for friction to hold them in the ice once placed and are removed easily. Safe use of ice pitons requires placement in pairs. Used singularly, ice pitons are a strong anchor but are easily removed, decreasing the perceived security of the anchor. Follow the instructions below for placing ice pitons in pairs.

(1) Cut a horizontal recess into the ice, and also create a vertical surface (two clean surfaces at right angles to each other).

(2) Drive one piton into the horizontal surface and another into the vertical surface so that the two pitons intersect at the necessary point (Figure 10-16).

(3) Connect the two rings with a single carabiner, ensuring the carabiner is not cross-loaded. Webbing or rope can be used if the rings are turned to the inside of the intersection.

(4) Test the piton pair to ensure it is secure. If it pulls out or appears weak, move to another spot and replace it. The pair of pitons, when placed correctly, are multidirectional.

Figure 10-16. Ice piton pair.

(5) The effective time and or strength for an ice piton placement is limited. The piton will heat from solar radiation, or the ice may crack or soften. Solar radiation can be nearly eliminated by covering the pitons with ice chips once they have been placed. If repeated use is necessary for one installation, such as top roping, the pitons should be inspected frequently and relocated when necessary. When an ice piton is removed, the ice that has accumulated in the tube must be removed before it freezes in position, making further use difficult.

c. **Ice Screws.** The ice screw is the most common type of ice protection and has replaced the ice piton for the most part (Figure 10-17). Some screws have longer "hangers" or handles, which allow them to be easily twisted into position by hand. Place ice screws as follows:

(1) Clear away all rotten ice from the surface and make a small hole with the ax pick to start the ice screw in.

(2) Force the ice screw in until the threads catch.

Figure 10-17. Placement of ice screw using the pick.

(3) Turn the screw until the eye or the hanger of the ice screw is flush with the ice and pointing down. The screw should be placed at an angle 90 to 100 degrees from the

lower surface. Use either your hand or the pick of the ice ax to screw it in. If you have a short ax (70 centimeters or less), you may be able to use the spike in the eye or hanger to ease the turning. (Remember that you may only have use of one hand at this point depending on your stance and the angle of the terrain.)

(4) As with ice pitons, melting of the ice around a screw over a period of time must be considered. The effective time and or strength of an ice screw placement is limited. The screw will heat from solar radiation, or the ice may crack or soften. Solar radiation can be nearly eliminated by covering the screw with ice chips once it has been emplaced. If repeated use is necessary for one installation, such as top roping, the screws should be inspected frequently and relocated when necessary. When an ice screw is removed, the ice that has accumulated in the tube, must be removed before further use.

d. **Horseshoe or Bollard Anchor.** This is an artificial anchor shaped generally like a horseshoe (Figure 10-18). It is formed from either ice or snow and constructed by either cutting with the ice ax or stamping with the boots. When constructed of snow, the width should not be less than 10 feet. In ice, this width may be narrowed to 2 feet, depending on the strength of the ice. The length of the bollard should be at least twice the width. The trench around the horseshoe should be stamped as deeply as possible in the snow and should be cut not less than 6 inches into the ice after all rotten ice is removed. The backside of the anchor must always be undercut to prevent the rope from sliding off and over the anchor.

(1) This type of anchor is usually available and may be used for fixed ropes or rappels. It must be inspected frequently to ensure that the rope is not cutting through the snow or ice more than one-third the length of the anchor; if it is, a new anchor must be constructed in a different location.

(2) A horseshoe anchor constructed in snow is always precarious, its strength depending upon the prevailing texture of the snow. For dry or wind-packed snow, the reliability of the anchor should always be suspect. The backside of the bollard can be reinforced with ice axes, pickets, or other equipment for added strength.

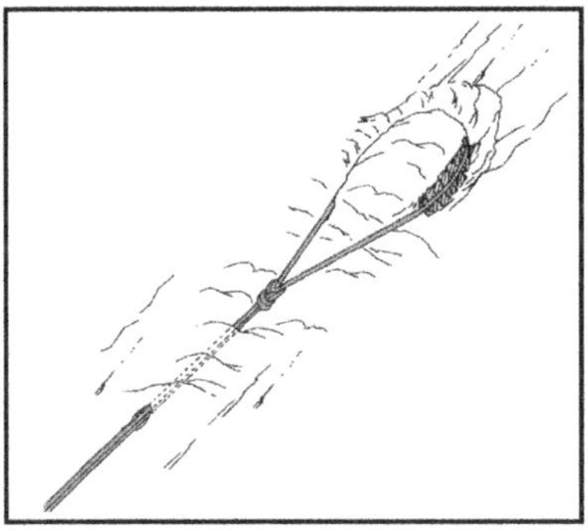

Figure 10-18. Horseshoe or bollard anchor.

e. **Pickets and Ice Axes.** Pickets and ice axes may be used as snow anchors as follows.

(1) The picket should be driven into the snow at 5 to 15 degrees off perpendicular from the lower surface. If the picket cannot be driven in all the way to the top hole, the carabiner should be placed in the hole closest to the snow surface to reduce leverage. The picket may also be tied off with a short loop of webbing or rope as in tying off pitons.

(2) An ice ax can be used in place of a picket. When using an ice ax as a snow anchor, it should be inserted with the widest portion of the ax shaft facing the direction of pull. The simplest connection to the ax is to use a sling or rope directly around the shaft just under the head. If using the leash ensure it is not worn, frayed, or cut from general use; is strong enough; and does not twist the ax when loaded. A carabiner can be clipped through the hole in the head, also.

(3) Whenever the strength of the snow anchor is suspect, especially when a picket or ax cannot be driven in all the way, the anchor may be buried in the snow and used as a "dead man" anchor. Other items suitable for dead man anchor construction are backpacks, skis, snowshoes, ski poles, or any other item large enough or shaped correctly to achieve the design. A similar anchor, sometimes referred to as a "dead guy," can be made with a large sack either stuffed with noncompressible items or filled with snow and buried. Ensure the attaching point is accessible before burying. The direction of pull on long items, such as a picket or ax, should be at a right angle to its length. The construction is identical to that of the dead man anchor used in earth.

f. **Equalized Anchors.** Snow and ice anchors must be constantly checked due to melting and changing snow or ice conditions.

(1) Whenever possible, two or more anchors should be used. While this is not always practical for intermediate anchor points on lead climbs or fixed ropes, it should be mandatory for main anchors at all belay positions, rappel points, or other fixed rope installations. (Figure 10-19, page 10-18, shows an example of three snow pickets configured to an equalized anchor.)

(2) As with multipoint anchors on rock, two or more snow or ice anchors can be joined together with a sling rope or webbing to construct one solid, equalized anchor. A bowline on a bight tied into the climbing rope can also be used instead of sling ropes or webbing.

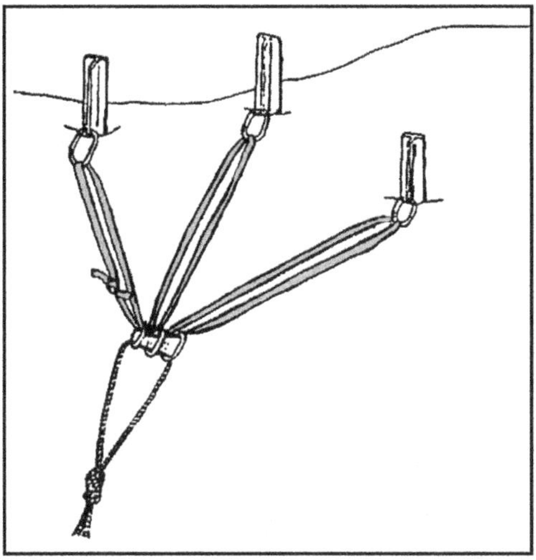

Figure 10-19. Equalized anchor using pickets.

10-6. ROPED CLIMBING ON ICE AND SNOW

When climbing on ice or snow team members tie into a climbing rope the same as when they climb on rock. When crevasses are expected, a three-man rope team is recommended.

a. **Tie-In Method.** For climbing on snow and ice, the tie-in procedure is normally the same as for rock climbing; however, when moving over snow-covered glaciers, the tie-in is modified slightly. (See paragraph 10-7, Movement on Glaciers, for more information).

b. **Movement.** For movement on gentle or moderate slopes where there is little chance of a serious fall, all climbers move simultaneously. Normally the climbers move in single file using the steps created by the lead climber and improving them when necessary. The rope between the climbers should be fully extended and kept reasonably tight. Should any member fall, he immediately yells "FALLING." The other rope team members immediately drop into a self-arrest position. The fallen climber also applies the self-arrest procedure. By using this method, called the "team arrest," the entire team as a whole arrests the fall of one member. On steeper slopes, and when crossing snow-covered crevasses where the snow bridges appear weak, the climbers use belayed climbing techniques as in rock climbing.

c. **Belaying on Snow and Ice.** The principles of belaying on ice and snow are the same as on rock. Generally, the high-force falls found in rock climbing are not present on snow and ice unless the pitch being climbed is extremely steep.

(1) ***Boot-Ax Belay.*** This belay can be useful in areas where the full length of the ice ax can penetrate the snow. The holding strength of the boot-ax belay is directly related to the firmness of the snow and to the strength of the ice ax shaft. The shaft of the ax is

tilted slightly uphill and jammed into the snow. The belayer places his uphill foot against the downhill side of the ax for support. A bight formed in the rope is placed over the boot and around the shaft of the ice ax. The brake is applied by wrapping the rope around the heel of the boot (Figure 10-20).

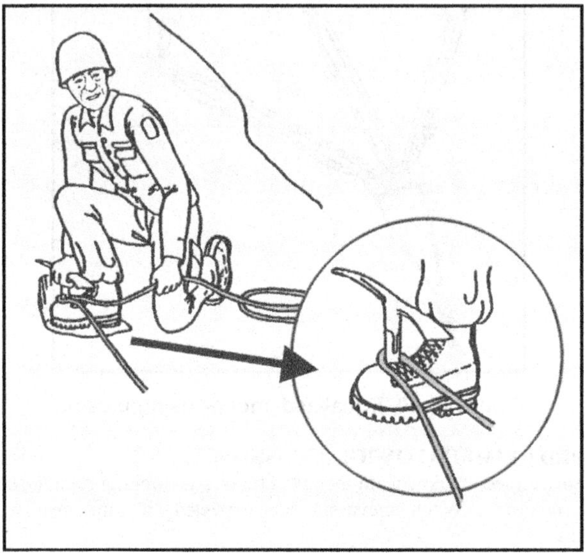

Figure 10-20. Boot-ax belay.

(2) ***Body Belay.*** The body belay can be used on snow and ice, also. The principles are the same as for belays on rock—solid anchors must be used and a well-braced position assumed. The position can be improved by digging depressions into the snow or ice for a seat and footholds. A strong platform should be constructed for the standing body belay.

(3) ***Munter Hitch.*** This belay technique is also used on snow and ice. When using the hitch off of the anchor, a two-point equalized anchor should be constructed as a minimum.

d. **Fixed Ropes.** The use of fixed ropes on ice is recommended for moving units through icefall areas on glaciers or other steep ice conditions. The procedures for emplacing fixed ropes on ice are basically the same as on rock with the exception that anchors need more attention, both in initial placement and in subsequent inspection, and steps may have to be cut to assist personnel.

10-7. MOVEMENT ON GLACIERS

Movement in mountainous terrain may require travel on glaciers. An understanding of glacier formation and characteristics is necessary to plan safe routes. A glacier is formed by the perennial accumulation of snow and other precipitation in a valley or draw. The accumulated snow eventually turns to ice due to metamorphosis. The "flow" or movement of glaciers is caused by gravity. There are a few different types of glaciers identifiable primarily by their location or activity.

- Valley glacier—resides and flows in a valley.
- Cirque glacier—forms and resides in a bowl.
- Hanging glacier—these are a result of valley or cirque glaciers flowing and or deteriorating. As the movement continues, portions separate and are sometimes left hanging on mountains, ridgelines, or cliffs.
- Piedmont glacier—formed by one or more valley glaciers; spreads out into a large area.
- Retreation glacier—a deteriorating glacier; annual melt of entire glacier exceeds the flow of the ice.
- Surging glacier—annual flow of the ice exceeds the melt; the movement is measurable over a period of time.

a. **Characteristics and Definitions.** This paragraph describes the common characteristics of glaciers, and defines common terminology used in reference to glaciers. (Figure 10-21 shows a cross section of a glacier, and Figure 10-22 depicts common glacier features.)

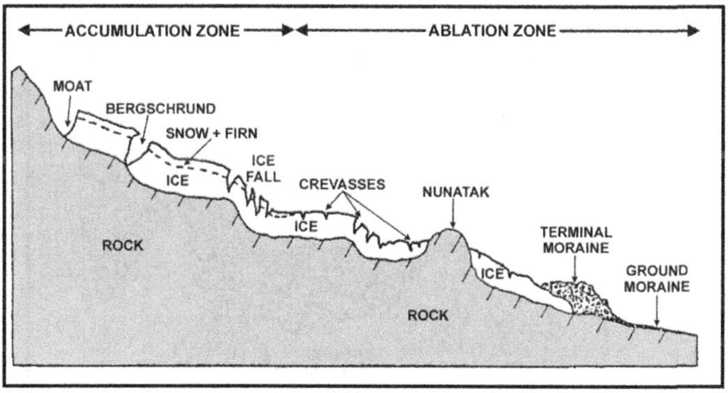

Figure 10-21. Glacier cross section.

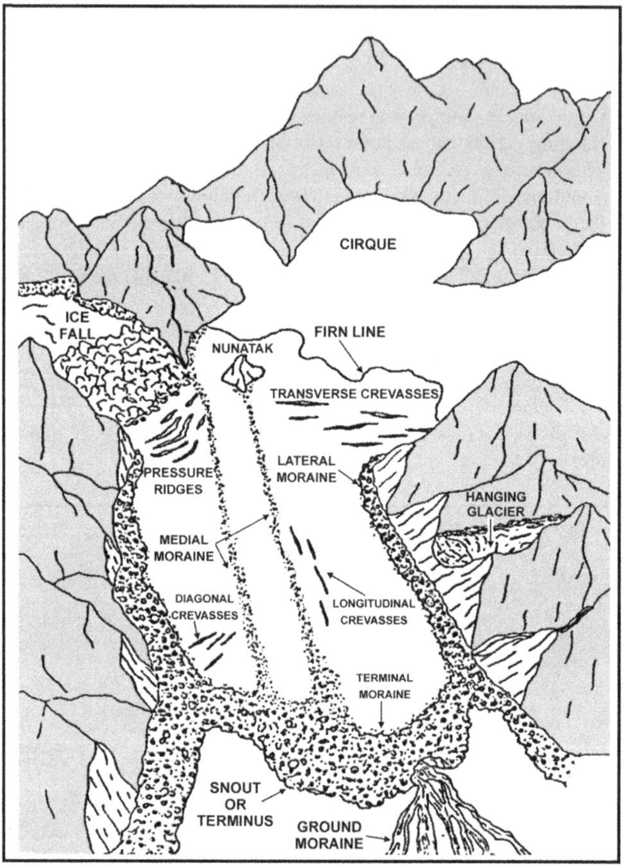

Figure 10-22. Glacier features.

(1) Firn is compacted granular snow that has been on the glacier at least one year. Firn is the building blocks of the ice that makes the glacier.

(2) The accumulation zone is the area that remains snow-covered throughout the year because of year-round snowfall. The snowfall exceeds melt.

(3) The ablation zone is the area where the snow melts off the ice in summer. Melt equals or exceeds snowfall.

(4) The firn line separates the accumulation and ablation zones. As you approach this area, you may see "strips" of snow in the ice. Be cautious, as these could be *snow bridges* remaining over crevasses. Remember that snow bridges will be weakest lower on the glacier as you enter the accumulation zone. The firn line can change annually.

(5) A bergschrund is a large crevasse at the head of a glacier caused by separation of active (flowing) and inactive (stationary) ice. These will usually be seen at the base of a major incline and can make an ascent on that area difficult.

(6) A moat is a wall formed at the head (start) of the glacier. These are formed by heat reflected from valley wall.

(7) A crevasse is a split or crack in the glacier surface. These are formed when the glacier moves over an irregularity in the bed surface.

(8) A transverse crevasse forms perpendicular to the flow of a glacier. These are normally found where a glacier flows over a slope with a gradient change of 30 degrees or more.

(9) Longitudinal crevasses form parallel to the flow of a glacier. These are normally found where a glacier widens.

(10) Diagonal crevasses form at an angle to the flow of a glacier. These are normally found along the edges where a glacier makes a bend.

(11) A snow bridge is a somewhat supportive structure of snow that covers a crevasse. Most of these are formed by the wind. The strength of a snow bridge depends on the snow itself.

(12) Icefalls are a jumble of crisscross crevasses and large ice towers that are normally found where a glacier flows over a slope with a gradient change of 25 degrees or more.

(13) Seracs are large pinnacles or columns of ice that are normally found in icefalls or on hanging glaciers.

(14) Ice avalanches are falling chunks of ice normally occurring near icefalls or hanging glaciers.

(15) The moraine is an accumulation of rock or debris on a glacier caused by rockfall or avalanche of valley walls.

(16) The lateral moraine is formed on sides of glacier.

(17) The medial moraine is in the middle of the glacier. This is also formed as two glaciers come together or as a glacier moves around a central peak.

(18) The terminal moraine is at the base of a glacier and is formed as moraines meet at the snout or terminus of a glacier.

(19) The ground moraine is the rocky debris extending out from the terminus of a glacier. This is formed by the scraping of earth as the glacier grew or surged and exposed as the glacier retreats.

(20) A Nunatak is a rock projection protruding through the glacier as the glacier flows around it.

(21) An ice mill is a hole in the glacier formed by swirling water on the surface. These can be large enough for a human to slip into.

(22) Pressure ridges are wavelike ridges that form on glacier normally after a glacier has flowed over icefalls.

(23) A glacier window is an opening at the snout of the glacier where water runs out of the glacier.

b. **Dangers and Obstacles.** The principle dangers and obstacles to movement in glacial areas are crevasses, icefalls, and ice avalanches. Snow-covered crevasses make movement on a glacier extremely treacherous. In winter, when visibility is poor, the difficulty of recognizing them is increased. Toward the end of the summer, crevasses are widest and covered by the least snow. Crossing snow bridges constitutes the greatest potential danger in movement over glaciers in the summer. On the steep pitch of a glacier, ice flowing over irregularities and cliffs in the underlying valley floor cause the ice to break up into ice blocks and towers, criss-crossed with crevasses. This jumbled cliff of ice is known as an icefall. Icefalls present a major obstacle to safe movement of troops on glaciers.

(1) Moving on glaciers brings about the hazard of falling into a crevasse. Although the crevasses are visible in the ablation zone in the summer (Figure 10-23), the accumulation zone will still have hidden crevasses. The risk of traveling in the accumulation zone can be managed to an acceptable level when ropes are used for connecting the team members (Figure 10-24, page 10-24). Crampons and an ice ax are all that is required to safely travel in the ablation zone in the summer.

Figure 10-23. Ablation zone of glacier in summer.

Figure 10-24. Rope teams moving in the accumulation zone of a glacier.

(2) When conditions warrant, three to four people will tie in to one rope at equal distances from each other. To locate the positions, if three people are on a team, double the rope and one ties into the middle and the other two at the ends. If four people are on a team, form a "z" with the rope and expand the "z" fully, keeping the end and the bight on each "side" of the "z" even. Tie in to the bights and the ends.

(3) Connect to the rope with the appropriate method and attach the Prusik as required. The rope should be kept relatively tight either by Prusik belay or positioning of each person. If the team members need to assemble in one area, use the Prusik to belay each other in.

(4) If a team member falls into a crevasse, the remaining members go into team arrest, assess the situation, and use the necessary technique to remove the person from the crevasse. The simplest and most common method for getting someone out of a crevasse is for the person to climb out while being belayed.

(5) All items should be secured to either the climber or the rope/harness to prevent inadvertent release and loss of necessary items or equipment. Packs should be secured to the rope/harness with webbing or rope. If traveling with a sled in tow, secure it not only to a climber to pull it, but connect it to the rope with webbing or rope also.

(6) If marking the route on the glacier is necessary for backtracking or to prevent disorientation in storms or flat-light conditions, use markers that will be noticeable against the white conditions. The first team member can place a new marker when the last team member reaches the previous marker.

c. **Roped Movement.** The first rule for movement on glaciers is to rope up (Figure 10-25). A roped team of two, while ideal for rock climbing, is at a disadvantage on a snow-covered glacier. The best combination is a three-man rope team. Generally, the rope team members will move at the same time with the rope fully extended and reasonably tight between individuals, their security being the team arrest. If an individual should break through a snow bridge and fall into a crevasse, the other members immediately perform self-arrest, halting the fall. At points of obvious weakness in the snow bridges, the members may decide to belay each other across the crevasse using one of the established belay techniques.

Figure 10-25. Preparation for roped movement.

(1) Even with proper training in crevasse rescue techniques, the probability exists that an individual may remain suspended in a crevasse for a fairly lengthy amount of time while trying to get himself out or while awaiting help from his rope team members. Because of this, it is strongly recommended that all personnel wear a seat/chest combination harness, whether improvised or premanufactured.

(2) Rope team members must be able to quickly remove the climbing rope from the harness(es) during a crevasse rescue. The standard practice for connecting to the rope for glacier travel is with a locking carabiner on a figure-eight loop to the harness. This allows quick detachment of the rope for rescue purposes. The appropriate standing part of the rope is then clipped to the chest harness carabiner.

(3) If a rope team consists of only two people, the rope should be divided into thirds, as for a four-person team. The team members tie into the middle positions on the rope, leaving a third of the rope between each team member and a third on each end of the rope. The remaining "thirds" of the rope should be coiled and either carried in the rucksack, attached to the rucksack, or carried over the head and shoulder. This gives each

climber an additional length of rope that can be used for crevasse rescue, should one of the men fall through and require another rope. If necessary, this excess end rope can be used to connect to another rope team for safer travel.

Note: The self-arrest technique used by one individual will work to halt the fall of his partner on a two-man rope team; however, the chance of it failing is much greater. Crevasse rescue procedures performed by a two-man rope team, by itself, may be extremely difficult. For safety reasons, movement over a snow-covered glacier by a single two-man team should be avoided wherever possible.

d. **Use of Prusik Knots.** Prusik knots are attached to the climbing rope for all glacier travel. The Prusiks are used as a self-belay technique to maintain a tight rope between individuals, to anchor the climbing rope for crevasse rescue, and for self-rescue in a crevasse fall. The Prusik slings are made from the 7-millimeter by 6-foot and 7-millimeter by 12-foot ropes. The ends of the ropes are tied together, forming endless loops or slings, with double fisherman's knots. Form the Prusik knot on the rope in front of the climber. An overhand knot can be tied into the sling just below the Prusik to keep equal tension on all the Prusik wraps. Attach this sling to the locking carabiner at the tie in point on the harness.

Note: An ascender can replace a Prusik sling in most situations. However, the weight of an ascender hanging on the rope during movement will become annoying, and it could be stepped on during movement and or climbing.

e. **Securing the Backpack/Rucksack.** If an individual should fall into a crevasse, it is essential that he be able to rid himself of his backpack. The weight of the average pack will be enough to hinder the climber during crevasse rescue, or possibly force him into an upside down position while suspended in the crevasse. Before movement, the pack should be attached to the climbing rope with a sling rope or webbing and a carabiner. A fallen climber can immediately drop the pack without losing it. The drop cord length should be minimal to allow the fallen individual to reach the pack after releasing it, if warm clothing is needed. When hanging from the drop cord, the pack should be oriented just as when wearing it (ensure the cord pulls from the top of the pack).

f. **Routes.** An individual operating in the mountains must appreciate certain limitations in glacier movement imposed by nature.

(1) Additional obstacles in getting onto a glacier may be swift glacier streams, steep terminal or lateral moraines, and difficult mountain terrain bordering the glacier ice. The same obstacles may also have to be overcome in getting on and off a valley glacier at any place along its course.

(2) Further considerations to movement on a glacier are steep sections, heavily crevassed portions, and icefalls, which may be major obstacles to progress. The use of current aerial photographs in conjunction with aerial reconnaissance is a valuable means of gathering advance information about a particular glacier. However, they only supplement, and do not take the place of, on-the-ground reconnaissance conducted from available vantage points.

g. **Crossing Crevasses.** Open crevasses are obvious, and their presence is an inconvenience rather than a danger to movement. Narrow cracks can be jumped, provided the take off and landing spots are firm and offer good footing. Wider cracks will have to be circumvented unless a solid piece of ice joins into an ice bridge strong enough to support at least the weight of one member of the team. Such ice bridges are often formed in the lower portion of a crevasse, connecting both sides of it.

(1) In the area of the firn line, the zone that divides seasonal melting from permanent falls of snow, large crevasses remain open, though their depths may be clogged with masses of snow. Narrow cracks may be covered. In this zone, the snow, which covers glacier ice, melts more rapidly than that which covers crevasses. The difference between glacier ice and narrow snow-covered cracks is immediately apparent; the covering snow is white, whereas the glacier ice is gray.

(2) Usually the upper part of a glacier is permanently snow covered. The snow surface here will vary in consistency from dry powder to consolidated snow. Below this surface cover are found other snow layers that become more crystalline in texture with depth, and gradually turn into glacier ice. It is in this snow-covered upper part of a glacier that crevasses are most difficult to detect, for even wide crevasses may be completely concealed by snow bridges.

h. **Snow Bridges.** Snow bridges are formed by windblown snow that builds a cornice over the empty interior of the crevasse. As the cornice grows from the windward side, a counter drift is formed on the leeward side. The growth of the leeward portion will be slower than that to the windward so that the juncture of the cornices occurs over the middle of the crevasse only when the contributing winds blow equally from each side. Bridges can also be formed without wind, especially during heavy falls of dry snow. Since cohesion of dry snow depends only on an interlocking of the branches of delicate crystals, such bridges are particularly dangerous during the winter. When warmer weather prevails the snow becomes settled and more compacted, and may form firmer bridges.

(1) Once a crevasse has been completely bridged, its detection is difficult. Bridges are generally slightly concave because of the settling of the snow. This concavity is perceptible in sunshine, but difficult to detect in flat light. If the presence of hidden crevasses is suspected, the leader of a roped team must probe the snow in front of him with the shaft of his ice ax. As long as a firm foundation is encountered, the team may proceed, but should the shaft meet no opposition from an underlying layer of snow, a crevasse is probably present. In such a situation, the prober should probe closer to his position to make sure that he is not standing on the bridge itself. If he is, he should retreat gently from the bridge and determine the width and direction of the crevasse. He should then follow and probe the margin until a more resistant portion of the bridge is reached. When moving parallel to a crevasse, all members of the team should keep well back from the edge and follow parallel but offset courses.

(2) A crevasse should be crossed at right angles to its length. When crossing a bridge that seems sufficiently strong enough to hold a member of the team, the team will generally move at the same time on a tight rope, with each individual prepared to go into self-arrest. If the stability of the snow bridge is under question, they should proceed as follows for a team of three glacier travelers:

(a) The leader and second take up a position at least 10 feet back from the edge. The third goes into a self-belay behind the second and remains on a tight rope.

(b) The second belays the leader across using one of the established belay techniques. The boot-ax belay should be used only if the snow is deep enough for the ax to be inserted up to the head and firm enough to support the possible load. A quick ice ax anchor should be placed for the other belays. Deadman or equalizing anchors should be used when necessary.

(c) The leader should move forward, carefully probing the snow and evaluating the strength of the bridge, until he reaches firm snow on the far side of the crevasse. He then continues as far across as possible so number two will have room to get across without number one having to move.

(d) The third assumes the middle person's belay position. The middle can be belayed across by both the first and last. Once the second is across, he assumes the belay position. Number one moves out on a tight rope and anchors in to a self-belay. Number two belays number three across.

(3) In crossing crevasses, distribute the weight over as wide an area as possible. Do not stamp the snow. Many fragile bridges can be crossed by lying down and crawling to the other side. Skis or snowshoes help distribute the weight nicely.

i. **Arresting and Securing a Fallen Climber.** The simplest and most common method for getting someone out of a crevasse is for the person to climb out while being belayed. Most crevasse falls will be no more than body height into the opening if the rope is kept snug between each person.

(1) To provide a quick means of holding an unexpected breakthrough, the rope is always kept taut. When the leader unexpectedly breaks through, the second and third immediately go into a self-arrest position to arrest the fall. A fall through a snow bridge results either in the person becoming jammed in the surface hole, or in being suspended in the crevasse by the rope. If the leader has fallen only partially through the snow bridge, he is supported by the snow forming the bridge and should not thrash about as this will only enlarge the hole and result in deeper suspension. All movements should be slow and aimed at rolling out of the hole and distributing the weight over the remainder of the bridge. The rope should remain tight at all times and the team arrest positions adjusted to do so. It generally is safer to retain the rucksack, as its bulk often prevents a deeper fall. Should a team member other than the leader experience a partial fall, the rescue procedure will be same as for the leader, only complicated slightly by the position on the rope.

(2) When the person falls into a crevasse, the length of the fall depends upon how quickly the fall is arrested and where in the bridge the break takes place. If the fall occurs close to the near edge of the crevasse, it usually can be checked before the climber has fallen more than 6 feet. However, if the person was almost across, the fall will cause the rope to cut through the bridge, and then even an instantaneous check by the other members will not prevent a deeper fall. The following scenario is an example of the sequence of events that take place after a fall by the leader in a three-person team. (This scenario is for a team of three, each person referred to by position; the leader is number 1.)

(a) Once the fall has been halted by the team arrest, the entire load must be placed on number 2 to allow number 3 to move forward and anchor the rope. Number 3 slowly releases his portion of the load onto number 2, always being prepared to go back into self-arrest should number 2's position begin to fail.

(b) Once number 2 is confident that he can hold the load, number 3 will proceed to number 2's position, using the Prusik as a self belay, to anchor the rope. In this way the rope remains reasonably tight between number 2 and number 3. Number 3 must always be prepared to go back into self-arrest should number 2's position begin to fail.

(c) When number 3 reaches number 2's position he will establish a bombproof anchor 3 to 10 feet in front of number 2 (on the load side), depending on how close number 2 is to the lip of the crevasse. This could be either a deadman or a two-point equalized anchor, as a minimum.

(d) Number 3 connects the rope to the anchor by tying a Prusik with his long Prusik sling onto the rope leading to number 1. An overhand knot should be tied into the long Prusik sling to shorten the distance to the anchor, and attached to the anchor with a carabiner. The Prusik knot is adjusted toward the load.

(e) Number 2 can then release the load of number 1 onto the anchor. Number 2 remains connected to the anchor and monitors the anchor.

(f) A fixed loop can be tied into the slack part of the rope, close to number 2, and attached to the anchor (to back up the Prusik knot).

(g) Number 3 remains tied in, but continues forward using a short Prusik as a self-belay. He must now quickly check on the condition of number 1 and decide which rescue technique will be required to retrieve him.

(3) These preliminary procedures must be performed before retrieving the fallen climber. If number 3 should fall through a crevasse, the procedure is the same except that number 1 assumes the role of number 3. Normally, if the middle person should fall through, number 1 would anchor the rope by himself. Number 3 would place the load on number 1's anchor, then anchor his rope and move forward with a Prusik self-belay to determine the condition of number 2.

j. **Crevasse Rescue Techniques.** Snow bridges are usually strongest at the edge of the crevasse, and a fall is most likely to occur some distance away from the edge. In some situations, a crevasse fall will occur at the edge of the snow bridge, on the edge of the ice. If a fall occurs away from the edge, the rope usually cuts deeply into the snow, thus greatly increasing friction for those pulling from above. In order to reduce friction, place padding, such as an ice ax, ski, ski pole, or backpack/rucksack, under the rope and at right angles to the stress. Push the padding forward as far as possible toward the edge of the crevasse, thus relieving the strain on the snow. Ensure the padding is anchored from falling into the crevasse for safety of the fallen climber.

(1) ***Use of Additional Rope Teams.*** Another rope team can move forward and assist in pulling the victim out of a crevasse. The assisting rope team should move to a point between the fallen climber and the remaining rope team members. The assisting team can attach to the arresting team's rope with a Prusik or ascender and both rope teams' members can all pull simultaneously. If necessary, a belay can be initiated by the fallen climber's team while the assisting team pulls. The arresting team member closest to the fallen climber should attach the long Prusik to themselves and the rope leading to the fallen climber, and the assisting team can attach their Prusik or ascender between this long Prusik and the arresting team member. As the assisting team pulls, the Prusik belay will be managed by the arresting team member at the long Prusik.

Note: Safety in numbers is obvious for efficient crevasse rescue techniques. Additional rope teams have the necessary equipment to improve the main anchor or establish new ones and the strength to pull a person out even if he is deep in the crevasse. Strength of other rope teams should always be used before establishing more time-consuming and elaborate rescue techniques.

(2) ***Fixed Rope.*** If the fallen climber is not injured, he may be able to climb out on a fixed rope. Number 1 clips number 3's rope to himself. He then climbs out using number 3's rope as a simple fixed line while number 2 takes up the slack in number 1's rope through the anchor Prusik for a belay.

(3) ***Prusik Ascending Technique.*** There may be times when the remaining members of a rope team can render little assistance to the person in the crevasse. If poor snow conditions make it impossible to construct a strong anchor, the rope team members on top may have to remain in self-arrest. Other times, it may just be easier for the fallen climber to perform a self-rescue. (Figure 10-26 shows the proper rope configuration.) The technique is performed as follows:

(a) The fallen climber removes his pack and lets it hang below from the drop cord.

(b) The individual slides their short Prusik up the climbing rope as far as possible.

(c) The long Prusik is attached to the rope just below the short Prusik. The double fisherman's knot is spread apart to create a loop large enough for one or both feet. The fallen climber inserts his foot/feet into the loop formed allowing the knot to cinch itself down.

(d) The individual stands in the foot loop, or "stirrup," of the long sling.

(e) With his weight removed from the short Prusik, it is slid up the rope as far as it will go. The individual then hangs from the short Prusik while he moves the long Prusik up underneath the short Prusik again.

(f) The procedure is repeated, alternately moving the Prusiks up the rope, to ascend the rope. Once the crevasse lip is reached, the individual can simply grasp the rope and pull himself over the edge and out of the hole.

(g) Besides being one of the simplest rope ascending techniques, the short Prusik acts as a self-belay and allows the climber to take as long a rest as he wants when sitting in the harness. The rope should be detached from the chest harness carabiner to make the movements less cumbersome. However, it is sometimes desirable to keep the chest harness connected to the rope for additional support. In this case the Prusik knots must be "on top" of the chest harness carabiner so they can be easily slid up the rope without interference from the carabiner. The long Prusik sling can be routed through the chest harness carabiner for additional support when standing up in the stirrup.

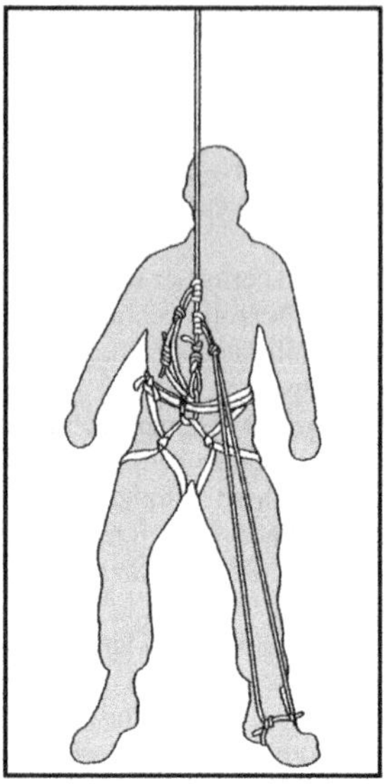

Figure 10-26. Prusik ascending technique.

(4) ***Z-Pulley Hauling System.*** If a fallen climber is injured or unconscious, he will not be able to offer any assistance in the rescue. If additional rope teams are not immediately available, a simple raising system can be rigged to haul the victim out of the crevasse. The Z-pulley hauling system is one of the simplest methods and the one most commonly used in crevasse rescue (Figure 10-27, page 10-32). The basic Z rig is a "3-to-1" system, providing mechanical advantage to reduce the workload on the individuals operating the haul line. In theory, it would only take about 33 pounds of pull on the haul rope to raise a 100-pound load with this system. In actual field use, some of this mechanical advantage is lost to friction as the rope bends sharply around carabiners and over the crevasse lip. The use of mechanical rescue pulleys can help reduce this friction in the system. The following describes rigging of the system. (This scenario is for a team of three, each person referred to by position; the leader is number 1.)

(a) After the rope team members have arrested and secured number 1 to the anchor, and they have decided to install the Z rig, number 2 will attach himself to the anchor without using the rope and clear the connecting knot used. Number 3 remains connected to the rope.

(b) The slack rope exiting the anchor Prusik is clipped into a separate carabiner attached to the anchor. A pulley can be used here if available.

(c) Number 3 will use number 2's short Prusik to rig the haul Prusik. He moves toward the crevasse lip (still on his own self-belay) and ties number 2's short Prusik onto number 1's rope (load rope) as close to the edge as possible.

(d) Another carabiner (and pulley if available) is clipped into the loop of the haul Prusik and the rope between number 3's belay Prusik and the anchor is clipped (or attached through the pulley). Number 3's rope becomes the haul rope.

(e) Number 3 then moves towards the anchor and number 2. Number 2 could help pull if necessary but first would connect to the haul rope with a Prusik just as number 3. If the haul Prusik reaches the anchor before the victim reaches the top, the load is simply placed back on the anchor Prusik and number 3 moves the haul Prusik back toward the edge. The system is now ready for another haul.

CAUTION

The force applied to the fallen climber through use of the Z-pulley system can be enough to destroy the harness-to-rope connection or injure the fallen climber if excess force is applied to the pulling rope.

Notes: 1. The Z-pulley adds more load on the anchor due to the mechanical advantage. The anchor should be monitored for the duration of the rescue.

2. With the "3-to-1" system, the load (fallen climber) will be raised 1 foot for every 3 feet of rope taken up during the haul.

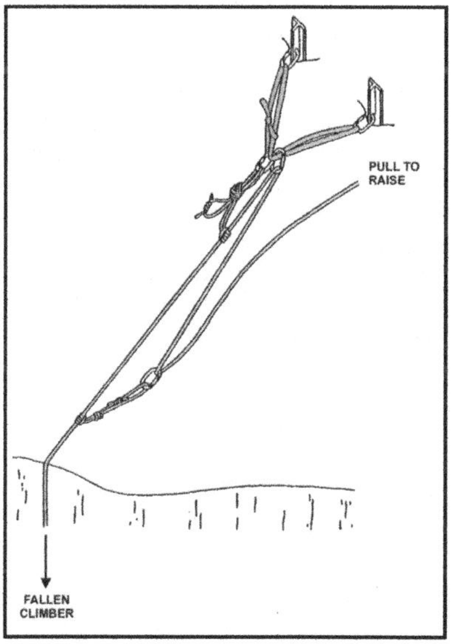

Figure 10-27. Z-pulley hauling system.

10-8. GLACIER BIVOUAC PROCEDURES

When locating a bivouac site or a gathering area where the team might need or want to unrope, at least one person will need to "probe" the area for hidden crevasses. The best type of probe will be the manufactured collapsing probe pole, at least eight feet in length. Other items could be used but the length and strength of the probe is most important. Other rope team members will belay the probers. The prober is "feeling" for a solid platform to place the tent by pushing the probe as hard and deep as possible into the surface. Probing should be in 2-foot intervals in all directions within the site.

a. If the probe suddenly has no resistance while pushing down, a crevasse is present. Attempts to outline the crevasse can be futile if the crevasse is large. Normally, the best decision is to relocate the proposed bivouac area far enough away to avoid that crevasse. (Sometimes only a few feet one way or the other is all that's needed to reach a good platform.) Probe the tent site again after digging to the desired surface. Mark boundaries with wands or other items such as skis, poles, and so on.

b. Occasionally while probing, increased pressure will be noticed without reaching a solid platform. The amount of snowfall may be such that even after digging into the snow, the probe still doesn't contact a hard surface. Try to find a solid platform.

c. There should be no unroped movement outside the probed/marked areas. If a latrine area is needed, probe a route away from the bivouac site and probe the latrine area also. If a dugout latrine is necessary, probe again after digging.

d. Multiple tent sites can be connected, which keeps tents closer together. Probe all areas between the tents if you plan to move in those areas. Closer tents will make communicating between tent groups and rope teams easier.

e. If there is a chance for severe storms with high winds, snow walls may be constructed to protect the tent site from wind. The walls can be constructed from loose snow piled on the perimeter, or blocks can be cut from consolidated snow layers. In deep soft snow, digging three or four feet to find a consolidated layer will result in enough snow moved to build up decent walls around the tent site.

(1) For block construction, move the soft snow from the surface into the wall foundation areas (down to a consolidated layer of snow).

(2) Cut blocks approximately 1 by 1 by 2 feet, and construct the walls by interlocking the blocks with overlapping placements. The walls should be slightly higher than the tent. At a minimum, build walls on the windward side of the tent site.

(3) Snow walls can also provide shelter from wind for food preparation.

CHAPTER 11

MOUNTAIN RESCUE AND EVACUATION

Steep terrain and adverse weather are common in mountainous environments. Under these conditions, relatively minor injuries may require evacuation. The evacuation technique chosen is determined by the type of injury, distance to be moved, terrain, and existing installations. Air evacuation is preferred; however, the weather, tactical situation, or operational ceiling of the aircraft may make this impossible. It is, therefore, imperative that all personnel are trained in mountain evacuation techniques and are self-sufficient. Casualties should be triaged before evacuation. Triage is performed by the most experienced medical personnel available (physician, physician's assistant, medic).

Performing a rescue operation can be a significant emotional event. Rescue scenarios must be practiced and rehearsed until rescue party members are proficient in the many tasks required to execute a rescue. To perform most of the high-angle rescues, Level I and Level II mountaineers are required with a Level III supervising.

11-1. CONSIDERATIONS

The techniques of evacuation are proven techniques. They are, however, all subject to improvement and should be discarded or modified as better methods of handling victims are developed.

a. When evacuating a victim from mountainous areas keep in mind that the purpose of a rescue operation is to save a life, and physical risk to the rescuers must be weighed against this purpose. However, there is no excuse for failing to make the maximum effort within this limitation. Work and expense should be no deterrent when a life is at stake.

b. Rescues will be unplanned (improvised) or planned rescue operations. For a planned rescue, equipment that is especially suited and designed for rescue should be used. For training missions always have a medical plan developed before an emergency arises (plan for the worst and hope for the best). Ensure that the MEDEVAC plan is a comprehensive plan and must be thought out and understood by all that may be involved in a potential rescue.

c. The following actions will be done immediately at the rescue scene.

(1) Assume command. One person, and one person only, is overall in charge at all times.

(2) Prevent further injuries to the victim and to others. Use reasonable care in reaching the victim.

(3) Immediately ensure the victim has an open airway, resume victim's breathing, control serious bleeding, and maintain moderate body warmth. If the victim is unconscious, continually monitor pulse. Protect the patient from environmental hazards.

(4) Do not move the victim until you have ascertained the extent of injuries, unless it is necessary to prevent further injuries or the victim is located in a dangerous location (for example, avalanche run-out zone, hanging glacier, possibility of falling rocks).

(5) Do nothing more until you have thoroughly considered the situation. Resist the urge for action. Speed is less important than correct action.

(6) Decide whether to evacuate with available facilities or to send for help. Speed in getting to a hospital must be balanced against the probability of further injury if working with inexperienced people, lack of equipment or wrong equipment, and terrain at hand.

(7) When the evacuation route is long and arduous, a series of litter relay points or stations should be established. These stations must be staffed with the minimum medical personnel to provide proper emergency treatment. When a victim develops signs of shock or worsens while being evacuated, he should be treated and retained at one of these stations until his condition allows evacuation.

(8) Helicopters or heated vehicles, if available, should be used for evacuation. While the use of aircraft or vehicles is preferred and can expedite a rescue operation, evacuation of a seriously wounded soldier should never be delayed to await aircraft, vehicle, or a change in weather.

11-2. PLANNING RESCUE OPERATIONS

Every commander should have a medical evacuation plan before undertaking an operation. This plan should have contingencies included so as not to rely on a single asset.

a. When rescuing a casualty (victim) threatened by hostile action, environmental hazard, or any other immediate hazard, the rescuer should not take action without first determining the extent of the hazard and his ability to handle the situation. THE RESCUER MUST NOT BECOME A CASUALTY.

b. The rescue team leader must evaluate the situation and analyze the factors involved. This evaluation can be divided into three major steps:

- Identify the task.
- Evaluate the circumstances of the rescue.
- Plan the action.

c. The task must be identified. In planning a rescue, the rescuer tries to obtain the following information:

- Who, what, where, when, why, and how the situation happened.
- Number of casualties by precedence (urgent, priority, routine, tactical immediate),
- number of casualties by type (litter or ambulatory), and the nature of their injuries.
- Terrain features and location of the casualties.
- Tactical situation.
- If adequate assistance is available to aid in security, rescue, treatment, and evacuation.
- If treatment can be provided at the scene; if the victims require movement to a safer location.
- Equipment required for the rescue operation.

d. Circumstances of the rescue are as follows:

(1) After identifying the task, relate it to the circumstances of the situation.

- Are additional personnel, security, medical, or special rescue equipment needed?
- Are there circumstances, such as aircraft accidents (mass casualties), that may require specialized skills?

- What is the weather condition?
- Is the terrain hazardous?
- How much time is available?

(2) The time element may cause a rescuer to compromise planning stages or treatment (beyond first aid). Make a realistic estimate of time available as quickly as possible to determine the action time remaining. The key elements are the casualty's condition and environment.

(3) Mass casualties are to be expected on the modern battlefield. All problems or complexities of rescue are now multiplied by the number of casualties. Time becomes the critical element.

(4) Considerations for the main rescue group for a planned rescue are as follows:

(a) Carry all needed equipment, hot food and drinks, stove, sleeping bags, tents, bivouac sacks, warm clothes, ropes, and stretchers.

(b) Prepare the evacuation route (ground transport to hospital, walking trails, fixed lines, lowering lines, anchor points, and rescue belay points). If the victim is airlifted out, attach a paper with the medical actions that were performed on the ground (for example, blood pressure, pulse rate, drugs started, and so on).

(c) When performing all rescues, the rescuers are always tied in for safety. With all rescue techniques, remember to think things through logically for safety and to prevent the rescuer from accidentally untying himself or the fallen climber.

(d) Constantly inform the casualty (if they are conscious) as to what you are doing and what he must do.

e. The rescue plan should proceed as follows:

(1) In estimating time available, the casualties' ability to endure is of primary importance. Age and physical condition may vary. Time available is a balance of the endurance time of the casualty, the situation, and the personnel and equipment available.

(2) Consider altitude and visibility. Maximum use of secure, reliable trails or roads is essential.

(3) Ensure that blankets and rain gear are available. Even a mild rain can complicate a normally simple rescue. In high altitudes, extreme cold, or gusting winds, available time is drastically reduced.

(4) High altitudes and gusting winds reduce the ability of fixed-wing or rotary-wing aircraft to assist in operations. Rotary-wing aircraft may be available to remove casualties from cliffs or inaccessible sites, and to quickly transport casualties to a medical treatment facility. Relying on aircraft or specialized equipment is a poor substitute for careful planning.

11-3. MASS CASUALTIES

When there are mass casualties, an orderly rescue may involve further planning.

a. To manage a mass casualty rescue or evacuation, separate stages are taken.

- FIRST STAGE: Remove personnel who are not trapped among debris or who can be easily evacuated.
- SECOND STAGE: Remove personnel who may be trapped by debris, but whose extraction only requires the equipment on hand and little time.

- THIRD STAGE: Remove the remaining personnel who are trapped in extremely difficult or time-consuming situations, such as moving large amounts of debris or cutting through a wall.
- FOURTH STAGE: Remove dead personnel.

b. Evacuation of wounded personnel is based on the victim's condition and is prioritized as follows:

- PRIORITY ONE: Personnel with life-threatening injuries that require immediate emergency care to survive; first aid and stabilization are accomplished before evacuation.
- PRIORITY TWO: Personnel with injuries that require medical care but speed of evacuation is not essential.
- PRIORITY THREE: Injured personnel who can evacuate themselves with minimal assistance.
- PRIORITY FOUR: The logistics removal of dead personnel.

11-4. SPECIAL TRAINING

Before receiving training in basic mountain evacuation, litter teams should receive instruction in military mountaineering and basic first aid. Litter bearers and medics must know the use and care of rope as an item of equipment. The members of litter teams must be proficient in the techniques of belaying and choosing belay points. Proper support and protection must be given to victims and litter bearers when evacuating over steep, difficult terrain.

11-5. PREPARATION FOR EVACUATION

Although the wounded soldier's life may have been saved by applying first aid, it can be lost through carelessness, rough handling, or inadequate protection from the elements. Therefore, before trying to move the wounded soldier, the type and extent of his injury must be evaluated. Dressings over wounds must be reinforced, and fractured bones must be properly immobilized and supported. Based upon the evaluation of the type and extent of the soldier's injury, the best method of manual transportation is selected.

11-6. MANUAL CARRIES

Personnel who are not seriously injured but cannot evacuate themselves may be assisted by fellow soldiers. Personnel who are injured and require prompt evacuation should not be forced to wait for mobile evacuation or special equipment.

a. **One-Man Carries.** The basic carries taught in the Soldier's Manual of Common Tasks (fireman's carry, two-hand, four-hand, saddleback, piggyback, pistol belt, and poncho litter) are viable means of transporting injured personnel; however, the mountainous terrain lends itself to several other techniques. One-man carries include the sling-rope carry and the rope coil carry.

(1) ***Sling-Rope Carry***. The sling-rope carry (Figure 11-1) requires a 4.5-meter sling rope and two men—one as the bearer and the other as an assistant to help secure the casualty to the bearer's back. Conscious or unconscious casualties may be transported this way.

(a) The bearer kneels on all fours.

(b) The assistant places the casualty face down on the bearer's back ensuring the casualty's armpits are even with the bearer's shoulders.

(c) The assistant then finds the middle of the sling rope and places it between the casualty's shoulders.

(d) The assistant runs the ends of the sling rope under the casualty's armpits, crosses the ends, and runs the ends over the bearer's shoulders and back under the bearer's arms.

(e) The assistant runs the ends of the rope between the casualty's legs, around the casualty's thighs, and back around to the front of the bearer. The rope is tied with a square knot with two overhand knots just above the bearer's belt buckle.

(f) The rope must be tight. Padding, when available, should be placed where the rope passes over the bearer's shoulders and under the casualty's thighs.

Figure 11-1. Sling-rope carry.

(2) ***Rope Coil Carry.*** The rope coil carry requires a bearer and a 36 1/2-meter coiled rope. It can be used to transport a conscious or unconscious victim.

(a) Place the casualty on his back.

(b) Separate the loops on one end of the coil, forming two almost equal groups.

(c) Slide one group of loops over the casualty's left leg and the other group over the right leg. The wraps holding the coil should be in the casualty's crotch with the loops on the other end of the coil extending upward toward the armpits.

(d) The bearer lies on his back between the casualty's legs and slides his arms through the loops. He then moves forward until the coil is extended.

(e) Grasping the casualty's arm, the bearer rolls over (toward the casualty's uninjured side), pulling the casualty onto his back.

(f) Holding the casualty's wrists, the bearer carefully stands, using his legs to lift up and keeping his back as straight as possible.

(g) A sling rope around both the casualty and bearer, tied with a joining knot at chest level, aids in keeping an unconscious victim upright. This also prevents the coils from slipping off the carrier's chest.

Note: The length of the coils on the rope coil and the height of the bearer must be considered. If the coils are too long and the bearer is shorter, the rope must be uncoiled and recoiled with smaller coils. If this is not done, the casualty will hang too low on the bearer's back and make it a cumbersome evacuation. A sling-rope harness can be used around the victim's back and bearer's chest, which frees the bearer's hands.

b. **Buddy Rappel.** The carrier can also conduct a seat-hip rappel with a victim secured to his back. In this case, the rappeller faces the cliff and assumes a modified L-shape body position to compensate for the weight of the victim on his back. The victim is top-rope belayed from above, which provides the victim with a point of attachment to a secured rope. The methods for securing a victim to a rappeller's back are described below.

(1) To secure the victim to the carrier's back with a rope, the carrier ties a standard rappel seat (brake hand of choice, depending on the injury) and rests his hands on his knees while the victim straddles his back.

(2) A 4.2-meter sling rope is used. A 45-centimeter tail of the sling is placed on the victim's left hip. (This method describes the procedure for a seat-hip rappel with right-hand brake.)

(3) The remaining long end of the sling rope is routed under the victim's buttocks, and passed over the victim's and carrier's right hip. The rope is run diagonally, from right to left, across the carrier's chest, over his left shoulder, and back under the victim's left armpit.

(4) The rope is then run horizontally, from left to right, across the victim's back. The rope is passed under the victim's right armpit and over the carrier's right shoulder.

(5) The rope is run diagonally, from right to left, across the carrier's chest and back across the carrier's and victim's left hip.

(6) The two rope ends should now meet. The two ends are tied together with a square knot and overhand knots.

(7) The knot is positioned on the victim's left hip. The carrier's shoulders may need to be padded to prevent cutting by the rope.

(8) An alternate method is to use two pistol belts hooked together and draped over the carrier's shoulders. The victim straddles the carrier, and the belay man secures the loose ends of the pistol belts under the victim's buttocks. Slack in the pistol belt sling should be avoided, since the carrier is most comfortable when the victim rests high on his back (see FM 8-35).

(9) A large rucksack can be slit on the sides near the bottom so that the victim can step into it. The victim is belayed from the top with the carrier conducting a standard rappel. The carrier wears the rucksack with the victim inside.

(10) A casualty secured to a carrier, as described above, can be rappelled down a steep cliff using a seat-shoulder or seat-hip rappel. The casualty's and rappeller's shoulders should be padded where the sling rope and rappel lines cross if a seat-shoulder

rappel is used. The buddy team should be belayed from above with a bowline tied around the victim's chest under his armpits. The belay rope must run over the rappeller's guide hand shoulder.

11-7. LITTERS

Many types of litters are available for evacuating casualties in rough mountain terrain. Casualties may be secured to litters in many different ways, depending on the terrain, nature of injuries, and equipment available. **All casualties must be secured.** This should be done under medical supervision after stabilization. It is also important to render psychological support to any victim awaiting evacuation.

If the litter must be carried, belayed, and then carried again, a sling rope should be wound around the litter end and tied off in a l-meter-long loop. This enables the carriers to hook and unhook the litter from the belay. Slings are available to aid the soldiers with litter carrying. Utility rope or webbing 6 meters long may be used. The rope is folded in half, and the loose ends are tied together with an overhand knot. These slings are attached to the litter rails (two or three to a side, depending on the number of litter bearers) by a girth hitch, and then routed up along the handling arm, over the shoulder, behind the neck, and then down along the other arm. The knot can be adjusted to help the outside arm grip the webbing. These slings help distribute the load more evenly, which is important if a great distance must be traveled.

a. **Manufactured Litters.** The following litters are readily available to mountaineering units.

(1) The poleless, nonrigid litter (NSN 6530-00-783-7510) is best issued for company medics since it is lightweight, easy to carry, and readily available. Casualties should be secured with the chest strap and pelvic straps, which are sewn on one side. This litter may be used when rappelling, on traverse lines, and on hauling lines in the vertical or horizontal position. It can be improvised with poles.

(2) The poleless semi-rigid litter (NSN 6530-00-783-7600) may be used the same as the nonrigid litter. It offers more victim protection and back support because of the wooden slats sewn into it.

(3) The mountain basket-type rigid litter (NSN 6530-00-181-7767) is best suited for areas where several casualties are to be transported. All other litters may be placed inside this litter basket and transported across traverse lines. This litter is rectangular and has no vertical leg divider so that it will accommodate other litters. It is also known as a modified Stokes litter.

(4) The Stokes metal litter (NSN 6530-00-042-8131) is suited for situations as above; however, the casualty must be moved in and out of the litter since no other litter will fit inside it. Some Stokes litter frames have a central weld on the frame end, which is a potential breaking point. Winding the rope around the frame end will distribute the force over a wider area and stabilize the system. (See FM 8-10-6 or USAF TO 00-75-5 for additional information on the Stokes litter.)

(5) The standard collapsible litter (NSN 6530-00-783-7905) (rigid pole folding litter) is most readily available in all units and, although heavy and unsuited to forward deployment, may be rigged for movement over rough or mountainous terrain. The folding aluminum litter (NSN 6530-00-783-7205) is a compact version of the pole litter and is better suited for forward deployment.

(6) The UT 2000 is manufactured in Austria and is specifically designed for mountaineering operations. The litter consists of two parts that join together to form a rigid litter. Each part has shoulder and waist straps that can be used to man-pack the litter making it extremely light and portable. When joined together the shoulder and waist straps are used to secure the casualty to the litter. Strapping is also provided to make a secure hoist point for aircraft extraction and high-angle rescues. Wheel sets are another accessory to the UT 2000 litter (either two wheels or one); they attach to the litter for use during a low-angle rescue.

(7) The patient rescue and recovery system (NSN 6530-01-260-1222) provides excellent patient support and protection (Figure 11-2). However, it is not a spinal immobilization device. A backboard must be used with this system for patients who have injuries to the shoulder area. This system will accommodate long and short backboards, scoop stretchers, and most other immobilization equipment.

Figure 11-2. Rescue and recovery system (NSN 6530-01-260-1222).

b. **Field-Expedient Litters.** A litter can be improvised from many different things. Most flat-surface objects of suitable size can be used as litters. Such objects include boards, doors, window shutters, benches, ladders, cots, and poles. Some may need to be tied together to obtain the required size. If possible, these objects should be padded.

(1) Litters can also be made by securing poles inside blankets, ponchos, shelter halves, tarpaulin, jackets, shirts, sacks, bags, or mattress covers. Poles can be improvised from strong branches, tent supports, skis, and other similar items.

(2) If poles cannot be found, a large item, such as a blanket, can be rolled from both sides toward the center. Then the rolls can be used to obtain a firm grip to carry the victim. If a poncho is used, the hood must be up and under the victim, not dragging on the ground.

(3) A rope litter is prepared using one rope (Figure 11-3). It requires 20 to 30 minutes to prepare and should be used only when other materials are not available. Four to six bearers are required to carry the litter. The rope litter is the most commonly used field-expedient litter.

Note: Above the tree line, little material exists to construct litters.

Figure 11-3. Rope litter.

(a) Make 24 bights about 45 to 61 centimeters long, starting in the middle of the rope so that two people can work on the litter at one time.

(b) With the remainder of the rope, make a clove hitch over each bight. Each clove hitch should be about 15 centimeters from the closed end of the bight when the litter is complete.

(c) Pass the remainder of the rope through the bights outside of the clove hitches. Dress the clove hitches down toward the closed end of the bight to secure the litter and tie off the ends of the rope with clove hitches.

(d) Line the litter with padding such as clothing, sleeping bags, empty boxes.

(e) Make the rope litter more stable by making it about 6 inches wider. After placing the clove hitches over the bights, slide them in (away from the closed end) about 15 centimeters. Take two 3- to 4-meter poles, 8 centimeters in diameter at the butt ends, and slide each pole down through the bights on each side. Dress down the clove hitches against the poles. Take two 1-meter poles, and tie them off across the head and foot of the litter with the remaining tails of the climbing rope.

Note: The above measurements may have to be altered to suit the overall length of rope available.

11-8. RESCUE SYSTEMS

Rescue systems are indispensable when conducting rescue operations. A large number of soldiers will not always be available to help with a rescue. Using a mechanical advantage rescue system allows a minimal amount of rescuers to perform tasks that would take a larger number of people without it.

a. **Belay Assist.** This system is used to bring a climber over a section that he is unable to climb, but will continue climbing once he is past the difficult section.

(1) First, tie off the following climber at the belay with a mule knot.

(2) Tie a Prusik knot with short Prusik cord about 12 inches below the mule knot, and place a carabiner into the loop. Place the tail from the mule knot into the carabiner in the Prusik cord.

(3) Untie the mule knot without letting the following climber descend any more than necessary. Do not remove the belay.

(4) Maintain control of the brake side of the rope and pull all of the slack through the carabiner in the Prusik cord.

(5) Pull up on the rope. The rope will automatically feed through the belay.

(6) If the leader has to pull for a distance he can tie another mule knot at the belay to secure the second climber before sliding the Prusik down to get more pulling distance.

(7) After the climber can continue climbing, the leader secures the belay with a mule knot.

(8) Remove the Prusik cord and carabiner, then untie the mule knot and continue belaying normally.

Notes: With all rescue techniques make sure that you always think everything through, and double check all systems to ensure that you don't accidentally untie the fallen climber or find yourself without back-up safety. Do not compound the problem! When doing any rescue work the rescuers will always be tied in for safety.

b. **Belay Escape.** The belay escape is used when a climber has taken a serious fall and cannot continue. The belayer is anchored and is performing an indirect belay, and must assist the injured climber or go for assistance. To accomplish this he must escape the belay system. The belayer will remain secured to an anchor at all times.

(1) After a climber has been injured, tie off the belay device on your body using a mule knot. To improve this system, clip a nonlocking carabiner through the loop in the overhand knot and clip it over the rope.

(2) Attach a short Prusik cord to the load rope and secure it to the anchor with a releasable knot.

(3) Using a guard knot or Munter mule, attach the climbing rope from the belay device.

(4) Untie the mule knot in the belay device attached to the harness and slowly lower the climber, transferring the weight of the climber onto the Prusik.

(5) Remove the climbing rope from the belay device attached to the harness.

(6) Release the mule knot securing the Prusik, transferring the weight to the anchor.

(7) At this point the climber is secured by the rope to the anchor system and the belayer can now assist the injured climber.

11-9. LOW-ANGLE EVACUATION

Cliffs and ridges, which must be surmounted, are often encountered along the evacuation path. Raising operations place a greater load on all elements of the system than do lowering operations. Since all means of raising a victim (pulley systems, hand winches, and power winches) depend on mechanical advantage, it becomes easy to overstress and break anchors and hand ropes. Using mechanical raising systems tends to reduce the soldier's sensitivity to the size of the load. It becomes important to monitor the system and to understand the forces involved.

a. **Raising Systems, Belays, and Backup Safeties.** Raising systems, belays, and backup safeties are of special importance in any raising operation. The primary raising system used is the Z-pulley system, which theoretically gives three pounds of lift for each pound of force expended. In practice, these numbers decrease due to rope-pulley friction, rope-edge friction, and other variables. A separate belay rope is attached to the litter and belayed from a separate anchor. Backup Prusik safeties should be installed in case any part of the pulley system fails.

(1) ***Raising System***. When considering a raising system for evacuations, the Z-pulley system is the most adaptable. It can be rigged with the equipment on hand, and can be modified and augmented to handle heavier loads. Although the vertical or horizontal hauling lines can also be used, the Z-pulley system offers a mechanical advantage that requires less exertion by the transport team.

(2) ***Belays***. Whenever ropes are used for an evacuation operation, the overriding safety concern is damage to the ropes. This is the main reason for two-rope raising systems (raising rope and belay rope).

(3) ***Backup Safeties.*** Because the stresses generated by the Z-pulley system can cause anchors to fail, backup safety Prusiks are used to safeguard the system. These should be attached to alternate anchor points, if possible.

b. **Raising the Litter.** The litter is prepared as already described.

(1) The raising ropes and belay ropes are secured to top anchors and are thrown down to the litter crew.

(2) Padding is placed at the cliff edge and over any protrusions to protect the ropes from abrasion.

(3) The litter attendants secure the ropes to the litter.

(4) The raising crew sets up the Z-pulley system.

(5) One member of the crew secures himself to an anchor and moves to the edge of the cliff to transmit signals and directions. (This is the signalman or relay man.)

Note: If the load is too heavy at this time, another pulley is added to the system to increase the mechanical advantage.

(6) Attendants guide the litter around obstacles while the crew continues to raise the system.

(7) As the litter nears the cliff edge, the signalman assists the attendant in moving the litter over the edge and onto the loading platform, taking care not to jar the casualty.

c. **Descending Slopes.** When descending a moderately steep slope that can be down-climbed, the litter and victim are prepared as described earlier (Figure 11-4, page 11-12).

(1) One man serves as the belay man and another takes his position on the rope in front of the belay man, assisting him in lowering the litter. The litter bearers take their positions and move the litter down with the speed of descent controlled by the belay man.

Figure 11-4. Low-angle evacuation—descending.

(2) The extra man may assist with the litter or precede the team to select a trail, clearing away shrubs and vines. He reconnoiters so that the team need not retrace its steps if a cliff is encountered.

(3) The most direct, practical passage should be taken utilizing available natural anchors as belay positions.

11-10. HIGH-ANGLE EVACUATION

Evacuation down cliffs should be used only when absolutely necessary and only by experienced personnel. The cliffs with the smoothest faces are chosen for the route. Site selection should have the following features: suitable anchor points, good loading and unloading platforms, clearance for the casualty along the route, and anchor points for the A-frame, if used. There are many ways to lower a casualty down a steep slope. As long as safety principals are followed, many different techniques can be used. One of the easiest and safest techniques is as follows (Figure 11-5):

a. Use multiple anchors for the litter and litter tenders.

b. Secure the litter to the lowering rope with a minimum of four tie-in points (one at each corner of the litter). Lengths of sling rope or 7-millimeter cordage work best. Make the attached ropes adjustable with Prussik knots so that each corner of the litter can be raised or lowered to keep the litter stable during descent. Tie the top of the ropes with loops and attach to the lowering rope with a pear shaped locking carabiner.

c. Two litter tenders will descend with the litter to control the descent and to monitor the casualty. They can be attached to separate anchors and either self-belay themselves or be lowered by belayers.

d. Once the steep slope has been negotiated, continue the rescue with a low-angle evacuation.

Figure 11-5. Cliff evacuation descent.

APPENDIX A

LEVELS OF MILITARY MOUNTAINEERING

Military mountaineering training provides tactical mobility in mountainous terrain that would otherwise be inaccessible. Soldiers with specialized training who are skilled in using mountain climbing equipment and techniques can overcome the difficulties of obstructing terrain. Highly motivated soldiers who are in superior physical condition should be selected for move advanced military mountaineering training (Levels 2 and 3) conducted at appropriate facilities. Soldiers who have completed advanced mountaineering training should be used as trainers, guides, and lead climbers during collective training. They may also serve as supervisors of installation teams and evacuation teams. Properly used these soldiers can drastically improve mobility and have a positive impact disproportionate to their numbers. Units anticipating mountain operations should strive to have approximately ten percent of their force achieve advanced mountaineering skills.

A-1. LEVEL 1: BASIC MOUNTAINEER

The basic mountaineer should be a graduate of a basic mountaineering course and have the fundamental travel and climbing skills necessary to move safely and efficiently in mountainous terrain. These soldiers should be comfortable functioning in this environment and, under the supervision of qualified mountain leaders or assault climbers, can assist in the rigging and use of all basic rope installations.

a. On technically difficult terrain, the basic mountaineer should be capable of performing duties as the "follower" or "second" on a roped climbing team, and should be well trained in using all basic rope systems. These soldiers may provide limited assistance to soldiers unskilled in mountaineering techniques.

b. Particularly adept soldiers may be selected as members of special purpose teams led and supervised by advanced mountaineers. Figure A-1 lists the minimum knowledge and skills required of basic mountaineers.

c. In a unit training program, Level 1 qualified soldiers should be identified and prepared to serve as assistant instructors to train unqualified soldiers in basic mountaineering skills. All high-risk training, however, must be conducted under the supervision of qualified Level 2 or 3 personnel.

• Characteristics of the mountain environment (summer and winter). • Mountaineering safety. • Use, care, and packing of individual cold weather clothing and equipment. • Care and use of basic mountaineering equipment. • Mountain bivouac techniques. • Mountain communications. • Mountain travel and walking techniques. • Hazard recognition and route selection. • Mountain navigation. • Basic medical evacuation.	• Rope management and knots. • Natural anchors. • Familiarization with artificial anchors. • Belay and rappel techniques. • Use of fixed ropes (lines). • Rock climbing fundamentals. • Rope bridges and lowering systems. • Individual movement on snow and ice. • Mountain stream crossings (to include water survival techniques). • First aid for mountain illnesses and injuries.

Figure A-1. Level 1, basic mountaineering tasks.

A-2. LEVEL 2: ASSAULT CLIMBER

Assault climbers are responsible for the rigging, inspection, use, and operation of all basic rope systems. They are trained in additional rope management skills, knot tying, and belay and rappel techniques, as well as using specialized mountaineering equipment. Assault climbers are capable of rigging complex, multipoint anchors, and high-angle raising/lowering systems. Level 2 qualification is required to supervise all high-risk training associated with Level 1. At a minimum, assault climbers should possess the additional knowledge and skills shown in Figure A-2.

• Use specialized mountaineering equipment. • Perform multipitch climbing: – Free climbing and aid climbing. – Leading on class 4 and 5 terrain. • Conduct multipitch rappelling. • Establish and operate hauling systems. • Establish fixed ropes with intermediate anchors.	• Move on moderate angle snow and ice. • Establish evacuation systems and perform high-angle rescue. • Perform avalanche hazard evaluation and rescue techniques. • Be familiar with movement on glaciers.

Figure A-2. Level 2, assault climber tasks.

A-3. LEVEL 3: MOUNTAIN LEADER

Mountain leaders possess all the skills of the assault climber and have extensive practical experience in a variety of mountain environments in both winter and summer conditions. Level 3 mountaineers should have well-developed hazard evaluation and safe route finding skills over all types of mountainous terrain. Mountain leaders are best qualified to advise commanders on all aspects of mountain operations, particularly the preparation and leadership required to move units over technically difficult, hazardous, or exposed terrain. The mountain leader is the highest level of qualification and is the principle trainer for conducting mountain operations. Instructor experience at a military mountaineering center or as a member of a special operations forces (SOF) mountain team is critical to acquiring Level 3 qualification. Figure A-3 outlines the additional knowledge and skills expected of mountain leaders. Depending on the specific AO, mountain leaders may need additional skills such as snowshoeing and all-terrain skiing.

• **Recognizing and evaluating peculiar terrain, weather, and hazards.** • **Preparing route, movement, bivouac, and risk management plans for all conditions and elevation levels.** • **Using roped movement techniques on steep snow and ice.** • **Performing multipitch climbing on mixed terrain (rock, snow, and ice).**	• **Performing glacier travel and crevice rescue.** • **Establishing and operating technical high-angle, multipitch rescue and evacuation systems.** • **Usiing winter shelters and survival techniques.** • **Leading units over technically difficult, hazardous, or exposed terrain in both winter and summer conditions.**

Figure A-3. Level 3, mountaineer leader tasks.

APPENDIX B

MEASUREMENT CONVERSION FACTORS

MULTIPLY	BY	TO OBTAIN
Millimeters	.03937	Inches
Centimeters	.3937	Inches
Centimeters	.03281	Feet
Meters	39.37	Inches
Meters	3.281	Feet
Meters	1.0936	Yards
Kilometers	.62137	Miles
Knots	1.1516	MPH

APPENDIX C

AVALANCHE SEARCH AND RESCUE TECHNIQUES

The effect of an avalanche can be disastrous. Chances of survival after burial by an avalanche are approximately 90 percent if the victim is located within the first 15 minutes. Probability of survival drops rapidly and, after two hours, chances of survival are remote. Suffocation accounts for 65 percent of avalanche fatalities, collision with obstacles such as rocks and trees accounts for 25 percent, and hypothermia and shock accounts for 10 percent.

The best chance of survival in snow country is to avoid an avalanche; but, if a member of your group ***is*** *in an avalanche, they are depending on you for rescue!*

C-1. IMMEDIATE ACTION

Survivors at the avalanche site organize into the first rescue team and immediately start rescue operations. If any indication of the location of the victim is found, random probing starts in that vicinity. The tip and edges of the slide are also likely areas to search. A human body is bulky and is apt to be thrown toward the surface or the sides.

C-2. GENERAL PROCEDURES

Establish from witnesses where the victim was located just before the avalanche to determine the point where the victim disappeared—the "last seen" point. Using this and any other information, establish a probable victim trajectory line leading to high priority search areas. Make a quick but systematic check of the slide area and the deposition area, and mark all clues. Look for skis, poles, ice axes, packs, gloves, hats, goggles, boots, or any other article the person may have been carrying—it might still be attached to the victim.

a. Organize initial searchers and probers. If using avalanche beacons, immediately select personnel to begin a beacon search. Ensure all other beacons are turned off or to receive to eliminate erroneous signals. All personnel should have a shovel or other tool for digging or, if enough personnel are available, a digger can be standing by to assist when needed. If the initial search reveals items from the victim, make an initial probe search in that area. This probing should take only a few seconds.

b. Make a coarse probe of all likely areas of burial, and repeat it as long as a live rescue remains possible. Resort to the fine probe only when the possibility of a live rescue is highly improbable. Unless otherwise indicated, start the coarse probe at the deposition area.

C-3. ESTABLISHING THE VICTIM'S MOST PROBABLE LOCATION

In many respects, a moving avalanche resembles a liquid. A human body, with a higher density than the flowing snow, would be expected to sink deeper and deeper into the avalanche; however, several factors influence the body's location. Turbulence, terrain, and the victim's own efforts to extricate himself all interact to determine the final burial position. Study of a large number of case histories leads to the following conclusions.

- The majority of buried victims are carried to the place of greatest deposition, usually the toe of the slide.
- If two points of the victim's trajectory can be established, a high probability exists that the victim will be near the downhill flow line passing through these two points.
- Any terrain features that catch and hold avalanche debris are also apt to catch a victim.
- If an avalanche follows a wandering gully, all debris deposit areas are likely burial spots. The likelihood of a victim being buried in a particular bend is proportional to the amount of debris deposited there.
- Vegetation, rocks, and other obstacles act as snares. The victim tends to be retained above the obstacle. An obstacle may simply delay the victim's motion, leading to final burial down flow from the obstacle.
- Maximum speed of the flowing snow occurs at the avalanche center. Friction reduces flow velocity along the edges. The closer the victim's trajectory is to the center of the slide, the greater will be his burial depth.
- Efforts of the victim to extricate himself by vigorous motion and "swimming" definitely minimize burial depth. Conversely, the limp body of an unconscious victim is likely to be buried deeply.
- An occasional exception to the above is emphasized. The victim may not be buried but may have been hurled away from the avalanche by wind blast. In the case of large and violent avalanches, a search of the surrounding terrain is advisable. Victims have been located in tree tops outside the slide area.

Use of avalanche transceivers is the most efficient method of searching for an avalanche victim, but only if the victim is wearing an active transceiver. Many models of transceivers are available, each with its own manufacturer's instructions for proper use and care. All currently available transceivers are compatible, although they may operate differently in the search mode.

C-4. PROBING FOR AVALANCHE VICTIMS

Probing offers the advantage of requiring simple equipment that can be operated by personnel with no previous training. Although the probers do not need previous training the search leader must be familiar with the technique to ensure proper execution of the probe line.

a. **Probe Poles.** Rigid steel tubing approximately 3/4-inch in diameter and approximately 10 feet long is recommended for the primary probe pole. Longer poles are difficult to manage, especially in a high wind. Although this type of pole performs best, it is difficult to transport to the avalanche site because of its length and weight.

(1) Each person operating in avalanche areas should carry folding sectional poles. These poles are similar to folding tent poles, but are stronger and are connected with cable instead of bungee cord. These poles should be carried on the outside of the pack for immediate access.

(2) If no probing poles are available, initial probing attempts can be started using ski poles in one of two ways: the ski pole can be reversed, probing with the wrist strap down; or the basket can be removed so that the point is down (the preferred method), which allows the ski pole to penetrate the snow more easily.

b. **Probing Lines.** For the probing operation to be effective, probing lines must be orderly and properly spaced. To ensure systematic and orderly probing, the number of personnel per line should be limited. Twenty per line is satisfactory, while thirty is normally the upper limit. The number of probers in the line will be dictated by not only the width of the area to be probed but the number of personnel available. A string may be used to keep the probe lines aligned, but will require added time to maintain.

(1) The probe line maintains a steady advance upslope. Advancing uphill automatically helps set the pace and permits easy probing to the full length of the probe. Probing does not come to a halt when a possible contact is made. The probe is left in contact and the line continues. A shovel crew follows up on the strike by digging down along the pole. Extra probes are carried by the shovel crew to replace those left in contact. Such a plan of operation is especially important when more than one victim is buried.

(2) Striking a body gives a distinct feel to the probe, which is easily recognizable in soft snow but less recognizable in hard compacted snow. A common problem is encountering debris within the snow that can be mistaken for the victim. The only sure check is by digging.

c. **Probing Techniques.** Two distinct probing methods are recognized: coarse probe and fine probe. As evidenced by their names, coarse probing implies a wider spacing of probe pole insertions with emphasis on speed. Fine probing involves close-spaced probing with emphasis on thoroughness. Coarse probing is used during initial phases of the search when live recovery is anticipated. Fine probing is the concluding measure, which almost guarantees finding the body. The coarse probe technique has a 70-percent chance of locating the victim on a given pass, while the fine probe has, essentially, a 100-percent chance of locating the body.

(1) The coarse probe functions as follows:

(a) Probers are spaced along a line 30 inches center to center, with feet about 15 inches apart.

(b) A single probe pole insertion is made at the center of the straddle span.

(c) On command of the probe line commander, the group advances 20 inches and repeats the single probe.

(d) Three commands are used for the complete sequence:

- "DOWN PROBE."
- "UP PROBE."
- "STEP FORWARD."

By using these commands, the leader can maintain closer control of the advancing probe line. It is important that the commands be adjusted to a rhythm that enforces the maximum reasonable pace. A string may also be used along the probe line to keep the probers dressed, although this requires the use of two soldiers to control the string. Strict discipline and firm, clear commands are essential for efficient probing. The probers themselves work silently.

(2) The fine probe functions as follows:

(a) Probers are spaced the same as for the coarse probe. Each man probes in front of his left foot, then in the center of his straddled position, and finally in front of his right foot.

(b) On command, the line advances 1 foot and repeats the probing sequence. Each probe is made 10 inches from the adjacent one.

(c) The commands for the fine probe are:

- "LEFT PROBE."
- "UP PROBE."
- "CENTER PROBE."
- "UP PROBE."
- "RIGHT PROBE."
- "UP PROBE."
- "STEP FORWARD."

(d) Good discipline and coordinated probing is even more important in fine probing than with the coarse probe. Careless or irregular probing can negate the advantages of fine probing. Use of a string to align the probers is especially important with the fine probe. The three insertions are made along the line established by the string, which is then moved ahead 1 foot.

GLOSSARY

AI	alpine ice
ALICE	all-purpose, lightweight, individual carrying equipment
AMS	acute mountain sickness
APS	anchor pulley system
ATC	air traffic controller
BDU	battle dress uniform
BTC	bridge team commander
C	Celsius
CE	community Europe
CEN	combined European norm
cm	centimeter
CMS	chronic mountain sickness
CPR	cardio-pulmonary resuscitation
CTA	common table of allowance
DA	Department of the Army
ECWCS	extreme cold weather clothing system
EN	European norm
F	Fahrenheit
FM	field manual
ft	foot; feet
GPS	Global Positioning System
GTA	graphic training aid
HACE	high-altitude cerebral edema
HAPE	high-altitude pulmonary edema
IAW	in accordance with
LBV	load-bearing vest
LCD	liquid crystal diode
LCE	load-carrying equipment
MEDEVAC	medical evacuation
MPH	miles per hour
MPS	moveable pulley system
MRE	meal ready-to-eat

NSN	national stock number
OP	observation post
PABA	para-amino benzoic acid
PCD	progress capture device
PFD	personal flotation device
RURP	realized ultimate reality piton
SE	southeast
SLCD	spring-loaded camming device
SOF	special operations forces
SOP	standard operating procedure
SOSES	shape, orientation, size, elevation, slope
SPF	sun protection factor
TOE	table of organization and equipment
UIAA	Union des International Alpine Association
U.S.	United States
USAF	United States Air Force
USN	United States Navy
UV	ultraviolet
UVA	ultraviolet A (radiation wavelengths between 320 and 400 nanometers)
UVB	ultraviolet B (radiation wavelengths between 295 and 320 nanometers)
WI	waterfall ice
YDS	Yosemite Decimal System

INDEX

ROPE LOG *(USAGE AND HISTORY)* For use of this form, seeFM 3-97.61; the proponent agency is TRADOC			UNIT ID MARKING
NSN	DOCUMENT NUMBER	SERIAL NUMBER	MFR LOT NUMBER
DATE OF MFR	ISSUE DATE	DATE IN SERVICE	LENGTH
DIAMETER	FIBER	COLOR	CONSTRUCTION

INSPECT ROPE FOR DAMAGE OR EXCESSIVE WEAR EACH TIME IT IS DEPLOYED AND AGAIN AFTER EACH USE. IMMEDIATELY RETIRE ALL SUSPECT ROPES.

DATE USED	LOCATION	TYPE OF USE	ROPE EXPOSURE	INSPECTOR'S INITIAL/DATE	ROPE CONDITION AND COMMENTS

DA FORM 5752-R, MAY 89

DATE USED	LOCATION	TYPE OF USE	ROPE EXPOSURE	INSPECTOR'S INITIAL/DATE	ROPE CONDITION AND COMMENTS

MANUFACTURER ______________________________

____

REVERSE OF DA FORM 5752-R, MAY 89

Field Manual
No. 4-02.2

Headquarters
Department of the Army
Washington, DC, 8 May 2007

MEDICAL EVACUATION

Contents

Distribution Restriction: Approved for public release; distribution is unlimited.

***This publication supersedes Chapters 1–7 and Appendixes A, B, D–F, K, L, and N of FM 8-10-6 dated 14 April 2000, and FM 8-10-26 dated 16 February 1999.**

Figures

Tables

Preface

This field manual (FM) provides doctrine, as well as techniques and procedures for conducting medical evacuation and medical regulating operations. Medical evacuation encompasses both the evacuation of Soldiers from the point of injury (POI) or wounding to a medical treatment facility (MTF) staffed and equipped to provide essential care in theater and further evacuation from the theater to provide definitive, rehabilitative, and convalescent care in the continental United States (CONUS) and the movement of patients between MTFs or to staging facilities. Medical evacuation entails the provision of en route medical care; supports the joint health service support (JHSS) system; and links the continuum of care. In addition, it discusses the difference between medical evacuation and casualty evacuation (CASEVAC), as well as coordination requirements for and the use of nonmedical transportation assets to accomplish the CASEVAC mission. This publication is intended for use by medical commanders and their staffs, command surgeons, and nonmedical commanders involved in medical evacuation operations.

Users of this publication are encouraged to submit comments and recommendations to improve this publication. Comments should include the page, paragraph, and line(s) of the text where the change is recommended. The proponent for this publication is the United States (US) Army Medical Department (AMEDD) Center and School (USAMEDDC&S). Comments and recommendations should be forwarded, in letter format, directly to **Commander, USAMEDDC&S, ATTN: MCCS-FCD-L, 1400 East Grayson Street, Fort Sam Houston, Texas 78234-5052** or by using the e-mail address: medicaldoctrine@amedd.army.mil.

This publication applies to the Active Army, the Army National Guard (ARNG)/Army National Guard of the United States (ARNGUS), and the U.S. Army Reserve (USAR), unless otherwise stated.

Unless this publication states otherwise, masculine nouns and pronouns do not refer exclusively to men.

Use of trade or brand names in this publication is for illustrative purposes only and does not imply endorsement by the Department of Defense (DOD).

The staffing and organization structure presented in this publication reflects those established in base tables of organization and equipment (TOE) and are current as of the publication print date. However, such staffing is subject to change to comply with manpower requirements criteria outlined in Army Regulation (AR) 71-32. Those requirements criteria are also subject to change if the modification table of organization and equipment (MTOE) is significantly altered.

This publication implements the following North Atlantic Treaty Organization (NATO) International Standardization Agreements (STANAGs) and American, British, Canadian, and Australian (ABCA) Quadripartite Standardization Agreements (QSTAGs):

TITLE	STANAG	QSTAG
Stretchers, Bearing Brackets, and Attachment Supports	2040	
Medical Employment of Air Transport in the Forward Area	2087	
Medical and Dental Supply Procedures	2128	
Minimum Labeling Requirements for Medical Materiel		436
Documentation Relative to Medical Evacuation, Treatment, and Cause of Death of Patients	2132	470
Road Movements and Movement Control	2454	

TITLE	STANAG	QSTAG
Orders for the Camouflage of the Red Cross and Red Crescent on Land in Tactical Operations	2931	
Aeromedical Evacuation	3204	

When amendment, revision, or cancellation of this publication is proposed which will affect or violate the international agreements concerned, the preparing agency will take appropriate reconciliatory action through international standardization channels. These agreements are available on request from the Standardization Documents Order Desk, 700 Robins Avenue, Building 4, Section D, Philadelphia, Pennsylvania 19111-5094.

The AMEDD is in a transitional phase with terminology. This publication uses the most current terminology; however, other FM 4-02-series and FM 8-series may use the older terminology. Changes in terminology are a result of adopting the terminology currently used in the joint and/or NATO and ABCA Armies publication arenas. Also, the following terms are synonymous and the current terms are listed first:

- Medical logistics (MEDLOG); health service logistics (HSL); and combat health logistics (CHL).
- Roles of care, echelons of care, and level of care.
- Combat and operational stress control (COSC) and combat stress control (CSC).
- Behavioral health (BH) and mental health (MH).
- Chemical, biological, radiological and nuclear (CBRN) and nuclear, biological, and chemical (NBC).
- Stability operations; stability, security, transition, and reconstruction (SSTR) operations; and stability operations and support operations.

Introduction

The Army Health System (AHS) is a complex system of interrelated and interdependent systems which provides a continuum of medical treatment from the POI or wounding through successive roles of health care to definitive, rehabilitative, and convalescent care in the CONUS, as required. *Medical evacuation* is the system which provides the vital linkage between the roles of care necessary to sustain the patient during transport. This is accomplished by providing en route medical care and emergency medical intervention, if required, and to enhance the individual's prognosis and to reduce long-term disability.

Medical evacuation occurs at the tactical, operational, and strategic levels and requires the synchronization and integration of service component medical evacuation resources and procedures with the DOD worldwide evacuation system operated by the United States Transportation Command (USTRANSCOM).

Army medical evacuation is a multifaceted mission accomplished by a combination of dedicated ground and air evacuation platforms synchronized to provide direct support (DS), general support (GS), and area support within the joint operations area (JOA). At the tactical level, organic or DS medical evacuation resources locate, acquire, treat, and evacuate Soldiers from the POI or wounding to an appropriate MTF where they are stabilized, prioritized, and, if required, prepared for further evacuation to an MTF capable of providing required essential care within the JOA.

Although the most recognized mission of Army medical evacuation assets is the evacuation and provision of en route medical care to battlefield wounded, the essential and vital functions of medical evacuation resources encompass many additional missions and tasks that support the JHSS system. Medical evacuation resources are used to transfer patients between MTFs within the JOA and from MTFs to United States Air Force (USAF) mobile aeromedical staging facilities (MASFs) or aeromedical staging facilities (ASFs); emergency movement of Class VIII, blood and blood products, medical personnel and equipment; and serve as messengers in medical channels.

Medical regulating provides the interface with the DOD worldwide medical evacuation system by determining the patient's destination (the MTF best suited to provide the required care) and scheduling the means to transport the patient with the required en route medical care. Formal medical regulating begins at Role 3, however technological advances in information management (IM)/information technology (IT) are permitting this capability to be used at Role 1 and Role 2 MTFs in some situations.

Chapter 1

Overview of Army Health System Operations and Medical Evacuation

The AHS is the Army component of the military health system (MHS). Its capabilities are focused on delivering health care across the continuum of military operations—from the POI or wounding, through the JOA, to the CONUS support base. The two missions of the AHS are to provide health service support (HSS) (casualty care, medical evacuation, and medical logistics) and force health protection (FHP) (casualty prevention). The AHS is focused on promoting wellness, preventing casualties due to disease and nonbattle injuries (DNBIs), and providing timely and effective casualty care and management. Medical evacuation is the key factor in ensuring the continuity of care provided to our Soldiers by providing en route medical care during evacuation, facilitating the transfer of patients between MTFs to receive the appropriate specialty care, and ensuring that scarce medical resources (personnel, equipment, and supplies [to include blood]) can be rapidly transported to areas of critical need on the battlefield.

SECTION I — ARMY HEALTH SYSTEM

1-1. The provision of AHS is governed by well established and time-tested principles and rules which ensure the care provided to our Soldiers is timely and effective. For an in-depth discussion of these principles, rules, and roles of medical care, refer to FM 4-02.

1-2. The AHS is comprised of 10 medical functions. They are—

- Medical command, control, communications, computers, and intelligence.
- Medical treatment.
- Medical evacuation.
- Hospitalization.
- Dental services.
- Preventive medicine (PVNTMED) services.
- Combat and operational stress control (COSC).
- Veterinary services.
- Medical logistics.
- Medical laboratory support.

1-3. Army health system resources are arrayed across the battlefield in successive levels of support. These successive levels have increased medical care capabilities at each higher level. Medical evacuation and the provision of en route medical care ensures an uninterrupted continuum of care is maintained while Soldiers are moved through the roles of medical care to the MTF best suited to treat the patient's specific injuries.

PRINCIPLES OF ARMY HEALTH SYSTEM

1-4. The principles of AHS provide a framework in which medical planners can ensure that comprehensive plans are developed to support the tactical commander's operation plan (OPLAN). The principles of AHS are conformity, continuity, control, proximity, flexibility, and mobility. For this publication, the discussion of these principles has been focused toward the medical evacuation mission. For a general discussion of how these principles relate to the overall AHS mission, refer to FM 4-02.

CONFORMITY

1-5. Participating in the development of the OPLAN or the operation order (OPORD) ensures that the medical planner conforms to strategic, operational, and tactical plans. This is the most fundamental element for effectively providing AHS and ensures medical influence over the execution of medical evacuation operations. Only by participating in the orders process and developing a medical evacuation plan, will the medical planner ensure that medical evacuation support is arrayed on the battlefield in the right place at the right time and synchronized across operational commands to maximize responsiveness and effectiveness.

CONTINUITY

1-6. En route medical care provided during medical evacuation must be effective and continuous to prevent interruptions in the continuum of care. An interruption in medical treatment may result in an increase in morbidity, mortality, and disability. No patient is evacuated any farther than his physical condition or the military situation requires.

1-7. Medical evacuation resources provide the linkage between the roles of care within the JOA. They also provide interface with other deployed elements of the MHS operated by other services to enhance and facilitate the continuum of care from the POI to the CONUS support base.

CONTROL

1-8. Medical planners must ensure medical control is exercised over the execution of ground medical evacuation operations and that medical influence (technical and operational supervision) is exercised over the execution of aeromedical evacuation (AE) operations. Furthermore, medical planners must ensure the medical evacuation system is responsive to changing requirements and tailored to effectively support the forces within an assigned area of operations (AO). Since medical evacuation resources are limited, it is essential that medical control and influence be retained at the highest level consistent with the tactical situation.

1-9. A thorough and comprehensive medical evacuation plan is essential to establishing and maintaining control of medical evacuation operations characterized by decentralized execution of the plan. The medical evacuation plan complies with the combatant command guidance and intent and maximizes the use of scarce medical evacuation resources. When directed by the combatant command, Army air and ground ambulances may support operations conducted by other services, allied and coalition partners, and the host nation (HN).

PROXIMITY

1-10. The location of medical evacuation assets in support of combat operations is dictated by orders and the tactical situation (mission, enemy, terrain and weather, troops and support available, time available, and civil considerations [METT-TC]). Accurately determining time and distance factors and the availability of evacuation resources are critical to determining the disposition of evacuation assets. The speed with which medical evacuation is initiated is extremely important in reducing morbidity, mortality and disability. Medical evacuation time must be minimized by the effective disposition of resources, ensuring close proximity of both supported elements and MTF. Medical evacuation assets cannot be located so far forward that they interfere with the conduct of combat operations. Conversely, they must not be located so far to the rear that medical treatment is delayed due to lengthy evacuation routes.

1-11. Medical evacuation resources, both ground and air, are arrayed on the battlefield to best support both the tactical commander and the AHS. Depending upon the situation, evacuation resources may be placed in a DS role to support maneuvering forces or GS, during stability operations which are centrally located to accomplish an area support mission.

FLEXIBILITY

1-12. Changes in tactical plans or operations may require redistribution or reallocation of medical evacuation resources. Therefore, the medical evacuation plan must be designed to ensure flexibility and agility as well as enhance the ability to rapidly task-organize and relocate medical evacuation assets to meet changing battlefield requirements. Medical planners must also ensure medical control and influence is exercised through the orders process and facilitates the synchronization of air and ground evacuation assets to rapidly clear the battlefield.

MOBILITY

1-13. Medical evacuation assets must have the same mobility and survivability capability (such as armor protection) as the forces supported. This mobility and survivability ensures that medical evacuation resources can rapidly respond and that evacuation routes do not become too lengthy. Medical evacuation assets also enhance the mobility of forward deployed MTFs by rapidly evacuating their patients to the next role of medical care.

BATTLEFIELD RULES

1-14. The AMEDD has developed a set of battlefield rules to aid in establishing priorities and to resolve competing priorities within AHS activities. These rules are intended to guide the medical planner to resolve system conflicts encountered in designing and coordinating AHS operations. Although medical personnel seek always to provide the full scope of AHS services and support in the best possible manner, during every combat operation there are inherent possibilities of conflicting support requirements. The planner or operator applies these rules to ensure that the conflicts are resolved appropriately. These battlefield rules are depicted in Table 1-1.

Table 1-1. Army Medical Department Battlefield Rules

AMEDD BATTLEFIELD RULES
BE THERE (MAINTAIN A MEDICAL PRESENCE WITH THE SOLDIER)
MAINTAIN THE HEALTH OF THE COMMAND
SAVE LIVES
CLEAR THE BATTLEFIELD OF CASUALTIES
PROVIDE STATE-OF-THE-ART MEDICAL CARE
ENSURE EARLY RETURN TO DUTY OF THE SOLDIER

BE THERE

1-15. Ensure that medical evacuation assets are in close proximity to supported elements to enhance response time, increase Soldier confidence and be a combat multiplier. This is accomplished by complementing organic medical evacuation assets with medical evacuation assets placed in DS, GS, and area support roles.

MAINTAIN THE HEALTH OF THE COMMAND

1-16. Ensure that medical evacuation assets are in close proximity to the MTF. The closer evacuation assets are to an MTF, the more rapidly care may be rendered.

SAVE LIVES

1-17. Ensure medical evacuation assets are the primary means of evacuating patients. Medical evacuation assets provide en route medical care that provides a continuum of care that is instrumental in preserving life and reducing long-term disability. The use of CASEVAC should only be used in extreme emergencies or when the medical evacuation system is overwhelmed. Refer to paragraphs 1-27 through 1-33 for a discussion on the differences between medical evacuation and CASEVAC operations.

PROVIDE STATE-OF-THE ART MEDICAL CARE

1-18. Ensure sufficient medical evacuation assets are available to facilitate the movement of patients between MTFs (intratheater evacuation) and the emergency movement of Class VIII, blood and blood products, and medical personnel and equipment. This capability permits the medical planner to maximize the use of scarce medical resources and exercise economy of force without sacrificing state-of-the-art medical care.

EARLY RETURN TO DUTY OF THE SOLDIER

1-19. Ensure seamless integration with the intertheater medical evacuation system. This allows rapid evacuation and continuous medical treatment throughout the continuum of care resulting rapid return to duty (RTD) rates.

SECTION II — MEDICAL EVACUATION

PURPOSE

1-20. An efficient and effective medical evacuation system—

- Minimizes mortality by rapidly and efficiently moving the sick, injured, and wounded to an MTF.
- Serves as a force multiplier as it clears the battlefield enabling the tactical commander to continue his mission with all available combat assets.
- Builds the morale of Soldiers by demonstrating that care is quickly available if they are wounded.
- Provides en route medical care that is essential in improving the prognosis and reducing disability of the wounded, injured, or ill Soldiers.
- Provides medical economy of force.
- Provides connectivity of the AHS as appropriate to the MHS.

Tactical	Operational	Strategic
Corps Division BCT Emergency Class VIII Resupply Emergency Movement of Medical Personnel/Equipment Stability Operations	Coalition Joint Class VIII Resupply Joint Blood Program DOD Civilians/Contractors Medical Personnel/Equipment Military Working Dogs Homeland Security Operations Shore-to-Ship Stability Operations	Host Nation Nongovernment Organizations Private Volunteer Organizations DOD support to Stability Operations International Disaster Relief

Figure 1-1. Army medical evacuation as it supports full spectrum operations

1-21. The AHS is established in roles of increasing capability from the POI to definitive care. On the battlefield, casualties are evacuated rearward from one role of care to the next, the sequencing of this movement is dependent upon METT-TC factors. In a contiguous battlefield, well established lines of communication (LOC), large numbers of casualties, and wide array of MTFs can result in very deliberate evacuation from one sequential role to the next higher. However in many situations, such as noncontiguous battlefield, the array of medical resources across the battlefield, the availability of medical evacuation resources, and the number of patients being evacuated may facilitate procedures that permit bypassing roles of care in order to ensure the timely treatment and care of casualties. This evacuation plan will be established by the appropriate level of command in coordination with the command surgeon to ensure the best treatment is provided to all casualties.

ATTRIBUTES

DEDICATED RESOURCES

1-22. The Army medical evacuation system is comprised of dedicated air and ground evacuation platforms. These platforms have been designed, staffed, and equipped to provide en route medical care to patients being evacuated and are used exclusively to support the medical mission. The focus of the medical evacuation mission coupled with the dedicated platforms permit a rapid response to calls for support. The dedicated nature of this mission dictates that Army medical evacuation unit's posture themselves in a "ready alert" status, ready to rapidly respond to evacuation missions and not diverted to perform any other task. Medical evacuation resources are protected under the provisions of the Geneva Conventions from intentional attack by the enemy. (For a discussion of the Geneva Conventions in relation to air and ground evacuation operations, refer to Appendix A.)

EN ROUTE MEDICAL CARE

1-23. En route care is provided on all Army medical evacuation platforms. This care is essential for minimizing mortality, enhancing survival rates, and reducing disability of wounded, injured, or ill Soldiers. Refer to paragraphs 1-27 through 1-33 for a discussion on the differences between medical evacuation and CASEVAC operations.

CONNECTIVITY

1-24. The Army provides connectivity to the MHS. The synchronized employment of medical evacuation resources provides and maintains the seamless continuum of care from the POI through successive roles of essential care within the theater. In addition to evacuating patients and providing en route medical care, medical evacuation resources provide for the emergency movement of scarce medical resources such as critical Class VIII, blood, medical personnel, and medical equipment. Further, medical evacuation resources are used to transfer patients from one MTF to another within the theater, to facilitate specialty care as well as transferring patients from an MTF to a MASF to facilitate intertheater evacuation. Figure 1-1 depicts the connectivity provided by medical evacuation platforms between the roles of care within theater. In addition to providing connectivity within the AHS, Army medical evacuation resources provide medical evacuation support and interface with MTFs of the other services deployed in the theater. In joint and multinational operations, the geographic combatant commander (GCC) may direct Army assets to provide this support and connectivity to joint, allied, and coalition forces within the JOA.

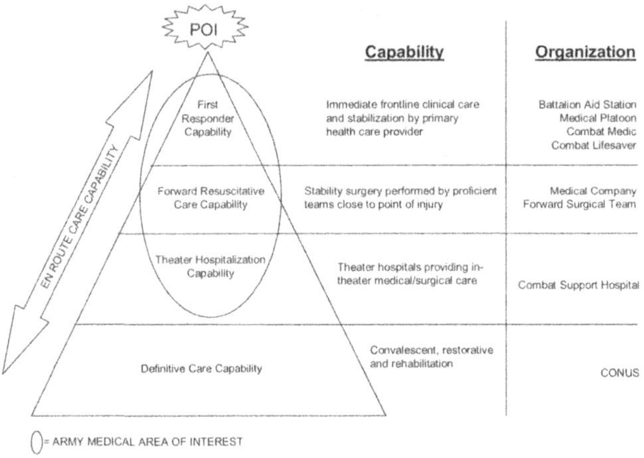

Figure 1-2. Army air and ground evacuation platforms provide connectivity to assure a seamless continuum of medical care

MEDICAL ECONOMY OF FORCE

1-25. The flexibility and versatility which medical evacuation resources afford the AHS to respond to urgent requirements throughout the battlefield is essential in ensuring the provision of seamless medical care. Medical economy of force is achieved for low density, high demand medical specialties (such as a neurosurgeon), medical supplies, and medical equipment (such as computer tomography scans or magnetic resonance imaging equipment) by having the capability to move the patient to the required care over long distances and by permitting the emergency of cross-leveling on medical supplies reducing the need for large Class VIII stockpiles. By providing medical economy of force, the deployed medical footprint is reduced without negatively impacting on the care provided to the Soldier.

FORCE MULTIPLIER

1-26. A highly effective, efficient, and responsive AHS increases Soldiers' confidence that if wounded in battle, they will receive timely and appropriate medical care to enhance their prognosis, speed their recovery, and reduce long-term disabilities. Medical evacuation resources locate and acquire wounded Soldiers and facilitate their entry into the AHS. These resources clear the battlefield of casualties which enables the tactical commander the ability to capitalize on and exploit battlefield opportunities with all available assets in the execution of his warfighting mission.

SECTION III — MEDICAL EVACUATION VERSUS CASUALTY EVACUATION

MEDICAL EVACUATION

1-27. Medical evacuation is performed by dedicated, standardized medical evacuation platforms, with medical professionals who provide the timely, efficient movement and en route care of the wounded, injured, or ill persons from the battlefield and/or other locations to MTFs. Medical evacuation is an AMEDD function that supports and is an integral part of the AHS. The provision of en route care on medically equipped vehicles or aircraft greatly enhances the patient's potential for recovery and may reduce long-term disability by maintaining the patient's medical condition in a more stable manner.

1-28. Medical evacuation ground/air ambulance platforms are defined as: Platforms especially for the medical evacuation mission with allocated medical equipment specifically designed for the purpose of enroute care and by trained medical personnel.

1-29. The gaining MTF in coordination with the losing MTF, is responsible for arranging for the evacuation of patients from the lower role of care. For example, Role 2 medical units are responsible for evacuating patients from Role 1 MTFs.

1-30. Medical evacuation begins when a request is initiated and continues until the patient is released from the AHS by medical authorities.

CASUALTY EVACUATION

1-31. Casualty evacuation is a term used to refer to the movement of casualties aboard nonmedical vehicles or aircraft.

WARNING

Casualties transported in this manner may not receive proper en route medical care or be transported to the appropriate MTF to address the patient's medical condition. If the casualty's medical condition deteriorates during transport, or the casualty is not transported to the appropriate MTF, an adverse impact on his prognosis and long-term disability or death may result.

1-32. If dedicated medical evacuation platforms (ground and air) are available, casualties should be evacuated on these conveyances to ensure they receive proper en route medical care.

1-33. Since CASEVAC operations can reduce combat power and degrade the efficiency of the AHS, units should only use CASEVAC to move Soldiers with less severe injuries when medical evacuation assets are overwhelmed. Medical planners should ensure CASEVAC operations are addressed in the OPLAN/OPORD as a separate operation, as these operations require preplanning, coordination, synchronization, and rehearsals. The CASEVAC plan should ensure casualties with severe or life-threatening injuries are prioritized for evacuation and are evacuated on dedicated medical evacuation platforms.

1-34. When possible, nonmedical vehicles/aircraft transporting casualties should be augmented with a trauma specialist or combat lifesaver (CLS). (On nonmedical aircraft, sufficient space may not be available to permit a caregiver to accompany the casualties.) The type of en route monitoring and medical care/first aid provided is limited by the following factors:

- Skill level of the individual providing care. (The combat medic is military occupational specialty [MOS]-qualified [MOS 68W] to provide emergency medical treatment [EMT]; the CLS is trained to provide enhanced first aid.) The combat medic can provide emergency medical intervention, whereas the CLS can only monitor the casualty and ensure that the basic lifesaving first-aid tasks are accomplished.
- Medical equipment available.
- Number of casualties being transported.
- Accessibility of casualties—If nonstandard evacuation vehicle is loaded with the maximum number of casualties, the combat medic or CLS may not be able to attend to the casualties while the vehicle is moving. If the condition of a casualty deteriorates and emergency measures are required, the vehicle will have to be stopped to permit care to be given.

SECTION IV — THEATER EVACUATION POLICY

ESTABLISHING THE THEATER EVACUATION POLICY

1-35. The theater evacuation policy is established by the Secretary of Defense, with the advice of the Joint Chiefs of Staff and upon the recommendation of the theater commander. The policy establishes, in number of days, the maximum period of noneffectiveness (hospitalization and convalescence) that patients may be held within the theater for treatment. This policy does not mean that a patient is held in the theater for the entire period of noneffectiveness. A patient who is not expected to be ready for RTD within the number of days established in the theater evacuation policy is evacuated to the CONUS or other safe haven. This is done providing that the treating physician determines that such evacuation will not aggravate the patient's disabilities or medical condition. For example, a theater evacuation policy of seven days does not mean that a patient is held in the theater for six days and then evacuated. Instead, it means that a patient is evacuated as soon as possible after the determination is made that the Soldier cannot be returned to duty within seven days following admission to a Role 3 hospital.

TEMPORARY REDUCTIONS

1-36. To the degree that unplanned for increases in patients occur (due perhaps to an epidemic or heavy combat casualties), a temporary reduction in the policy may be necessary. This reduction is used to adjust the volume of patients being held in the theater hospital system. A reduction in the evacuation policy increases the number of patients requiring evacuation out-of-theater and increases the requirement for evacuation assets. This action is necessary to relieve the congestion caused by the patient increases. A decrease in the theater evacuation policy increases the evacuation asset requirements.

ESSENTIAL CARE IN THEATER AND LENGTH OF POLICY

1-37. Due to the reduction of the medical footprint within the theater, health care delivery to deployed forces is now provided under the concept of essential care in theater. This concept provides for essential and stabilizing care being rendered within the theater, with the patient being evacuated to CONUS or other safe haven for definitive, rehabilitative, and convalescent care. In current and ongoing operations, if the patient cannot be treated and returned to duty within seven days of admission to a combat support hospital (CSH), the patient is evacuated from the theater for continued care.

1-38. The time period established by the theater evacuation policy starts on the date the patient is admitted to the first hospital. The total time a patient is hospitalized in the theater (including transit time between MTFs) for a single, uninterrupted episode of illness or injury should not exceed the number of days stated in the theater evacuation policy. Though guided by the evacuation policy, the actual selection of a patient for evacuation is based on clinical judgment as to the patient's ability to tolerate and survive the movement to the next role of care.

EXCEPTION TO POLICY

1-39. An exception to the theater evacuation policy may be required with respect to SOF personnel. This exception may be required to retain low density MOS skills within the theater. Retaining these personnel within the theater for an extended period of time is possible if the medical resources are available within the theater to treat their injuries and provide for convalescence and rehabilitation. If retention within the theater would result in a deterioration of their medical condition or would adversely impact on their prognosis for full recovery, they are evacuated from the theater for definitive care.

EVACUATION POLICY CONSIDERATIONS

1-40. Paragraphs 1-41 through 1-45 discuss the various considerations of the theater evacuation policy.

Physicians and Dentists

1-41. To physicians and dentists engaged in direct patient treatment and decisions relating to patient disposition, it means that there is a maximum period which clinical staffs may complete the necessary treatment needed to return the patient to full duty within the theater. If the theater policy is seven days and full RTD can be predicted within that time, the patient is retained in the theater. If the patient cannot be returned to full duty within seven days, the patient is evacuated out-of-theater as early as clinically prudent.

Medical Planners and Staff

1-42. The medical staff computes mix, number, and distribution of hospital beds required in the theater.

1-43. The medical operator has a management tool which, when properly adjusted and used, provides the balance between patient care and tactical support requirements. The medical staff is able to tailor a medical package specifically designed to handle the patient workloads, with maximum benefit to the patients and with maximum economy of available resources.

Logisticians

1-44. The nonmedical logistician is able to estimate his total obligation to support this system.

United States Air Force Planner

1-45. The USAF planner can accurately plan the USAF AE requirements for both intra- and intertheater patient movements.

FACTORS DETERMINING THE THEATER EVACUATION POLICY

1-46. To fully understand how the theater evacuation policy affects AHS operations, the medical operator should be aware of the factors that influence the establishment of this policy.

NATURE OF THE OPERATION

1-47. A major factor in determining the theater evacuation policy is the nature of the tactical operations. For example: Will the operations be of short duration and with a low potential for conflict? Will there be operations of long duration with significant combat operations? Will chemical, biological, radiological, nuclear, and high yield explosives (CBRNE) be employed? Will only conventional weapons be used? Is a static combat situation expected? Is there a significant threat of terrorist activities? Are the majority of patients anticipated to be DNBI patients or those with combat-related trauma?

Number and Type of Patients

1-48. Another factor affecting the policy is the number and types of patients anticipated and the rate of patient RTD. Admission rates vary widely in different geographical areas of the world and in different types of military operations.

Evacuation Means

1-49. The means (quantity and type of transportation) available for evacuation of patients from the theater to CONUS is an essential factor impacting on the evacuation policy.

Availability of Replacements

1-50. The capability of CONUS to furnish replacements to the theater is another consideration. For each patient who is evacuated from the theater to CONUS, a fully trained and equipped replacement must be provided. During a small-scale conflict overseas, the CONUS replacement capability is much greater than when compared to a large-scale conflict such as World War II.

Availability of In-Theater Resources

1-51. Limitations of all AHS resources (such as insufficient number and types of medical units in the division/corps to support the brigade combat teams [BCTs] and an insufficient amount of medical and nonmedical logistics) will have an impact on the theater evacuation policy. The more limitations (or shortages), the shorter the theater evacuation policy.

IMPACT OF THE EVACUATION POLICY ON ARMY HEALTH SYSTEM REQUIREMENTS

Shorter Evacuation Policy

1-52. A shorter theater evacuation policy—

- Results in fewer hospital beds required in the theater and a greater number of beds required elsewhere.
- Creates a greater demand for intertheater USAF evacuation resources. (A shortened intratheater evacuation policy would likewise increase the number of airframes required in the theater.)
- Increases the requirements for replacements to meet the rapid personnel turnover which could be expected, especially in combat units. (The impact this would have on both intra- and intertheater transportation and other requirements must also be considered.)

Longer Evacuation Policy

1-53. A longer theater evacuation policy—

- Results in a greater accumulation of patients and a demand for a larger AHS infrastructure in the theater. It decreases bed requirements elsewhere.
- Increases the requirements for medical logistics (medical supplies, equipment, and equipment maintenance) and nonmedical logistics support.
- Increases the requirements for hospitals, engineer support, and all aspects of base development for deployed AHS force. (It demands the establishment of a larger number of hospitals within the theater and may require medical specialty augmentation.)
- Provides for a greater proportion of patients to RTD within the theater and, thus, reduces the loss of experienced manpower.
- A longer intratheater evacuation policy may decrease the demand on the intratheater evacuation assets and system.

1-54. The concept of essential care in theater does not support longer evacuation policies as the deployed hospitals are not designed to provide definitive, rehabilitative, and convalescent care/services. If the theater evacuation policy is extended in theaters operating under the essential care in theater concept, augmentation of medical specialty resources will be required.

PATIENT STABILIZATION

1-55. The evacuation policy has no impact on the patient stabilization period for movement. This period is known as the *evacuation delay*. It is the period of time planned for between the time of patient reporting and the time of medical evacuation of the patient to the next role of care. Evacuation delays normally range from 24 to 72 hours and are designated by the Army service component command (ASCC) surgeon.

Chapter 2

Medical Evacuation Resources

This chapter discusses the mission, functions, and capabilities of medical evacuation units and elements, as specified in the unit's TOE. It also discusses the command and control (C2) headquarters to which they are assigned. The discussion of each organization also includes the basis of allocation (BOA), limitations, and dependencies.

SECTION I — MANEUVER BATTALION MEDICAL PLATOON AMBULANCES

MEDICAL PLATOON AMBULANCE SQUADS

2-1. The ground ambulance team is the basic medical evacuation element used within the BCTs and at the corps and theater AOs. These ground ambulance teams provide medical evacuation from the POI to supporting MTFs while ensuring the continuity of care en route. Ambulance squads consisting of two ambulance teams organic to the medical platoons of movement and maneuver and fires units and to the evacuation platoon of the brigade support medical companies (BSMCs).

2-2. The primary mission of the maneuver medical platoon ambulance squads is to provide ground ambulance evacuation support from supported infantry/armor companies or from POI back to a casualty collection point (CCP) or to the Role 1 MTF/battalion aid station (BAS). Maneuver medical platoon ambulance teams are assigned to heavy BCT (HBCT), infantry BCT, Stryker BCT (SBCT), reconnaissance squadrons, and reconnaissance and target acquisition (RSTA) squadrons. They also provide area support to other elements (which do not have organic medical evacuation resources) operating in their AO. Ground ambulance support provided is consistent with evacuation precedence, tactical standing operating procedures (TSOPs), and other operational considerations.

2-3. For definitive information on maneuver battalion medical platoon operations, see FM 4-02.4, FM 4-02.6, and FM 4-02.21.

MANEUVER UNIT MEDICAL PLATOON

ORGANIZATION

2-4. The medical platoon consists of a platoon headquarters section, a treatment squad, combat medic section, and an ambulance squad. The number of ambulance squads in an ambulance platoon varies and is based on the type of parent organization. The infantry, airborne, and air assault battalions' ambulance platoons have two ambulance squads equipped with high mobility multipurpose wheeled vehicle (HMMWV) ambulances. The mechanized infantry and armor combat maneuver battalions' ambulance platoons have four ambulance squads equipped with M-113 track ambulances. The SBCT infantry battalion's and RSTA squadron's ambulance platoons have two ambulance squads equipped with Stryker wheeled armored ambulances (referred to as the Stryker medical evacuation vehicle [MEV]).

Medical Platoon Headquarters

2-5. The medical platoon leader is a physician and also serves as the battalion surgeon. He is assisted by the medical operations officer (field medical assistant) in the operational, administrative, and logistical support aspects of the platoon.

Ambulance Section

2-6. The ambulance section of the medical platoon is organized into ambulance squads and is supervised by the platoon sergeant. Each squad contains a noncommissioned officer (NCO) squad leader, three combat medics, and two ambulances.

Ambulance Squad Leader

2-7. The ambulance squad leader ensures that ambulance team members perform operator maintenance on assigned vehicles. An ambulance team consists of one vehicle and two medical personnel (an aide/evacuation NCO and a combat medic).

Ambulance Team

2-8. Ambulance team members are also responsible for performing operator maintenance on assigned ambulances. Specific duties of the ambulance team are to—

- Administer EMT as required.
- Initiate the Department of Defense (DD) Form 1380, US Field Medical Card (FMC), as required.
- Prepare patients for ground and air medical evacuation.
- Provide medical evacuation of wounded or injured Soldiers from the POI to supporting MTF or ambulance exchange point (AXP).
- Provide patient care en route.
- Provide emergency movement of medical personnel and emergency delivery of blood, medical supplies, and medical equipment.
- Assist in the care of COSR casualties.
- Maintain operational readiness of the ambulance.
- Resupply company and platoon trauma specialists, when required.
- Operate the vehicle and maintain contact with supported and supporting elements.
- Find and collect the wounded.
- Serve as messengers within medical channels.

PRIMARY FUNCTIONS OF AMBULANCE TEAMS

2-9. The primary function of the ambulance team is to collect and treat the sick, injured, and wounded Soldiers on the battlefield and to provide medical evacuation support from the POI, CCP, or AXP to the supporting MTF.

SECTION II — AMBULANCE/EVACUATION PLATOON—FORWARD SUPPORT, BRIGADE SUPPORT, OR AREA SUPPORT MEDICAL COMPANY

MEDICAL COMPANIES

2-10. Medical companies provide Role 2 MTF and medical evacuation. Medical companies assigned to or in DS of a BCT are referred to as either a forward support medical company (FSMC) or a BSMC. Area support medical companies (ASMCs) are assigned to multifunctional medical battalions and operate Role 2 MTF and provide medical evacuation support for division, and theater units. In addition, all medical companies provide Role 1 area medical support to units without organic medical personnel and may augment/reinforce maneuver medical platoons/sections with treatment and medical evacuation support.

Forward Support Medical Company

2-11. The FSMC, an Army of Excellence (AOE) designed unit is organized to provide triage and management of mass casualties, advanced trauma management (ATM), initial resuscitation and stabilization, care for patients with DNBI, and battle wounded and injured Soldiers. The FSMC also provides intervention for combat and operational stress disorders to include COSR and preparation of patients for further medical evacuation. The FSMC is organized into a company headquarters, a treatment platoon, an ambulance platoon, a PVNTMED section, and a BH section.

2-12. The ambulance platoon of FSMC-Light and Airborne units employ a platoon headquarters with four-wheeled ambulance squads (eight ambulances); the FSMC-Air Assault employs a platoon headquarters with three-wheeled ambulance squads (six ambulances); the FSMC-Heavy employs a platoon headquarters with three-tracked ambulance squads (six ambulances) and two-wheeled ambulance squads (four ambulances); and the FSMC-Airborne employs a platoon headquarters with four-wheeled ambulance squads (eight ambulances). A complete discussion on the organization, mission, and functions of the FSMC is provided in FM 4-02.6.

Brigade Support Medical Company

2-13. The new BSMC is designed to support the modular BCT. The BOA for the BSMC is one per maneuver brigade supported. The mission of the BSMC is to provide HSS to all BCT units operating within the brigade AO. The BSMC locates and establishes its company headquarters and also operates a Role 2 MTF and may operate Role 1 MTFs on an area support basis for units that do not have organic medical assets. The company provides C2 for its organic and attached/operational control (OPCON) medical augmentation elements. The BSMC may be augmented with a forward surgical team (FST) providing the company surgical capability, see FM 4-02.25. The BSMC is organized into a company headquarters, a PVNTMED section, a BH section, a treatment platoon, and an evacuation platoon.

2-14. The evacuation platoon performs ground medical evacuation and en route patient care for the supported units. The evacuation platoon consists of a platoon headquarters, a GS evacuation section, and a DS evacuation section. The platoon employs five HMMWV evacuation squads (or ten evacuation teams). A complete discussion on the organization, mission, and functions of the BSMC is provided in FM 4-02.6 and Field Manual Interim (FMI) 4-90.1.

Area Support Medical Company

2-15. The ASMC performs functions similar to those of the BCT medical companies. The ASMCs are employed primarily in support of division, corps, and echelons above corps (EAC) units. They are deployed to a geographical area to provide area HSS, or may be deployed to provide HSS for designated units. The ASMC also establishes its Role 2 MTF in a secure location centrally located for supported units. Medical treatment squads/teams of the ASMC may be deployed to establish Role 1 MTF and provide HSS support to concentrations of nondivisional units that do not have organic medical capabilities. The ASMC is organized into a company headquarters, a treatment and ambulance platoon, and a BH section.

2-16. The ambulance platoon performs ground medical evacuation and en route patient care for supported units. The ambulance platoon consists of a platoon headquarters, four ambulance squads (or eight ambulance teams), one HMMWV control vehicle, and eight HMMWV ambulances. A complete discussion on the organization, mission, and functions of the ASMC is provided in FM 4-02.24.

Organizations

2-17. The BSMC/FSMC normally establishes its Role 2 MTF in the BSA, while the ASMC normally establishes its Role 2 MTF in echelons above brigade (EAB) and echelons above division sustainment areas and at theater level. The medical company provides DS/GS ground ambulance evacuation for supported units. It coordinates/ requests air ambulance support for air evacuation of patients from supported Role 1 MTFs/BASs, POI, CCP, and AXPs. The organizational structure, number and type of ambulances assigned to an ambulance/evacuation platoon may differ depending on their location on the battlefield and the types of units supported. For example, a BSMC in support of a HBCT has a mix of armored and wheeled evacuation vehicles while a BSMC in support of the SBCT has only wheeled evacuation vehicles.

2-18. The company headquarters section is organized into a command element; a support element; a unit supply element; a medical supply and medical maintenance element; and an operations and communications element.

Treatment Platoon

2-19. The treatment platoon is composed of a platoon headquarters, treatment squads, and an area support section, which consists of an area support squad, an area support treatment squad, and an area patient holding squad.

2-20. The platoon headquarters is the C2 element of the platoon. The platoon headquarters:

- Determines and directs the disposition of patients and submits requests through the company command post (CP) for their evacuation of patients to supporting hospitals.
- Directs, coordinates, and supervises platoon operations.
- Directs the activities of the Role 2 MTF, monitors Class VIII supplies, blood usage and inventory levels, and keeps the commander informed.
- Manages platoon operations, operations security (OPSEC), communications, medical administration, organizational training, supply transportation, patient accountability, statistical reporting functions, and blood situation reporting.

2-21. The treatment squad element can contain up to four treatment squads, depending on the type company assigned. The treatment squads have the capabilities to:

- Provide emergency and routine sick call treatment services to Soldiers assigned to supported units.
- Operate as separate treatment teams (Teams A and B) for limited periods of time.
- Operate up to four treatment stations.
- Reinforce or reconstitute similar treatment squads and can operate for up to 48 hours while separated from their parent unit.

2-22. For medical companies deployed in the BSA, these squads can further be used to—

- Provide augmentation to maneuver battalion medical platoons.
- Facilitate the movement of the treatment platoon MTF from one location to another or echelons forward to establish an MTF at a new location. A treatment squad can be echeloned forward to establish an MTF at a new location. The echeloning of elements allows the old treatment site to remain operational until the new site is established.
- Augment an AXP, CCP, or positioned at designated points in the ambulance shuttle system, when required.

2-23. The area support section of the treatment platoon is composed of an area treatment squad, an area support squad, and a patient-holding squad. These squads form the medical company Role 2 MTF. The area support treatment squad provides trauma care and routine sick call services to personnel assigned to units located in the supported area of the medical company. The area support squad provides operational dental care that includes emergency and essential dental services, limited laboratory and radiological services, and blood support commensurate with Role 2 MTF. The patient-holding squad provides up to 40 patient beds (40 patient beds for heavy divisions and 20 patient beds for light divisions) for patients requiring minimal treatment. Patients held in the patient-holding area are those patients who are expected to be RTD within 72 hours from the time they are held for treatment.

2-24. The area treatment squad is identical in personnel and equipment as the treatment squads of the treatment section. It is the base medical treatment element of a Role 2 MTF. It provides sick call services and initial resuscitative treatment for supported units. For communications, the squad employs frequency modulated (FM) radios and is deployed in the medical radio and wires communications nets.

Ambulance/Evacuation Platoon

2-25. The ambulance/evacuation platoon provides C2 for ambulance platoon operations. The ambulance platoon headquarters element maintains communications to direct ground ambulance evacuation of patients. It provides ground ambulance evacuation support for supported maneuver battalions and for supported units operating in the sustainment areas. The ambulance headquarters element performs route reconnaissance and develops and issues graphic overlays to all its ambulance teams. It also coordinates and establishes AXPs for both air and ground ambulances, as required.

2-26. Ambulance squads provide ground ambulance evacuation of patients from supported BASs/unit aid stations back to the Role 2 MTF that is located in the brigade support area (BSA). An ambulance squad consists of two ambulance teams (two ambulances, wheel or tracked vehicles). Ambulance squad personnel—

- Perform EMT, evacuate patients, and provide for their continued care en route.
- Operate and maintain assigned communication and navigational equipment.
- Perform preventive maintenance checks and services (PMCS) on ambulances and associated equipment.
- Maintain supply levels for the ambulance medical equipment sets (MES).
- Ensure that appropriate property exchange of medical items (such as litters and blankets) is made at sending and receiving MTF.
- Prepare patients for ground and air medical evacuation.
- Provide medical evacuation of wounded or injured Soldiers from the POI to supporting MTF, CCP or AXP.
- Initiate the DD Form 1380, FMC, as required.
- Provide emergency movement of medical personnel and emergency delivery of blood, medical supplies, and medical equipment.
- Assist in the care of COSR casualties.
- Maintain operational readiness of the ambulance.
- Resupply company and platoon trauma specialists, when required.
- Operate the vehicle and maintain contact with supported and supporting elements.
- Find and collect the wounded.
- Serve as messengers within medical channels.

Preventive Medicine Section

2-27. Preventive medicine sections are organic to the divisions and the BSMC of the BCT. The PVNTMED section is also found in the multifunctional medical battalion (MMB) headquarters. The PVNTMED section assigned to medical companies has a primary responsibility for supervising the unit's PVNTMED program as described in AR 40-5. The section—

- Ensures measures are implemented to protect personnel against food, water, and arthropodborne diseases, as well as environmental injuries.
- Provides advice and consultation in the area of health threat assessment, FHP, environmental sanitation, epidemiology, sanitary engineering, and pest management.
- Assists the higher headquarters in determining requirements for medical intelligence assessments, particularly with respect to CBRN and disease prevalence.
- Coordinates with supporting veterinary teams for conducting and implementing food safety and quality assurance surveillance and assisting in foodborne and zoonotic disease surveillance and control.

Behavioral Health Section

2-28. Combat and operational stress is controlled through vigorous prevention, consultation, and restoration programs by the BSMC BH section. These programs are designed to maximize the RTD rate of

COSR Soldiers by providing rest/restoration within or near their unit areas. The BSMC BH section has a primary responsibility for assisting commanders in controlling combat stress and serves as a consultant to the commander, staff, and others involved with providing prevention and intervention services to unit Soldiers and their families. For additional information on the operations and functions of BSMCs, and ASMCs, refer to FMs 4-02.6 and 4-02.24.

SECTION III — MEDICAL COMPANY (GROUND AMBULANCE)

OPERATIONAL INFORMATION

2-29. The AHS is a continuum of increasing roles of care extending from the POI through the CONUS base. All sick, injured, and wounded patients must be evacuated from the battlefield in the shortest possible time to the MTFs that can provide the required treatment. Ground ambulances serve as one of the primary means of evacuating patients from the battlefield.

MISSION

2-30. The mission of the medical company, ground ambulance, TOE 08453A00, is to provide ground evacuation within the theater.

ASSIGNMENT

2-31. The medical company, ground ambulance is normally assigned or attached to MMB (TOE 08485G00) or a medical brigade (MEDBDE) for C2.

EMPLOYMENT

2-32. The medical company, ground ambulance, provides DS to BCTs and is employed in the division, corps, and EAC to provide area support. It is tactically located where it can best control its assets and execute its patient evacuation mission.

BASIS OF ALLOCATION

2-33. The BOA within the combat zone is one per division supported and one per two divisions supported within the theater.

CAPABILITIES

2-34. At Level I, the unit is capable of providing:

- A single-lift capability for evacuation of 96 litter patients or 192 ambulatory patients.
- Medical evacuation from BCT/division medical companies and ASMCs to supporting hospitals.
- Reinforcement of BCT/division medical company evacuation assets.
- Reinforcement of covering force and unassigned area operations.
- Movement of patients between hospitals and aeromedical staging squadrons/MASFs, railheads, or seaports in the division, corps, and EAC.
- Area evacuation support beyond the capabilities of the ASMC.
- Emergency movement of medical personnel and supplies.
- Preparation of patients for ground and air medical evacuation.
- Medical evacuation of wounded or injured Soldiers from the POI to supporting MTF, CCP, or AXP.
- Field feeding and vehicle refueling support when collocated with the MMB.

DEPENDENCY

2-35. This unit is dependent upon the appropriate elements of the division, corps, EAC, or ASCC for—

- Religious, financial management, legal, personnel, mortuary affairs, and administrative services.
- Viable communications systems for C2 and adequate road networks.
- Laundry, shower, and clothing repair.
- Generator, communication equipment, and communication security (COMSEC) equipment maintenance.
- Health service support to include hospitalization.

ORGANIZATION AND FUNCTIONS

2-36. This company is organized into a company headquarters section, two ambulance platoon headquarters, and two evacuation sections (Figure 2-1).

2-37. The company headquarters provides C2, communications, administration, field feeding, and logistical support (to include maintenance) for the subordinate ambulance platoons.

2-38. The ambulance platoon headquarters provides C2 for the subordinate ambulance squads.

2-39. The evacuation section operates ambulances and provides en route medical care for patients. Each evacuation section consists of 12 ambulances, each with a two-man crew. The members of the squad maintain the level of expendable Class VIII supplies in the ambulance MES by reconstituting supplies from medical companies or hospitals when they pick up or drop off patients. They are also responsible for performing operator maintenance on assigned vehicles.

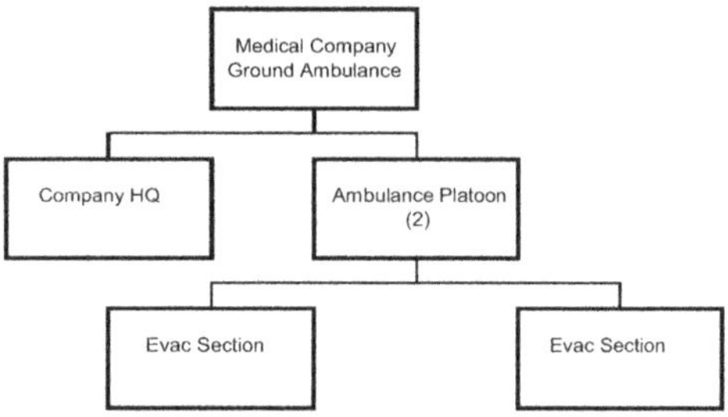

Figure 2-1. Medical company, ground ambulance

SECTION IV — MULTIFUNCTIONAL MEDICAL BATTALION

MODULAR FORCE

2-40. A modular force is a key aspect in the transformation of the Army. The modular force model is based on brigade-sized elements that are more responsive than division-sized elements and can perform joint and expeditionary-type missions. Using modular units, Army planners can tailor force structure, reduce strategic lift requirements, and create flexible forces with specialized capabilities based on everchanging mission requirements. To facilitate the synchronization and provision of AHS services and support, the AMEDD C2 organizations have been redesigned into a modular design. The MMB will replace the functional battalions (area support, evacuation, and medical logistics).

2-41. The new modular design of medical units gives the medical command (MEDCOM) the ability to—

- Assist in the employment of the optimal mix of medical capabilities within an expeditionary to campaign framework.
- Ensure seamless, state-of-the-art care, regardless of delivery location.
- Provide tested and proven systems to the battlefield and ensure the provision of the right care at the right place and time.
- Promote scalability through easily tailored, capabilities-based packages that result in improved tactical mobility, reduced footprint, and increased modularity for flexible task organization.
- Provide and enable the joint force commander (JFC) to choose among augmentation packages, thus enabling rapid synchronization of desired medical capabilities.

MULTIFUNCTIONAL MEDICAL BATTALION

MISSION

2-42. The mission of the MMB, TOE 08486G000, is to provide a scalable, flexible and modular C2, administrative assistance, logistical support, and technical supervision capability for assigned and attached medical organizations (companies, detachments, and teams) task-organized for support of deployed forces.

ASSIGNMENT

2-43. This unit is assigned to a MEDCOM or a MEDBDE.

EMPLOYMENT

2-44. The MMB is normally employed in a MEDCOM or MEDBDE AO.

BASIS OF ALLOCATION

2-45. The BOA is one per three to six subordinate company-size units plus the blood detachment.

CAPABILITIES

2-46. At Level I, this unit provides—

- Command and control, staff planning, supervision of operations, medical and general logistics support as required, and administration of the assigned and attached units conducting AHS operations in its supported AO.
- Task organization of medical assets to meet the projected patient workload.
- Advice to senior commanders in the AO on the medical aspects of their operations.
- Coordination of medical regulating and patient movement within the AO.
- Monitoring, planning, and coordinating ground and air evacuation within the battalion AO. Coordinating air evacuation support requirements and synchronizating the air evacuation plan into the overall medical evacuation plan with the supporting aviation unit.
- Consultation and technical advice on PVNTMED (medical entomology; medical, and occupational and environmental health [OEH] surveillance; and sanitary engineering), pharmacy procedures, COSC and BH, medical records administration, veterinary services, nursing practices and procedures, dental services, medical laboratory procedures, and automated medical information systems to supported units.
- Guidance for facility site selection and area preparation, as required.
- Monitoring and supervising medical logistics operations, to include Class VIII supply/resupply, medical equipment maintenance and repair support, optical fabrication and repair support, and blood management.
- Planning and coordination for Role 1 and Role 2 AHS operations, to include staff advice to units without organic medical assets within the AO.

- Unit level maintenance for wheeled vehicles and power generation equipment and wheeled vehicle recovery operations support to assigned or attached units.
- Organizational communications equipment maintenance support for the battalion.
- Food service support for staff and other assigned/attached medical units.
- Consolidated property book maintenance.
- Religious support for MMB staff, assigned/attached medical units, and casualties in MMB subordinate unit MTFs.

DEPENDENCY

2-47. This unit is dependent on the—

- Corps for legal, administrative, financial management, human resources, and transportation services, area damage control, CBRN decontamination assistance, mortuary affairs, and laundry and bath services.
- Quartermaster supply company or equivalent for Class I rations.
- Engineer company or equivalent for site selection, waste disposal, and minor construction.
- Movement control battalion or equivalent for supplemental transportation requirements.

ORGANIZATION AND FUNCTIONS

2-48. Figure 2-2 depicts the organization of the MMB. Specific information concerning the early entry and campaign elements is provided below.

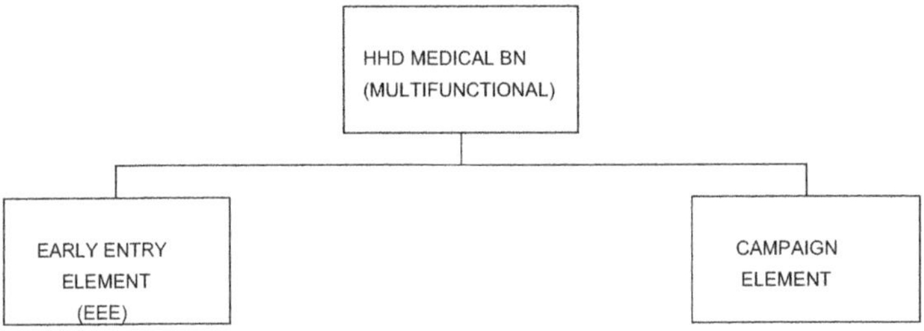

Figure 2-2. Multifunctional medical battalion

EARLY ENTRY ELEMENT, MULTIFUNCTIONAL MEDICAL BATTALION

2-49. Deployed units will begin operations immediately upon arrival in the theater avoiding the delays experienced in past conflicts. There is a need to deploy an EEE medical force to assess, plan, and coordinate AHS operations in the AO and provide medical surveillance, PVNTMED, and area medical support early on. One of the keys to providing early medical care is the development of modular medical elements. These modular AHS elements will perform specific medical functions and will allow medical planners to bring necessary medical units into theater as required by the battlefield mission.

MISSION

2-50. The mission of the EEE of the MMB, TOE 08486GA00, is to provide a scalable, flexible and modular C2, administrative assistance, logistical support, and technical supervision capability for assigned and attached medical organizations (companies, detachments, and teams) task-organized for support of deployed forces.

Assignment

2-51. This EEE is assigned to an MMB.

Employment

2-52. Normally employed in a MEDCOM/MEDBDE AO. It can also be deployed to provide C2 to expeditionary forces in early entry operations and facilitates the reception, staging, onward movement, and integration (RSOI) of medical forces.

Basis of Allocation

2-53. The BOA is one per MMB.

Capabilities

2-54. At Level I, this unit provides—

- Command and control, staff planning, supervision of operations, medical and general logistics support as required, and administration of the assigned and attached units conducting AHS operations in its supported AO.
- Task organization of medical assets.
- Advice to senior commanders in the AO on the medical aspects of their operations.
- Coordination of medical regulating and patient movement within the AO.
- Monitoring, planning, and coordinating ground and air evacuation within the battalion AO. Coordinating air evacuation support requirements with the supporting aviation unit, and synchronizing the air evacuation plan into the overall medical evacuation plan.
- Consultation and technical advice on PVNTMED (medical entomology, medical and OEH surveillance, and sanitary engineering), COSC and BH, medical records administration, veterinary services, nursing practices and procedures, dental services, and automated medical information systems to supported units.
- Guidance for facility site selection and area preparation, as required.
- Monitoring and supervising medical logistics operations, to include Class VIII supply/resupply.
- Consolidated property book maintenance.
- Unit-level maintenance for wheeled vehicles and power generation equipment and wheeled vehicle recovery operations support to assigned or attached units.
- Organizational communications equipment maintenance support for the battalion.
- Food service support for staff and assigned/attached medical units.

Dependency

2-55. This unit is dependent on the—

- Corps for legal, administrative, financial management, human resources, religious support, transportation services, area damage control, CBRN decontamination assistance, mortuary affairs, and laundry and bath services.
- Quartermaster supply company or equivalent for Class I rations.
- Engineer company or equivalent for site selection, waste disposal, and minor construction.
- Movement control battalion or equivalent for supplemental transportation requirements.

ORGANIZATION AND FUNCTIONS

2-56. The battalion command section provides C2 of assigned and attached medical companies, detachments, and teams. (See Figure 2-3.)

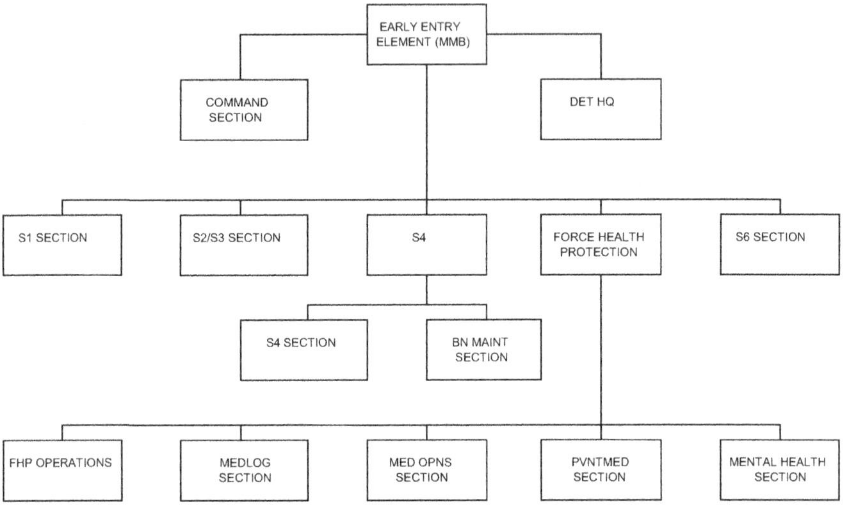

Figure 2-3. Early entry element, multifunctional medical battalion

2-57. The Adjutant (US Army) (S1) operates according to METT-TC and is the principle coordinating staff officer responsible for the delivery of human resources support. The S1 is responsible for the execution of all of the human resources (HR) core competencies to include: personnel readiness management (PRM); personnel accountability and strength reporting (PASR); personnel information management (PIM); reception, replacement, RTD, rest and recuperation (R&R), redeployment (R5) operations; casualty operations; essential personnel services (EPS); and postal operations. Morale, welfare, and recreation (MWR) operations, HR planning; and staff operations. The S1 also coordinates with elements of supporting agencies for finance, legal, religious, and administrative services.

2-58. The Intelligence Officer (US Army)/Operations and Training Officer (US Army) (S2/S3) section is responsible for security, plans and operations, deployment, relocation, and redeployment of the battalion and its assigned and attached units. The battalion's primary net control station is in this paragraph.

2-59. The Supply Officer (US Army) (S4) section is responsible for coordination, control, and management of logistics for assigned and attached units. The battalion maintenance section is under the staff supervision of the battalion S4. This section provides unit level maintenance for wheeled vehicles assigned to the HHD and assigned or attached units without unit level maintenance capability.

2-60. The FHP operations section is responsible for the planning, coordination, and execution of the AHS mission within the battalion's AO. This section supervises the operations of the medical logistics, operations, PVNTMED, and BH sections as well as coordinates medical regulation and patient movement within the AO. The medical operations section is responsible for—

- Management of area medical support (Roles 1 and 2).
- Planning, coordination, and execution of the Class VIII mission within the battalion's AO. This includes blood and medical maintenance management.

- Situational awareness for locations of ground and air ambulance assets.
- Planning and coordination for evacuation of patients using ground and air ambulance assets.
- Planning and coordination between air evacuation elements and the maneuver brigade S3. Information provided to the maneuver brigade S3 should include, but not be limited to the location of—
  - Medical treatment facilities, CCPs, and AXPs.
  - Forward air refueling equipment.
  - Location of supported units and liaison requirements. Air evacuation corridors, ground evacuation routes, and recommendations on usage.
- The PVNTMED section is responsible for the planning, coordination, and execution of the PVNTMED mission within the battalion's AO. This includes management of PVNTMED and veterinary assets.
- The BH section is responsible for the planning, coordination, and execution of the COSC mission within the battalion's AO. This section also collects and records social and psychological data.

2-61. The S6 section is responsible for all aspects of information management, automation, and communications-electronic (CE) support to assigned and attached units. The battalion's alternate net control station is in this paragraph.

2-62. The detachment headquarters provides for billeting, field feeding, discipline, security, training, and administration for personnel assigned to the EEE.

2-63. The battalion maintenance section is under the staff supervision of the battalion S4. This section provides unit level maintenance for wheeled vehicles assigned to the HHD and assigned or attached units without unit level maintenance capability.

CAMPAIGN SUPPORT ELEMENT, MULTIFUNCTIONAL MEDICAL BATTALION

MISSION

2-64. The mission of the campaign support element, MMB, TOE 08486GB00, is to complete the staffing of the headquarters of the MMB to enhance the delivery of campaign quality health care to deployed forces. To provide a scalable, flexible, and modular C2, administrative assistance, logistical support, and technical supervision capability for assigned and attached medical organizations (companies, detachments, and teams) task-organized for support of deployed forces.

ASSIGNMENT

2-65. This unit is assigned to an MMB.

EMPLOYMENT

2-66. Normally employed with the MEDBDE or a MEDCOM. If the EEE is deployed first, the campaign element falls in on this deployed module to provide a more robust C2 element and additional operational and planning capabilities, increased medical logistical staff, and a more diverse and robust clinical staff.

BASIS OF ALLOCATION

2-67. The BOA is one per MMB.

CAPABILITIES

2-68. At Role 1, this unit provides—

- Augmentation of C2 in personnel, logistics, HSS, FHP, medical operations, PVNTMED, BH, automation, maintenance, religious support, and food service to the EEE.
- Unit maintenance on all organic equipment except COMSEC equipment.

DEPENDENCY

2-69. This unit is dependent on the HHD, MMB for legal, administrative, financial management, human resources, food service, transportation services, area damage control, CBRN decontamination assistance, mortuary affairs, and laundry and bath services.

ORGANIZATION AND FUNCTIONS

2-70. The command section provides C2 for assigned and attached medical companies, detachments, and teams. (See Figure 2-4.)

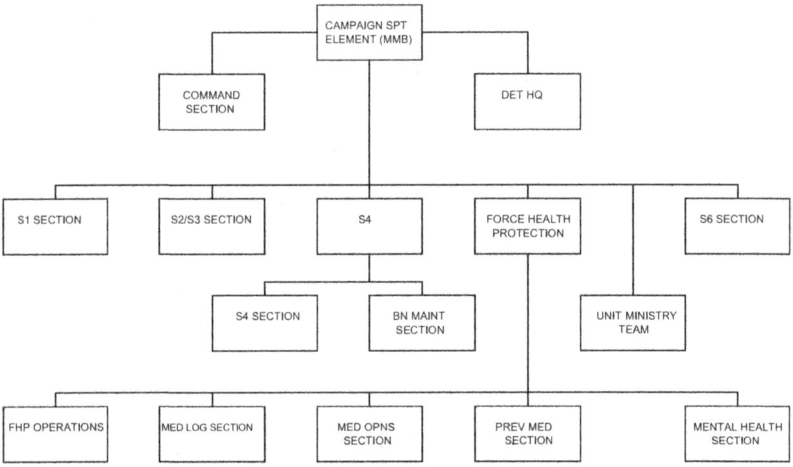

Figure 2-4. Campaign support element, multifunctional medical battalion

2-71. The S1 operates according to METT-TC and is the principle coordinating staff officer responsible for the delivery of HR support. The S1 is responsible for the execution of all of the HR core competencies to include: PRM, PASR, PIM, R5 operations, and casualty operations, EPS, postal operations, MWR operations, and HR planning and staff operations. The S1 also coordinates with elements of supporting agencies for finance, legal, religious, and administrative services.

2-72. The S2/S3 section coordinates with the EEE S2/S3 section of the MMB in the plans and operations support to subordinate organizational elements.

2-73. The S4 section is responsible for coordination, control, and management of logistics for assigned and attached units.

2-74. The FHP operations section plans and coordinates the HSS/FHP mission within the battalion's AO when collocated with the EEE. This section also provides advice on the operation of medical logistics, operations, PVNTMED, BH, medical regulating, and patient movement to supported organizations.

2-75. The medical logistics section plans and organizes Class VIII supply support (including blood and medical equipment management) in coordination with the medical logistics section of the EEE.

2-76. The medical operations section is responsible for—

- Management of area medical support (Roles 1 and 2).
- Situational awareness for locations of ground and air ambulance assets.
- Planning and coordination for evacuation of patients using ground and air ambulance assets.
- Planning and coordination between air evacuation elements and the maneuver brigade S3. Information provided to the maneuver brigade S3 should include, but not be limited to the location of—
  - Medical treatment facilities, CCPs, and AXPs.
  - Forward air refueling equipment.
  - Supported units and liaison requirements.
  - Evacuation corridors and evacuation routes and recommendations on usage.

2-77. The PVNTMED section plans and coordinates PVNTMED/veterinary support to subordinate organizations.

2-78. The BH section plans and coordinates BH support to subordinate organizations. This section also collects and records social and psychological data when collocated with the EEE.

2-79. The S6 section provides advice and support to subordinate organizations for CE support and information management.

2-80. The unit ministry team provides and coordinates religious support for all MMB staff, assigned/attached medical units and casualties in MMB subordinate unit MTFs.

2-81. The detachment headquarters provides coordination with the EEE in support of the field feeding and administration mission of the unit.

2-82. The battalion maintenance section provides unit level maintenance management support to assigned and attached units in coordination with the battalion S4 and the battalion maintenance section of the EEE.

SECTION V — GENERAL SUPPORT AVIATION BATTALION, COMBAT AVIATION BRIGADE

GENERAL SUPPORT AVIATION BATTALION

MISSION

2-83. The mission of the general support aviation battalion (GSAB) (Figure 2-5) is to provide the combat aviation brigade with aerial sustainment and maneuver support.

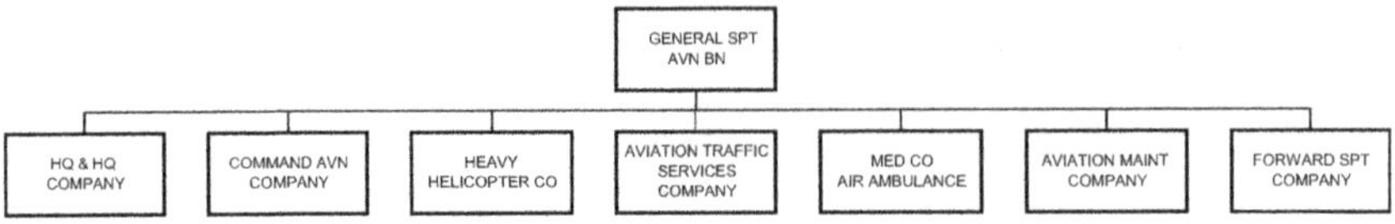

Figure 2-5. General support aviation battalion

ASSIGNMENT

2-84. Organic to the division combat aviation brigade, theater assault aviation brigade, theater general support aviation brigade, and the Army National Guard (ARNG) aviation expeditionary brigade (TOE 01300G400).

BASIS OF ALLOCATION

2-85. The BOA is one per division combat aviation brigade, one per ARNG aviation expeditionary brigade, one per theater assault aviation brigade, and three per theater GSAB (TOE 01300G400).

CAPABILITIES

2-86. This organization provides the following when all of its subordinate units are organized at Level I as shown in their respective TOEs:

- Battle command on the move.
  - Employment of Army airspace command and control (A2C2) system platforms in support of the division command group, maneuver, and aviation commanders' C2 requirements.
  - Radio relay and transport of liaison officers as required.
- Air assault.
  - Airlift for air assault operations and air movement of tactical forces (other than assault), field artillery or other fire support assets, tactical air defense (AD) systems, engineer equipment and personnel, and other resources.
  - Casualty evacuation when medical aircraft are not readily available.
  - Evacuation of downed aircraft and personnel.
- Sustainment.
  - Division sustainment with heavy-lift helicopter company.
  - General aviation support to the aviation brigade.
  - Support aircraft to the aviation support battalion.
  - Personnel recovery (PR).
  - Downed aircraft recovery team operations.
  - Casualty evacuation operations.
  - Air movement of general supplies, equipment, and personnel, including logistics over-the-shore (LOTS) when applicable.
  - Aerial noncombatant evacuation operations when directed by applicable commander/ authority.
  - Air movement of nuclear and other special munitions.
  - Self-deployment of organic CH-47 helicopters to a theater of operations when the aircraft have been configured with extended range fuel systems.

    - Retail level aircraft refueling when aircraft are configured with extended range fuel system (commonly referred to as "Fat Cow").
- Air traffic services support.
- Aeromedical evacuation.
    - Patient evacuation.
    - Patient movement between MTFs (patient transfers).
    - Class VIII resupply.
    - Joint blood program support.
    - Medical C2.
    - Movement of medical personnel and medical equipment.
    - Air crash rescue support.

2-87. Due to the CH-47's capabilities, heavy helicopter units can perform two unique missions:

- High-altitude operations.
- Oversized, heavy, and special munitions movement.

2-88. This GSAB is dependent upon the following organizations:

- Appropriate elements of the division aviation brigade for legal, HSS, FHP, financial management, and personnel and administrative services.
- United States Air Force weather team in the headquarters and headquarters company of the combat aviation brigade (TOE 01302G200) for air weather service support.
- Aviation support company (TOEs 01918G200 or 01919G000 [ARNG]) for intermediate level field maintenance support, DS maintenance for assigned aviation ground power units (AGPU), line item number (LIN) P44627, and support for aviation and common Class IX repair parts.

MEDICAL COMPANY, AIR AMBULANCE (HH-60), GENERAL SUPPORT AVIATION BATTALION, COMBAT AVIATION BRIGADE

OPERATIONAL INFORMATION

2-89. The medical company, air ambulance (Figure 2-6), provides aeromedical evacuation for all categories of patients consistent with evacuation precedence and other operational considerations within the division. In today's ongoing efforts in building a modular force, medical company, air ambulance will fall under the GSAB which will provide aircraft maintenance and logistics support, aviation communications, and real-time operational picture associated with today's combat environment.

MISSION

2-90. The mission of the medical company, air ambulance (HH-60) (TOE 08443G00) is to provide air ambulance evacuation support within the division and theater.

ASSIGNMENT

2-91. The medical company, air ambulance, (HH-60) is organic to the GSAB (TOEs 01125G000, 01225G100, and 01305G200) for C2.

EMPLOYMENT

2-92. This unit is employed in the division and corps area of responsibility. It is tactically located where it can best control its assets and execute its patient evacuation mission.

BASIS OF ALLOCATION

2-93. The BOA is one per GSAB and two per theater of operations (TOEs 01125G100, 01225G100, and 01305G100).

CAPABILITIES

2-94. At Level I, this unit provides—

- Twelve helicopter ambulances to evacuate critically wounded or other patients consistent with evacuation priorities and operational considerations, from points as far forward as possible, to division, corps, and EAC hospitals. Single patient lift capability for the company is 72 litter patients or 84 ambulatory patients, or some combination thereof.
- Four forward support medical evacuation teams (three helicopters each) that can be deployed as a team or an independent aircraft.
- Air crash rescue support.
- Expeditious delivery of whole blood, biological, and medical supplies to meet critical requirements.
- Rapid movement of medical personnel and accompanying equipment/supplies to meet the requirements for mass casualty (MASCAL), reinforcement/reconstitution, or emergency situations.
- Movement of patients between hospitals, ASFs, hospital ships, casualty receiving and treatment ships, seaports and railheads in the theater of operations.
- Military working dog evacuation and aeromedical evacuation support to a combat search and rescue task force.

DEPENDENCY

2-95. This unit is dependent upon—

- Appropriate elements of the division or corps for religious, legal, financial management, personnel and administrative services, and CE and COMSEC maintenance.
- The headquarters and headquarters company of the GSAB for religious, legal, HSS, FHP, personnel and administrative service, and unit CBRN support.
- The forward support company of the GSAB for Class III, automotive and generator maintenance, and field feeding.
- The aviation support company of the GSAB for unit level field maintenance of organic aircraft, including unit level supply support for aircraft Class IX.
- United States Air Force weather team in the headquarters and headquarters company of the division aviation brigade for air weather service support.

ORGANIZATION AND FUNCTIONS

2-96. The company headquarters provides C2, logistical support, and administration for the company.

2-97. The four forward support MEDEVAC teams (FSMTs) provide a task-organized means for medical evacuation at the DS and GS level. They also, provide emergency movement of medical personnel and emergency delivery of whole blood, biological, and medical supplies and equipment.

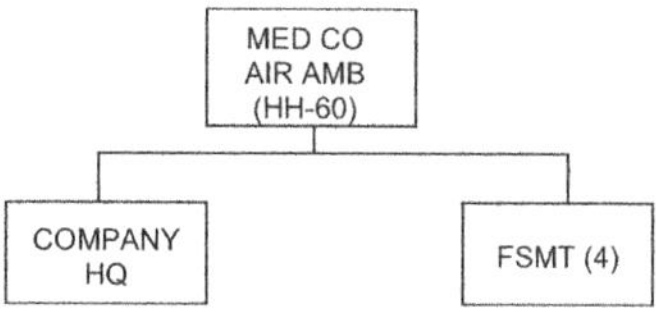

Figure 2-6. Medical company, air ambulance (HH-60)

AVIATION MAINTENANCE COMPANY, AVIATION UNIT MAINTENANCE, GENERAL SUPPORT AVIATION BATTALION, COMBAT AVIATION BRIGADE

MISSION

2-98. The mission of the aviation maintenance company is to perform aviation unit maintenance for the GSAB.

ASSIGNMENT

2-99. This unit is organic to a GSAB (TOE 01305G100).

EMPLOYMENT

2-100. The aviation maintenance company is employed as part of combined forces and provides unit level field maintenance support to the GSAB. The aviation maintenance company is located and its operations are normally conducted immediately behind the main battle area (MBA) in the EAB. The aviation maintenance company operations in forward areas are usually limited to daily inspections, minor adjustments, emergency and contact maintenance, and aircraft recovery. Contact teams and slice elements could be utilized as necessary.

BASIS OF ALLOCATION

2-101. The BOA is one per GSAB (TOE 01305G100).

CAPABILITIES

2-102. The aviation maintenance company provides unit level field maintenance support for 8 UH-60, 12 HH-60, and 12 CH-47 aircraft. Individuals of this organization can assist in the coordinated defense of the unit's area or installation. This unit performs unit level field maintenance on only the supported aircraft and some associated maintenance equipment.

DEPENDENCY

2-103. This unit depends on—

- Appropriate elements of the division or corps for religious, legal, HSS, FHP, financial management, personnel and administrative services, and logistical support (including supplemental transportation).
- Headquarters and headquarters company of the GSAB (TOE 01306G100) for religious, legal, HSS, FHP, financial management, personnel, and administrative services, unit-level logistical support, and communications maintenance.

- Forward support company (TOE 63317G200) for Class III, automotive and generator maintenance, and field feeding service support.
- Aviation maintenance battalion (TOE 01918G200), for unit level field maintenance support, backup unit level field maintenance support, DS maintenance for assigned AGPU, and support for aviation and common Class IX repair parts.

ORGANIZATION AND FUNCTIONS

2-104. This unit inspects aircraft systems, subsystems, components, and the maintenance performed by crew chiefs and repairers to determine airworthiness and compliance with applicable technical publications; performs operational checks and troubleshoots aircraft systems to isolate malfunctions; participates in maintenance test flights and maintains historical maintenance files.

2-105. This unit provides company maintenance operations. It maintains and regulates uniform flow of aircraft and associated components. Further, it maintains the company's aircraft logbooks, provides unit level Class IX support, and performs maintenance on aviation life support equipment (ALSE).

2-106. This unit provides unit level maintenance on UH/HH-60/CH-47 helicopters and associated equipment; services and lubricates helicopter systems and subsystems; and diagnoses and troubleshoots operational malfunctions.

MEDICAL COMMAND

2-107. The MEDCOM serves as the senior medical command within the theater, and works directly for the theater commander. With augmentation, MEDCOM is joint capable. In the absence of a theater commander the MEDCOM works for the regional combatant commander during a major combat operation (MCO). It possesses a centralized capability to effectively and efficiently task-organize medical elements based on theater specific medical requirements. The MEDCOM is depicted in Figure 2-7.

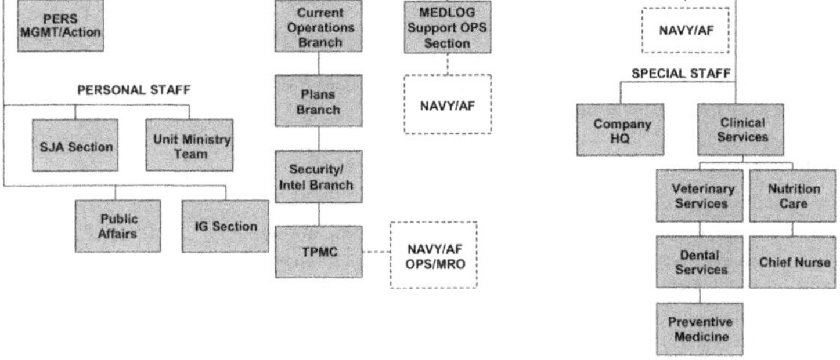

Figure 2-7. Medical command

2-108. The MEDCOM serves as the medical force provider (Figure 2-8) for the corps and EAC and provides the requisite medical C2 necessary to provide campaign-quality health care to deployed forces. The MEDCOM also focuses on theater medical operations plans and medical contingency plans.

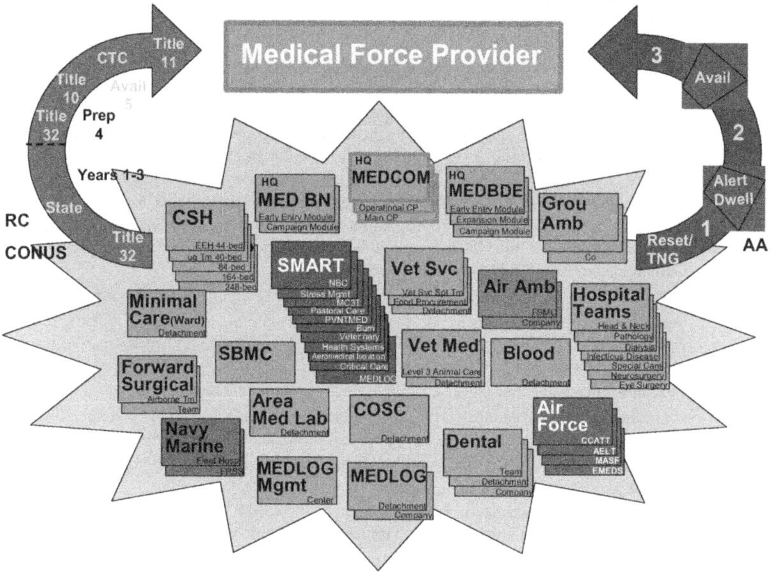

Figure 2-8. Medical force pool

2-109. The ability of the MEDCOM staff to develop accurate patient estimates based on real-time medical data ensures the right mix of medical specialties and resources are in the right place and at the right time to treat casualties on the battlefield. Care must be continuous and seamless from the POI or wounding through a phased system providing essential care in theater, with evacuation for definitive and rehabilitative care in CONUS.

2-110. The MEDCOM is a dedicated, regionally focused command with a basis of allocation of one per theater.

2-111. The MEDCOM is a versatile, modular medical C2 structure composed of a main command post (MCP) and an operational command post (OCP) which are 100 percent mobile. The MEDCOM is comprised of two standard requirements code (SRC) identified modules to provide a scalable medical C2 capability to the corps commander. Further, the MEDCOM is capable of—

- Providing subordinate medical organizations to operate under the subordinate MEDBDE and/or MMB and providing FSTs and/or other augmentation required by BCTs, and echelons above brigade (EAB) medical companies.
- Operating as the medical force provider for provisioning of required assets to the MEDBDEs.
- Advising the corps and other senior level commanders on the medical aspects of their operations.
- Providing staff planning, supervision of operations, and administration of assigned and attached medical units.
- Providing reach to the MHS.

- Providing medical Army support to other Services (ASOS)/Title 10 responsibilities of the ASCC commander.
- Coordinating with the USAF theater patient movement requirements center (TPMRC) for medical regulating and movement of patients from MEDCOM MTFs.
- Providing consultation services, technical advice, and clinical policy development in all aspects of medical and surgical services.
- Containing the functional staff to coordinate hospitalization, nursing, PVNTMED (entomology, epidemiology, OEH surveillance, potable water inspection, pest management, food facility inspection, and control of medical and nonmedical waste), medical evacuation, medical regulating, MEDLOG to include blood management, dental services, veterinary services (zoonotic disease control, investigation and inspection of subsistence, and animal medicine), pharmacy, optometry, nutrition care, and COSC, neuropsychiatric (NP), and BH services, medical laboratory services, and area medical support to supported units.
- Providing technical advice and consultation on medical automated information systems and programs such as the Theater Medical Information Program (TMIP) and Medical Communications for Combat Casualty Care (MC4).
- Functioning as the senior theater joint medical C2 element when augmented with joint augmentation package.
- Peacetime planning, training for and with the MEDBDEs, and participation in theater engagement plans.
- Providing advice and assistance in facility selection and preparation.
- Coordinating and orchestrating MEDLOG operations to include Class VIII supply, distribution, medical maintenance and repair support, optical fabrication, and blood management.
- Planning for the single integrated medical logistics manager (SIMLM), mission when designated.
- Planning and coordination of religious support for MEDCOM staff, assigned and attached medical units, and casualties in MEDCOM subordinate unit MTFs.

Operational Command Post

2-112. The MEDCOM OCP is a deployable, versatile module, which provides medical C2, policy development, and technical guidance to subordinate medical units and provides interface and liaison with supported theater forces in EAC, corps, and subordinate BCTs. The MEDCOM OCP has an assigned SRC to facilitate the placement of personnel in the CP and to integrate the CP into the time-phased force deployment list (TPFDL). The MEDCOM OCP is depicted in Figure 2-9.

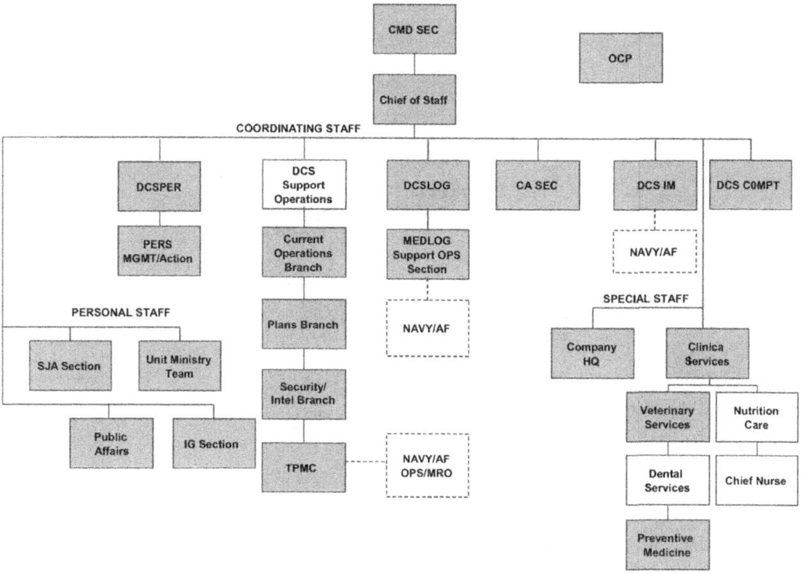

Figure 2-9. Operational command post, medical command

2-113. As the force continues to build within the theater, the MEDCOM's OCP can be incrementally expanded or augmented with additional planning and professional expertise. The deployment of personnel from the MCP model can then be built into the medical C2 required structure in order to provide campaign-quality health care to the deployed force.

Main Command Post

2-114. The MEDCOM MCP is a deployable augmentation module, which completes the staffing of the MEDCOM headquarters to enhance the delivery of campaign-quality health care to deployed forces in EAC, corps, and supported BCTs. It facilitates medical C2, policy development, and technical guidance to subordinate MEDBDEs and provides interface and liaison with supported theater forces in EAC, corps, and subordinate BCTs. If the OCP is deployed and needs additional personnel or clinical skills, the MEDCOM MCP can deploy personnel to the deployed OCP. This will provide a more robust medical C2 element and additional operational and planning capabilities; increased medical logistics staff; automated medical information systems; a more diverse clinical staff; and more depth in the command judge advocate (CJA), inspector general (IG), civil affairs (CA), and public affairs sections. The MEDCOM MCP has an assigned SRC to facilitate the placement on and the integration into the TPFDL. The MCP is depicted in Figure 2-10.

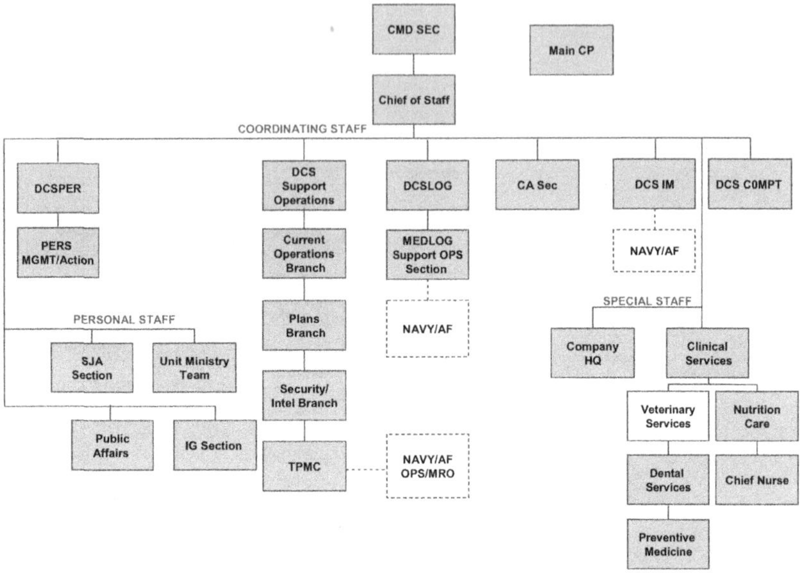

Figure 2-10. Main command post, medical command

MEDICAL BRIGADE

2-115. The MEDBDE provides a scalable expeditionary medical C2 capability for assigned and attached medical functional organizations task-organized in support of the deployed division/corps forces. The MEDBDE brings all requisite medical C2 and planning capabilities to provide responsive and effective HSS/FHP throughout the corps AO. The MEDBDE is depicted in Figure 2-11.

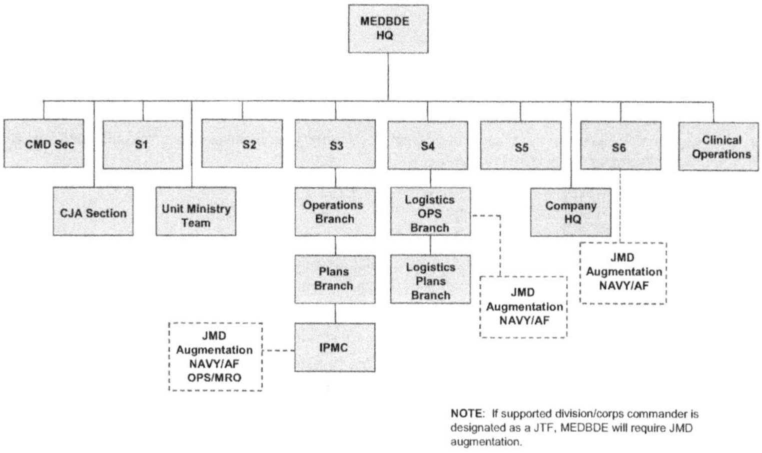

Figure 2-11. Medical brigade

2-116. The MEDBDE ensures the division/corps commander has the right mixture of medical professional (operational, technical, and clinical) expertise to synchronize the complex system of AHS functions required to maintain the health of his command by promoting fitness, preventing casualties from DNBI, and promptly treating and evacuating those injured on the battlefield.

2-117. The design and flexibility of the MEDBDE permits the division/corps commander the ability to meet expeditionary medical support requirements in support of early entry forces. As the supported corps commander's forces grow in both size and complexity, the MEDBDE can deploy additional modules that build upon one another to support full spectrum operations. The MEDBDE also provides the corps commander the appropriate medical C2 to continue to build his medical force capabilities through the integration of Army, joint, and multinational medical forces to ensure he can identify and counter any health threats in his supported areas. This will permit him to transition from expeditionary medical support operations to providing campaign-quality health care in his AO.

2-118. The MEDBDE has the capability to provide SRC elements for an early entry module (EEM), an expansion module, and the campaign module (CM), thus enabling its capability to be tailored to METT-TC factors of a specific division/corps operation.

2-119. The MEDBDE provides medical C2, administrative assistance, and technical supervision of assigned and attached medical units providing AHS in support of a division/corps. Further, it is capable of—

- Providing a rapidly responsive early entry medical C2 module that can quickly integrate into the early entry deployment sequence for crisis management.
- Providing full spectrum continuous medical C2 in support of all Army corps and joint (when augmented) forces.
- Providing operational medical augmentation to Role 2 BCT medical companies.
- Advising corps and BCT commanders on the medical aspects of their operations.
- Providing medical staff planning, operational and technical supervision, and administrative assistance for MMBs and CSHs task-organized under the MEDBDE.
- Coordinating with supporting TPMRC for medical regulating and medical evacuation from MMBs and hospitals to supporting EAC MTFs and CONUS.

- Providing medical consultation services and technical advice in—
  - Preventive medicine (medical and OEH surveillance, environmental health, sanitary engineering, and medical entomology).
  - Nursing services.
  - Dental services.
  - Behavioral health (to include COSC) and NP care.
  - Veterinary services (including food safety and inspection, animal medical care, and veterinary PVNTMED services).
  - Nutrition care and medical food service.
  - Medical laboratory support.
- Providing advice and recommendations for conduct of civil-military operations (CMO).
- Control and supervision of Class VIII supply and resupply movement to include blood management. When designated by the combatant command, serves as the SIMLM.
- Providing joint capable medical C2 capability when augmented with appropriate joint assets.
- Serves as the executive agent for veterinary services according to DODD 5200.31E.
- Coordinates ASOS for ship-to-shore/shore-to-ship medical evacuation mission.
- Plans and coordinates religious support for MEDBDE staff, assigned and attached medical units, and casualties in MEDBDE subordinate unit MTFs.

Chapter 3

Operational and Tactical Evacuation Planning

A comprehensive medical evacuation plan is essential to ensure effective, efficient, and responsive medical evacuation is provided to all wounded, injured, and ill Soldiers in the JOA. The Army medical evacuation plan flows from the GCC's guidance and intent and incorporates all missions and tasks directed by the GCC to be accomplished and is synchronized with supporting and supported units. In some scenarios, Army air and ground evacuation resources may be directed to provide support to sister Services, allied and coalition partners, and HN forces. Examples of a medical evacuation OPLAN and OPORD are contained in Appendix D.

SECTION I — THEATER MEDICAL EVACUATION PLANNING RESPONSIBILITIES

JOINT PLANNING

3-1. All medical evacuation planning within the theater reflects the GCC's guidance and intent. Planning at the joint level is governed by Joint Publications (JPs) 5-0, 5-00.1, and 5-00.2. Guidance on medical evacuation in joint operations is addressed in JP 4-02. When directed by the GCC, Army medical evacuation assets may be tasked to support other than Army forces engaged in the execution of the joint mission. These additional support missions will be clearly articulated in the GCC's OPLAN and OPORD. The ASCC surgeon, with the advice of the senior medical evacuation planner, will coordinate and synchronize these support operations with the combatant command surgeon, joint task force (JTF) surgeon, and the other Services and/or multinational partners as required to ensure that a comprehensive and effective, efficient, and responsive plan is developed and implemented.

MEDICAL COMMAND AND CONTROL ORGANIZATIONS

3-2. Army evacuation assets support the full spectrum of all Army operations and are capable of supporting Army forces engaged in the execution of the joint mission, as directed. Major combat operations supported by an ASCC will have a deployed MEDCOM. The C2 coordination and orders flow for AHS is illustrated in Figure 3-1. Evacuation planning should occur at each major level of command in order to effectively execute command, control, and influence over evacuation planning.

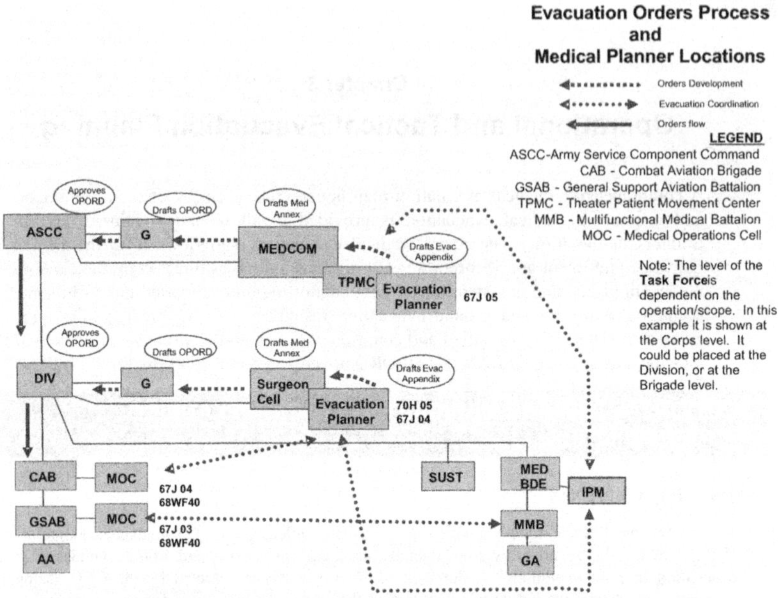

Figure 3-1. Command and control organizations for coordination and orders flow

3-3. Under the new Army modular design, some early entry operations may not require the deployment of a theater or a brigade C2 structure. In these cases, medical evacuation planning, coordination, and synchronization are initiated by the most senior medical command/staff level (such as the division surgeon, medical brigade, GSAB, and/or MMB). The senior leaders responsible for air and ground medical evacuation provide estimates and recommendations for the plan. This plan is normally provided as an appendix to the HSS annex and is disseminated down through the chain of command.

3-4. Dedicated Army rotary-wing aircraft (air ambulances) are under the C2 of the GSAB within the transformed Army aviation structure. Ground ambulances remain under the C2 of medical units/elements that are organic to movement and maneuver, fires, and sustainment units. The medical company, ground ambulance is under the C2 of the MMB.

3-5. To ensure an integrated medical evacuation system, synchronization of air with ground ambulance assets is essential. To affect this synchronization, the medical C2 headquarters must maintain influence over the execution of the medical evacuation plan. This influence is accomplished through the orders process.

Theater Patient Movement Center

3-6. In the design of the modular MEDCOM, the medical reengineering initiative (MRI) MEDCOM's medical regulating office (MRO) staff was reevaluated, realigned, and augmented with medical evacuators to more effectively coordinate and synchronize the various ground and air evacuation aspects of the operation and to supervise and facilitate the medical regulating functions. This staff section was renamed as the theater patient movement center (TPMC). The addition of medical evacuators enables this section to develop, coordinate, and synchronize the medical evacuation portion of the MEDCOM AHS

hour coordination and management responsibility for patient regulating and administration within the MEDCOM AO and directs patient evacuation between facilities within its AO as required to avoid patient overflow. This staff is responsible for preparing patient statistical reports for the command. This section also coordinates with the supporting TPMRC for the transfer of patients out of the MEDCOM/theater. The TPMRC regulates the flow of patients out of the theater to an MTF capable of providing the required specialty care and arranges for the transport of these patients by USAF aeromedical evacuation assets.

INTRATHEATER PATIENT MOVEMENT CENTER

3-7. As with the modular MEDCOM, the MRO of the modular MEDBDE was also redesigned in a similar fashion to enhance the capabilities for medical evacuation planning within the MEDBDE AO. This staff section was renamed the intratheater patient movement center (IPMC). This office is responsible for coordinating and synchronizing medical evacuation operations with the TPMC and the division surgeon cell. It regulates patients within its subordinate hospitals and arranges for transportation to transfer patients between facilities to relieve surgical backlogs, to obtain specialty care, and to ensure bed availability prior to start of operations, as appropriate. It also coordinates with the MEDCOM TPMC and the TPMRC for the evacuation of patients out of theater. This office maintains a 24-hour coordination and management responsibility for patient regulating and administration with the MEDBDE AO. It also prepares patient statistical reports for the MEDBDE commander. In the absence of the TPMC, IPMC assumes duties of the TPMC.

SECTION II — PLANNING PROCESS

EVACUATION PLANS AND ORDERS

CHARACTERISTICS

3-8. Operation plans and orders are the means by which commanders express their visualization, commander's intent, and decisions. They focus on results the commander expects to achieve. These OPLANs and OPORDs provide the basis for ensuring that medical evacuation operations are comprehensively planned, synchronized, responsive, and ensure a seamless continuum of care.

3-9. Evacuation plans and orders help form the basis commanders use to synchronize military operations. They encourage initiative by providing the *what* and *why* of a mission, and leave the *how to accomplish the mission* to subordinates. They give subordinates the operational and tactical freedom to accomplish the mission by providing the minimum restrictions and details necessary for synchronization and coordination. As medical evacuation operations are complex and may cross command and service boundaries, it is essential that the medical evacuation annex to the medical OPLAN/OPORD clearly delineates responsibilities, communications interoperability and procedures, and coordination requirements.

PLANS AND ORDERS

3-10. Publishing the commander's OPLAN/OPORD permits subordinate commanders to prepare supporting plans and orders. They can discern and implement instructions derived from a higher commander's plan and order. Appendix D provides an example of operational planning and a sample OPORD. Additionally, the commander's OPLAN/OPORD—

- Focuses subordinates' activities.
- Provides tasks and activities, constraints, and coordinating instructions necessary for mission accomplishment.
- Encourages agility, speed, and initiative during execution.
- Conveys instructions in a standard, recognizable, clear, and simple format.

3-11. The amount of detail provided in an evacuation plan or order depends on several factors, to include the experience and competence of subordinate commanders, cohesion and tactical experience of subordinate units, and the complexity of the operation.

3-12. Commanders balance these factors with their guidance and commander's intent, and determine the type of plan or order to issue. To maintain clarity and simplicity, plans and orders include annexes only when necessary and only when they pertain to the entire command. Annexes contain the details of support and synchronization necessary to accomplish the mission.

3-13. For a complete discussion of the military decision-making process and development of OPLANs/OPORDs refer to FM 5-0 and FMI 5-0.1.

MEDICAL EVACUATION PLANNING

3-14. Medical evacuation planning which supports the HSS/FHP plan is an ongoing process and is accomplished by medical operations element at battalion, brigade, division, corps, and theater levels. The HSS/FHP operations planning process must take under consideration all issues that could affect or influence HSS/FHP support for tactical operations. These issues should be considered in the initial developmental stages of theater, corps, division, or brigade plan as the supporting HSS/FHP annex is developed.

3-15. Planning is a dynamic and continuous process. The framework established in a robust initial plan can be built on in subsequent phases of an operation. The evacuation plan for a given AO will require adjustment and change during different phases of an operation. For example, transitioning from offensive operations to stability operations will generally dictate a significant change in evacuation coverage priorities.

3-16. Medical evacuation planning considerations, although essentially the same at all levels of command, may entail a broader scope, complexity, and detail at the higher levels of command due to the coordination and synchronization requirements which exist across command lines and service or multinational boundaries. Table 3-1 provides medical evacuation planning considerations at various levels of command based on METT-TC.

Table 3-1. Medical evacuation planning considerations

Mission	1. Define the protocol for a valid evacuation mission (mission statement). This protocol will serve as the authority to execute the mission by each medical evacuation unit/element/crew. 2. Use the HSS/FHP annex and mission analysis to: • Task evacuation units where needed. This may involve tasking part or all of DS units to perform a GS or area support mission. 3. Based on the type of operation: • Define trigger points for changes in the evacuation support plan. • Develop a concept of operations for all of the corps evacuation resources. 4. Define task organization of units assigned to the corps (and when directed to the JTF) and support relationships, if possible. Lines of coordination and mutual support relationships should be clearly delineated. 5. Area support and GS coverage areas will be

	defined with flexibility built into the plan. Coverage areas along division boundaries, for example, should not restrict or hinder evacuation of neighboring units that are geographically close. 6. Define pilot certifications for shore-to-ship medical air evacuation. 7. Develop standing operating procedure (SOP) requirements for single versus dual ship medical air evacuation. Address theater-specific procedures for in-flight link-up between medical evacuation aircraft and nonmedical evacuation aircraft for the purpose of dual ship requirements and/or security of pick-up zone. 8. Delineate any nonstandard air evacuation missions and allocate resources accordingly, such as support to Army special operations, joint and coalition forces. 9. Coordinate with any units that will have shared responsibilities for evacuation in a specific area and/or a specific mission. 10. Evaluate medical evacuation specific A2C2 considerations. Specifically, coordinating the crossing of brigade/division/corps boundaries/ area of operations. Medical evacuation must be afforded special expedited procedures for medical evacuation missions. Additionally, METT-TC dependent air corridors may have to be developed for the medical evacuation mission.
Enemy	1. Define the level or air-to-air threat and the threat faced by ground evacuation resources and the impact of this threat on the overall evacuation plan. This will be a major consideration in dictating which means of evacuation will be the primary means. 2. Based on the estimate of the number of enemy prisoners of war (EPW) or other categories of detainees involved in the operation— • Define procedures for evacuating wounded, injured, or ill persons through medical channels to include responsibilities of the echelon commander for designating guards for those evacuated directly from the point of capture to an MTF. • Allocate assets as necessary to provide this support and to support detainee collection points and holding areas.
Terrain and Weather	Define the types of terrain in the theater and use information as a major consideration in dictating the primary means of evacuation (ground or air in

	each sector) and any factors that may impair evacuation efforts.
Troops and Support Available	1. Disseminate locations of facilities used in evacuation, MTFs, MASFs, and aerial ports of debarkation (APOD). 2. Disseminate the capabilities and availability of joint evacuation assets and how to request their assistance. 3. Collect contact information from all evacuation units, assemble and publish for theater use. 4. Standardize a mandatory evacuation mission data collection format and collection method and schedule.
Time Available	1. Define the acceptable limits of evacuation time based on the distance from evacuation units. 2. Determine the proper allocation of resources to support the entire theater requirements. 3. Plan for trigger points (in time or distance) for changes in the evacuation plan to occur. 4. Define phases of the operation that reflect the commander's intent. Major changes in evacuation coverage usually occur when changes occur in the operational tempo (OPTEMPO). 5. Define briefing levels for launch authority. Consider developing a uniform theaterwide aviation risk assessment sheet.
Civil Considerations	1. Based on the HSS/FHP estimate and plan, determine requirements for the evacuation of HN civilians and others. 2. Define and disseminate protocols and procedures for evacuating civilians. 3. Allocate resources to support this mission, when directed.

3-17. For an effective HSS/FHP plan, medical staffs must coordinate and share operations information in order to synchronize support operations for mission accomplishment. See FMs 4-02, 4-02.1, 4-02.6, 4-02.12, 4-02.17, 4-02.25, 8-42 (4-02.42), and 8-55 (4-02.55), for doctrinal guidance on HSS/FHP operations. After the HSS/FHP plan is completed, it is incorporated into the C2 headquarters plan. After all annexes and appendixes of the plan are approved by the commander, it is incorporated into the OPORD.

SURGEON'S RESPONSIBILITIES

3-18. The surgeon is responsible for supervision and development of AHS input for the OPORD. The HSS plan serves as the base document for this input. The HSS plan is revised or updated based on mission analysis or changes in HSS requirements. The surgeon is tasked by the Assistant Chief of Staff (Operations and Plans) (G3)/S3 for HSS/FHP input to the OPORD for support of operations. The G3/S3 indicates timeline requirements. The surgeon and his staff are involved in all stages of the planning process, to identify all HSS requirements. Information on development of running estimates and the OPLAN/OPORD are discussed below. The medical plan/operations cell develops a HSS/FHP plan based on guidance received from the commander and the surgeon and provides HSS/FHP operational planning updates to the surgeon. The plan is briefed to the commander for approval, as required and provided to the G3/S3 according to the format of the in FM 8-55 (4-02.55).

MEDICAL EVACUATION TOOLS

3-19. It is essential that the evacuation plan for all combat operations be well planned, coordinated, and disseminated. In designing the medical evacuation plan, the medical planner should consider the following:

- Casualty collection points normally are predesignated along the axis of advance or evacuation routes.
- Forward of the BAS, the trauma specialist, CLS, and combat troops take casualties to the CCPs. These points facilitate acquisition by supporting ambulance teams and reduce evacuation time.
- When used by the BAS, CCPs help preserve BAS mobility, preclude carrying casualties forward, and reduce evacuation time to the sustainment area. When designating a CCP, the designating authority makes a decision whether or not to provide medical staff at the location. This decision is based upon the assessment of risk versus the availability of personnel.
- Normally, the role of care designating the point is responsible for staffing. Medical personnel may not be available to staff these points, and CLSs and ambulatory patients may be required to perform self-aid, buddy aid, or enhanced first aid. Casualty collecting points should be identified on operational overlays.

Note. A CCP staffed by a trauma treatment team is designated as a BAS (minus) rather than as a CCP.

3-20. Ambulance exchange points are a predetermined position where patients are exchanged from one evacuation platform to another.

3-21. These points are normally preplanned and are a part of the HSS/FHP annex to the OPLAN. In the forward area, the threat of enemy ground activities, large concentrations of lethal weapons systems, and effective use of antiaircraft weapons may dictate that the AXP be a predetermined rendezvous point for the rapid transfer of patients from one evacuation platform to another. The location of AXPs should be frequently changed to preclude attracting enemy fires.

3-22. Ambulance exchange points are established for many different reasons:

For example, the ambulance platoon of the heavy BSMC possesses a mixture of wheeled and tracked ambulances.

- The brigade tracked ambulances are provided so that they may keep up with maneuver elements. These vehicles carry the patients from the BAS to an AXP where the brigade wheeled ambulances take over for the relatively longer trip to the sustainment area.
- Ambulance exchange points are not limited to ground evacuation assets. Another example is a situation where the threat air defense artillery capability is such that air ambulances cannot fly as far forward as the BASs. However, an AXP could be established a few kilometers to the rear, still well forward of the BSA.
- The brigade combat team (BCT) tracked or wheeled ambulances could then transfer the patients to the air assets, thereby facilitating the rapid evacuation of patients and realizing a significant time savings.

3-23. By using AXPs evacuation assets are returned to their supporting positions faster. This facilitates evacuation as the returning crews are familiar with the road network and the supported unit's tactical situation. In the case of air ambulance (AA) assets, it is important because of the requirements for integration into the A2C2 system at each role and the enhancement to survivability provided by current threat and friendly air defense information.

3-24. The ambulance shuttle system is an effective and flexible method of employing ambulances during combat. It consists of one or more ambulance loading points, relay points, and when necessary, ambulance control points, all leveled forward from the principal group of ambulances, the company location, or basic relay points as tactically required. The ambulance—

- Loading point is a point in the shuttle system where one or more ambulances are stationed ready to receive patients for evacuation.
- Relay point is a point in the shuttle system where one or more empty ambulances are stationed. They are ready to advance to a loading point or to the next relay post to replace an ambulance that has moved from it. As a control measure, relay points are generally numbered from front to rear.
- Control points consists of a Soldier (from the ambulance company or platoon) stationed at a crossroad or road junction where ambulances may take one more directions to reach loading points. The Soldier, knowing from which location each loaded ambulance has come, directs empty ambulances returning from the sustainment area. The need for control points is dictated by the situation. Generally, they are more necessary in forward areas.

3-25. In the establishment of the ambulance shuttle system, once the relay points are designated, the required numbers of ambulances are stationed at each point. If the tactical situation permits, the ambulances may be delivered to the relay points by convoy.

3-26. Staffing of relay, loading, and ambulance control points. Important points may be manned to supervise the blanket, litter, and splint exchange and to ensure that messages and medical supplies to be forwarded are expedited.

3-27. Advantages of the ambulance shuttle system are that the system—

- Places ambulances at CCPs and BASs as needed.
- Permits a steady flow of patients through the system to MTFs.
- Avoids unnecessary massing of transport in forward areas.
- Minimizes the danger of damage to ambulances by the enemy.
- Permits the commander or platoon leader to control his elements and enables him to extend their activities without advancing the headquarters.
- Facilitates administration and maintenance.
- Maximizes the use of small battle command elements (sections or platoons) to operate the ambulance shuttle without employing the entire parent unit.
- Provides for flexible use of other ambulance assets for specific situations.

OBSTACLES MARKING

3-28. Ambulance crews must know and recognize the standard land/gap marking patterns. Unit TSOPs on the types of materials used should be available to ambulance crews. Refer to FM 3-34.2 for additional information.

Chapter 4
Army Medical Evacuation

Medical evacuation operations are planned to provide comprehensive, responsive, flexible, and agile support to the tactical commander in conformity with the commander's intent and OPLANs. This chapter discusses the employment of medical evacuation resources and the coordination and synchronization required to effectively execute medical evacuation operations (to include the transfer of patients between MTFs and to staging facilities) by air and ground evacuation assets.

SECTION I — MEDICAL EVACUATION SUPPORT

EVACUATION PRECEDENCE

4-1. Casualties requiring evacuation are prioritized to ensure the most seriously injured or ill receive timely medical intervention consistent with their medical condition. As with medical treatment, medical urgency is the only factor used to determine the medical evacuation precedence. (Appendix A provides an in-depth discussion of the provisions of the Geneva Conventions.)

This paragraph implements STANAG 3204.

4-2. The determination to request medical evacuation and assignment of a precedence is made by the senior military person present or, if available, the senior medical person at the scene. This decision is based on the advice of the senior medical person at the scene (if available), the patient's condition, and the tactical situation. Assignment of a medical evacuation precedence is necessary. The precedence provides the supporting medical unit and controlling headquarters with information that is used in determining priorities for committing their evacuation assets. For this reason, correct assignment of precedence cannot be overemphasized; over classification may result in an increase in evacuation which could burden the HSS system.

4-3. The patient's medical condition is the overriding factor in determining the evacuation platform and destination facility. The AA operates wherever needed on the battlefield, dependent on risk and METT-TC factors. The crew of the AA, assisted by onboard patient monitoring and diagnostic equipment, is trained in aeromedical procedures to provide optimum en route patient care. It is the preferred method of evacuation for most categories of patients. Air ambulances are a low-density, high-demand resource and must be managed accordingly. To conserve these valuable resources, medical planners should plan to use AA to primarily move Priority I, URGENT and Priority IA, URGENT-SURG patients with the other categories on a space-available basis. Health service support planners must plan for a synchronized air and ground evacuation plan. Depending on the length of time required for an AA to be dispatched and arrive at the POI, it may be prudent to evacuate the casualty by ground evacuation assets to a BAS and/or Role 2 MTF for stabilization by a physician. Further evacuation could then be accomplished by AA.

4-4. Patients will be picked up as soon as possible, consistent with available resources and pending missions. Table 4-1 depicts the categories of evacuation precedence and the criteria used to determine the appropriate precedence.

Table 4-1. Categories of evacuation precedence

Priority I—URGENT	Is assigned to emergency cases that should be evacuated as soon as possible and within a maximum of 2 hours in order to save life, limb, or eyesight, to prevent complications of serious illness, or to avoid permanent disability.
Priority IA—URGENT-SURG	Is assigned to patients who must receive far forward surgical intervention to save life and to stabilize them for further evacuation.
Priority II—PRIORITY	Is assigned to sick and wounded personnel requiring prompt medical care. This precedence is used when the individual should be evacuated within 4 hours or his medical condition could deteriorate to such a degree that he will become an URGENT precedence, or whose requirements for special treatment are not available locally, or who will suffer unnecessary pain or disability.
Priority III—ROUTINE	Is assigned to sick and wounded personnel requiring evacuation but whose condition is not expected to deteriorate significantly. The sick and wounded in this category should be evacuated within 24 hours.
Priority IV—CONVENIENCE	Is assigned to patients for whom evacuation by medical platform is a matter of medical convenience rather than necessity.
The NATO STANAG 3204 has deleted the category of Priority IV—CONVENIENCE; however, it will still be included in the US Army evacuation priorities as there is a requirement for it on the battlefield.	

SECTION II — MEDICAL EVACUATION REQUESTS

4-5. This section discusses requests for medical evacuation. Specific procedures, frequencies, and security requirements for transmittal of medical evacuation requests are delineated through the orders process and are made a part of the unit/command SOPs. Each sector based on the METT-TC may be designated with a different method of evacuation as the primary means to effect evacuation. In sectors which have a high ground-to-air or air-to-air threat may rely on ground evacuation assets to move the majority of patients. In other sectors where the ground threat is high and comprised of small arms, improvised explosive devices, and bombs, medical evacuation operations may be more efficiently and effectively executed by AAs. An additional consideration in planning medical evacuation operations is to determine whether armed escorts are required for either the ground or AA mission. Those missions that require armed escort must be thoroughly coordinated and synchronized between the medical assets and force protection assets that will accompany them.

4-6. Medical evacuation requests often are sent from the POI, through intermediaries, such as higher headquarters, who then transmit the request up to the nearest medical evacuation unit. The unit relaying the request must ensure that it relays the exact information originally received. The radio call sign and frequency relayed (Line 2 of the request) should be that of the requesting unit and not that of the relaying unit. However, the intermediaries contact information can be given as additional information if a callback is necessary to clarify details of the mission. Figure 4-1 depicts the communication flow for evacuation requests.

4-7. The necessity for secure communication in transmitting a 9-line request is METT-TC dependent. A technically capable enemy may be able to intercept a nonsecure 9-line request and strike at the pickup zone causing additional harm to the patients and the medical evacuation crew. In combat and when directed by higher headquarters, always use secure communications when requesting medical evacuation.

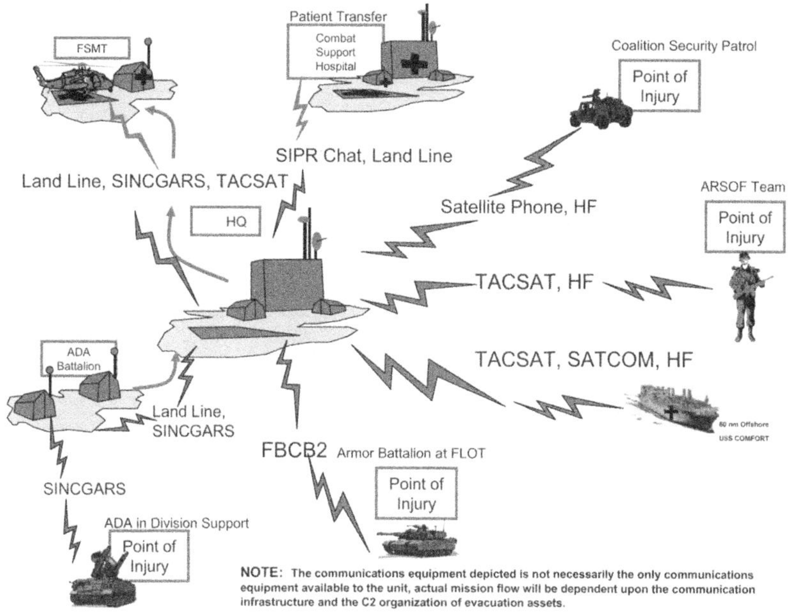

Figure 4-1. Flow of communication for evacuation request

GROUND EVACUATION REQUEST

4-8. Figure 4-2 provides an example of a request from the POI for ground evacuation. The patient is a ROUTINE precedence patient requiring evacuation from the POI to the supporting BAS (Role 1 MTF) with follow-on evacuation to an AXP for further evacuation to the rear. Should the patient require further treatment, either an air or ground evacuation could be used.

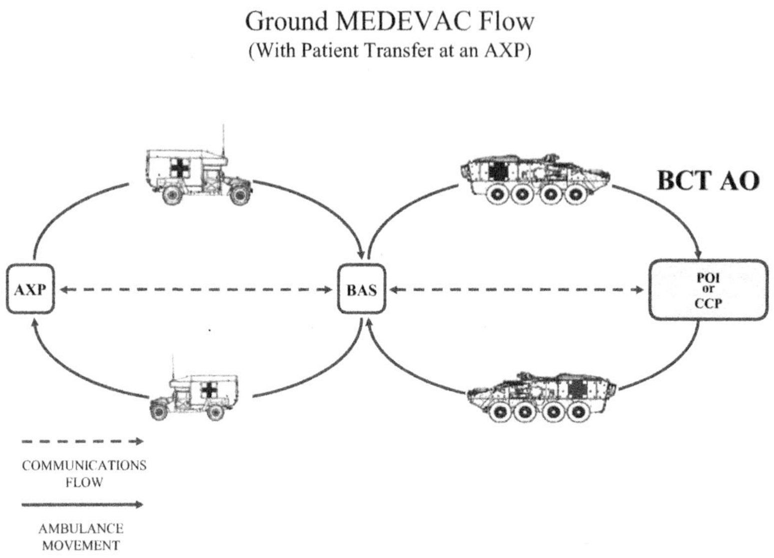

Figure 4-2. Ground evacuation request in a maneuver unit

4-9. Ground medical evacuation assets also provide DS and GS to supported units in the EAC. Figure 4-3 and Figure 4-4 depict the processes of providing DS to an ASMC and the GS support role of transferring a patient from a corps hospital to a MASF for further evacuation out of the theater.

4-10. In the corps area, the ASMC provides Roles 1 and 2 medical support on an area basis. The ASMC has organic evacuation assets and receives augmentation when required from the medical company ground ambulance assigned to the MMB.

4-11. Medical evacuation resources are also used to transfer a patient between hospitals within the theater and from a hospital to MASF for further evacuation out of theater. In Figure 4-3, this mission is being accomplished by a ground ambulance. However, depending upon the distance from the hospital to the MASF and/or APOD, this mission can also be accomplished by AA assets.

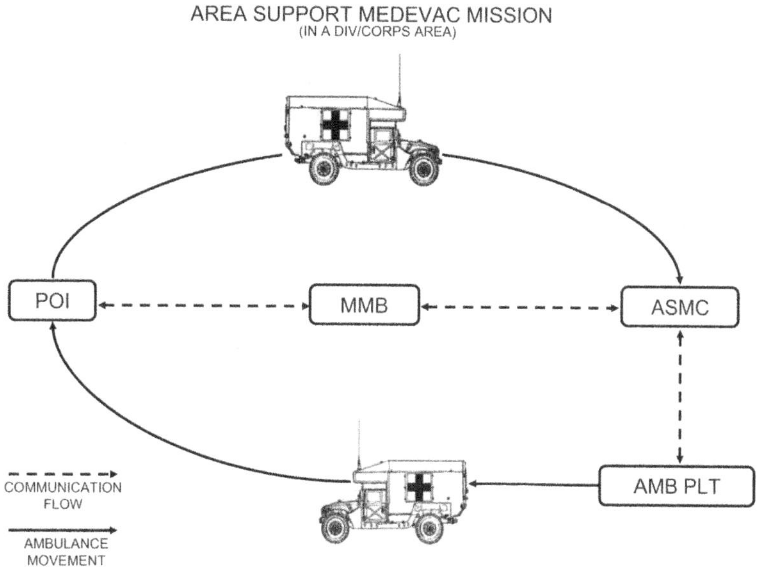

Figure 4-3. Ground evacuation request in the corps

AIR EVACUATION REQUEST

4-12. Air ambulances may fly as far forward as possible on the battlefield. Although evacuation by AA is the preferred means for all casualties, when high evacuation workloads exist, evacuation by AA should be the primary means used for URGENT and URGENT-SURG patients. Figure 4-5 depicts the zones of evacuation. The GSAB/CAB will place AA assets where they can best support the tactical commander's plan and facilitate the timely and responsive evacuation. To coordinate and synchronize the ground and air medical evacuation plans, a staff element was designed within the GSAB headquarters. This cell consists of the medical planners on the GSAB and CAB staff. The medical operations cell (MOC) will:

- Establish flight procedures specific to air evacuation missions within the CAB. This may include special routes or corridors as well as procedures for escort aircraft linkup.
- Ensure lines of communication (LOC) to supported units and higher levels of medical command are available. Further the MOC ensures supported units understand medical evacuation procedures and capability.
- Establish medical evacuation briefing and launch procedures. Ensure there is 24-hour access to those able to launch high and very high-risk missions.
- Maintain awareness of tactical and medical situation. Coordinate with MROs at higher levels to efficiently conduct the general support mission and to work in concert with adjacent units.
- Assist the AA CO and GSAB/CAB staff conduct medical evacuation operations.

- Coordinate missions with supported command surgeons. To ensure situational awareness and coordination of medical evacuation efforts, this staff keeps command surgeons informed on air evacuation missions being executed in their AO.
- Consults and coordinates with supported command surgeons when AAs cannot be launched to execute a requested mission. Ensures that the appropriate command surgeon is notified to ensure mission can be accomplished by ground evacuation assets.

4-13. Figure 4-4 depicts a request for air medical evacuation. The scenario shows the request being sent to the FSMT, the AA company, and the GSAB. The routing of the request would be dependant on how our forces are arrayed. In a similar situation where ground evacuation may not be readily available or the tactical situation precludes its use the requesting unit could transmit their medical evacuation request directly to a supporting FSMT as a primary means for a medical evacuation request or the AA company as a secondary option with a third being sent directly to the GSAB.

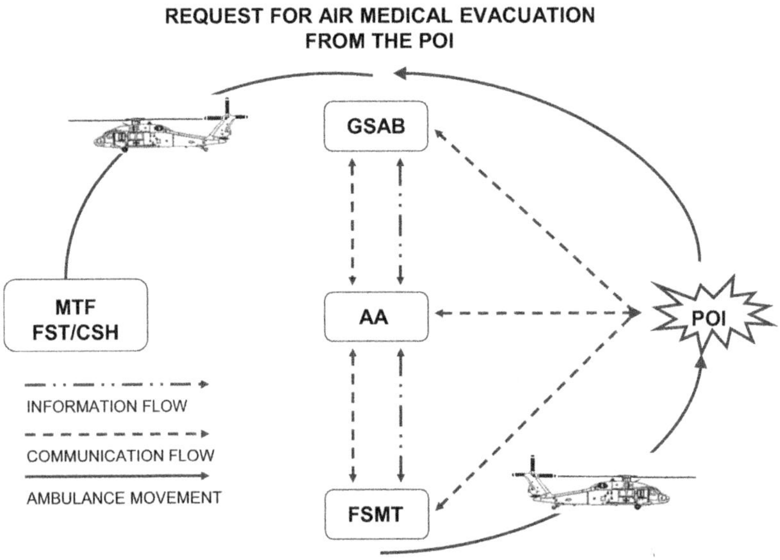

Figure 4-4. Air evacuation request in the corps

4-14. Figure 4-5 illustrates evacuation zones during a noncontiguous conflict. Here, no forward line of own troops (FLOT) is clearly visible.

- Forward support medical evacuation teams are placed where they are most needed. This can be with troops most often engaged in combat, high population density areas, area of famine or disease with high civilian casualties, refugee areas, geographically centralized location, and such.

- Area or general support FSMTs accomplish the patient transfer mission that develops between FSMTs and MTFs, between MTFs, and between MTFs and intertheater movement locations. These FSMT must also be responsible for the DS mission in their immediate vicinity.

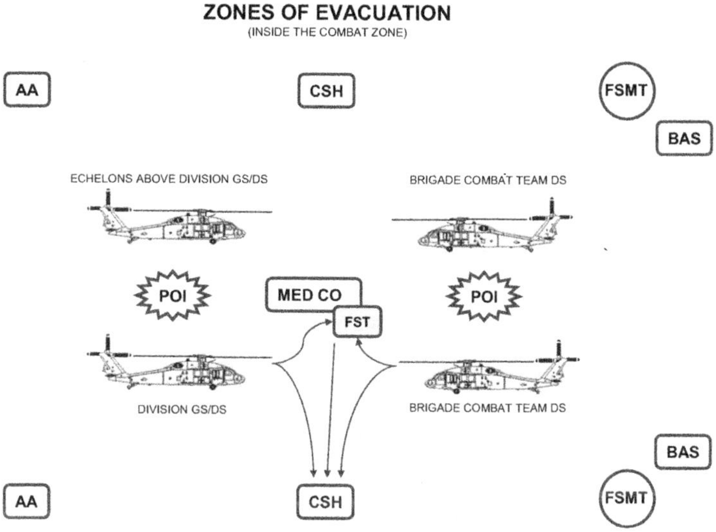

Figure 4-5. Air ambulance zones of evacuation

4-15. Figure 4-6 illustrates a comparison in aviation mission planning/execution cycle for medical evacuation and typical aviation.

- In contract to a typical aviation mission cycle, continuous aeromedical evacuation coverage results in extended operational duty days that often exceed 24 hours in length in order to provide seamless coverage to our forces. Medical evacuation units must plan and develop a detailed battle rhythm that addresses composite risk management, mission execution processes, and tactics, techniques, and procedures unique to 24-hour continuous evacuation operations.
- A medical evacuation crew cycle begins much like any other mission planning however they will not know the location or the time of execution for their mission. Since during these cycles it is critical the crew responds as rapidly as safely possible when the last information is received, it is imperative that crews are afforded the ability to manage rest while ensuring the procedures are in place to maintain battlefield situational awareness and facilitate a rapid, safe execution when a 9-line mission is received. This includes but is not limited to crew rest cycles, briefing procedures, and aviation and logistical support to remote/split-based crews to ensure the safe and timely execution of missions.

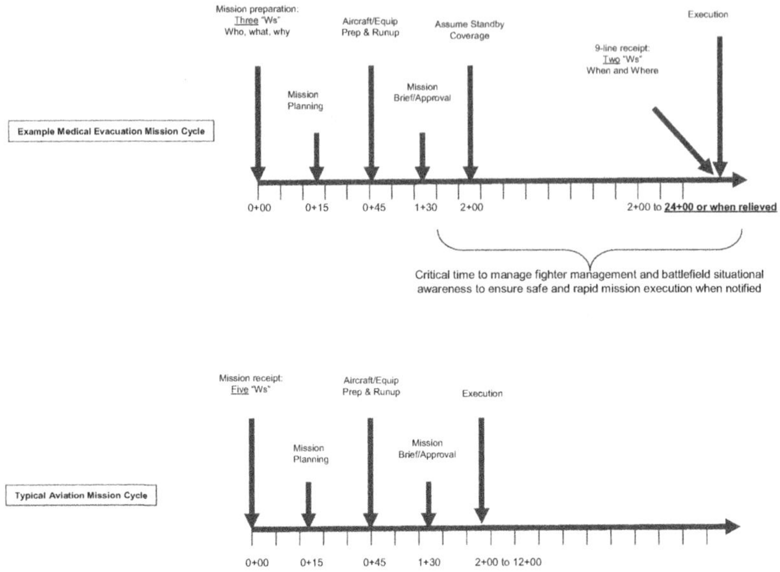

Figure 4-6. Aviation mission planning/execution cycle

JOINT INTERCONNECTIVITY

4-16. When directed by the GCC, Army evacuation assets may be used to evacuate patients from the other services or may be placed in direct support of other service, allied, or coalition units participating in the operation. Prior to the initiation of the operation, communications, and procedural issues should be synchronized to ensure the seamless execution of the operation. Interoperability issues can result in slower response times and may adversely effect medical evacuation operations. Figure 4-7 depicts the communications and patient flow in a joint scenario.

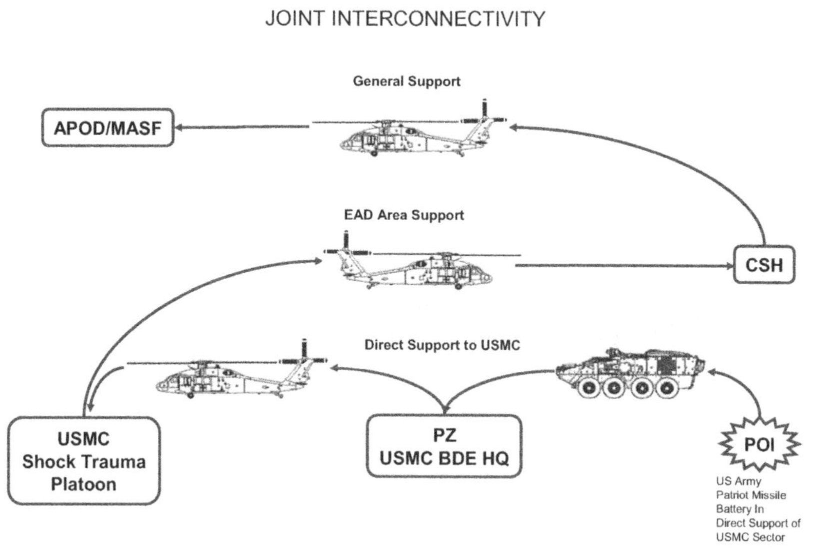

Figure 4-7. Air and ground ambulance evacuation in a joint environment

SECTION III — MEDICAL EVACUATION UNITS, ELEMENTS, AND PLATFORMS CONSIDERATIONS

MEDICAL EVACUATION SUPPORT PROTOCOL

4-17. Medical evacuation is performed by the higher role of medical care supporting forward and evacuating from the lower role. Medical evacuation support for theater, corps, and division is provided by organic ambulance teams, squads, and platoons. In addition, it is provided by the MMB medical company, ground ambulance medical company (GAMC), ASMC, and the GSAB medical company, AA.

4-18. The MMB maximizes the effective employment of ground ambulances resources whether in support of a division, corps, or at theater level. It coordinates through the combat aviation brigade's GSAB for effective use of AA support of its ASMCs. The medical brigade, MMBs, GSAB, sustainment brigade, BCTs and division surgeon section (DSS) coordinate and synchronize medical evacuation operations for the division AOs. The DSS coordinates with the MEDBDE for the evacuation of patients from division MTFs to supporting CSHs. The MEDBDE coordinates and arranges the required intratheater medical evacuation of patients to and from supporting CSHs. The MEDCOM coordinates for evacuation of patients out of the theater. To ensure that patients are evacuated to the appropriate MTFs, MROs are organic to the MEDCOM, and MEDBDE.

MEDICAL EVACUATION PLATFORMS FACTORS

4-19. Evacuation platforms must be capable of keeping pace with the troops supported. The patient's medical condition is the overriding factor in determining the evacuation platform and destination MTF. The AA operates wherever needed on the battlefield, dependent on risk and METT-TC factors. Use of hardened armored MEV may be the vehicle of choice for some missions for short evacuation to an MTF or to a secure AXP for transfer to an air or wheeled ground ambulance. The criteria for an efficient and effective evacuation system were identified in Chapter 1.

4-20. The crew of the AA, assisted by onboard patient monitoring and diagnostic equipment, is trained in aeromedical procedures to provide optimum en route patient care. It is the platform of choice for most categories of patients. However, insufficient numbers of AAs are available to evacuate all patients expected in a battle zone. To conserve these valuable resources, medical planners should plan to use AAs to primarily move Priority I, URGENT, Priority IA, URGENT-SURG, and Priority II patients with other categories on a space-available basis.

SECTION IV — MEDICAL EVACUATION AT UNIT LEVEL

CONSIDERATIONS FOR MEDICAL EVACUATION MISSIONS

4-21. As METT-TC factors affect all medical evacuation missions and the employment of ambulance assets. The medical commander must consider the basic tenets that influence the employment of medical evacuation assets. These factors include the patient's medical condition and the:

- Tactical commander's plan for employment of combat forces.
- Enemy's most likely course of action.
- Anticipated patient load.
- Expected areas of patient density.
- Availability of medical evacuation resources to include ground and air crews.
- Availability, location, and type of supporting MTFs.
- Protection afforded medical personnel, patients, and medical units, vehicles, and aircraft under the provisions of the Geneva Conventions.
- Army airspace command and control plan.
- Engineer obstacle plans.
- Fire support plan (to ensure medical evacuation assets are not dispatched onto routes and at the times affected by the fire support mission).
- Road network/dedicated medical evacuation routes (contaminated and clean).
- Weather conditions.

PATIENT ACQUISITION

4-22. Units with organic medical evacuation assets have the primary responsibility for patient acquisition. Methods of employment and evacuation techniques differ depending upon the nature of the operation.

4-23. Units without organic ambulance assets are provided medical evacuation support on an area support basis. Units must develop techniques which facilitate the effective employment of their combat medics/company/platoon medics, enhance the ability to acquire patients in forward areas, and rapidly request medical evacuation support. The techniques developed should be included in the unit TSOP. As a minimum, the TSOP should include the—

- Vehicle assignment for the organic medical personnel.
- Vehicles designated to be used for casualty transport and/or patient evacuation.
- Procedures for requesting medical evacuation support (during routine operations or during mass casualty situations).
- Role of the first sergeant (1SG), platoon sergeants, and CLSs in medical evacuation.

Maneuver Battalion Medical Platoons

4-24. The medical platoon leader (a physician) along with the medical platoon field medical assistant should be included in all battalion tactical planning. This officer—

- Maintains knowledge of the concept of operations, commander's intent, and the anticipated medical requirements.
- Develops the HSS/FHP plan (FM 8-55 [FM 4-02.55] and FM 4-02.4) and provides HSS/FHP overlays with preplanned evacuation routes and CCPs, to ambulance squads or teams for inclusion in the battalion OPLAN.
- Ensures that the platoon sergeant provides strip maps or other navigational tools to the ambulance drivers, if needed.
- Requests augmentation support from the supporting medical company in advance of the operation, if required. When elements of other maneuver battalions are attached as part of a task force, then ensures that adequate medical elements are included in the support package.
- Ensures that orientation and support are provided for the medical personnel. This precludes taxing the medical elements of the receiving unit. These responsibilities are normally delegated to the medical operations officer (field medical assistant).

4-25. The ambulance/evacuation section squad leader/NCO ensures that the ambulance teams have a working knowledge of the terrain features in the AO. Whenever possible, he familiarizes with primary and secondary medical evacuation routes through route reconnaissance conducted by the squad leaders. This platoon sergeant and squad leader manages the employment of the ambulance teams and monitors the communications net to remain abreast of the tactical situation.

4-26. The following factors should be considered when selecting ambulance routes:

- Tactical mission.
- Coordinating evacuation plans and operations with the unit movement officer.
- Security of routes and security escort.
- Availability of routes.
- Physical characteristics of roads and cross-country routes (to include natural obstacles).
- Requirements to traverse roads in urban areas and potential obstructions from rubble and debris.
- Traffic density.
- Time and distance factors.
- Proximity of possible routes to areas that may be subject to enemy fire.
- Lines of patient drift.
- Cover, concealment, and available defilade for moving and stationary vehicles.
- Engineer obstacle plans.
- Fire support plan (to ensure medical evacuation assets are not dispatched onto routes and at the times affected by the fire support mission).

4-27. Depending upon the combat situation, the modes of evacuation may include walking soldiers who are wounded, manual and litter carries, nonmedical transportation assets, or dedicated medical evacuation platforms. Evacuation in the battalion area normally depends on the organic ambulances assigned. Evacuation by AA is dependent upon the patient's medical condition, availability of air assets, tactical situation, air superiority, and weather conditions.

4-28. The ambulance team is normally deployed to each maneuver company trains and one remains with the Role 1 MTF/BAS in the combat trains. Ambulances operate as far forward as the tactical situation permits. An ambulance team operating in a maneuver company AO, is normally under the tactical control of the maneuver company executive officer or 1SG. This ambulance team, however, remains under the OPCON of the medical platoon.

4-29. The medical operations officer ensures that the ambulances are located close to the anticipated patient workload. An ambulance team consists of one ambulance and two combat medics (on track vehicles, a third emergency care specialist is required to provide en route medical care). One or two of these teams serve in DS of a maneuver company. To become familiar with the specific terrain and battlefield situation, the team maintains contact with the supported maneuver company during most combat operations. Maneuver medical platoon ambulances not in DS are positioned strategically throughout the battalion area or are sited at the BAS to—

- Evacuate patients from the company aid posts, CCPs to the BAS.
- Reinforce the forward positioned ambulance teams.
- Support the combat forces held in reserve and/or scout, mortar and forward observer platoons.

4-30. Another employment option is to forward site the additional ambulance teams at company CCPs, as well as at the BASs. Many times the ambulance team finds battlefield casualties who have not been seen by a trauma specialist. In these cases, the team members dismount and then find, treat, and evacuate the patients. Ambulance teams not specifically dedicated to support combat elements can be used as messengers in medical channels and to provide transport of emergency medical personnel, equipment, and supplies.

4-31. During static situations where the maneuver company is not in enemy contact or is in reserve, the ambulance team returns to the BAS to serve as reinforcement to other elements in contact. However, during movement to contact, the ambulance team immediately deploys with its supported unit. In moving patients back to the CCP, the team may be assisted by nonmedical personnel. Specific duties of the ambulance team are to—

- Maintain contact with supported elements.
- Find and collect the wounded.
- Administer EMT.
- Initiate or complete the FMC.
- Evacuate patients to the BAS.
- Direct or guide ambulatory patients to the BAS.
- Resupply company/platoon medics.
- Serve as messengers in medical channels.
- Route recons from Role 1 MTF to Role 2 MTF or the higher role of medical care.

Casualty Collection Points

4-32. During the offense, CCPs may be used to avoid hampering the movement of the maneuver elements. In fast-moving situations, preplanned CCPs are included in the HSS plan and activated based on the crossing of phase lines, upon the occurrence of predetermined events, or on the execution of other control measures. It may be necessary to set up multiple CCPs for each phase of an operation. Rotating the use of these points precludes the enemy from using them to pinpoint maneuver elements or from attracting enemy fires. When the situation permits, medical evacuation of patients from CCPs, or AXPs may be accomplished by AAs.

4-33. Ambulance teams move using available terrain features for cover and concealment. They avoid prominent terrain features and likely targets. When stationary, the ambulance crew should conceal the vehicle as much as possible. When a casualty occurs in a tank, Bradley infantry fighting vehicle (BIFV), a Stryker wheeled armored vehicles, the ambulance team moves as close to the armored vehicle as possible. Assisted by the armored crew, if possible, the casualty is extracted from the vehicle and then administered EMT. The ambulance team moves the patient to the Role 1 MTF/BAS or to a CCP point to await further evacuation.

4-34. The senior company medic/trauma specialist normally remains with the company combat trains, but may be used anywhere in the company area, even assisting the ambulance teams in some situations. He may be used to direct ambulance teams to locations where vehicle crews need assistance or where injured or wounded crew members have been left. In some situations, crew members may have to rely on self-aid or buddy aid until the CLS or the platoon medic/trauma specialist arrives.

Area Medical Evacuation Support

4-35. Medical evacuation on an area basis is required at all roles in the AHS system. The division units without organic medical evacuation resources, such as combat engineers, will require evacuation support on an area basis. To ensure that these elements receive adequate support, the medical planner must include their requirements into the OPLAN. Prior coordination is essential to ensure that the locations of CCPs, AXPs, and BASs are disseminated to these elements and that any unique support requirements are included in the OPLAN. Further, the triggers for relocating these assets and their next locations should also be included.

MEDICAL EVACUATION SUPPORT FROM ROLE 1 AND ROLE 2 MEDICAL TREATMENT FACILITIES

4-36. Evacuation from Role 1 MTFs operated by treatment squad or team is normally provided by the BSMC/FSMC ambulance platoon and the GSAB AA company. Further, these ambulance assets provide medical evacuation support on an area basis to other units in the brigade area, as required.

Ground Ambulances

4-37. The ambulance teams from the ambulance platoon are normally collocated with the BSMC/FSMC treatment platoon for mutual support. They establish contact and locate one ambulance team with the medical platoon of each maneuver battalion. The remaining ambulances are used for BCT operations and area support. The ambulances are positioned at AXPs or CCPs, or are field-sited with Role 1 MTFs operated by the BSMC/FSMC treatment teams.

4-38. A FSMT from the supporting GSAB medical company, AAs provides DS AA support for the BSMC/FSMC Role 2 MTF. The teams normally remain under the operational control of the GSAB. The DS relationship provides authority to the BSMC/FSMC in coordination with the brigade support battalion/forward support battalion (BSB/FSB) distribution support section, brigade surgeons section to direct the integrated air and ground medical evacuation support for the BCT. The section leader of the FSMT should be included in the brigade tactical planning process.

Air Ambulances

4-39. The AA team evacuates Priority I, URGENT patients from as far forward as possible to the BSMC/FSMC. Further, when a FST is colocated with a BSMC/FSMC, AAs evacuate Priority IA, URGENT-SURG to this facility if the patient can not be flown to a Role 3 MTF condition due to distance or patient condition. External lift capabilities of aeromedical evacuation helicopters add an important dimension to its role on the battlefield in moving medical supplies and equipment.

4-40. The GSAB AAs may operate from the sustainment area and BSAs providing around-the-clock, immediate response, evacuation aircraft, as well as provide the BSMC/FSMC commander flexibility and agility in the movement of treatment teams and equipment to the forward battle area. It also provides the capability to rapidly resupply Class VIII supplies to combat units.

4-41. The medical elements within the BSB/FSB will request AA support through the brigade aviation element (BAE). The FSMT leader will provide coordination between the BSB/FSB, BAE, or the BCT S3.

4-42. In the BSMC/FSMC, the executive officer (XO) is the principal assistant to the commander for the tactical employment of the company assets. The XO should be included in all brigade tactical planning. Officer preparation will be required to reinforce or reconstitute forward HSS elements and to request augmentation through the brigade surgeon section (BSS), if required. The FSMT leader keeps the BSMC/FSMC XO apprised of the operational capability. This enables the XO to effect timely reinforcement or augmentation. The BSMC/FSMC XO must be familiar with the specific terrain and battlefield situation. Further, the XO should have a thorough understanding of the division and brigade commanders' ground tactical plan.

Brigade Medical Evacuation Plan

4-43. The BSS is responsible for developing the BCT medical evacuation plan. The BSMC/FSMC commander is responsible for the execution of brigade medical evacuation plan, to include the use of both ground and air assets. The BSS medical planner should include the medical company commander, XO, brigade S1, brigade XO, medical platoon leaders, FSMT leader, the BSB/FSB medical plans and operations officer, and the BAE company commander (if available) in the planning process.

Ground Ambulance Evacuation Brigade Role 2 Medical Treatment Facilities

4-44. Evacuation from the BSMC/FSMC is provided by ground ambulances from the GAMC or from AAs from the GSAB AA company. The GAMC provides support to the BSMCs and medical elements in the division AO. They usually evacuate only those patients who will not RTD within 72 hours.

Ground Ambulance Support for Echelons Above Brigade/Division Units

4-45. The ASMC ambulance platoon normally collocates with the treatment platoon for mutual support and area taskings. It performs ground evacuation and en route patient care for supported units in the sustainment area. The ASMC may also evacuate patients from the supported Role 2 MTFs in the sustainment area to the ASMC Role 2 MTF. The ambulance platoon is mobile in operations as its assets may be totally deployed at one time. The platoon normally forward stations a portion of its teams in support of those units in the sustainment area.

4-46. The remaining teams are used for GS support to division sustainment, movement and maneuver and fires units, and for reinforcing support or ambulance shuttles. Platoons or squads from the GAMC will be in DS, or OPCON to, and collocated with supported division units.

SECTION V — EXCHANGE OF PROPERTY

PROPERTY EXCHANGE AND PATIENT MOVEMENT ITEMS

4-47. United States Army medical evacuation operations require that whenever a patient is evacuated from one MTF to another, or is transferred from one ambulance to another, medical items of equipment (casualty evacuation bags [cold weather-type bags], blankets, litters, and splints) remain with the patient. To prevent rapid and unnecessary depletion of supplies and equipment, the receiving Army element exchanges like property with the transferring element. This reciprocal procedure will be practiced to the fullest extent possible through all phases of evacuation from the most forward element through the most rearward hospital.

4-48. With USAF aeromedical evacuation operations, a major factor in the evacuation of patients is that specific medical equipment and durable supplies designated as patient movement items (PMI) must be available to support the patient during the evacuation. Examples of PMI include—

- Ventilators.
- Litters.
- Patient monitors.
- Pulse oximeters.

These items will be available for exchange at the supporting ASFs and MASFs. Refer to JP 4-02 and FM 4-02.1 for additional information on PMI.

This subparagraph implements STANAG 2128 and QSTAG 436.

4-49. Medical property of allied nations (NATO and ABCA armies) accompanying patients of allied nations will be returned to the parent nation at once, if possible. If it is not possible, like items will be exchanged as in the paragraph above.

4-50. When medical property of coalition forces without ratified standardization agreements accompanies patients of coalition forces, it is returned as soon as practical. Commanders should consult with their staff judge advocate early in the planning process to ensure appropriate policy and procedures are developed and disseminated.

SECTION VI — MEDICAL EVACUATION SUPPORT FOR THE OFFENSE, THE DEFENSE, AND STABILITY OPERATIONS

MEDICAL EVACUATION SUPPORT FOR THE OFFENSE

4-51. The offense is the decisive form of war, the commander's only means of attaining a positive goal or of completely destroying an enemy force (FM 3-0). The offense is characterized by rapid movement, deep penetrations, aggressive action, and the ability to sustain momentum regardless of counterfires and countermeasures.

4-52. When considering the evacuation plans to support an offensive action, the medical planner must consider many factors (FM 8-55 [FM 4-02.55]). The forms of maneuver, as well as the enemy's capabilities, influence the character of the patient workload and its time and space distribution. The analysis of this workload determines the allocation of AHS resources and the location or relocation of MTFs. Evacuation support of offensive operations must be responsive to several essential characteristics. As operations achieve success, the areas of casualty density move away from the supporting facilities. This causes the routes of medical evacuation to lengthen. Heaviest patient workloads occur during disruption of enemy main defenses, at terrain or tactical barriers, during the assault on final objectives, and during enemy counterattacks. The accurate prediction of these workload points by the medical planner is essential if medical evacuation operations are to be successful.

4-53. In traditional combat operations, the major casualty area of the division BCTs is normally the zone of the main attack. As the main attack accomplishes the primary task of the division, it receives first priority in the allocation of combat power. The allocation of combat forces dictates roughly the areas which are likely to have the greatest casualty density. As a general rule, all BCT MTFs are located initially as far forward as combat operations permit. This allows the maximum use of these facilities before lengthening evacuation lines force their displacement forward. In operations that feature deep battles with CBRN weapons targeted at supporting logistical bases, MASCAL operations may be conducted in sustainment area.

4-54. As advancing combat formations extend control of the battle area forward, supporting medical elements overtake patients. This facilitates the acquisition of the battle wounded and reduces the vital time elapsed between wounding and treatment. In offensive operations, two basic problems confront the supporting evacuation units.

- First, contact with the supported unit must be maintained. Responsibility for the contact follows the normal HSS pattern of rear to front. The contact is maintained by forward deployed air and ground evacuation resources.
- Secondly, the mobility of the MTFs supporting the combat formations must be maintained. Periodically, BCT medical companies and collocated FSTs, are cleared of their patients so that they may move forward. In some scenarios, CSHs may also have to evacuate patients to reduce patient overflow, surgical backlogs, or to increase bed availability.

4-55. This requirement for prompt evacuation of patients from forward MTFs requires available ambulances to be leveled well forward from the outset. The requirement for periodic movement of large numbers of patients from divisional and corps facilities further stresses the evacuation system.

TYPES OF OFFENSIVE OPERATIONS

Movement to Contact/Attack

4-56. Medical evacuation support in movement to contact/attack is key to the tactical plan. Prior deployment of evacuation resources with parent and supported units permits uninterrupted and effective evacuation support.

Exploitation and Pursuit

4-57. Evacuation support of exploitation and pursuit operations resembles those discussed for the envelopment (paragraph 4-59). Since exploitation and pursuit operations can rarely be planned in detail, evacuation operations must adhere to TSOPs and innovative C2. These actions are often characterized by—

- Decentralized operations.
- Unsecured ground evacuation routes.
- Exceptionally long distances for evacuation.
- Increased reliance on convoys and AAs.

MEDICAL EVACUATION SUPPORT FOR DIFFERENT FORMS OF OFFENSIVE MANEUVER

Penetration and Attack

4-58. In this tactic, the attack passes through the enemy's principal defensive position, ruptures it, and neutralizes or destroys the enemy forces. Of all forms of offensive maneuver, the penetration of main enemy defenses normally produces the heaviest medical evacuation workload. Patient acquisition starts slowly, but becomes more rapid as the attack progresses. The evacuation routes lengthen as the operation progresses. The penetration maneuver is often preceded by heavy preparatory fires which may evoke heavy return fire. These enemy fires may modify the decision to place evacuation assets as far forward as possible. Patient evacuation may be slow and difficult due to damage to roads or the inaccessibility of patients. Evacuation support problems multiply when some combat units remain near the point of original penetration. This is done to hold or widen the gap in enemy defenses while the bulk of division forces exploit or pursue the enemy. Treatment elements are placed near each shoulder of the penetration; ground evacuation cannot take place across an avenue of heavy combat traffic. Besides the heavy traffic, the area of the penetration is normally a target for both conventional and nonconventional CBRN weapons.

Envelopment

4-59. In the envelopment, the main or enveloping attack passes around or over the enemy's principal defensive positions. The purpose is to seize objectives which cut the enemy's escape routes and subject him to destruction in place from flank to rear. Since the envelopment maneuver involves no direct breach of the enemy's principal defensive positions, the medical evacuation system may not be confronted with a heavy workload in the opening phase—

- Ambulances are echeloned well forward in all levels of AHS to quickly evacuate the patients generated by suddenly occurring contact.
- Medical treatment facilities moving with their respective formations overtake patients during evacuation and reduce delays in treatment. After triage and treatment, the patients are evacuated to division/corps-level facilities by accompanying division/corps ground ambulance assets or AAs.

- When the isolated nature of the envelopment maneuver precludes prompt evacuation, the patients are carried forward with the treatment element. Again, nonmedical vehicles may be pressed into emergency use for this purpose. When possible, nonmedical vehicles should be staffed by a combat medic or CLS.
- When patients must be carried forward with the enveloping forces, AHS commanders use halts at assembly areas and phase lines to arrange combat protection for ground ambulance convoys through unsecured areas.
- Further, the commander may take advantage of friendly fires and suppression of enemy air defenses to call for prearranged AA support missions, or emergency use of medium-lift helicopter backhaul capabilities.

Infiltration

4-60. Infiltration is a form of maneuver used during offensive operations.

- The division can attack after infiltration or use it as a means of obtaining intelligence and harassing the enemy.
- Though it is not restricted to small units or dismounted actions, the division employs these techniques with a portion of its units, in conjunction with offensive operations conducted by the remainder of its units.

Army Health System of Infiltration/Maneuver

4-61. The AHS of infiltration/maneuver is restricted by the amount of medical equipment, supplies, and transportation assets that can be introduced into the attack area. No deployment of division-level medical units without their organic transportation should be attempted.

4-62. Elements of unit-level AHS should be accompanied by their organic vehicles, and ambulances should receive priority for deployment. It may be necessary to man-carry enough BAS equipment into the attack area to provide EMT and ATM; however, this results in degrading mobility. When the element is committed without its ambulances, patients are evacuated to the BAS by litter bearer teams. This requires reinforcement of the medical platoon by division or corps medical personnel or improvisation of litter teams using combat troops (if available and approved by the tactical commander). Patient evacuation from the BAS and medical resupply of the force may be provided by litter bearers, depending upon distances and degree of secrecy required.

4-63. When airborne and air assault forces are used, infiltrating elements may land at various points within the enemy's rear area and proceed on foot to designated attack positions. As in surface movement, the amount of medical equipment taken may be limited. In airborne operations, the evacuation of patients will be by litter bearers or frontline ambulances to CCPs or the BAS and then by division-level ambulances to the Role 2 MTF. In air assault operations, the evacuation is by litter bearers to CCPs or to the BAS and then by AAs to a Role 2 MTF. Once the combat element begins the assault on the objective, secrecy is no longer important and its isolated location requires HSS characteristic to airborne and air assault operations until ground linkup.

Turning Movement

4-64. The turning movement is a variant to the envelopment in which the attacker attempts to avoid the defense entirely; rather, the attacker seeks to secure key terrain deep in the enemy's rear and along his LOC. Faced with a major threat to his rear, the enemy is thus "turned" out of his defensive positions and forced to attack rearward at a disadvantage.

Note. General MacArthur's invasion at Inchon during the Korean War is an example of a classic turning movement. Casualties were initially light as the main defenses were avoided; however, as the invasion developed, resistance stiffened and higher casualty rates were experienced. Further, as fighting occurred in a populated area (Seoul), significant civilian casualties resulted. The lack of Korean health care providers caused many of these civilians to seek medical aid from US field medical units.

Medical Evacuation for Turning Movement

4-65. Medical evacuation support to the turning movement is provided basically in the same manner as to the envelopment. As the operation is conducted in the enemy's rear area, LOC and evacuation routes may be unsecured resulting in delays in resupply and evacuation. In the Inchon example, a hospital ship was located off the coast to accept patients evacuated from the fighting. However, due to the precarious tides, evacuation and resupply were often delayed for hours and sometimes days since the harbor could not be navigated by small vessels. It was not until Kimpo Airfield fell that timely evacuation could occur. The deployed HSS units must be able to quickly clear the battlefield of patients, evacuate them from the forward areas, and sustain the patients in rear areas until evacuation routes are established.

MEDICAL EVACUATION SUPPORT FOR THE DEFENSE

TYPES OF DEFENSE

4-66. There are three forms of the defense:

- **Area defense** concentrates on denying enemy access to designated terrain for a specific period of time, rather than on the outright destruction of the enemy.
- **Mobile defense** focuses on denying the enemy force by allowing him to advance to a point where he is exposed to a decisive counterattack by the striking force. The primary defeat mechanism, the counterattack, is supplemented by the fires of the fixing force.
- **Retrograde** is an organized movement to the rear and away from the enemy. The enemy may force these operations or a commander may execute them voluntarily. Within the retrograde operation there are three forms: delay, withdrawal, and retirement.

4-67. Support is generally more difficult to provide in the defense. The patient load reflects lower casualty rates, but forward area patient acquisition is complicated by enemy actions and the maneuver of combat forces.

4-68. Medical personnel generally are permitted much less time to reach the patient, complete vital EMT, and remove him from the battle site. Increased casualties among exposed medical personnel further reduce the medical treatment and evacuation capabilities. Heaviest patient workloads, including those produced by enemy artillery and CBRN weapons, may be expected during the preparation or initial phase of the enemy attack and in the counterattack phase.

4-69. The enemy attack may disrupt ground and air routes and delay evacuation of patients to and from treatment elements. The depth and dispersion of the defense create significant time and distance problems for evacuation assets. Combat elements may be forced to withdraw while carrying their remaining patients to the rear. The enemy exercises the initiative early in the operation which may preclude accurate prediction of initial areas of casualty density. This makes the effective integration of air assets into the evacuation plan essential. The use of AAs must be coordinated and synchronized with the supporting GSAB to ensure the synchronized execution of evacuation operations occurs.

4-70. The support requirements for retrogrades may vary widely depending upon the tactical plan, the enemy reaction, and the METT-TC factors. Firm rules that apply equally to all types of retrograde operations are not feasible, but considerations include:

- Requirement for maximum security and secrecy in movement.

- Influence of refugee movement that may impede medical evacuation missions conducted in friendly territory.
- Integration of evacuation routes and obstacle plans should be accomplished.
- Difficulties in controlling and coordinating movements of the force which may produce lucrative targets for the enemy.
- Movements at night or during periods of limited visibility.
- Time and means available to remove patients from the battlefield. In stable situations and in the advance, time is important only as it affects the physical well-being of the wounded. In retrograde operations, time is more important. As available time decreases, medical managers at all levels closely evaluate the capability to collect, treat, and evacuate all patients.
- Medical evacuation routes will also be required for the movement of troops and materiel. This causes patient evacuation in retrograde movements to be more difficult than in any other type of operation.
  - Command, control, and communications may be disrupted by the enemy.
  - Successful medical evacuation requires including ambulances on the priority list for movement.
  - Providing for the transportation of the slightly wounded in cargo vehicles and providing guidance to subordinate commanders defining their responsibilities in collecting and evacuating patients.
  - Special emphasis must be placed on the triage of patients and consideration given to the type of transportation assets available for evacuation.
- When the patient load exceeds the means to move them, the tactical commander must make the decision as to whether patients are to be left behind. The medical staff officer keeps the tactical commander informed in order that he may make a timely decision. Medical personnel and supplies must be left with patients who cannot be evacuated (refer to FM 4-02 for additional information).

Air Ambulance Evacuation Support to Covering Forces in Defensive Operations

4-71. The division covering force uses combat power to screen the FLOT and force the early deployment of the enemy's main force. Organic AHS elements of the covering forces establish aid stations of minimal size or operate in the split-team mode (FM 4-02.4). Ambulances are deployed well forward to promptly clear patients from combat units. Medical care is limited to EMT and ATM, followed by rapid evacuation whenever possible.

4-72. The covering force surgeon maintains communications with attached aviation elements and uses these assets (augmented by medical personnel to provide en route medical care when feasible) to provide backhaul of casualties, rather than use ground ambulances. The early evacuation of patients from BASs ensures their mobility for rearward displacement. The BCT or armor cavalry regiment (ACR) medical company/troop is equipped to receive patients generated in the covering force area. Depending upon the fluidity of the defensive situation, the FSMT (if available) may have to rapidly evacuate patient's further rearward to a BSMC/FSMC and collocated FST or a CSH.

4-73. The wide dispersion of units and the manner in which they withdraw make patient acquisition difficult. When covering forces withdraw, patients are transported to the rear by the medical element that acquired them. Seriously wounded or injured patients are given priority for evacuation by air. There are usually few COSRs during covering force operations, although delayed symptoms may occur once the element is withdrawn to a safe area. Ground ambulances augmented by nonmedical transportation assets evacuate the remaining patients.

MEDICAL EVACUATION SUPPORT FOR STABILITY OPERATIONS

4-74. Stability operations promote and protect US national interests by influencing the threat, political, and information dimensions of the operational environment. They include developmental, cooperative activities during peacetime and coercive actions in response to crisis.

4-75. Some of the purposes for which the US Forces are employed to conduct stability operations are as follows:

- Civil control.
- Civil security.
- Provide essential services.
- Support economic and infrastructure development.
- Governance.

4-76. Stability operations may complement the main effort or they may themselves constitute the main effort. Table 4-2 identifies the types of stability operations and the types of missions the BCT may conduct.

Table 4-2. Types of stability operations

Types Of Operations	*Types Of Missions*
Civil Control	Establish police forces training.
	Integrate trained police into operations.
	Counter organized crime.
	Establish judicial system.
	Transition to indigenous police forces.
Civil Security	Initiate security forces training.
	Establish National Security Forces.
	Integrate security forces into operations.
	Isolate population from insurgency.
	Defeat insurgency.
	Transition to National Security Forces.
Essential Services	Sewage treatment plants operating.
	Establish trash disposal.
	Water treatment plants operating.

Table 4-2. Types of stability operations

	Restore electrical power. Reopen hospitals and clinics.
	Reopen schools and universities.
Economic/Infrastructure Development	Implement Employment Programs.
	Secure vital natural resources.
	Repair/rebuild distribution infrastructure.
	Prioritize reconstruction projects.
	Essential banking services available.
	Implement public works projects.
Governance	Identify and recruit local leaders.
	Facilitate establishment of sector representation.
	Facilitate establishment of neighborhood councils.
	Facilitate establishment of district councils.
	Support and secure elections.

4-77. Medical evacuation support to forces deployed in stability operations is dependent upon the specific type of operation, anticipated duration of the operation, number of forces deployed, theater evacuation policy, medical troop ceiling, and anticipated level of violence. In most situations, medical evacuation support follows the traditional support provided to combat forces. If there is a shortened theater evacuation policy, a limited medical troop ceiling, and limited hospitalization assets within the AO, organic and DS ambulance support is provided from the POI to the supporting Roles 1 or 2 MTF and, once the patient is stabilized for further evacuation, from the treatment element to an airfield for evacuation out of the theater.

4-78. Reconstruction operations use Army forces to assist civil authorities, foreign or domestic, as they prepare for or respond to crises and relieve suffering. In reconstruction operations, Army forces provide essential support, services, assets, or specialized resources to help civil authorities deal with situations beyond their capabilities. The purpose of reconstruction operations is to meet the immediate needs of designated groups for a limited time, until civil authorities can do without Army assistance.

4-79. Humanitarian and civic assistance operations can include a number of activities such as disaster relief, civil support, refugee assistance, the provision of medical care to isolated populations, and refeeding programs resulting from famines or natural disasters. Medical evacuation assets may be used to evacuate the injured from disaster sites, to provide the emergency transport of critically needed medical supplies and personnel to remote locations, or to perform emergency rescues during times of flooding, wildfires, or other natural disasters.

SECTION VII — MEDICAL EVACUATION SUPPORT FOR ENABLING/SHAPING OPERATIONS

PASSAGE OF LINES

4-80. Passage of lines presents a challenge for the medical planner. There will be a number of medical evacuation units using the same air and road networks. Coordination and synchronization are essential if confusion and over evacuation are to be avoided. The information required to operate in the BCT or division AO includes:

- Radio frequencies and call signs.
- Operations plans and TSOPs.
- Location of MTFs.
- Location of CCPs and AXPs.
- Main supply route, forward arming and refueling points (FARPs), and A2C2 data.

SECURITY OPERATIONS

4-81. In security operations, the covering forces are dependent upon organic resources found in the maneuver battalion medical platoon for initial support. The level of command for the covering force (division or corps) determines the responsibility for the subsequent evacuation plan. In a corps covering force, for example, the corps AHS structure has the responsibility for establishing and operating the medical evacuation system to support the forward deployed corps forces. This is done to prevent the divisions following the covering forces from becoming overloaded with patients prior to the hand off and passage of lines. The use of CCPs, AXPs, and nonmedical transportation assets (CASEVAC) to move the wounded is essential. The covering force battle may be extremely violent. Patient loads will be high and the distance to MTFs may be much longer than usual. The effectiveness of the medical evacuation system depends upon the forward positioning of a number of ground ambulances and the effective integration of AAs into the evacuation plan.

ADVANCE, FLANK, AND REAR GUARDS

4-82. In advance, flank, and rear guards, these forces normally receive medical evacuation support through the attachment of evacuation teams. The teams evacuate patients to predesignated CCPs along a main axis of advance or to the nearest treatment element providing area support. Employment of AAs provides a measure of agility and flexibility.

RIVER CROSSING OPERATIONS

4-83. The river barrier itself exerts decisive influence on the use of division and BCT medical units. Attack across a river line creates an AHS delivery problem comparable to that of the amphibious assault. Army health system protection elements cross as soon as combat operations permit.

4-84. Early crossing of treatment elements reduces turnaround time for all crossing equipment that is used to load patients on the far shore. Maximum use of AA assets is made to prevent excessive patient buildup in far shore treatment facilities. Near shore MTFs are placed as far forward as assault operations and protective considerations permit to reduce ambulance shuttle distances from off-loading points. For detailed information on river crossing operations, refer to FM 90-13.

RECONNAISSANCE OPERATIONS

4-85. The reconnaissance in force is an attack to discover and test the enemy's position and strength or to develop other intelligence. The division usually probes with multiple combat units of limited size, retaining sufficient reserves to quickly exploit known enemy weaknesses.

4-86. Ambulances are positioned well forward at both unit and division levels. Ambulances are moved at night to enhance secrecy. The echeloning of ambulances is an indication to the enemy that an attack is imminent due to the forward placement of HSS. Role 2 MTFs are not established until a significant patient workload develops.

4-87. Patients received at BASs of reconnoitering units are evacuated to Role 2 MTFs as early as practical or are carried forward with the force until a suitable opportunity for evacuation presents itself. Maximum possible use of AA assets is made to cover extended distances and to overcome potentially unsecured ground evacuation routes.

RETROGRADE OPERATIONS

4-88. A retrograde operation is a movement to the rear or away from the enemy. This type of operation may be forced by enemy action or may be executed voluntarily.

AEROMEDICAL EVACUATION SUPPORT IN RETROGRADE OPERATIONS

4-89. Aeromedical evacuation support in retrograde operations may vary widely depending upon the enemy reaction and the situation. Firm rules that apply equally to all types of retrograde operations are impossible to establish. Factors to consider in planning aeromedical evacuation support for retrograde operations include—

- Patient's condition/status.
- Mission, enemy, terrain and weather, troops and support available, time available and civil considerations.
- Requirement for maximum efforts in secrecy and employment of aircraft survivability equipment.
- Influence of refugee movement (which may impede military medical movements conducted in friendly territory).
- Integration of evacuation routes and other Army aviation routes.
- Difficulty in controlling and coordinating movements of the force which may produce lucrative targets for the enemy.

4-90. During retrograde operations, battalion, brigade, and division surgeons must evaluate their capability to collect, treat, and evacuate patients. Three factors influence the number of patients evacuated from the battlefield and the time and place of patient treatment:

- Time available prior to MTF displacement.
- Location of the enemy advance.
- Means of evacuation.

4-91. Due to congestion on main supply routes and general evacuation routes, the AAs may provide the best means of evacuation during retrograde operations. Seriously wounded patients should be evacuated by the fastest and most comfortable means and should receive medical care en route. Proper sorting (triage) and rapid evacuation of patients lessen the need for establishing a complete Role 2 MTF operation. Therefore, the FSMT (if available) leader should be prepared to surge his efforts in the retrograde. Combat and operational stress reaction casualty rates are usually low as compared to wounded in action (WIA) rates during controlled retrograde operations, but increase after safety is reached.

4-92. Commanders must decide to displace medical evacuation units before they hamper the maneuver forces conducting the retrograde movement. During mission planning, the medical evacuation units/elements integrate with the supported unit's displacement plan. The medical evacuation element leaders must coordinate with the supported unit commander and the BSMC for displacement decision points and locations.

WITHDRAWAL OPERATIONS

4-93. A withdrawal operation is one in which a deployed force disengages from an enemy force. It may be forced by enemy pressure or conducted voluntarily.

4-94. Although the deployed force disengages from the enemy, contact is maintained by security elements while the main force moves to the sustainment area, forms march columns, and moves to a predesignated location. Aeromedical evacuation elements must be prepared to support the withdrawing force as well as any forces left in contact. Time available to acquire, treat, and evacuate patients from the line of contact is limited.

AEROMEDICAL EVACUATION SUPPORT IN WITHDRAWAL OPERATIONS

4-95. The AA medical company element supporting the withdrawing force must ensure it maintains sufficient station capability at its present location to support the patients while taking action to move to a new location.

4-96. In addition to the evacuation support provided to the force left in contact, there is a requirement to support the force moving to the sustainment area. If the main force infiltrates to the sustainment area, patients are carried by their parent units to the assembly area or MTFs in the sustainment areas. Medical assets are positioned in the assembly area to receive patients.

4-97. The division establishes Role 2 MTFs at a minimal size and well to the rear of the first line of alternate positions. If combat and environmental conditions indicate a light patient workload and the road network permits rapid ambulance movement, the division may use another Role 2 MTF to provide GS to all withdrawing brigades.

4-98. Role 2 MTFs conduct leapfrog rearward, occupying successive positions placed along the withdrawal route to minimize the requirement for multiple displacements by any one MTF. This avoids unnecessary interference with combat operations while providing continuous medical support. With the exception of an ambulance squad to support the covering force the division normally does not employ ambulances further forward than the BAS. AA elements will normally evacuate to sustainment area and corps positions to avoid impeding the withdrawal of brigades.

4-99. Preparation for the withdrawal operation includes distribution of extra medical consumable supplies and nonexpendable exchange items to each medical element. This is done to overcome the potential effects of isolated MTFs or the possible intermittent operation of the medical evacuation system. The AA medical company may receive the mission to assist with this distribution.

4-100. If withdrawal is made under enemy pressure, the provision of aeromedical evacuation is modified. Since the time available is critical, aeromedical resources cannot remain in the forward areas. The assets are usually withdrawn as a unit. Patients occurring during the withdrawal are carried to the sustainment areas by the parent unit, normally using nonmedical vehicles.

DELAYING OPERATIONS

4-101. Delaying operations occur when forces are insufficient to attack or to defend and when the defensive plan calls for drawing the attacker into an unfavorable situation. The delay maneuver may include friendly maneuver to successive positions across a broad front.

4-102. Delaying brigades split their combat power, moving their less mobile forces directly to the next defensive position while the elements remaining in contact fight to the sustainment area.

4-103. Aeromedical support for a delaying operation is conducted like withdrawal operations.

RETIREMENT OPERATIONS

4-104. A retirement operation is a rearward movement of a force not in contact with the enemy. It is conducted according to the force's OPLAN and without pressure by enemy forces.

4-105. Because the division is no longer in contact with the enemy, it can march (in multiple columns) directly to the rear. The AHS requirements for this type of operation are similar to those in a movement to contact. The patient workload is light. The FSMT (if available) leader will coordinate with the BSMC commander, and the AA medical company CP will coordinate with the DSS for guidance in the positioning of company assets to support this operation.

SECTION VIII — SPECIAL FORCES OPERATIONS

4-106. Army special operations forces (ARSOF) have limited organic AHS resources. They are, therefore, dependent upon the theater AHS for the majority of their health care requirements. To ensure that ARSOF receive comprehensive and timely AHS and logistics, medical planners must consider each medical function and determine the support organic and theater AHS assets will provide. The significant challenge for ARSOF medical planners is to ensure the efficient delivery and use of the conventional Class VIII (medical materiel) system for support of ARSOF missions.

4-107. Army special operations forces consist of special forces (SF), rangers, special operations aviation (SOA), psychological operations (PSYOP), and CA. The SOA, CA, and PSYOP have limited medical capabilities and are very dependent upon the theater HSS assets and the organizations they are supporting. Special forces and rangers often operate far removed from conventional AHS and must be more self-reliant and sustaining than conventional forces. Accordingly, SF and ranger medical personnel receive enhanced medical training above that provided for a trauma specialist. The Army educates the SOF medic to be an independent care practitioner qualified to provide ATM to combat casualties. When deployed on independent operations, the two SF medics are the sole source of medical care for their operational detachment, the indigenous forces, and their families that the detachment supports. They can train the indigenous populace in basic medical skills and establish an austere AHS system. Nonmedical ARSOF personnel receive medical training at the combat lifesaver level. Special operation combat medics (SOCMs) are trained in emergency medical treatment and in combat trauma management, but they do not possess the invasive surgical skills of the SF medical sergeant. The SOCMs provide medical care in combat environments where resuscitative surgery is more rapidly accessible than in the small unit missions of SF. Physicians and physician assistants assigned to units with SOCMs can provide ATM. Army health system planning for missions where the risk of penetrating trauma is high and further complicated by time and extended distance requires extensive planning and coordination. The capability for medical personnel and units that can perform resuscitative surgery and invasive stabilization must be provided regardless of the ARSOF unit involved.

4-108. Although the ARSOF health care provider receives enhanced medical training exceeding the level and scope found in conventional forces, he depends heavily on the conventional AHS system to conserve the combat strength of the ARSOF (particularly in the area of medical evacuation where the ARSOF does not have a dedicated system).

4-109. Ideally, medical evacuation for ARSOF personnel should follow the doctrinal flow sequence. The ARSOF medical planner must be innovative and follow the tenets of immediate far forward stabilization. He directs evacuation to the appropriate MTF when the condition of the patient warrants it, with whatever means of transportation are available.

4-110. Medical evacuation of ARSOF casualties is an operational matter. That is, it must reflect the commander's concept of the operation. It can only succeed when the medical planner integrates the medical evacuation plan with the tactical plan and logistics airflow.

4-111. Conventional medical evacuation assets are not normally used when operations occur in hostile and denied areas. Use the conventional medical evacuation system once the casualty is extracted from the hostile or denied area. For additional information on AHS to ARSOF, refer to FM 3-05, FM 4-02.6. FM 8-42 (FM 4-02.42), and FM 4-02.43.

SECTION IX — URBAN OPERATIONS

TERRAIN AND ENVIRONMENT

4-112. Throughout history, battles have been fought on urbanized terrain. Some recent examples urban operations (UO) include Hue, Beirut, Panama City, and Baghdad. Military operations on urbanized terrain are those military actions planned and conducted on a terrain where man-made structures impact on the tactical options available to the commander. This terrain is characterized by a three-dimensional battlefield, having considerable rubble, ready-made fortified fighting positions, and an isolating effect on all movement and maneuvers and fires, and sustainment units. In this environment, the requirement for a sound and understandable evacuation plan cannot be overstated. Of concern to medical and tactical planners is the need to plan, train, prepare, and equip for evacuation from under, above, and at ground level.

TERRAIN CONSIDERATIONS

4-113. Medical evacuation in the urban environment is a labor-intensive effort. Due to rubble, debris, barricades, and destroyed roadways, much of the evacuation effort will be by manual litter teams. When this occurs, establish a litter or ambulance shuttle system. The shuttle system reduces the distance that litter teams have to carry the wounded or injured Soldiers. This enhances the litter teams' effectiveness by providing brief respites, reducing fatigue. Further, the litter teams stay in the forward areas. They are familiar with the geography of the area and what areas have or have not been searched for casualties. In moving patients by litter, you should:

- Use covered evacuation routes such as subways, whenever possible.
- Use easily identifiable points for navigation CCPs.
- Rest frequently by alternating litter teams.

EVACUATION PLANNING CONSIDERATIONS FOR URBAN OPERATIONS

4-114. Conducting medical evacuation operations in the urban environment challenges the medical planner. He must ensure that the HSS/FHP plan includes special or unique materiel requirements or improvised use of standard equipment. The plan must be sufficiently flexible to support unanticipated situations.

4-115. Special equipment requirements include, but are not be limited to:

- Axes, crowbars, and other tools used to break through barriers.
- Special harnesses; portable block and tackle equipment; grappling hooks; collapsible litters; lightweight collapsible ladders; heavy gloves; and casualty blankets with shielding. This equipment, using pulleys, is for lowering casualties from buildings or moving them from one building to another at some distance above the ground.
- Equipment for the safe and quick retrieval from craters, basements, sewers, and subways. Casualties may have to be extracted from beneath rubble and debris.

4-116. Air ambulances equipped with a rescue hoist may be able to evacuate casualties from the roofs of buildings or may be able to insert needed medical personnel and supplies.

COMMUNICATIONS IN THE URBAN ENVIRONMENT

4-117. The urban environment degrades communications. Line of sight radios will be ineffective and individual Soldiers will not have access to radio equipment. The medical evacuation teams will have difficulty in locating injured or wounded Soldiers due to their isolation within buildings, or by being hidden in rubble and debris. Injured Soldiers can display alternate forms of communications, such as markers, panels, or field expedients (field/uniform jackets or T-shirts) indicating where they may be found.

Establishing Casualty Collection Points

4-118. Preplan and establish CCPs at relatively secure areas accessible to both ground and AAs. The location of these points are depicted on the HSS/FHP overlay to the OPLAN. Casualty collection points should—

- Offer cover from enemy fires.
- Be as far forward as the tactical situation permits.
- Be identifiable by an unmistakable feature (natural or man-made).
- Allow rapid turnaround of ambulances.

4-119. The tactical commander must approve route markings to the MTF and if to display of the Geneva Red Cross at the facility. (Camouflaging or not displaying the Geneva Red Cross can forfeit the protections, for both medical personnel and their patients, afforded under the Geneva Conventions. Refer to Appendix A and FM 4-02 for additional information.) The location of the MTF must be as accessible as possible, but well separated from fuel and ammunition depots, motor pools, reserve forces, or other lucrative enemy targets, as well as civilian hazards such as gas stations or chemical factories.

Ground Evacuation For Urban Operations

4-120. When using ground evacuation assets in support of UO, the medical planner must be aware that built-up areas may have significant obstacles to vehicular movement. Factors requiring consideration include the following:

- Rubble and other battle damage complicate and canalize transportation operations within the urban terrain.
- Bypassed pockets of resistance and ambushes pose a constant threat along evacuation routes.
- Land navigation with most tactical maps proves to be difficult. Using commercial city maps when available can aid in establishing evacuation routes.
- Ambulance teams must dismount from the ambulance, search for, and rescue casualties.
- Movement of patients becomes a personnel intensive effort. There are insufficient medical personnel to search for, collect, and treat the wounded. Medical units may require assistance in the form of litter bearers and search teams from supported units.
- Refugees may hamper movement into and around urban areas.
- Dislocated civilians, EPW, and detainees receive medical treatment according to the command policy and the Geneva Conventions.

Air Evacuation for Urban Operations

4-121. When using AA assets in support of UO, the medical planner must consider enemy air defense capabilities and terrain features, both natural and man-made, within and adjacent to the built-up areas. Air ambulances are the preferred means of evacuation in UO. Considerations in the use of AAs include the following:

- Movement is highly restricted and is canalized over secured areas, down wide roads, and open areas.
- Telephone and electrical wire and communication antennas hinder aircraft movement.
- Secure landing zones (LZs) must be available.
- Landing zones may include buildings with helipads on their roofs or sturdy buildings, such as parking garages.
- Snipers with air defense capabilities may occupy upper stories of the urban area's taller buildings.

Specialty Training

4-122. Medical personnel require special training in the tactics, techniques, and procedures required to operate in an UO environment. If they are to survive in this environment, they must know how to:

- Cross open areas safely.
- Avoid barricades and mines.
- Enter and depart safely from buildings.
- Recognize situations where booby traps or ambushes are likely and would be advantageous to the enemy. Detailed information on the conduct of combat operations in the urban environment is contained in FM 3-06 and FM 3-06.11.

Note. Medical personnel do not engage in offensive-type actions. They must rely on the supported unit to provide covering fires and to clear rooms and buildings prior to entry.

4-123. Units can modify and apply many of the extraction and patient evacuation techniques used in a mountainous terrain for evacuation in UO.

SECTION X — HOMELAND SECURITY

4-124. To preserve the freedoms guaranteed by the Constitution of the US, the Nation must have a homeland that is secure from threats and violence, including terrorism. Homeland security (HS) is the Nation's first priority, and it requires a national effort; in which the DOD has a key role in that effort. The National Strategy for Homeland Security (NSHS) complements the National Security Strategy of the US by providing a comprehensive framework for organizing the efforts of federal, state, local, and private organizations whose primary functions are often unrelated to national security. Critical to understanding the overall relationship is an understanding of the distinction between the role that DOD plays with respect to securing the Nation and HS, and the policy in the NSHS, which has the Department of Homeland Security (DHS) as the lead. Homeland security at the national level has a specific focus on terrorist threats. The DOD focus in supporting HS is broader. Military application of the NSHS calls for preparation, detection, deterrence, prevention, defending, and responding to threats and aggression aimed at the homeland. The DOD also provides military assistance to civil authorities (MACA), including consequence management activities. The Armed Forces of the US support the NSHS through two distinct but interrelated mission areas–homeland defense (HD) and civil support. The United States Northern Command (USNORTHCOM) is the unified command that has the DOD responsibility for HD and civil support.

HOMELAND DEFENSE

4-125. Homeland defense is the protection of US territory, sovereignty, domestic population, and critical infrastructure against external threats and aggression. The purpose of HD is to protect against and mitigate the impact of incursions or attacks on sovereign territory, the domestic population, and defense critical infrastructure. For HD missions, as directed by the President of the US and/or the Secretary of Defense, DOD serves as the lead federal agency.

Homeland Defense Characteristics

4-126. The DOD is the lead, supported by other agencies, in defending against traditional external threats/aggression (such as air and missile attack). However, against internal asymmetric, nontraditional threats (such as terrorism), DOD may be in support of DHS. When ordered to conduct HD operations within US territory, DOD will coordinate closely with other federal agencies or departments.

Types of Homeland Defense Operations

4-127. The types of homeland defense operations that the future brigade combat teams is expected to perform include—

- Offensive operations.
- Defensive operations.
- Humanitarian and civic assistance.
- Support to counterdrug operations.
- Show of force.
- Domestic support operations.
- Security operations (screen, guard, area security, and local security).
- Reconnaissance operations (zone, route, and area reconnaissance).

CIVIL SUPPORT

4-128. Civil support missions consist of military support to US civil authorities for domestic emergencies, and for designated law enforcement within the scope of restrictions required by the Posse Comitatus Act and other support approved by the Secretary of Defense (see FM 3-07).

4-129. Employment of military forces within the US, its territories, and possessions, under the auspices of civil support, typically falls under the broad mission of MACA. The MACA missions consist of three mission subsets. These mission subsets consist of military support to civil authorities, military support to civilian law enforcement agencies, and military assistance for civil disturbances.

Military Support to Civil Authorities

4-130. Military support to civil authorities (MSCA) is the most widely recognized form of DOD civil support because it usually consists of support for high-profile emergencies such as natural or man-made disasters that often invoke Presidential or state emergency/disaster declarations:

- **Natural Disasters.** In the event of a natural disaster or emergency, there will be a concerted US Government effort to support the affected areas. The DOD may be asked to provide assistance to DHS/Federal Emergency Management Agency (FEMA) in an attempt to save lives, protect property, and lessen the threat of a catastrophe in the US. Examples of natural disasters include, but are not limited to severe weather, wildland fire fighting, and animal disease eradication. When natural disasters occur and military assistance is anticipated, DHS/FEMA will request a defense coordinating officer (DCO) who serves as the single DOD point of contact within the disaster area. The DCO will be under the OPCON of the designated supported combatant command. Units supporting the event are normally OPCON to the supported combatant command.
- **Special Events.** The DOD provides support to a special event, a planned program of athletic competition or related activities involving participants from the US and/or other nations. Historic examples of these events are the Olympic Games and the Pan American Games.
- **Man-made Disasters.** Manmade disasters can be accidental or intentional. An example of an accidental event is an oil spill. The implications of a deliberate or unintentional large release of a CBRNE are severe. A catastrophic CBRNE event or attack may occur with little or no warning, resulting (either immediately or over time) in mass casualties and producing a mass exodus of evacuees. The response capabilities and resources of the local jurisdiction (to include mutual aid from surrounding jurisdictions and response support from the state) may be insufficient and quickly, if not immediately, overwhelmed.

Domestic Support Operations

4-131. Domestic disasters occur due to various causes and many may require the assistance of DOD and more specifically, the Army medical evacuation assets to fill response shortfalls. In the post-9/11 environment, terrorism or other intentional acts are recognized as a more likely cause of contingencies within the US than many had acknowledged since the civil defense era of the forties and fifties. This threat increases the likelihood that a large-scale disaster will occur that requires the augmentation of civil authorities' resources by DOD. The US DHS and USNORTHCOM have been established to coordinate the preparedness for and response to such events for civil authorities and DOD respectively.

4-132. Domestic support operations are typically conducted to prevent loss of life and property. Certain operations may be conducted in the form of immediate and automatic response by US military commanders as defined in DOD Directive 3025.15, to save lives, prevent human suffering, or to mitigate great property damage under imminently serious conditions. Other requests for support for civil disasters are made by civil authorities or a DCO (in a Stafford Act declared emergency) through the Joint Directorate of Military Support per the directive and work down through command channels from the Secretary of Defense.

4-133. Homeland Security Presidential Directive 5 recognized that there currently exists an artificial separation of local, state, and federal response to an incident and the role of DOD is not well streamlined or defined. This recognition led the White House and the DHS to initiate the transition from the Federal Response Plan to the National Response Plan, which serves as the doctrine for how the US responds to disasters no matter the cause.

4-134. Medical evacuation support plans to augment civil authorities are necessary for a variety of contingencies. Domestic support planning must address a range of problems such as—

- Preidentifying medical evacuation capabilities, units and personnel available to support various contingencies large enough to require civil support.
- Command and control relationships between civil authorities and DOD forces especially when DOD units are divided.
- Support for deployed DOD forces when no DOD logistics operations are deployed, including medical support.
- Cost capture and reimbursement from civil authorities to DOD in non-Stafford Act emergencies.

SECTION XI — OTHER TYPES OF MEDICAL EVACUATION SUPPORT MISSIONS

EVACUATION OF MILITARY WORKING DOGS

4-135. Military working dogs (MWDs) when injured or ill may be evacuated on any transportation means available. The using unit is responsible for the evacuation of the animal. Use of dedicated medical evacuation assets (air or ground ambulances) is authorized based on mission priority and availability. When possible, the handler should accompany the animal during the evacuation. Using units should include the location of veterinary treatment facility/support units on operational overlays. See FM 4-02.18 for information on veterinary support for MWDs.

PERSONNEL RECOVERY OPERATIONS

4-136. Air ambulances do participate in PR operations in a support role and are not protected from attack while engaged in a PR mission over contested or denied territory. If the involvement in these operations consists solely of evacuating wounded crewmembers from a crash site in friendly territory, AAs retain the protection accorded to them under the provisions of the Geneva Conventions. Air ambulances flying in contested or denied areas are not protected from attack unless flying at times and routes agreed to by the enemy and may be summoned to land. Air ambulances must obey a summons to land. Personnel retain their protections under the Conventions. However, if AAs participate in the actual search and rescue phases of the operation, they are not solely engaged in the provision of AHS and are, therefore, not afforded the protections. Further, because the mission does not fall within the protected activities, the AAs participating in the PR operation must remove the Geneva Conventions emblem from the aircraft.

SHORE-TO-SHIP EVACUATION OPERATIONS

4-137. It is imperative that Army AA units be able to interface on the first day of battle with US Navy (USN) air-capable ships. Lessons learned from past operations, have shown that US Army helicopters should be able to operate to and from USN air-capable ships. An interservice agreement between the Army and the Navy allows for deck-landing qualification of Army pilots (refer to FM 1-564 for additional information). It is important that units having contingency missions requiring Navy support establish training requirements to obtain naval-operations orientation, water egress training, water survival, and deck-landing qualification. This enhances the successful accomplishment of the aeromedical evacuation mission to naval vessels.

COMMUNICATIONS

4-138. In past joint operations, communications have been burdensome for both Army and Navy elements. It is essential to establish commonality of communication requirements during training exercises and to establish communication equipment and frequencies for medical evacuation to Navy vessels. This will provide smooth integration of Army helicopters into the Navy airspace management system during actual operations.

NAVIGATION

4-139. As Navy vessels may operate relatively long distances from the ground combat operations, Army AA units need to be proficient in over water navigation. The use of navigational aids (NAVAIDS) from the Navy element in support of the operation is the first priority for over water navigation. Basic dead-reckoning remains a secondary measure.

MEDICAL EVACUATION OF ENEMY PRISONERS OF WAR AND DETAINEES

4-140. Sick, injured, and wounded EPW and detainees are treated and evacuated in military police (MP) channels when possible. They must be physically segregated from US, allied, and coalition patients. Guards for these prisoners are provided according to the BCT, division or corps TSOP and are from other than medical resources. The echelon commander is normally responsible for this support. The US provides the same standard of medical care for wounded, sick, and injured EPW and detainees as that given to US, allied, and coalition Soldiers. Wounded, sick, or injured EPW and detainees in the CZ may be treated and returned to MP channels for evacuation; or the EPW and detainees may be stabilized and moved through medical channels to theater hospitals for treatment. Enemy prisoners of war and detainees are not evacuated from the theater for medical treatment.

4-141. When EPW are evacuated through medical channels, medical personnel—

- Report this action through medical channels to the next higher headquarters.
- Request disposition instructions from the MEDBDE IPMC.
- The MEDBDE IPMC is responsible for—
  - Coordinating the transportation means.
  - Identifying the MTF to which the EPW/detainees will be taken.
  - Coordinating, in conjunction with the hospital commander, with the Detainee Reporting System to account for EPW within medical channels.

SECTION XII — MEDICAL EVACUATION REQUEST

4-142. Procedures for requesting medical evacuation support must be institutionalized down to the unit level. Procedural guidance and standardization of request procedures are provided below. The same format used to request aeromedical evacuation is also used for requesting ground evacuation.

Table 4-3. Procedures for information collection and medical evacuation request preparation

Line	*Item*	*Explanation*	*Where/How Obtained*	*Who Normally Provides*	*Reason*
1	Location of pickup site	Encrypt the grid coordinates of the pickup site. When using the DRYAD Numeral Cipher, the same "SET" line will be used to encrypt the grid zone letters and the coordinates. To preclude misunderstanding, a statement is made that the grid zone letters are included in the message (unless unit SOP specifies it's use at all times)	From map	Unit leader(s)	Required so evacuation vehicle knows where to pickup patient. Also, so that the unit coordinating the evacuation mission can plan the route for the evacuation vehicle (if the evacuation vehicle must pick up from more than one location).
2	Radio frequency, call sign and suffix	Encrypt the frequency of the radio at the pickup site, not a relay frequency. The call sign (and suffix if used) of person to be contacted at the pickup site may be transmitted in the clear.	From SOI	RTO	Required so that evacuation vehicle can contact requesting unit while enroute (obtain additional information or change in situation or directions).
3	Number of patients by precedence	Report only applicable information and encrypt the brevity codes. A—URGENT B—URGENT-SURG C—PRIORITY D—ROUTINE E—CONVENIENCE If two or more categories must be reported in the same request, insert the word "BREAK" between each category.	From evaluation of patient(s)	Medic or senior person present	Required by unit controlling vehicles to assist in prioritizing missions.

4	Special equipment required	Encrypt the application brevity codes. A—None B—Hoist C—Extraction equipment D—Ventilator	From evaluation of patient/situation	Medic or senior person present	Required so that the equipment can be placed on board the evacuation vehicle prior to the start of the mission.
5	Number of patients by type	Report only applicable information and encrypt the brevity code. If requesting medical evacuation for both types, insert the word "BREAK" between the litter entry and ambulatory entry. L+# of patients–Litter A+# of patients–Ambulatory (sitting)	From evaluation of patient(s)	Medic or senior person present	Required so that the appropriate number of evacuation vehicles may be dispatched to the pickup site. They should be configured to carry the patients requiring evacuation.
6	Security of pickup site (wartime)	N—No enemy troops in area P—Possibly enemy troops in area (approach with caution). E—Enemy troops in area (approach with caution). X—Enemy troops in area (armed escort required).	From evaluation of situation	Unit leader	Required to assist the evacuation crew in assessing the situation and determining if assistance is required. More definitive guidance can be furnished to the evacuation vehicle while it is enroute (specific location of enemy to assist an aircraft in planning its approach).
6	Number and type of wound, injury or illness (peacetime)	Specific information regarding patient wounds by type (gunshot or shrapnel). Report serious bleeding, along with patient's blood type if known.	From evaluation of patient(s)	Medic or senior person present	Required to assist evacuation personnel in determining treatment and special equipment needed.
7	Method of marking pickup site	Encrypt the brevity codes. A—Panels B—Pyrotechnic signal C—Smoke signal D—None E—Other	Based on situation and availability of materials	Medic or senior person present	Required to assist the evacuation crew in identifying the specific location of the pick up. Note that the color of the panel or smoke should not be transmitted until the evacuation vehicle contacts the unit (just prior to its arrival). For security, the crew should identify the color and the unit verifies it.
8	Patient	The number of patients in each category need not	From evacuation	Medic or	Required to assist in

	nationality and status	be transmitted. Encrypt only the applicable brevity codes. A—US military B—US citizen C—Non-US military D—Non-US citizen E—enemy prisoner of war (EPW)	platform	senior person present	planning for destination facilities and need for guards. Unit requesting support should ensure that there is an English speaking representative at the pickup site.
8	CBRN contamination (wartime)	Include this line only when applicable. Encrypt the applicable brevity codes. C—Chemical B—Biological R—Radiological N—Nuclear	From situation	Medic or senior person present	Required to assist in planning for the mission. (determine which evacuation vehicle will accomplish the mission and when it will be accomplished)
9	Terrain description (peacetime)	Includes details of terrain features in and around proposed landing site. If possible, describe relationship of site to prominent terrain feature (lake, mountain, tower).	From area survey	Personnel present	Required to allow evacuation personnel to assess route/avenue of approach into area. Of particular importance if hoist operation is required.

Chapter 5

Medical Evacuation in Specific Environments

This chapter addresses medical evacuation in specific environments or under special circumstances. The medical evacuation effort must be well-planned and its execution synchronized to be effective. Further, medical evacuation personnel must be flexible and ready to improvise, if needed, to meet the demands of unique situations.

SECTION I — MOUNTAIN OPERATIONS

5-1. In the past, armies have experienced great difficulty in evacuating patients from mountainous areas because they are extremely diverse in nature. Some mountains are dry and barren with temperatures ranging from extreme heat in the summer to extreme cold in the winter. In tropical regions, mountains are frequently covered by lush jungles and heavy seasonal rains occur. Many areas display high rocky crags with glaciated peaks and year-round snow cover. Elevations can also vary from as little as 1,000 feet to over 16,000 feet with drastic and rapidly occurring weather changes.

5-2. Operations in mountainous terrain require some procedure modifications. This is due to the environmental impact on personnel and equipment. Important physical characteristics and considerations that influence medical evacuation are—

- Rugged peaks, steep ridges, and deep valleys.
- Limited number of trafficable roads.
- Reduced communications ranges.
- Unpredictability of and severe changes in weather.
- Decreased partial pressure of oxygen.
- Limited availability of LZs.

5-3. In order to effectively support the tactical plan, the HSS/FHP plan must provide maximum flexibility. The HSS planner should consider using all means of evacuation. Due to the length of evacuation times and the limited means of ground evacuation, it is important to triage and prioritize patients prior to movement.

- The availability of improved, hard-surfaced roads is extremely limited, if they exist at all. Usually, improved roads are only found in valley corridors. Such roads are often dependent upon a system of narrow bridges spanning mountain streams and ravines. They may also twist along ridgelines and cling to steep shoulders.
- Secondary roads and trails may be primitive and scarce. However, they may provide the only routes capable of vehicular traffic. Cross-compartment travel between adjacent valleys may be impossible by ground vehicle. Off-road travel requires detailed planning, even for short distances.
- Because of rough terrain, the Role 2 MTF may not be able to reach the BAS by ground vehicle. An ambulance shuttle system is established with an AXP for air and ground evacuation vehicles to meet litter bearers. Litter bearers and beasts of burden may be the only means of evacuation available. Any available personnel may be used as litter bearers (nonmedical personnel from supported units may be required to augment the litter bearer teams). Close coordination between Role 2 MTFs and BASs in establishing CCPs or AXPs is necessary to—
  - Reduce distance traveled by litter bearers.
  - Reduce evacuation time.

    - Conserve personnel.
    - Locate the best potential LZs for AAs.
- In mountainous areas, evacuation of patients by air is the preferred means. Air ambulances permit the rapid movement of patients over rugged terrain. For example, to travel a distance of only 6 kilometers on foot could take up to 2 hours, while flying time could be less than 2 minutes.
- Frequency-modulated radios are the principal means of communication in this environment. The ability to transmit is hampered by the limitations of line of sight transmissions.
- The briefing of ambulance drivers needs to be extensive, including detailed strip maps and overlays. Further, specific instructions on what to do in various situations should be covered (such as if the vehicle breaks down or the unit moves).

5-4. The mountain environment, with its severe and rapidly changing weather, impacts on aircraft performance capabilities; accelerates crew fatigue; and requires special flying techniques. Having to rely on continuous aviation support for a successful mountain operation is risky.

- Flying in mountainous areas requires special training. Both the terrain and the weather influence basic flying techniques and operational planning. Rugged, mountainous terrain complicates flight route selection. Direct routes can seldom be flown without exposing the aircraft to detection and destruction by the enemy.
- Important considerations for AA operations in mountainous areas are—
    - Density altitude which is the most important factor affecting aircraft performance. Density altitude combines temperature, humidity, and pressure altitude, and provides the basis for lift capability. Density altitude can vary significantly between the pickup point and the LZ because of the time of day and changes in elevation. Frequent performance planning updates are essential.
    - Unpredictable winds which can produce significant turbulence, wind sheers, updrafts, and downdrafts. This further increases the risk of a catastrophe in a seemingly routine mission. Adverse winds along with high density altitude demand current and accurate performance planning. Pilots must plan for greater margins of safety.
    - Ice can clog intake ports, thus starving the engine of air, or it can collect on rotor blades resulting in a significant loss of lift. Asymmetrical shedding can cause severe out-of-balance rotor conditions.
    - Visibility can be reduced by low clouds or fog greatly decreasing the ability to navigate or to avoid obstacles.
    - Lack of LZs is critical because the characteristics of mountain terrain do not usually afford adequate LZs. The terrain may only allow the aircraft to hover while loading patients on board.
    - Hoist operations can be expected in mountainous terrain requiring the use of the internal or external rescue hoist. Mounting the rescue hoist on the aircraft as standard equipment in mountain operations may be required. When possible, orientation and training sessions with supported troops should be conducted to help minimize the difficulty of such missions. Depending on the terrain, the forest penetrator may also be needed to accomplish the mission.
    - Enemy air defenses must be considered because when enemy air defense capabilities preclude using AAs in forward areas, they should be used to evacuate patients from AXPs or from Role 2 MTFs.
    - Ambulatory patients may be reported as litter patients in mountainous terrain because these patients may be unable to move unassisted over the rugged terrain. Once placed on the AA, their status may be upgraded.
    - Additional crew training for ground and air evacuation crews should include training and orientation in mountaineering skills, handling patients, and survival skills. Such training might include cold weather survival training (cold injury prevention, mountain [rock] climbing, and the use of ropes and vertical rescue techniques). Training may also include individual and unit

movement at high altitudes and techniques of patient evacuation by litter, emphasizing the use of pack animals (if available from the host country), and the improvised travois litter.

- Patient loading can be hazardous and care must be taken when loading patients where there is a great deal of slope to the LZ. Emphasis on approaching and loading the aircraft from the down-slope side of the aircraft must be reinforced.

DANGER

Approaching the aircraft from the up-slope side is hazardous.

5-5. Troops operating in mountainous areas are exposed to other injuries and illnesses that frequently occur in this environment. These conditions include—

- An increased rate of fracture, sprain, and dislocation injuries.
- Incidents of acute mountain sickness, high-altitude pulmonary edema, and cerebral edema caused by rapid ascent to heights over 7,500 feet.
- Cold weather injuries.
- Dehydration and heat exhaustion.
- Sunburns and snow blindness.
- Aggravated sickle cell anemia. Although this condition is not considered a mountain illness, personnel with the sickle cell trait can be seriously affected by the decrease in barometric pressure and lower oxygen levels found at higher altitudes.

5-6. The proportion of litter cases to ambulatory cases is increased in mountainous terrain, for even the slightly wounded may be unable to move unassisted over rough terrain. Litter relay stations may be required along the evacuation route to conserve the energy of litter bearers and to speed evacuation. It is important to be able to predict the number of patients that can be evacuated with available personnel. When the average terrain grade exceeds 20 degrees, the four-man litter team is no longer efficient and should be replaced by a six-man team. The average mountain litter team should be capable of climbing 120 to 150 vertical meters of average mountain terrain and return with a patient in approximately 1 hour.

5-7. Mountain operations may require medical personnel to carry additional equipment. Items such as ropes, pulleys, pitons, piton hammers, and snap links are all necessary for evacuating patients and establishing BASs. All unnecessary items of equipment including those for which substitutes or improvisations can be made should be left behind. Heavy tentage, bulky chests, extra splint sets, excess litters, and nice-to-have medical supplies should be stored. Such medical supplies, if stored, should be readily available for airdrop or other means of transport. Medical items that are subject to freezing should be safeguarded; they should not be exposed to the low temperatures experienced in mountainous areas.

5-8. For further information on mountain operations, refer to FM 3-97.6.

SECTION II — JUNGLE OPERATIONS

5-9. Health service support elements in a jungle environment retain the same basic capabilities as in other environments. Jungle operations, however, subject personnel and equipment to effects not found in other environments. The jungle environment degrades the ability to maneuver. Security problems are also increased and affect medical evacuation operations as much as they do the combat forces.

5-10. In jungle operations, combinations of air and ground evacuation units are used to maximize the patient evacuation potential. Using this dual system of evacuation ensures that the inherent limitations of one system can be compensated for by the other. Jungle variations affect the organizing, positioning, and securing of HSS/FHP. Due to the terrain, aerial resupply is usually a common practice. The responsiveness provided by aerial resupply requires fewer supplies to be stockpiled in the combat trains.

5-11. Jungle combat operations are characterized by ambushes and other unconventional warfare operations. The security threat caused by infiltrators requires that LOC be patrolled often and that convoys be escorted. It is, therefore, essential that HSS be performed as far forward as the tactical situation permits. Positioning assets forward—

- Improves response time.
- Reduces road movement.
- Allows the HSS elements to take advantage of the security offered by combat units.

5-12. The thick foliage often makes evacuation by ground more difficult than in other types of terrain. Factors such as the threat, limited road network, and reliance on nonmedical personnel for convoy security make air evacuation the preferred means. By using the ambulance shuttle system, patients can be transferred from forward operating ground ambulances to either ground or AAs operating further to the rear. In situations where evacuation assets are delayed by various factors (weather or terrain), patients are held for longer periods of time at forward locations. This will dictate the need for additional medical supplies. Health service support planners must try to anticipate these delays whenever possible. The increased disease and infection incidences associated with the jungle environment may worsen the patient's condition; therefore, timely evacuation is essential.

5-13. In some remote and densely foliaged jungles, the only means of evacuation may be by litter. Ambulances may not be practical on trails, unimproved muddy roads, or in swamps. As in mountain operations, there is a higher proportion of litter cases than usual. In the jungle even a slightly wounded Soldier may find it impossible to walk through dense undergrowth. At best, litter teams can carry patients only a few hundred meters over rough jungle terrain before needing rest or relief. Litter carries should be kept as short as possible and medical elements prepositioned and retained forward.

5-14. Other special planning considerations in jungle operations include—

- Water is vital in the jungle and is plentiful. Water from natural sources, however, should be considered contaminated. Water purification procedures must be taught to all Soldiers. (Refer to FMs 21-10 and 4-25.12 for additional information.)
- Due to the tropical climate, units should pack hot weather clothing when deploying to jungle areas. Insect (mosquito) nets, insect repellent, permethrin treated uniforms and sunscreen should be issued to all Soldiers operating in this environment.
- The jungle environment is ideal for the transmission of a large number of diseases. The rate of DNBI casualties is potentially the highest in this climate. The heat, humidity, and terrain places the troops at high risk for dehydration, heat injury, skin diseases, endemic diseases, and immersion foot. Small wounds can rapidly become infected and lead to loss of effectiveness and possibly require evacuation. High standards of personal hygiene must be taught, encouraged, and maintained by the command. Mosquitoes and other arthropods that carry disease flourish under jungle conditions. Use of all possible personal protective measures must be ensured. Food- and waterborne diseases leading to diarrhea or other symptoms will abound. Food service sanitation measures must be strictly followed. The potential for contamination of food and water increases with each time they are handled, stored, or transported. Soldiers must be encouraged to consume adequate amounts of water that has been purified and to eat only approved foods. In the jungle it is necessary for the commander to pay meticulous attention to the details of PVNTMED measures to maintain an effective fighting force. For additional information on PVNTMED measures, refer to FM 21-10 and FM 4-25.12.
- Health service support personnel should be trained in survival and support techniques in jungle environments. For example, training should be conducted in—
  - Hot weather acclimatization and survival.
  - Prevention, early detection, and treatment of arthropod-, food-, and waterborne diseases.
  - Land navigation in a jungle environment.
  - Field sanitation and other PVNTMED measures.
  - Care and maintenance of equipment and supplies.

- Due to the increased heat and humidity, vehicles and equipment require additional maintenance. Equipment tends to rust quickly and must be cleaned and oiled more frequently. Canvas items rot and rubber deteriorates much faster than in more temperate climates.
- The range of FM communications in the jungle is significantly reduced due to the dense undergrowth, heavy rains, and hilly terrain. The range of a radio set operated in the jungle may be reduced by 10 to 25 percent. The heavy rain and high humidity of the tropics also reduce the range (about 20 percent) and reliability of wire communications. The transmission range can be extended by using additional radio relays and field expedient antennas.
- Utility helicopters are not able to lift the same size loads that can be lifted in more temperate areas. This results in a reduced patient load in some evacuation aircraft. Again, frequent and accurate performance planning is essential for mission accomplishment.
- There may be few suitable LZs. Many LZs will only be large enough to support one or two helicopters at a time.
- Hoist operations may be required more frequently in the thick jungle vegetation where LZs are not available. The forest penetrator should be carried on all operations.
- For aviation-specific information, refer to training circular (TC) 1-400.

SECTION III — DESERT OPERATIONS

5-15. Deserts are arid, barren regions of the earth incapable of supporting normal life due to a lack of fresh water. Although deserts are often thought of as hot climates, it is important to note that temperatures range from over 136 degrees (°) Fahrenheit (F) in some deserts, to bitter cold in others. Day to night fluctuations in temperature can exceed 70° F. Desert terrain can have mountains, rocky plateaus, or sandy dunes; some desert areas may contain all of these characteristics. Rain, when it falls, often causes flooding in low-lying areas. Winds can have a devastating effect upon AHS operations by destroying equipment and supplies and causing dust storms. Dust storms make navigation and patient treatment difficult. Since deserts vary considerably in the type of terrain and temperature, and in their cultural makeup, current medical intelligence should be obtained prior to deployment on operations conducted on desert terrain.

5-16. People have lived and fought in desert areas for thousands of years. However, the environmental effects on personnel can be extreme, especially for Soldiers not prepared for these operations.

- To be effective, Soldiers must be properly acclimatized to the desert. Two weeks are usually required to satisfactorily acclimatize troops to hot environments, using progressive degrees of heat exposure and physical exertion. Other potential acclimatization problems that may be encountered are the effects of dry air and altitude on the respiratory system. Since many desert areas are located in mountainous terrain, Soldiers must be acclimatized for both the altitude and the temperature. In some areas of the world, such as the Gobi Desert in East Asia, people must be acclimatized to the cold in addition to the dryness. (For additional information, refer to FM 21-10, FM 8-250, and TB MED 507.)
- Units deployed in desert areas typically have long LOC and are widely dispersed. As unit elements become more dispersed, commanders must rely more heavily on junior officers and NCOs to ensure that C2 is maintained and that discipline and PVNTMED programs are enforced. For a unit to be effective, a high level of discipline must exist at all levels of the organization.
- Water is the most basic need in a desert. Without it, Soldiers cannot function effectively for more than a few hours. Thirst is not an adequate indicator of the need for water. It is necessary for each commander to establish and enforce a supervised drinking program. Experience has shown many times that Soldiers do not drink enough fluids unless forced to do so. It is important to cool the water, if at all possible, to make it more appealing. Water supplies should be carefully guarded against accidental loss, sabotage, or contamination. Extra water must be carried by AHS vehicles for patients to drink and to cool heat casualties.
- Soldiers deployed in the desert are susceptible to endemic diseases and environmental injuries. Proper water discipline, vaccines, prophylactic measures, field sanitation measures, personal

hygiene, and other PVNTMED measures can reduce these risks. Cold weather injuries, heat injuries, and respiratory disease can also be prevalent. Proper clothing, equipment, and a water discipline program must have command emphasis in desert operations.

- Winds may very easily damage materiel such as aircraft, antennas, and tents. Equipment is protected by using covers, tie-downs, and shelters. Terrain helps shield equipment from the wind if site selection is done carefully. In some cases, special tools, such as extra long metal tent stakes, are necessary.
- The effects of wind and sand are interrelated. Desert sand starts to become airborne when the wind reaches about 20 knots. Sandstorms—
  - Restrict visibility.
  - Pose a hazard to eyes (especially for Soldiers wearing contact lenses).
  - Can contaminate water supplies (if they are not protected).
  - Make navigation difficult.
- The sun may cause sunburn of the skin and eyes (if protection is not used).

5-17. Seven characteristics of the desert environment that may adversely effect equipment are—

- Terrain. Trafficability varies with the type of terrain covered. Open, flat, and rocky terrain affords higher trafficability than do mountainous areas, lava beds, or salt marshes. Drivers must be well-trained in judging the terrain over which they are driving to select the best alternative routes.
  - Tracked vehicles are best suited for desert operations. However, they can throw tracks when traversing a rocky area. Their use is also limited in rough terrain with steep slopes.
  - Wheeled vehicles may be used in desert operations; however, they normally have a lower average speed than tracked vehicles and a higher incidence of damage and malfunction. Wheeled vehicles often bog down in sandy areas and cannot traverse many of the rougher areas.
  - In planning for desert operations, vehicles should carry extra repair parts (fan belts, tires, and other items apt to malfunction).
- Excessive heat causes vehicles to overheat, leading to greater than normal wear. The frequency of leaks on vehicles and aircraft is greater than in some other environments. Engine and transmission seals tend to dry out and crack; fuel lines wear out quickly; and water requirements for cooling vehicle engines are greater. Loss of water, through evaporation, must be included in logistical planning. Using vehicle and aircraft covers reduce the effects of heat while vehicles and aircraft are not in use. Refer to FM 1-202.
  - Batteries do not hold their charge efficiently in intense heat. Dry battery supplies should be increased to compensate for a higher usage rate.
  - Communications equipment must be protected from the heat in the desert. Dust covers are used on this type of equipment. If the equipment has ventilating ports, these should be cleaned regularly to avoid clogging.
  - Medical supplies must be protected from the heat to prevent deterioration. The shelf life of some medical supplies decreases when stored in hot climates.
  - Medical equipment must be protected from the effects of heat. It may be protected using the same techniques as those used to protect communications equipment.
- The sun burns unprotected skin and it may damage unprotected eyes. Soldiers should dress in loosely fitting clothing, use sunburn cream or oils to protect exposed skin, and wear sunglasses or goggles to protect their eyes. Soldiers should remain fully clothed. Removing clothing increases direct exposure of the skin to the sun and eliminates the beneficial cooling effects of the moisture trapped in clothing. Radiant light or its heat effects may be detrimental to plastics, lubricants, pressurized gasses, rubber, and some fluids. All vehicles and aircraft should be kept well-ventilated, and windshields should be covered to reduce heat buildup inside. Supplies of all types should be stored in a well-ventilated, shady area. Placing supplies in covered holes in the ground may reduce adverse heat effects.

- Dust and sand present one of the greatest dangers to the proper functioning of equipment. Sand mixed with lubricants forms an abrasive paste. Lubrication fittings, bearings, and filters should be inspected frequently and changed when required.
  - Aircraft should not be exposed to dust and sand any more than is absolutely necessary. Ground handling instead of hovering reduces sand ingestion. Dust and sand increase failure of microphone switches, signal distribution panels, and circuit breakers, and cause electrical motors and generators to burn out. Wheel and flight control bearings require more frequent cleaning; engines should be flushed frequently.
  - Medical and communications equipment may be adversely affected by dust and sand. Over a period of time, electrical insulation is damaged by windblown sand. When combined with the effects of lubricants on the insulation, dust and sand can become a major communications problem. Special care should be taken to brush dust off radio equipment and to keep ventilating ports and channels clear.
  - Sand can accumulate in airframes, on the bottom of armored vehicles, and in bearings on all types of equipment. This accumulation, combined with oil and condensation, adds extra weight to aircraft and vehicles as well as jamming their control linkages. Sand and grease buildups must be removed from bearings to ensure safe operation and control of aircraft and vehicles.
  - Dust trails created by hovering aircraft or ground vehicles can be seen in excess of 10 miles on a relatively flat desert. This exposes these assets to direct and indirect enemy fires. Ground vehicles should reduce their speed to the point that they do not create a dust signature.
- Humidity is a factor in some desert areas of the world, especially in the Middle East. Humidity can become a problem for short periods of time in other desert areas. Light coats of lubrication can help prevent rust; however, these benefits should be weighed against the dust-gathering qualities of oil. Demisting equipment is used on optics and night vision equipment to combat the effects of humidity.
- Temperature variation can cause condensation in humid desert areas affecting optics, fuel lines, air tanks, and weapons. Expansion and contraction of air and fluids cause tires to over inflate during the day and under inflate at night. Fuel tanks may overflow during the day causing a fire hazard. Oil fluid levels become overfull and cause leaks during the day, or insufficient lubrication occurs when the oil cools. Vehicle operators and crew chiefs must ensure that the effects of temperature variations do not become a significant problem.
- Static electricity is prevalent in the desert. This is important to remember during refueling operations and when oxygen is being used onboard vehicles or aircraft. Proper refueling procedures must be followed. Static electricity also causes severe shock to ground personnel in sling load and hoist operations.

CAUTION

During hoist and sling load operations, the load must touch the ground before the ground crew can handle it.

5-18. To ensure success in desert operations, detailed planning is required. Factors to consider include—

- Additional quantities of water are required for HSS/FHP operations for the survival of both medical personnel and their patients. Load plans for all vehicles and aircraft must include water. Water is as mission essential as any piece of unit equipment. It should be a priority item when load plans are developed.
- Prescribed load lists are expanded to carry sufficient quantities of repair parts easily degraded by the environmental factors. For example, rubber and plastic fittings and tubes, or spare parts for communications equipment.

- Plan for the effects of wind, sand, and sun on all equipment. All plastic and glass surfaces on vehicles, aircraft, and other equipment should be covered when not in use. Covers should be ordered or made prior to deployment.
- Fuel planning is critical due to power limitations, extended range requirements, and increased vulnerability of refueling sites in the relatively open desert terrain. Careful planning of FARPs is essential for mission accomplishment.
- Units should plan to pack both hot and cold weather clothing when deploying due to varying temperatures between day and night.
- Petroleum, oils, and lubricants (POL) should be of the proper viscosity for desert operations. Maintenance services are also performed more frequently on ground vehicles and aircraft, thus requiring a larger amount of POL than normal.
- Extra filters of all types are planned for due to a higher consumption rate.

5-19. Training for desert operations is not significantly different than training for operations in other areas except for the following:

- Because many desert areas are in mountainous terrain and because high temperatures increase density altitude, aeromedical evacuation units should conduct mountain training to prepare for contingencies in desert areas. Further, procedures and techniques for evacuation in mountainous terrain must be practiced by all AHS personnel. Special equipment requirements (paragraph 5-7) must also be considered.
- Navigation in desert terrain varies from simple to extremely difficult. Factors affecting navigation are the type of desert and the scale and quality of the available navigational charts. At times, aircraft may have to use dead-reckoning navigational techniques (time, distance, and heading). Ground vehicles must have compasses available, as they have to rely on compass headings and odometer readings to navigate. Ground and AA crews should be able to interpret navigational charts and maps of all types and scales. Use of convoys is a viable technique to improve security and to ensure that ground vehicles do not get lost. Aircraft may be used to assist in navigation by convoys in those areas in which there are poor road networks and the terrain offers no distinctive features by which to navigate.

5-20. In principle, medical evacuation operations in the desert do not differ greatly from these operations in other environments.

5-21. Helicopter landing sites should be chosen with care. Common mistakes made by many units when establishing the LZ are—

- Locating the pad relative to the patient and tents, vehicles, and other obstacles. A common tendency is to locate the helipad downwind of MTFs so that approaches may be made into the wind towards the facility. In mountainous deserts, winds normally channel down the valleys and are more predictable along valley floors. A better site selection for an LZ is with the MTF, alongside the approach and takeoff zone. Thus, the landing direction is up or down the valley, depending on the airflow, and the MTF is not over flown.
- Situating landing sites in washes, small confined areas between large rocks, or close to moving tracked vehicles. Map coordinates are rarely accurate unless the site is beside a major terrain feature. Therefore, LZs should be located next to major terrain features or on higher ground where they can be seen from the air at a distance of 2 to 3 kilometers, if possible. Lack of distinctive features in the open desert and on large-scale maps makes pinpoint navigation difficult, especially at night.
- Marking of helicopter LZs is done so that the pad can be seen from the air, but the markings should not be a hazard in themselves. If engineer tape is used, it should be firmly secured to prevent it from blowing loose. Panel markers are not a good tool to use as they are difficult to see. If panel markers are used, they need to be secured. If used, flares or marker smoke should not be deployed on or directly upwind from the pad. Smoke grenades or flares should not be thrown under the aircraft as it lands. Avoid using white smoke to mark the LZ. Colored smoke is probably the best daylight marking method. It is difficult to detect a smoke grenade more

than 2 to 3 kilometers away, but an aircraft in the general vicinity can normally see it. Radios are used to guide aircraft to the LZs, but this creates an electronic signature. Units requesting medical evacuation must be prepared to signal the evacuation aircraft upon its arrival. Normally, map coordinates will guide the aircraft to within 2 to 3 kilometers of the LZ. Even from nap-of-the-earth (NOE) altitudes, the aircrew may be able to see several units in the area. The requesting unit must signal the aircraft to ensure the designated LZ is used.

5-22. When planning for night flights, considerations include the following—

- Moonlight aids the medical evacuation pilot by providing him with the light to see with either unaided vision or night vision goggles (NVGs). When adequate ambient light exists, medical evacuation crews function almost as effectively at night as they do during daylight. The small arms threat is somewhat reduced at night, although it still exists from radar-guided weapons, infrared-sited weapons, and passive night vision device-equipped weapons systems. Flying into a bright moon with NVGs on can be compared to flying into the sun during the day. The goggles darken and visibility becomes extremely poor. Flight routes should not be planned to fly directly into a bright moon if NVGs are to be used.
- The lack of visual cues over sand is similar to that over water. It is very easy for pilots to become disoriented and fly into the ground. Reliance on radar altimeters is a must over flat sandy areas of the desert.
- Frequently, desert areas do not have sufficient ambient light to allow adequate night vision, even with the aid of NVGs. A pilot wearing NVGs is often unable to see the ground at an altitude of 100 feet using a landing light equipped with a pink light filter. Under these conditions, dead reckoning is the only effective navigation method unless global positioning satellites, Doppler equipment, or NAVAIDS are available. Unfiltered light can be used with or without NVGs; however, this increases the risk of exposing the aircraft's position to the enemy.
- Desert warfare is usually characterized by extended battle zones which increase evacuation distance and time. Combat health support units are located further to the sustainment area in the desert. Establishing an ambulance shuttle system or CCPs is useful. Combat health support units require a greater number of vehicles for operating in deserts than in other environments. Air evacuation by fixed and rotary-wing aircraft is the preferred method due to their speed and range. Further, using aircraft reduces the load on ground vehicles. Augmentation from higher echelon HSS may also be required to meet the extended evacuation needs.
- Smoke is used extensively on the modern battlefield by both sides. It can be effectively used to mask friendly actions to include medical evacuation. (Refer to Appendix E for further information.)
  - Smoke can be a major hazard, especially to medical evacuation helicopters. Smoke reduces visibility and forces an aircraft higher where it can be acquired by threat weapons systems. The phenomenon of inversion occurs often in the desert. When this happens, medical evacuation vehicles and aircraft may be able to work underneath the smoke using the smoke layer for overhead concealment.
  - Medical units must coordinate closely with supported organizations on smoke operations. Smoke can either help or hinder the evacuation mission, depending upon how it is used.
- Communications in the desert are affected by a number of factors. Atmospheric interference and the skip of signals occur frequently. Mineral deposits in the desert may unexpectedly disrupt communications. Many of these problems can be overcome by using additional radio relays, pre-established control measures, and visual signals.
- Artificial lights may be used at times in the desert. They are very easily detected. Even with blackout lights, vehicles using lights can be detected for miles with NVG. Serious consideration should be given to driving without using lights when the tactical situation dictates. Ground guides are used to help vehicles navigate through areas that are not clearly marked or through area where troops are present.
- Wind is one of the most significant environmental factors affecting medical evacuation in the desert. Wind can be destructive to both structures and equipment; tents, antennas, and aircraft

can be easily damaged. Wind direction and speed vary greatly within the space of a few miles. Velocity is substantially increased when wind channels between hills and direction changes due to interference of terrain features. The wind frequently makes aeromedical evacuation impossible by exceeding the operating limitations of the aircraft. At other times, it may limit the use of some potential LZs. Blowing sand and dust can slow down the evacuation system by making navigation by either ground or AA difficult, if not impossible. High winds are predictable to a certain extent. For example, at certain times of the year in the Mojave Desert high winds occur every day at dusk and last for 3 to 4 hours. At other times, high winds, based on frontal weather patterns, can remain for several days at a time. These factors should be considered by HSS planners, and medical evacuation assets should be massed or relocated accordingly.

- The desert provides little or no protection from enemy air defenses except in mountainous terrain. Aircraft may have to be flown above NOE altitudes to prevent a dust signature. These factors cause increased exposure and vulnerability of AAs to enemy air defenses and may limit their employment.

5-23. Further Information. Refer to FM 90-3 for additional information on desert operations.

SECTION IV — EXTREME COLD WEATHER OPERATIONS

5-24. Operations in the extreme cold have many of the limiting factors found in desert operations. The tundra and glacial areas are harsh, arid, and barren. Temperatures may reach lows of -80° F to -100° F which, combined with gale force winds, make exposure unsurvivable.

- The greatest environmental detriment to operations is blowing snow. This results in a loss of depth perception from total white conditions. Blowing snow is caused by the wind or by the rotor wash of helicopters; its effect reduces visibility to zero.
- Other environmental considerations are as extreme but easier to circumvent. Solid footing is suspect in both the dead of winter and in the summer. Snow and ice cover crevasses, holes, and otherwise unstable ground. In traversing suspect ground situations, consider linking Soldiers by rope. During the summer, ground transportation is more restricted than in any other environment due to the marsh and muskeg composition of the arctic tundra. Patients must be sustained for a longer duration due to terrain delays and the lack of direct lines of evacuation.
- Greater responsibility has to be placed on each Soldier, especially for maintenance of nutrition and water consumption. It is imperative to stress that leadership and training are important in the prevention of cold weather injury. Strict adherence to the guidelines found in FMs 21-10 and 31-70 assures an effective fighting force. Water conservation is essential; however, adequate consumption by the individual should be enforced.
- Factors to consider for conducting evacuation in arctic operations include the following:
  - Arctic warfare is usually characterized by extended battle zones that increase evacuation distance and time. Establishing an ambulance shuttle system or CCPs is useful. Augmentation from higher echelon AHS may also be required to meet the extended evacuation needs.
  - Additional supplies of water should be carried by ambulances and maintained at CCPs, if possible.
  - Due to the decreased temperature and frozen environment, ambulance maintenance requirements are increased. Lubricants must be of the correct viscosity for the temperature. In extreme cold, batteries perform less efficiently. Consult the specific aircraft maintenance manual for the recommended battery and procedures for extreme cold weather operations.

Note. All ambulances are considered deadlined without a functional heater for the patient compartment.

  - The proper storage of medical supplies is essential to prevent loss from freezing.

- There are few terrain features or road networks; therefore, evacuation routes must be surveyed and marked over open terrain. At extreme latitudes, operations during the winter months are conducted in extended hours of darkness. The use of NVG may be required. Compass accuracy is inconsistent due to a geomagnetic phenomenon. Beacons and homing devices are essential for air navigation.
- Weather is extremely unpredictable. There are too few observers to allow for accurate assessment of weather patterns. Unfavorable weather conditions cause unexpected delays; therefore, medical personnel must be prepared to provide survival measures for their patients and themselves.
- Landing zones must be chosen with extreme care in both winter and summer. Blowing snow mandates instrument-assisted takeoffs and running landings. Landing areas must be correspondingly larger. The full weight of the aircraft cannot be allowed to settle on the skis until after firm ground conditions are established. Movement of patients to and from the aircraft is difficult. Where an aircraft lands is where it stays. A rocking motion, to free the skis prior to lift off, is performed using the cyclic and antitorque controls.

- Thorough planning and strict preparation are the keys to survival. Factors to consider include the following:
  - Mud obstacles at noon may become an avenue of approach at midnight.
  - Snow complicates all work. Snow-covered terrain hampers reinforcements, muffles noise, makes cross-country driving hazardous, and creates different camouflage requirements.
  - Because of thermal sights, a complete reappraisal of concealment is required.
  - Tracks in the snow destroy concealment.
  - No Soldier is assigned to any job alone. The buddy system is used at all times.
  - Anticipate that all maintenance tasks will take twice as long.
  - Bare metal can stick to skin or wet garments in subfreezing temperatures.
  - Fuel spilled on skin or garments increases the freezing factor; it is one of the greatest causes of injury in winter operations.
  - When operating in the cold, anticipate increased POL needs. Fuel consumption can rise as much as 25 percent for vehicles operating in deep snow, slush, or mud.
  - The recommended fuel for Yukon stoves is diesel.
  - Make every effort to warm gearboxes and engines before starting.
  - The first consideration in the AO is heat; followed by shelter for sustained work.
  - Soldiers need to stand clear of taut cables; steel tends to be brittle and breaks in extremely cold temperature.
  - Fire extinguishers are winterized by adding 15 percent nitrogen to the carbon dioxide.
  - Degradation of battery life requires changes as much as six times more frequently than in a more temperate environment.
  - Radio sets are warmed up prior to transmission. The sets may be turned on but should not transmit for at least one-half hour.

Note. Single-channel ground and airborne radio system (SINCGARS) radios do not require a warm up period.

  - Frost shields (such as using the plastic bag in which the batteries are packed) should be placed over microphones.
  - Grounding rods have to be buried horizontally instead of pounded in vertically. Recovery of stakes and rods placed in the ground is significantly more difficult.
  - Flooring is needed in heated areas because of the thawing of the tundra.
  - Soldiers must take breaks for water and warmth.
  - Static electricity presents a serious safety hazard especially around flammable materials.

5-25. For additional information, refer to FM 31-70.

SECTION V — CHEMICAL, BIOLOGICAL, RADIOLOGICAL, AND NUCLEAR ENVIRONMENTS

5-26. Medical evacuation and treatment are conducted continuously throughout operations conducted in a CBRN contaminated environment. The AHS commander must have a comprehensive plan which is rehearsed on a periodic basis to ensure the timely evacuation and treatment of casualties in a CBRN environment. Techniques and procedures which are essential for operating in a contaminated environment should be contained in the unit TSOP. The number of casualties and their medical condition, type of contaminant, the size of the land area contaminated, the expected duration of operation, risk assessment and acceptable level of risk, and the number of AHS assets (medical personnel, medical units, and evacuation vehicles and aircraft) initially contaminated will determine the quantity and type of uncontaminated AHS resources, if any, which will be introduced into the contaminated environment to ensure timely medical treatment and evacuation occur.

5-27. The Commander must take into consideration the number of assets he is willing to commit during evacuation operations in a CBRN contaminated environment. Since the combinations of evacuation methods are nearly endless, the commander has greater flexibility in tailoring an evacuation plan to meet his particular tactical situation in a CBRN contaminated environment.

5-28. On the modern battlefield there are three basic modes of evacuating patients (personnel, ground vehicles, and aircraft).

- In using personnel to physically carry the casualties, the commander must realize the inherent stress involved. Cumbersome mission-oriented protective posture (MOPP) gear needed in a contaminated environment (added to climate, increased workloads, and the fatigue of battle) greatly reduces the effectiveness of unit personnel.
- If the commander must send evacuation personnel into a radiological contaminated area, he must establish operational exposure guidance (OEG) for the medical evacuation operation. Radiation exposure records are maintained by the unit CBRN NCO and are made available to the commander, staff, and surgeon. Based on OEG, the commander decides which medical evacuation assets to send into the contaminated environment.

5-29. Commanders should make every effort to limit the number of evacuation assets which are contaminated while still maintaining a timely and effective medical treatment and evacuation operation.

- It is expected that a certain number of both ground and AAs will become contaminated in the course of battle. The commander can, therefore, segregate the contaminated ones. This results in the smallest impact on his available assets and the greatest possibility for continuing the patient evacuation mission. Optimize the use of resources, medical or nonmedical, which are already contaminated before employing uncontaminated resources.
- Once a vehicle or aircraft has entered a contaminated area, it is highly unlikely that it will be able to be spared long enough to undergo a complete decontamination. This depends upon the contaminant, the tempo of the battle, and the resources available. Normally, contaminated vehicles (air and ground) have restricted use and are confined to dirty environments.
- Introducing uncontaminated aircraft into a contaminated area should be avoided, whenever possible. Ground ambulances should be used instead of AAs as long as their use does not adversely affect the patient's medical condition. Ground ambulances are more plentiful and are easier to decontaminate. This does not, however, preclude using aircraft in a contaminated environment or in the evacuation of contaminated patients.
- The relative positions of the contaminated area, FLOT, and threat air defense systems determine if and where helicopters are to be used. The commander may choose to restrict one or more helicopters to the contaminated areas and use ground vehicles to cross the line separating contaminated and clean areas. The ground ambulance can proceed to the receiving MTF with a

patient decontamination station. The patient can then be transferred to a clean ground or AA if further evacuation is required. The routes used by ground vehicles to cross between contaminated and clean areas are considered dirty routes and should not be crossed by clean vehicles. The effects of wind and time upon the contaminants must also be considered.

- The rotor wash of the helicopters must always be kept in mind when evacuating contaminated casualties. The intense winds disturb the contaminants in the area and further aggravate the condition by additionally spreading the contaminants. Ideally, the aircraft must be allowed to land and reduce to a flat pitch prior to bringing any patients near. This will be dictated by the tactical situation, but allows some reduction in the effects of the downwash. A helicopter must not land too close to a decontamination station (especially upwind) because any trace of contaminants in the rotor wash will compromise the decontamination procedure.
- Immediate decontamination of aircraft and ground vehicles should be accomplished to minimize crew exposure. Units should develop their own procedures for thorough decontamination and document them in their TSOPs. A sample aircraft decontamination station that may be tailored to a particular unit's needs is provided in FM 3-11.5.
- Evacuation of patients must continue even in a contaminated environment. The commander must recognize the constraints placed upon him by resources and plan and train to overcome deficiencies.
- Refer to FM 4-02.7 for additional information on AHS operations in a CBRN environment.

SECTION VI — SHORE-TO-SHIP EVACUATION OPERATIONS

5-30. It is imperative that Army aeromedical evacuation units be able to interface with USN air-capable ships. Lessons learned from past operations, such as Vietnam and Grenada, have shown that US Army helicopters should be able to operate to and from USN air-capable ships. An interservice agreement between the Army and the Navy allows for deck-landing qualification of Army pilots. (Refer to FM 1-564 for additional information.)

- It is important that units having contingency missions requiring Navy support establish training requirements to obtain naval-operations orientation, water egress training, water survival, and deck-landing qualification. This enhances the successful accomplishment of the aeromedical evacuation mission to naval vessels.
- In past joint operations, communications have been burdensome for both Army and Navy elements. Commonality of communication requirements should be established during training exercises. Communication equipment and frequencies for medical evacuation to Navy vessels must be established. This will provide smooth integration of Army helicopters into the Navy airspace management system during actual operations.
- As the Navy vessels may operate relatively long distances from the ground combat operations, Army aeromedical evacuation units need to be proficient in over water navigation. The use of NAVAIDS from the Navy element in support of the operation is the first priority for over water navigation. Basic dead-reckoning remains a secondary measure.

5-31. An important aspect of joint operations is the medical capabilities of Navy vessels servicing the CZ. Knowledge of ship's medical capabilities assists the MRO to direct patients to proper treatment sites. There are many classes of ships which can meet the medical needs of ground forces. Destroyer tenders and aircraft carriers have helicopter landing areas, one operating room (OR) and, at a minimum, one medical officer. Amphibious ships have the most extensive medical facilities of any Navy combat ship. The Navy has 59 amphibious ships in active commission plus two tank landing ships, which are operated by the Navy Reserve Force. The primary mission of the amphibious ships is to transport and support the Fleet Marine Force. The ships have the additional duty of casualty receiving and treatment ships (CRTS). During normal operations, the medical staff is kept to a minimum. The medical staff is augmented when expanded capabilities are needed. Current information regarding landing requirements and medical capabilities should be obtained during training periods with the Navy. Casualty care is secondary to the combat mission of all US combat ships.

- Amphibious assault ships "WASP" class are designated by the Navy as amphibious helicopter assault carrier dock (LHD) (followed by a number) and have the largest patient care facilities on any US combat ship. The WASP class ships have six main ORs, four dental ORs, bed capacity that can be expanded to 600, and it carries 1,500 pints of frozen blood. This ship can receive casualties from helicopters or landing craft.
- Amphibious assault ships "TARAWA" class are designated by the Navy as amphibious helicopter assault carrier (LHA) (followed by a number). The TARAWA class ships have three main ORs, two dental ORs, an overflow bed capacity of 300, and carries 1,500 pints of frozen blood. The ship can receive casualties from helicopters or landing craft.
- Amphibious assault ships "IWO JIMA" class are designated by the Navy as amphibious assault helicopter carrier—aviation (LPH) (followed by a number). These ships were specifically designed to operate helicopters. The IWO JIMA class ships have two ORs and an overflow bed capacity of 200.
- The amphibious transport dock (LPD) (followed by a number) has less medical capabilities than the LHD, LHA, or LPH ships. It can be designated as secondary casualty receiving ship.
- The older dock landing ship (LSD) (followed by a number) can be used as a secondary CRTS when augmented. The newer class of LSD currently under construction can be used as a casualty receiving ship with a capacity for 50 wounded.
- The tank landing ship (LST) (followed by a number) is another type of ship used in amphibious operations. It is designed with a helicopter platform and a stern ramp. Patients can be delivered by air or boat when required by tactical or mass casualty situations. When the LST is augmented with medical personnel and materiel, it can be used for the emergency treatment and evacuation of patients.
- The troop transport (AP) (followed by a number) is not in active service. When available, the troop transport can be outfitted with special medical facilities and carry sick, injured, and wounded personnel.

5-32. The Military Sealift Command operates two hospital ships. The United States Navy Ship (USNS) MERCY T-AH 19 and the USNS COMFORT T-AH 20. One ship is based on each coast and, when needed, will be assigned medical staffs from military hospitals, getting underway within 5 days. The hospital ships MTFs were designed for a total capacity of 1,000 casualties, including 500 acute care beds and 500 recuperation beds. The hospital ships have 50 trauma stations in the casualty receiving area; 12 operating rooms; a 20-bed recovery room; 80 intensive care beds; and 16 intermediate, light, limited care wards. The maximum patient flow rate, for which the helicopter facility and the casualty reception area were designed, is 300 patients per 24 hours. There is a limited capability to receive casualties from boats.

5-33. The US Army has the shore-to-ship medical evacuation mission on an area support basis for Marine forces deployed on land.

SECTION VII — AIRBORNE AND AIR ASSAULT OPERATIONS

5-34. Airborne and air assault operations are some of the most complicated military operations to undertake, not only from an operational perspective, but also from a sustainment and conservation of the fighting strength perspective. A number of personnel within the medical chain that need to coordinate, not only amongst the AMEDD, but also include the aviation community as what proponent own the aircraft (for example, BAE, BSS, medical platoon, and such). Manual evacuation techniques may have to be used and planners should plan for redundant communications, only ground medical evacuation may or may not be available.

5-35. The airborne and air assault operational forces are specialized forces employed to maximize their design characteristics. Airborne units are a flexible force that can be strategically or tactically deployed. They can be inserted rapidly anywhere in the world as either a deterrent or strike force. Air assault units are flexible and lethal fighting organizations. They are ideally suited for rapid employment to critical areas beyond the reach of ground forces.

5-36. After airborne forces have landed in the objective area, they reorganize and maneuver to seize objectives. When it is necessary for assault aircraft to land in the drop zone, they are parked and unloaded rapidly. Then, they may be used to transport Soldiers injured during the parachute assault. It must be understood that organic medical units may experience an overload of patients during the early phases of an airborne assault. These units have to hold the patients until either ground link-up is made or evacuation can be established at airheads. Aeromedical evacuation from the airhead is accomplished using tactical and strategic USAF aircraft.

5-37. The air assault division's organic aircraft have the ability to attack from any direction, over fly obstacles, and bypass enemy positions. Evacuation of patients in the assault phase is accomplished by AAs. Air ambulances may accompany the air assault task force (AATF) or respond from laager sites once the initial assault has taken place. If AAs are providing on-call support, it will be necessary to fly secure air avenues of approach.

5-38. When both airborne and air assault divisions have been employed and become a part of other conventional forces, their operations are similar to that of light infantry forces. During initial deployment, division medical evacuation assets may be used to evacuate patients to the airhead for air evacuation directly to division/corps hospitals.

Chapter 6
Medical Regulating

GENERAL

6-1. Medical regulating is the coordination and control of moving patients to MTFs which are best able to provide the required specialty care. This system is designed to ensure the efficient and safe movement of patients.

PURPOSE OF MEDICAL REGULATING

6-2. Medical regulating entails identifying the patients awaiting evacuation, locating the available beds, and coordinating the transportation means for movement. Careful control of patient evacuation to appropriate hospitals is necessary to—

- Effect an even distribution of cases.
- Ensure adequate beds are available for current and anticipated needs.
- Route patients requiring specialized treatment to the appropriate MTF.

6-3. The factors that influence the scheduling of patient movement include the following:

- Patient's medical condition (stabilized to withstand evacuation).
- Tactical situation.
- Availability of evacuation means.
- Locations of MTFs with special capabilities or resources.
- Current bed status of MTFs.
- Surgical backlogs.
- Number and location of patients by diagnostic category.
- Location of airfields, seaports, and other transportation hubs.
- Communications capabilities (to include radio silence procedures).

MEDICAL REGULATING TERMINOLOGY

6-4. As medical regulating may include coordination with other services, it is necessary to use the correct terminology. These terms include—

- Intracorps Medical Regulating—This is the system by which patients are transferred or evacuated from an FSB or main support battalion (MSB) to a combat support hospital (CSH).
- Intratheater Medical Regulating—This is the system by which patients are transferred or evacuated from one hospital to another within the AO. This includes evacuations between CZ hospitals, between EAC hospitals, or from CZ hospitals to EAC hospitals.
- Intertheater Medical Regulating—Moving patients between, into, and out of the different theaters of the geographic combatant commands and into the CONUS or another supporting theater. (JP 4-02.)
- Patient Administrator—The patient administrator (PAD) accomplishes the medical regulating function at the hospital level in addition to his normal duties. His medical regulating functions include consolidating all evacuation requests within the hospital and forwarding an evacuation request to his next higher headquarters for action. The PAD is also responsible for keeping his next higher MRO apprised of the current beds available and the OR status.

- Medical Regulating Officer—The MRO functions as the responsible individual at C2 headquarters for receiving and consolidating evacuation requests. These requests are initiated by the DSS or subordinate hospitals. The MRO also maintains the current patient status, bed status, and the surgical backlog at subordinate hospitals. His duties include—
  - Managing what patient classes are regulated into his facility.
  - Determining what resources are available to move the patients and coordinating for the use of these assets.
  - Maintaining accountability of patients within the MTFs.
  - Preparing reports as required.
- Theater Patient Movement Requirements Center—The TPMRC is a joint agency normally located at or near the unified theater headquarters. The theater surgeon supervises the functions of this office. These functions include—
  - Maintaining direct liaison with the Global Patient Movement Requirements Center (GPMRC), the MROs of component services, and the transportation agencies which furnish the means for evacuation.
  - Obtaining periodic reports of available beds from the services MROs providing hospitalization.
  - Selecting hospitals based on the reported bed availability to receive patients within EAC.
- Global Patient Movement Requirements Center—The GPMRC is a joint agency located in CONUS and established by the USTRANSCOM. The GPMRC receives requests from the TPMRCs. The primary role of the GPMRC is to apportion intertheater assets to the TPMRCs, collaborate and integrate proposed TPMRC intertheater plans and schedules, and communicate lift and bed requirements. The destination hospital is determined based on the patient's medical needs, the available transportation resources, and MTF capabilities.
- Joint Patient Movement Requirements Center (JPMRC)—Joint patient movement requirements center provides theater patient movement requirements center-type domain, automatic information system support and operations for a joint task force operating within a unified command area of responsibility. The theater patient movement requirements center maintains overall responsibility for theater patient movement operations, but the joint patient movement requirements center is responsible for patient movement operations within its area scope of responsibility and coordinates with the theater patient movement requirements center for intra-theater patient movement and the GPMRC for inter-theater patient movement. Also called JPMRC.
- Theater Aeromedical Evacuation System (TAES)—The TAES is a functional organization which is provided by the USAF and performs the mission of theater AE. Significant components of the TAES include the following:
  - Aeromedical evacuation command squadron.
  - Aeromedical evacuation control team (AECT).
  - Mobile aeromedical staging facility.
  - Contingency aeromedical staging facility (CASF) (under the C2 of the medical group commander).
  - Aeromedical evacuation crews.
  - Critical care air transport teams (CCATTs) (not necessarily Air Force).
- Aeromedical Evacuation Command Squadron—The AE command squadron provides C2 of all assigned TAES forces and can deploy in advance of other AE components to arrange support requirements for AE forces. The command squadron advises commanders and other personnel/agencies on AE operations, capabilities, and requirements and provides procedural and technical guidance for attached and transiting AE elements.
- Aeromedical Evacuation Control Team—The AECT is a CASF (under the C2 of the medical group commander) and is an USAF element responsible for linking validated patient movement requests with the appropriate airlift platform. The AECT is normally located within the Air

Mobility Division of the theater air operations center (AOC) for intratheater and the tanker airlift control center (TACC) for intertheater coordination of AE operations using USAF aircraft. The CONUS and strategic AOC for the USAF is the TACC located at Scott Air Force Base, Illinois. The AECT coordinates with and receives requirements from the PMRC for execution.

- Aeromedical Evacuation Liaison Team (AELT)–The AELT provides a direct communications link and immediate coordination between the user service and the TAES. The AELT is located at any level where Air Force fixed-wing requests are initiated, verifies and coordinates with the AECT/patient movement requirements center (PMRC) the physiology of flight issues and patient flight/movement requirements, assists with patient preparation for flight, and directs patient on-load activities.
- Mobile Aeromedical Staging Facility—The MASF is a USAF staging facility employed at forward airfields in the CZ to provide a temporary staging capability for preparation of patients being evacuated from corps to EAC hospitals. The MASF is employed to ensure patients are prepared for aircraft loading with the main focus of reducing AE aircraft ground time.
- Contingency Aeromedical Staging Facility—The CASF, similar to the MASF, is a USAF staging facility. The CASF, which provides greater capability and longer holding periods than the MASF, is a fixed facility located at theater AE strategic hubs and interfaces with the TAES. As opposed to the TAES elements, which are under the C2 of the AE command squadron, the CASF reports to the medical group commander.
- Military Sealift Command (MSC)—The MSC is the USN element responsible for coordinating movement of supplies, equipment, and personnel into the JOA by Navy ships. Further, it coordinates, through the TPMRC, the medical evacuation of patients by ship from the JOA to the support base or hospital ship, as required.
- Corps Movement Control Center (CMCC)—The CMCC is the corps movement control organization. It provides centralized movement control and highway regulation for movement of personnel and material into, within, and out of the corps area. When USAF capabilities are exceeded, the CMCC coordinates requests for additional air and ground resources. It also obtains the necessary clearances to support the medical evacuation mission from the CZ.
- Movement Control Battalion (MCB)—The MCB mission is to provide movement management services and highway traffic regulations and to coordinate for personnel and material movements into, within, and out of the theater. The MCB coordinates with allied and HN movement control agencies. It also coordinates with the USTRANSCOM and its subordinate units (such as the Air Mobility Command and MSC) and prepares movement and port clearance plans and programs.
- Joint Military Transportation Board (JMTB)—The JMTB is a joint staff composed of members of the Army, USAF, and USN that coordinates transportation requirements for patients requiring intertheater evacuation.
- Transportation Command Regulating and Command and Control Evacuation System—The Transportation Command Regulating and Command and Control Evacuation System is the single DOD automated information system that supports medical regulating, patient movement, and patient in-transit visibility requirements.
- Automated Patient Evacuation System—The Automated Patient Evacuation System is the system that automates the patient movement portion of medical evacuation.

MEDICAL REGULATING FOR THE DIVISION

6-5. Medical regulating in and from the division is the responsibility of the DSS. Medical regulating in the division is not as formalized as the rest of the medical regulating system. It is usually operated procedurally so as not to depend solely on communications to effect rapid evacuation. The medical regulating function in the DSS is concerned primarily with.

- Tracking the movement of patients throughout the division MTFs and into the corps facilities.
- Monitoring the use of ambulance assets.

6-6. Air and ground ambulances placed in GS of the BCTs can be field sited in the sustainment area when the mission directs. When these assets go forward to the BSMC or the ASMC to evacuate patients to corps MTFs, they have corps MTF destinations predetermined (blocks of beds). The DSS, in coordination with the medical group/brigade MRO, establishes the number of patients a supporting corps hospital can accept during a particular period of time. These blocks of available beds are then provided to the GS ambulances prior to the call for missions.

Note. Under the MRI force structure, the medical group functions were absorbed by the medical brigade.

- Once an evacuation mission is completed, the originating division MTF contacts the patient disposition section of the BSS and provides—
  - Patient numbers by category and precedence.
  - Departure times.
  - Modes of transportation.
  - Destination MTFs.
  - Any other information required by TSOP.
- The DSS, in turn, notifies the medical group/brigade MRO via the patient administration net, which is monitored by the corps MTFs. Since corps ground ambulances have no onboard communications ability and AAs have no amplitude-modulated high-frequency capability at present, all patient information is passed to the gaining MTFs via the patient administration nets. To reduce the turnaround time for ground ambulances and to move more serious patients to the CSHs in the sustainment area—
  - Air ambulances are given blocks of beds in the corps hospitals farther to the rear.
  - Ground ambulances are normally given blocks of beds in the more forward deployed CSHs.

6-7. Medical evacuation can be effected immediately, procedurally, and under conditions of communications silence without interrupting the continuum of care by—

- Preparing patient estimates.
- Prioritizing and task-organizing ambulance support.
- Assigning blocks of hospital bed designations prior to the start of the mission.
- The assistant chief of staff (personnel) (G1) is also involved in medical regulating as the G1 is responsible for personnel accountability and casualty reporting. Casualty liaison teams (CLTs) will normally be colocated at CSH to assist and coordinate patient tracking and casualty reporting.

MEDICAL REGULATING WITHIN THE COMBAT ZONE

6-8. The requirement to transfer patients from one hospital to another within the CZ occurs. This results from—

- Surgical backlogs.
- Mass casualty situations.
- Specialty care requirements.
- Planned movement of an MTF.

6-9. When it is necessary to transfer a patient, the attending physician notifies the hospital PAD. The PAD consolidates all such requests from the hospital and requests movement authority from the IPMC.

6-10. If the medical IPMC can transfer the patient or patients to its subordinate hospitals, he designates the hospitals to receive the patients and notifies both the requesting and receiving hospitals of the transfer. The IPMC also tasks subordinate medical evacuation units for the assets to transfer the patients.

6-11. If the MEDBDE cannot provide the needed hospitalization within its own resources, the MRO forwards the request to the medical TPMC for action. The TPMC then designates the receiving hospitals and notifies the subordinate IPMCs. The medical IPMCs disseminate the information to the hospital PAD and coordinate the evacuation resources for the transfer. The MRO also coordinates the regulation of patients to—

- Other US military service hospitals and naval hospital ships.
- Allied nations' military hospitals.
- Other authorized supporting facilities.

MEDICAL REGULATING FROM THE COMBAT ZONE TO ECHELONS ABOVE CORPS

6-12. Hospital attending physicians and oral and maxillofacial surgeons submit daily reports to the hospital PAD listing the patients requiring evacuation. The PAD assembles this information and transmits the report to the medical brigade headquarters. This report is a request for transportation, as well as a notification of the number of patients requiring evacuation. The report classifies the patients according to—

- Diagnostic category.
- Desired on-load points.
- When the patients will be available for evacuation.

6-13. The medical brigade TPMC consolidates these reports from each hospital attached to the medical brigade and forwards his report to the corps TPMC. The medical brigade MRO consolidates the reports and transmits the data to the theater TPMC.

6-14. If a JPMRC has been activated within the theater, the MEDCOM MRO consolidates all reports from the CZ medical brigades and forwards them to the JPMRC. The JPMRC designates hospitals in the EAC to receive the patients. The designation is based on the previously received bed status reports from all service components and available means of evacuation. The JPMRC then notifies the MEDCOM MRO of designated hospitals. The TPMC accomplishes this task if the JPMRC is not activated.

6-15. The primary means of moving patients from the CZ out of theater is USAF aircraft. With the elements of the TAES deployed, it is possible to find AELTs at each echelon and as far forward as Air Force fixed-wing requests are initiated. The PMRC monitors the MRO patient evacuation requests and passes requirements to the AELT. At the same time they pass airlift requirements to the AECT, seeking an aircraft to perform the evacuation mission. The AELT, through the MRO patient movement request, requests the PMRC/AECT to move patients. Included in the request are the originating medical facility (OMF) and the destination airfields. The airfields selected are those serving the hospitals designated to receive patients.

6-16. The AECT is a component of the TAES and performs the mission of coordinating the movement of and providing in-flight medical care to patients while under the USAF control. The AECT receives patient movement requirements from the TPMRC, then works with the airlift control team (ALCT) in the AOC to meet the evacuation requirements.

6-17. The AOC coordinates the forward movement of cargo and personnel aboard USAF aircraft with other USAF units, Army transportation representatives, and USN agencies. Certain of these aircraft are scheduled to evacuate patients on their return trips. These aircraft seldom go forward solely to evacuate patients; however, these missions may be used for retrograde airlift and mixed cargo/AE missions is the primary means of AE due to limited aircraft and the need to maximize the use of this asset.

6-18. After the schedules have been arranged, the AECT returns the detailed flight schedule to the JPMRC who passes the information to the TPMC/AELT and the parent AE element.

6-19. The MEDCOM MRO, in coordination with the AELT, issues these instructions to the IPMCs (with the authority to move patients in Army CZ facilities) and the receiving hospitals. The hospitals must prepare to receive the patients at the destination airfields. (This may be accomplished by collocating an ASMC at the airfield to receive the incoming patients.) The patients are sorted by destination hospital and moved by Army medical evacuation means. The instructions mentioned above include, as a minimum, the—

- Number of patients to be moved.
- On-load airfield.
- Destination airfield.
- United States Air Force aircraft mission number.
- Estimated time of arrival at the destination airfield.

6-20. The TPMC issues the flight and movement instructions to its subordinate medical IPMCs. The IPMCs then direct the evacuation units and hospitals within their AOs to move the patients to the on-load airfield according to the arrival time of the aircraft. This movement must be closely controlled, as a MASF can accommodate up to 50 patients. The patients cannot be delivered to the MASF earlier than 6 hours prior to arrival of the aircraft and no later than 1 hour prior to arrival.

MEDICAL REGULATING WITHIN ECHELONS ABOVE CORPS

6-21. Medical regulating within the EAC is similar to the system used within the CZ. Attending physicians within the Role 3 hospitals notify the hospital PAD of patients requiring evacuation to CSHs. The PAD then consolidates the requests from the hospital and forwards the consolidated requests to the IPMC. The IPMC, in turn, consolidates the requests and forwards them to the TPMC.

6-22. The TPMC, based on periodic bed status and availability reports from subordinate hospitals, designates specific hospitals to receive the patients. The hospitals are designated based on bed availability, to include specialty beds, to support the specific patient. The TPMC then notifies the requesting medical IPMC of the designated hospitals and, in turn, notifies the designated hospitals.

INTERTHEATER MEDICAL REGULATING

6-23. The patients who are evacuated to EAC are treated there and then further evacuated to CONUS. The attending physicians at the hospital notify the PAD. The PAD then consolidates these requests and forwards them to the IPMC. This IPMC forwards the consolidated request to the TPMC who, in turn, consolidates and forwards a request to the JPMRC (if established) or the TPMRC.

6-24. Upon request of the J/TPMRC for authority to evacuate patients to CONUS, the GPMRC directs the distribution of these patients into hospitals throughout the CONUS; advises the J/TPMRC of the destination hospital; and provides the authority for such movement. As a rule, the destination hospitals are military facilities. Civilian national disaster medical system member hospitals and other federal hospitals may also receive patients. The Department of Veterans Affairs hospitals, for example, may receive patients who are expected to be discharged from service. The GPMRC validates the TPMRC patient movement requirements and, if moving by air, tasks the TACC to plan, schedule, and execute the intertheater evacuation.

6-25. When the J/TPMRC receives the authorization to move patients, it notifies the TPMC of destination hospitals in CONUS. The TPMC coordinates with the JMTB to arrange movement of CONUS-bound patients. The TPMC then authorizes the movement to ASFs/aeromedical staging squadrons (ASTSs) that are located on or near air bases or airstrips capable of handling long-range aircraft. Transportation is arranged, within Army channels, to move patients from the hospitals to the staging facilities. The medical brigade, in coordination with the AELT, then notifies the subordinate CSHs of the flight schedule and the evacuation arrangements for movement to strategic airheads. At strategic airheads, there is an established ASF/ASTS. When the patients are delivered to the USAF, the responsibility for those patients is transferred from the Army hospital to the TAES. Upon arrival in CONUS further movement is the responsibility of the GPMRC.

6-26. All patients may not be able to be moved by air from the theater to CONUS. In that event, the MSC is used to move them by surface means. The movement authority also comes from the G/T/JPMRC or TPMC which has arranged with the Navy Service Component Command for the movement of patients by hospital ships. When the patients are moved by ships, the MEDCOM has to provide holding facilities at the port (collocating an ASMC can provide this support). Patients are delivered to these holding facilities and held there until loaded aboard the ships.

MOBILE AEROMEDICAL STAGING FACILITY

6-27. The MASF is a 13-person, mobile, tented, temporary staging facility deployed to provide supportive patient care and administration. Each MASF is capable of a throughput of 40 patients per 24 hours and should be able to sustain this tempo for 72 hours before augmentation is required and is not intended to hold patients overnight or for an extended period.

6-28. This theater system is used to evacuate patients from—

- United States Air Force operational locations within the CZ to hospital facilities outside the CZ.
- Airhead or airborne objective areas where airborne operations include USAF forward logistics support.

6-29. Mobile aeromedical staging facilities—

- Provide supportive medical care when not augmented by the CCATTs. (When augmented by CCATTs, continued stabilization of patients can be accomplished. Aeromedical evacuation crews and CCATTs fly airlift missions to provide in-flight patient care.)
- Confirm sending facility has prepared patients for evacuation.
- In coordination with the OMF, ensure patient evacuation manifests are completed. (If the OMF is an Role 2 facility [such as an FSMC augmented with an FST], the forms required to complete the manifest may not be available and the MASF will be required to complete the appropriate forms.)
- Identify patient baggage tags.

6-30. Upon deployment, the OMF provides an adequate quantity of medications for patients' transit time to the regulated destinations.

6-31. The MASF staff also establishes liaison with the OMF. The AELT is composed of one medical service corps officer for administrative assistance and a flight nurse. The team provides the initial interface between the user service and the TAES. The AELT is located at any level of the combat forces where Air Force fixed-wing requests are initiated.

LIMITATIONS OF THE UNITED STATES AIR FORCE THEATER AEROMEDICAL EVACUATION SYSTEM

6-32. There are a number of limitations that are inherent in the current system. These include the following:

- Absence of biological weapon and chemical weapon agent decontamination ability.
- The MASF does not have the capability to provide patient meals.
- The AECT ensures the initial 30-days medical resupply package arrives at the MASF. The MASF/AELT/aeromedical evacuation operations team rely on the user service for all other logistical support.
- It is the Army's responsibility to provide food and other logistical support required including moving patients back to Army facilities should USAF AE support be delayed.

ORIGINATING MEDICAL FACILITY'S RESPONSIBILITIES

6-33. Once the authorization to move the patient has been given, the OMF must complete the following administrative procedures prior to entering the patient into the TAES:

- The patient's baggage tag, patient evacuation manifest, and patient evacuation tag are the specified evacuation forms for all services and are completed as required by triservice regulation.

Note. If the OMF is a Role 2 facility (such as the BSMC), the forms may not be available; in which case, the forms will need to be completed by the MASF.

- All of the patient's medical records must be collected together and packaged. The dental records are forwarded separately in the event they are needed for identification.
- At the appropriate time, the OMF provides transportation to the MASF and assists in the offload.
- The OMF must provide the necessary medications, medical supplies, and equipment to support the patients' travel time to the regulated destination.
- Any requirements for armed guards must be met by the echelon commander.

Note. Medical units do not provide guards for prisoners or EPW in their care. When guards are required, they are provided by the echelon commander. The OMF will coordinate for this support when needed.

- A limited amount of personal baggage is authorized if each piece is properly tagged and delivered to the MASF with the patient. Patients will always be evacuated with CBRN-protective equipment, less the protective overgarment.
- Each patient must be clearly identified with a wristband or equivalent identification and properly classified as to his medical condition.
- The OMF must ensure that each patient is properly briefed and prepared for his evacuation prior to his arrival at the MASF.

MEDICAL REGULATING FOR ARMY SPECIAL OPERATIONS FORCES

6-34. As in medical evacuation, the medical regulating plan must be integrated with the ARSOF operational and logistic plan. Maximum use of opportune (operational and logistics) aircraft and command and logistics communications nets must be coordinated to expedite mission requests and ensure success.

6-35. The ARSOF medical planner must constantly coordinate with the battalion or group operations and logistics sections to obtain up-to-date information of opportune transportation assets to be used for evacuation. In an operation, or when the theater is not sufficiently developed to allow the TAES to be used effectively, the primary means of air evacuation will be those SOA or USAF SOF airframes conducting the clandestine mission. It is essential that coordination is made through the Theater Special Operations Command (TSOC) or the highest C2 element for flight medics or para rescue men to accompany the flight when backhauling the casualties. Otherwise, a medic from the SOF unit being supported may have to accompany the patient, leaving the mission without proper medical support, or the casualty may have to be transported without en route care.

6-36. For all other special operations, the supporting medical evacuation unit provides air and ground ambulances according to standard doctrinal procedures. United States Air Force MASFs or AELTs may be collocated at SOF support bases, or C2 bases, particularly during contingency operations where the build-up phase allows for prepositioning of assets.

6-37. During sustained special operations missions, the TSOC cannot afford to lose the services of low-density ARSOF skilled Soldiers who become casualties. Every effort must be made to preserve ARSOF capabilities and combat power in theater. A determination must be made as to who can be treated and returned to duty at hospitals within the EAC. As an exception to the theater evacuation policy, the combatant command may retain injured or wounded ARSOF in theater where they can be returned to limited duty without jeopardizing their recovery and health. There they can assume the support duties performed by other ARSOF Soldiers, freeing the latter for operational duties.

Appendix A

Geneva Conventions and Medical Evacuations

The conduct of armed hostilities on land is regulated by customary international law and lawmaking treaties such as The Hague and Geneva Conventions. The rights and duties set forth in the Conventions are part of the supreme Law of the Land. The US is obligated to adhere to these obligations even when an opponent does not. It is a DOD and Army policy to conduct operations in a manner consistent with these obligations. An in-depth discussion of the provisions applicable to medical units and personnel is provided in FM 4-02. This appendix discusses only those articles or actions which affect medical evacuation operations.

DISTINCTIVE MARKINGS AND CAMOUFLAGE OF MEDICAL FACILITIES AND EVACUATION PLATFORMS

This paragraph implements STANAG 2454.

A-1. All US medical facilities and units, except veterinary, display the distinctive flag of the Geneva Conventions. This flag consists of a Red Cross on a white background. It is displayed over the unit or facility and in other places as necessary to adequately identify the unit or facility as a medical facility.

AUTHORIZED EMBLEM

A-2. The Geneva Conventions authorizes the use of the following distinctive emblems on a white background: Red Cross; Red Crescent; and Red Lion and Sun. In operations conducted in countries using an emblem other than the Red Cross on a white background, US Soldiers must be made aware of the different official emblems. Although not specifically authorized as a symbol in lieu of the Red Cross, enemies of Israel in past wars have recognized the Red Star of David and have afforded it the same respect as the Red Cross. This showed compliance with the general rule that the wounded and sick must be respected and protected when they are recognized as such, even when not properly marked.

A-3. United States forces are legally entitled to only display the Red Cross. However, commanders have authorized the display of both the Red Cross and the Red Crescent to accommodate HN concerns and to ensure that confusion of emblems would not occur. Such use of the Red Crescent must be in a smaller size than the Red Cross.

This paragraph implements STANAG 2931.

Camouflage of the Distinctive Emblem

A-4. Camouflage of medical facilities (medical units, medical vehicles, and medical aircraft on the ground) is authorized when the lack of camouflage might compromise tactical operations. The marking of facilities and the use of camouflage are incompatible and should not be undertaken concurrently.

A-5. If failure to camouflage endangers or compromises tactical operations, the camouflage of medical facilities may be ordered by a NATO commander of at least brigade level or equivalent. Such an order is temporary and local in nature and is rescinded as soon as circumstances permit.

Note. There is no such thing as a *camouflaged* Red Cross. When camouflaging a medical unit or ambulance, either cover up the Red Cross or take it down. A *black cross* on an olive drab or any other background is not a symbol recognized under the Geneva Conventions.

MEDICAL AIRCRAFT

A-6. Medical aircraft exclusively employed for the removal of wounded and sick and for the transport of medical personnel and equipment shall not be attacked, but shall be respected by the belligerents, while flying at heights, times, and on routes specifically agreed upon between the belligerents concerned.

A-7. The medical aircraft shall bear, clearly marked, the distinctive emblem together with their national colors on their lower, upper, and lateral surfaces.

A-8. Unless agreed otherwise, flights over enemy or enemy-occupied territory are prohibited.

A-9. Medical aircraft shall obey every summons to land. In the event that a landing is thus imposed, the aircraft with its occupants may continue its flight after examination, if any.

A-10. In the event of involuntary landing in enemy or enemy-occupied territory, the wounded and sick, as well as the crew of the aircraft, shall be prisoners of war; medical personnel will be treated as prescribed in these Conventions.

SELF-DEFENSE AND DEFENSE OF PATIENTS

A-11. When engaging in medical evacuation operations, medical personnel are entitled to defend themselves and their patients. They are only permitted to use individual small arms.

A-12. The mounting or use of offensive weapons on dedicated medical evacuation vehicles and aircraft jeopardizes the protections afforded by the Geneva Conventions. These offensive weapons may include, but are not limited to machine guns, grenade launchers, hand grenades, and light antitank weapons.

A-13. Medical personnel are only permitted to fire in their personal defense and for the protection of the wounded and sick in their charge against marauders and other persons violating the Law of War.

ENEMY PRISONERS OF WAR/DETAINEES

A-14. Sick, injured, wounded enemy EPW, and detainees are treated and evacuated through normal medical channels but are physically segregated from US, allied, or coalition patients.

A-15. Personnel resources to guard EPW or detainee patients are provided by the echelon commander. Medical personnel do not guard EPW or detainee patients.

COMPLIANCE WITH THE GENEVA CONVENTIONS

A-16. The US is a party to the 1949 Geneva Conventions. Two of these Conventions afford protection for medical personnel, facilities, and evacuation platforms (to include aircraft on the ground). All HSS personnel should thoroughly understand the provisions of the Geneva Conventions that apply to medical activities.

CONSEQUENCES OF VIOLATIONS

A-17. Violation of these Conventions can result in the loss of protection afforded by them. Medical personnel should inform the tactical commander of the consequences of violating the provisions of these Conventions. The consequences may include—

- Medical evacuation assets subjected to attack and destruction by the enemy.
- Health service support capability degraded.
- Captured medical personnel becoming prisoners of war rather than retained persons. They may not be permitted to treat their fellow prisoners.
- Loss of protected status for medical unit, personnel, or evacuation platforms (to include aircraft on the ground).

PERCEPTION OF IMPROPRIETY

A-18. Because even the perception of impropriety can be detrimental to the mission and US interests, HSS commanders must ensure they do not give the impression of impropriety in the conduct of medical evacuation operations. For example, if a medical commander included in the unit's TSOP rules governing the use of automatic or crew-served weapons, it would give the impression that the unit possessed and intended to use these types of weapons. Under the provisions of the Geneva Conventions, medical units are only authorized individual small arms for use in the defense of the patients under their care and for themselves. Even though the unit does not possess these types of weapons, the entry in the TSOP could be misinterpreted and a case made that the commander intended to use these weapons in violation of the Geneva Conventions.

Appendix B
Legacy Units

Currently there are two units that are still active but will be transitioning out of the Army's inventory. They are the medical evacuation battalion and the medical company, AA (UH-1V).

SECTION I — MEDICAL EVACUATION BATTALION

B-1. The HHD, medical evacuation battalion, serves as the central manager of ground and air evacuation assets within the corps and EAC.

B-2. Assignment. The medical evacuation battalion is assigned to the MEDCOM in the EAC or to the medical brigade in the corps. It is normally further assigned to a medical group for C2.

Note. Under the MRI design, the medical group was replaced by the MEDBDE.

B-3. Air and ground ambulance companies assigned to the MEDCOM or MEDBDE are attached to the medical evacuation battalion for C2.

B-4. The basis of allocation is one medical evacuation battalion per a combination of three to seven of the following units:

- Medical companies, AA.
- Medical companies, ground ambulance.

B-5. Mission and capabilities.

- The mission of the medical evacuation battalion is to provide C2 of air and ground medical evacuation units within the AO. It tactically locates in the area where it can best control subordinate air and ground ambulance companies.
- The medical evacuation battalion is designed to focus on C2, planning, patient evacuation, subordinate unit support, and vehicle management. Specific capabilities are—
  - Command and control, planning and supervision of operations and training, and administration of a combination of air and ground ambulance companies.
  - Staff and technical supervision of aviation operations, safety, standardization, and aviation unit maintenance (AVUM)-level maintenance within the attached AA companies.
  - Coordination of medical evacuation operations and communications functions.
  - Coordination of logistics and service support to attached units.
  - Aviation medicine and unit-level AHS.
- This unit is dependent upon appropriate elements of the corps or ASCC for—
  - Personnel service support.
  - Health service support, to include hospitalization.
  - Mortuary affairs support.
  - Laundry, shower, and clothing repair.
  - Communications security equipment maintenance.
  - Military police support.

B-6. Organization and functions.

- Medical Evacuation Battalion (Figure B-1). The headquarters and headquarters detachment (HHD), medical evacuation battalion, is organized into a—
  - Battalion headquarters section.
  - Adjutant (US Army) section.
  - Intelligence Officer (US Army)/S3 section.
  - Supply Officer (US Army) section.
  - Detachment headquarters.
  - Treatment team.
- Battalion Headquarters Section. This section provides C2 of the assigned and attached air and ground ambulance companies. It also assists the commander on all military intelligence matters (to include the health threat), organization, training, operations, planning, personnel service support, and logistics support. Further, it provides information on the health of the command and aviation medicine expertise, as well as providing supervision over technical and flight aspects of administration, training, and safety within subordinate aviation units.
- Adjutant (US Army) Section. This section operates according to METT-TC and is the principle coordinating staff officer responsible for the delivery of human resources support. The S1 is responsible for the execution of all of the HR core competencies to include —PRM; PASR; PIM; R5 operations; casualty operations; EPS; and postal operations. It also includes MWR operations and HR planning and staff operations. The S1 also coordinates with elements of supporting agencies for finance, legal, religious, and administrative services.
- Intelligence Officer (US Army)/S3 Section. This section assists the S2/S3 officer in the execution of his duties and is capable of sustained 24-hour operations. This section remains abreast of the tactical situation and determines future medical evacuation requirements. It plans for ground and air evacuation operations, coordinates command post operations, and maintains the status of the AA units and plans for their employment. Further, this section maintains communications systems and nets, determines intelligence requirements, coordinates with movement control elements, and prepares orders and overlays.
- Supply Officer (US Army) Section. The S4 section assists the S4 officer in the execution of his duties. This section plans, coordinates, and supervises the requisitioning, receipt, storage, issue, and accounting for all classes of supply. Further, it monitors and keeps the commander informed on all matters pertaining to maintenance on assigned aircraft, ground vehicles, and medical equipment. This section also serves as the interface with the supporting medical battalion, logistics (forward/rear) for medical nonexpendable and durable item supply transactions.
- Detachment Headquarters Section. The detachment headquarters section provides, C2, administration, and logistics support for assigned personnel. It is also responsible for company supply and armament functions, food service operations, maintenance operations, and unit administration.
- Treatment Team. The treatment team provides unit-level AHS to assigned and attached elements collocated with the detachment headquarters and to adjacent units on an area support basis. The physician is a flight surgeon and provides staff assistance to the battalion commander on all matters pertaining to aviation medicine. The flight surgeon provides care and treatment for all assigned and attached aircrew members. This physician is dual-hatted as the battalion surgeon.

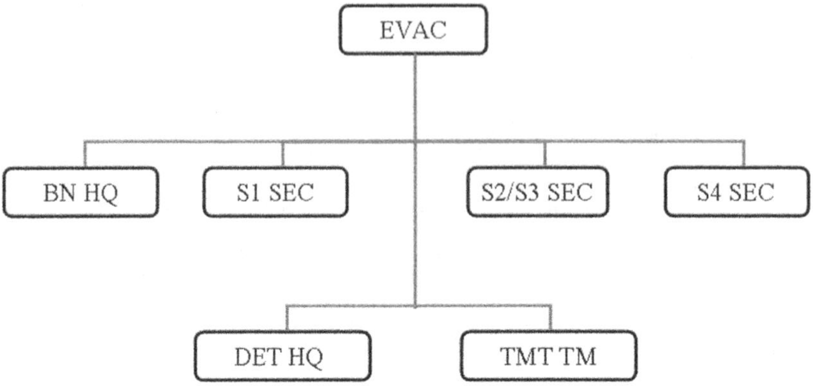

Figure B-1. Medical evacuation battalion

SECTION II — MEDICAL COMPANY, AIR AMBULANCE

B-7. The medical company, AA, provides aeromedical evacuation for all categories of patients consistent with evacuation precedence and other operational considerations. Medical evacuation is effected from as far forward as possible in the tactical AO to division- and corps-level MTFs.

B-8. Assignment.

- The medical company, AA, is normally assigned to the MEDCOM or MEDBDE and attached to the medical evacuation battalion for C2.
- The basis of allocation is one unit in support of each division or equivalent force supported. Further, one unit is in GS in the corps per two division or fraction thereof; or .333 units per separate brigades or ACRs.

B-9. Mission and capabilities.

- The mission of the medical company, AA, is to provide.
  - Aeromedical evacuation support within the theater, either DS to the divisions or GS to the corps.
  - Emergency movement of medical personnel, equipment, and supplies including whole blood, blood products, and biologicals.
- Specific capabilities of this unit are to operate on a 24-hour-a-day basis and evacuate patients based on operational capability (dependent on type of aircraft).
  - Operate 15 AAs (UH-60s). These ambulances are each capable of carrying six litter patients and one ambulatory patient, or seven ambulatory patients, or some combination thereof. Single patient lift capability is 90 litter patients, or 105 ambulatory patients, or some combination thereof. In-flight medical treatment and surveillance of patients is provided by a flight medic.
  - Operate 15 AAs (UH-1H/V). These ambulances are capable of carrying six litter, or nine ambulatory patients, or some combination thereof. Single patient lift capability is 90 litter, 135 ambulatory, or some combination thereof. In-flight medical treatment and patient surveillance are provided by a flight medic.
- Provide internal/external load capability for the movement of medical personnel and equipment.

- Perform AVUM on all organic aircraft and organizational maintenance on all organic avionics equipment. It also performs unit-level maintenance on all organic equipment less medical.
- Provide air crash rescue support, less fire suppression.
- Operate as an area support MEDEVAC section and three FSMTs to provide flexibility in supporting division, brigade, or brigade TF equivalent operations.

B-10. This unit is dependent upon:

- Support elements of corps or ASCC for—
  - Financial management, legal, and religious support.
  - Human resources support.
  - Logistics.
  - Health service support, to include medical supply and equipment.
  - Food service support.
  - Communications security equipment maintenance.
  - Mortuary affairs support.
  - Military police support.
  - Laundry, shower, and clothing repair.
  - Engineer support for heliport/landing strip construction and maintenance.
- The supporting aviation intermediate maintenance (AVIM) organization for AVIM support.

B-11. Organization and functions.

- The medical company, AA (Figure B-2), is organized into a—
  - Company headquarters.
  - Flight operations platoon consisting of a platoon headquarters, a flight operations section, and an airfield service section.
  - Aircraft maintenance platoon consisting of a platoon headquarters, a component repair section, and a maintenance section.
  - Air ambulance platoon consisting of a platoon headquarters, an area support MEDEVAC section, and three FSMTs.

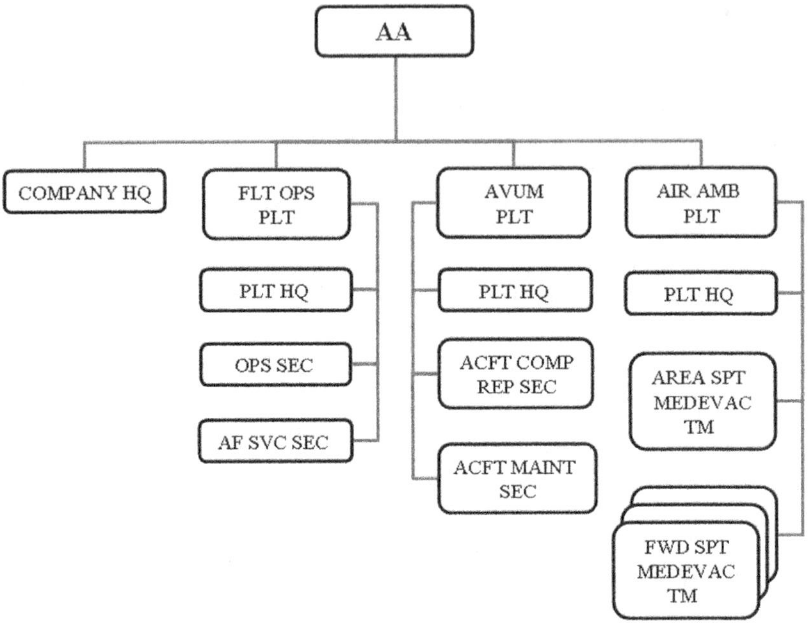

Figure B-2. Medical company, air ambulance

Appendix C

Tactical Standing Operating Procedures

This appendix provides a sample TSOP for an MMB engaged in medical evacuation missions. It should not be considered as all-inclusive. It may be supplemented with the information and procedures required for operating within a specific command or special operation.

PURPOSE

C-1. The TSOP prescribes policy, guidance, and procedures for routine support of tactical operations of a specific unit. It should cover broad areas of unit operations, but be sufficiently detailed to provide newly assigned personnel with the guidance required for them to perform their mission. A TSOP may be modified by the TSOP and OPLANs/OPORDs of higher headquarters. It applies to a specific unit and all subordinate units assigned and attached. Should a TSOP not be in conformity with the TSOP of the higher headquarters, the higher headquarters TSOP governs. The TSOP is periodically reviewed and updated as required.

FORMAT FOR THE TACTICAL STANDING OPERATING PROCEDURE

C-2. There is not a standard format for all TSOPs; however, it is recommended that a unit TSOP follow the format used by its higher headquarters. The TSOP can be divided into sections (specific functional areas or major operational areas) and further subdivided into annexes. An annex can be further subdivided into appendixes and then into tabs. Appendixes can be used to provide detailed information on major subdivisions of the annex, and tabs can be used to provide additional information (such as report formats or area layouts) addressed in the appendix.

C-3. Regardless of the format used, the TSOP follows a logical sequence in the presentation of material. It should discuss the chain of command, major functions and staff sections of the unit, operational requirements, required reports, necessary coordination with higher and subordinate elements for mission accomplishment, programs (such as command information, preventive medicine measures [PMM], and combat stress control) and other relevant topics.

C-4. Pagination of the TSOP can be accomplished by starting with page 1 and numbering the remaining pages sequentially. If the TSOP is subdivided into sections, annexes, appendixes, and tabs, a numbering system that clearly identifies the location of the page within the document can be used. Annexes are identified by letter and are listed alphabetically. Appendixes are identified by numbers and arranged sequentially within a specific annex. Tabs are identified by a letter and are listed alphabetically within a specific appendix. After numbering the initial sections using the standard numbering system (sequentially starting with page 1 through to the end of the sections), number the annexes and their subdivisions. They are numbered as the letter of the annex, the number of the appendix, the letter of the tab, and the page number. For example, page 4 of Annex D is written as "D-4"; page 2 of Appendix 3 to Annex D is written as "D-3-2"; page 5 of Tab A to Appendix 3 of Annex D is written as "D-3-A-5." This system of numbering makes the pages readily identifiable as to their place within the document as a whole.

C-5. In addition to using a numbering system to identify specific pages within the TSOP, descriptive headings should be used on all pages to identify the subordinate elements of the TSOP.

- The first page of the TSOP should be prepared on the unit's letterhead. The remaining pages of the major sections should include the unit identification in the upper right-hand corner of the page (for example: "Multifunctional Medical Battalion").
- A sample heading for an Annex is: "ANNEX B (Command Post) to Multifunctional Medical Battalion."

- A sample heading for an Appendix to Annex B is: "APPENDIX 3 (Command Post Security) to ANNEX B (Command Post) to Multifunctional Medical Battalion."
- A sample heading for a Tab to Appendix 3 to Annex B is: "Tab A (Tactical Operations Center [TOC] Security) to APPENDIX 3 (Command Post Security) to ANNEX B (Command Post) to Multifunctional Medical Battalion."

C-6. As the TSOP is developed there may be an overlap of material from one annex to another. This is due in part to similar functions that are common to two or more staff sections. Where overlaps occur, the material presented should not be contradictory. All discrepancies will be resolved prior to the authentication and publication of the TSOP. The TSOP is authenticated by the unit commander.

Sample Tactical Standing Operating Procedures (Sections)

C-7. The information contained in this paragraph can be supplemented. It is not intended to be an all-inclusive listing. Different commands will have unique requirements that need to be included. The first section of the TSOP identifies the specific unit/headquarters that developed it.

- Scope–This document establishes and prescribes procedures to be followed by the unit identified and its assigned, attached, or OPCON units/elements.
- Purpose–This document provides policy and guidance for routine tactical operations of this headquarters and its assigned, attached, or OPCON units.
- Applicability–Except when modified by policy guidance, TSOP, or OPLANs/OPORDs of the higher headquarters, this document applies to this unit and to all units assigned, attached, or under OPCON for combat operations. These orders, however, do not replace judgment and common sense. In cases of nonconformity, the document of the higher headquarters governs. Each subordinate element will prepare a unit TSOP, conforming to the guidance herein.
- References–This paragraph can include any pertinent regulations, policy letters, higher headquarters TSOP, or any other appropriate documents.
- General information–This paragraph discusses the required state of readiness of the unit; primary, secondary, and contingency missions; procedures for operating within another command's AO; and procedures for resolution of conflicts with governing regulations, policies, and procedures.

C-8. The second section of the TSOP discusses the specific organization.

- Organization–This paragraph furnishes specific information concerning the authority for establishing the unit, such as applicable MTOE or other staffing documentation.
- Succession of command–The guidance for determining the succession of command is discussed.
- Task organization–Task organization is contingent on the mission and will be approved by the headquarters ordering deployment.
- Joint, multinational, and interagency operations–This paragraph provides guidance on any issues concerning C2 and TF organization in joint, multinational, and interagency operations. It also provides guidance on liaison officer requirements.
- Organizational charts–Contained in Annex A.

C-9. The third section of the TSOP discusses the unit functions.

- Battalion headquarters–This paragraph discusses the C2 of the assigned, attached, and OPCON units.
- Headquarters detachment–This paragraph discusses the functions of the headquarters detachment element, such as supervising movements, internal arrangements, area layout, physical security, and operation of the headquarters and staff.
- Attached, assigned, and OPCON units–This paragraph discusses the missions and functions of these units (such as providing medical evacuation of patients, emergency movement of blood and blood products, biologicals, medical logistics, and transportation of medical personnel and equipment).

- Staff responsibilities–This paragraph lists the unit's key personnel and their duties as prescribed in FM 5-0 and any command-specific duties.

C-10. The fourth section of the TSOP pertains to staff operations and is subdivided into annexes.

Sample Tactical Standing Operating Procedures (Annexes)

C-11. Annexes are used to provide detailed information on a particular function or area of responsibility. The commander determines the level of specificity required for the TSOP. Depending upon the complexity of the material to be presented, the annex may be further subdivided into appendixes and tabs. If the annex contains broad guidance or does not provide formats for required reports, paragraphs may be used. The annex should not require further subdivision. However, as the material presented becomes more complex, prescribes formats, or contains graphic materials, the annex may require additional subdivision. This paragraph discusses the subdivision of the annex by appendixes. It does not contain examples of subdividing the information presented in the appendixes into tabs. Applicable references such as ARs, FMs, and technical manuals should be provided in each annex. The number of annexes and their subdivisions presented below are not to be considered as an all-inclusive listing. Different commands will have unique requirements; therefore, supplementation of the information presented is permitted.

C-12. Annex A. (Organizational Charts.)

C-13. Annex B. (Command Post.)

- General. The division may operate main and/or tactical command post CPs (TAC CP) depending on the mission and tactical situation. Characteristics of the CPs include—
  - Main CP–Normally located in the sustainment area. Personnel staffing is tailored to provide planning, coordinating, and C2 of assigned/attached/OPCON units. The area location for the main CP is selected by the battalion S3; the exact site is designated by the commander in coordination with the XO and battalion S3. The XO designates work areas within the main CP. The commander, HHD, serves as the headquarters commandant. His duties include coordinating for and obtaining construction, maintenance, and logistical services and support for the CP.
  - Tactical command post CP–Normally established in the objective area or at an intermediate staging base (ISB). The ISB is the forward designated staging base at which final preparation for assault operations are controlled and is generally a safe area for support operations. The TAC CP is tailored to extend the commander's span of control and to better assess the situation on the ground and coordinate the arrival and subsequent activities of AHS organizations within the AO.
- Battalion tactical operations center.
  - Definition–The TOC is the command element of the battalion containing communications and personnel required to command, control, and coordinate HSS operations. The TOC is located within a secure, controlled area whether at a main CP or TAC CP.
  - Purpose–The purpose of the TOC is to provide the commander, in a secure environment, current evaluated information and recommendations concerning HSS/FHP operations.
  - Responsibilities–The battalion commander has overall supervision and control over the TOC. The battalion XO operates the TOC and has primary staff responsibility in the absence of the commander.
  - Operations–The TOC will operate on a 24-hour basis. It is principally staffed by each primary staff section furnishing necessary manpower as required. Secure and nonsecure telephone communications connect the TOC to other staff sections within the CP area. Access to the TOC is strictly controlled by means of an access roster, and if available, security badges. Only essential personnel and authorized visitors are allowed to enter. Each staff section will maintain an SOP on the organization and operation of its element. All elements within the TOC will, when appropriate, maintain a current situational map of their specific operations. Discussion and portrayal of tactical plans outside of the security area is prohibited.

  - Composition of the TOC–This is a listing of those personnel comprising the TOC. It normally includes the commander, XO, command sergeant major, principal staff members, and other specific staff members such as the S3 (air) or the battalion aviation maintenance officer.
  - Tactical operations center configuration–This is a schematic representation of the physical layout of the TOC.
- Responsibilities for the main CP services–The HHD commander's main mission is to support the headquarters, medical evacuation battalion. He has the personnel and resources assigned to facilitate this mission. He plans and coordinates for providing shower, laundry, transportation, maintenance, and other required services. He is also responsible for establishing and maintaining security; supplying fuel and other POL products; establishing the support area for the main CP (orderly room, supply room, motor pool, billets, and dining facilities); providing and operating generators to meet electrical requirements.
- Camouflage–This appendix discusses what camouflage procedures are required, to include type and amount of required camouflage materials (such as nets and shrubs); display of the Geneva Conventions distinctive emblem on facilities, vehicles, and aircraft on the ground (STANAG 2931); and other pertinent policies, guidance, or procedures.
- Message center–This establishes procedures for the handling of classified messages; provides delivery and service of IMMEDIATE and FLASH messages to the appropriate staff section; establishes procedures for preparing outgoing messages; and establishes a delivery service to the servicing message center for transmission of outgoing messages.

C-14. Annex C. (Administration and Human Resources.) This annex outlines procedures relating to administrative and human resources matters and associated activities.

- Personnel accountability and strength reporting.
  - Personnel daily summary (PDS)–This provides the procedures for filling out and submitting a daily personnel status report. The instructions may include requirements for encrypting the report prior to transmission, and specific guidance on time of submission, corrections, or other administrative requirements.
- Personnel readiness management.
  - Maintain the personnel readiness of unit personnel to include projected gains and losses.
- Personnel information management.
  - Maintain 100 percent accountability of the unit's military personnel files.
  - Update records directly through the officer record brief/enlisted record brief.
  - Reception, replacement, RTD, R&R, and redeployment.
  - Track the flow and duty status of the five categories of Soldiers into and out of the unit.
  - Update deployed theater accountability software database as required.
  - Coordinate the movement of the five categories of Soldiers.
  - Replacements–Individual replacements will not be readily available during the initial phases of operations. The S1 will automatically initiate replacement requests for personnel who are reported on the PDS report as WIA, missing in action (MIA), or killed in action (KIA).
- Essential personnel services.
  - Issue identification cards and tags.
  - Personnel actions–All personnel actions will be channeled through the S1. Company XOs and 1SGs will be the company points of contact. Actions will be handled expeditiously and meet suspense dates (tactical situation permitting).
  - Efficiency reports–This paragraph provides pertinent information on the completion and submission of these reports.
  - Award recommendations–This paragraph delineates the responsibilities for and guidance concerning submitting recommendations for awards and for scheduling and conducting award ceremonies.
  - Promotions–This paragraph discusses the procedures for submitting recommendations for promotion and scheduling and conducting promotion ceremonies.

  - Correspondence–All correspondence addressed to higher headquarters will be submitted through the S1. Requirements for submission, preparation, and approval are also provided.
- Casualty operations.
  - Casualty Feeder Card–This report is submitted on DA Form 1156 (Casualty Feeder Card). Instructions on the completion of the form and submission requirements are included.
  - Tracking of casualties will require coordination with CLT and CSH.
- Postal operations.
  - This includes hours of operation and services available.
- Morale, welfare, and recreation operations.
  - Sporting activities and morale and welfare activities.
  - American Red Cross.
  - This includes hours of operation and services available.
  - Post Exchange services–This includes hours of operation and availability.
- Management, and HR planning and staff operations.
  - Preparation of the personnel estimate and personnel portion of plans or OPORD.
  - Coordinate external HR support with HR operations cell within the sustainment brigade.
- Other administrative support.
  - Financial management–This service includes disbursements and currency control, payday activities, currency conversion, check cashing, and the appointment of Class A agents.
  - Legal services–Information and specific guidance on administrative boards, courts-martial authority and jurisdiction, legal assistance, and general services should be provided.
  - Religious activities–Religious activities include chaplain support, services available from different faiths, schedule of services, and hospital visitations.
  - Distribution–Pick up and delivery schedules and any command-specific issues/procedures are provided.
- Graves registration (GRREG)–Commanders at all levels are responsible for the recovery, identification, and evacuation of US dead. This section discusses the responsibilities and procedures for unit-level GRREG activities for assigned and attached personnel.

Note. This activity is for unit members only. Medical evacuation units do not accept nor transport the remains of nonunit members. Remains will not be transported in ambulances under any circumstances.

  - Responsibilities–This paragraph discusses both unit and battalion requirements.
  - Disposition–Specific guidance on procedures, GRREG collection points, transportation requirements, and the handling of remains are provided.
  - Hasty burials–Specific requirements for conducting hasty burials, marking, and reporting of grave sites are included.
  - Personal effects–Guidance on accounting for personal effects and requirements for burial should a hasty burial be required is contained in this paragraph.
  - Disposition of civilian and EPW remains–The local civilian government is responsible for the burial of remains of its citizens. The remains of EPW should be accomplished in separate cemeteries from US and allied personnel. If this is not possible, separate sections of the same cemetery should be used.
  - Contaminated remains–This paragraph discusses handling and disposition requirements (to include protective clothing), procedures, and marking and reporting of burial site.
- Public information–This annex contains procedures for obtaining approval on the public release of information to include the hometown news release program. In overseas locations, specific guidance on the interaction of unit members with the news media should be provided.

- Maintenance of law, order, and discipline–This appendix should provide applicable regulations, policy, and command guidance on topics such as serious incident reports, notifications, and submission formats, straggler control, confinement of military prisoners, and EPW and detainees (also discussed below).
- Enemy prisoners of war–This appendix discusses the unit responsibilities of EPW and detainees surrendered to the medical unit. These procedures do not pertain to EPW and detainee patients captured by other units. Medical personnel do not guard, search, or interrogate EPW and detainees while in the AHS; guards are provided by nonmedical personnel designated by the tactical commander for these duties. Until EPW and detainee personnel can be evacuated to an EPW collection point, medical personnel should remember and enforce the basic skills: segregate, safeguard, silence, secure, speed, and tagging. (The speed portion of evacuating EPW and detainees to designated collection points is of paramount importance to medical units.)

Note. The treatment of EPW and detainees is governed by international and US law and the provisions of the Geneva Conventions. Personnel should be aware of these requirements and have ready access to the applicable regulations and policy guidance.

- Allied, coalition, and interagency personnel–This paragraph should provide guidance on the eligibility of allied, coalition, and interagency personnel for evacuation by Army ambulance and destination locations.

C-15. Annex D. (Intelligence and Security.) This annex pertains to intelligence requirements and procedures and OPSEC considerations.

- Intelligence–The S2 has the responsibility of collecting information to assist the commander in reaching logical decisions as to the best COA to pursue. Priority intelligence requirement (PIR) may include the location, type, and strength of the air defense threat; location, type, and strength of enemy air defense radars; known or suspected CBRN activity; the health threat; and issues which the commander considers to be PIR. In addition to PIR, the commander's critical information requirements (CCIR) are also considered.
- Intelligence reports–The S2 is responsible for disseminating all applicable estimates, analyses, periodic intelligence reports, and intelligence summaries generated within the battalion or received from higher headquarters. Information on submission of reports and suspense on intelligence products and reports should also be addressed in this appendix.
- Weather data–Weather conditions can disrupt ground evacuation efforts and planners must recognize their impact when coordinating AA operations.
- Reports–These include information acquired during the routine performance of duty by pilots, ambulance drivers, and medics.
- Counterintelligence.
  - Camouflage–When ordered or directed by the tactical commander, all units will initiate and continually strive to improve camouflage operations of positions, vehicles, and equipment. Noise and light discipline is emphasized at all times.
  - Communications security–These measures will be enforced at all times. The specific requirements and considerations are included.
  - Signs and countersigns–This paragraph lists the signs and countersigns to be used during hours of darkness. It also includes reporting requirements and procedures if the sign/countersign is lost or compromised.
  - Signal operating instructions.
  - Document security–This paragraph discusses the procedures for marking and safeguarding classified material, both work documents and completed documents. (Reporting requirements in the event of compromise are also included.)
- Captured or surrendered personnel, equipment, supplies, and documents–This appendix provides specific guidance on the handling of captured personnel, equipment, supplies, and

documents. The disposition of captured medical equipment and supplies is governed by the Geneva Conventions and is protected against intentional destruction.

- Security–This appendix discusses weapons security and checks, safeguarding of patient weapons, aircraft security, signal operating instructions (communications) security, TOC security, Sensitive Item Status Report, and escape and evasion.

C-16. Annex E. (Operations.) This annex establishes procedures for S3 operations within the medical evacuation battalion, and provides a basis for standardization of medical evacuation operations in a tactical environment. It is essential that these procedures be standardized to ensure common understanding, facilitate control and responsiveness, and enhance mission accomplishment. Information on readiness levels, threat levels, warning levels, camouflage, security, area damage control, operations, and leader checklists is also included.

- Operational situation report–Requirements for preparation, format, and submission of this report are discussed in this appendix.
- Operations security–This appendix provides the guidance and procedures for secure planning and conduct of combat operations.
  - Priority intelligence requirements and security classification.
  - Responsibilities–The commander is ultimately responsible for denying information to the enemy. The S3 is responsible to the commander for the overall planning and execution of operations. The S2 has the principle staff interest in assuming the required degree of OPSEC and has the primary staff responsibility for coordinating the efforts of all other staff elements in this regards. The OPSEC officer is responsible for the preparation of the PIR and providing classification guidance. Additionally, the OPSEC officer identifies the priorities for OPSEC analysis and develops OPSEC countermeasures. The S2 coordinates with the S3 in planning OPSEC.
  - Hostile intelligence threat–The different sources of intelligence (human intelligence, signal intelligence, and so forth) are discussed.
  - Operational security program–This includes physical security, information security, and signal security.
  - Document downgrading/declassification and classification authority.
- Terrorist threat and countermeasures–This appendix provides guidance on security measures to counter the terrorist threat, both on an individual and unit level.
- Minefield threat–This appendix provides guidance on the potential of minefields being established in the AO and delineates reporting requirements. (Specific minefield extraction techniques are contained in Annex L of this TSOP.)
- Operations security and countermeasures–This appendix discusses camouflage, light discipline, noise discipline, physical security, information security, and signal security.
- Unit location update–his appendix provides timely information on the location of main and forward CPs, location of subordinate unit CPs, location of helipads, and location of POL points.
- Flight operations–This appendix provides information concerning Army aviation LZs throughout the corps area and division areas; required reports; airdrop information; and mission debriefing.
- Communications-electronics–This appendix establishes communications policies, procedures, and responsibilities for the installation, operation, and maintenance of CE equipment.
  - Responsibilities of the battalion CE noncommissioned officer in charge.
  - Concept of operations.
  - Command and control.
  - Radio communications.
  - Radio teletypewriter communications.
  - Message/communications center service.
  - Message handling procedures.
  - Wire communications.

- Switchboard operations.
- Communications security and operations.
- Intelligence security.
- Meaconing, intrusion, jamming, and interference (MIJI) reporting and electronic communications countermeasures.
- Security violations–This prescribes procedures for reporting any event of action which may have jeopardized the security of communications.
- Destruction of material.
- Daily shift inventory.
- Physical security.
- Transmission security.
- Security areas–This discusses access procedures and rosters, access approval requirements, and prohibited items.
- Inventory of classified materials.
- Communications security officers and custodians–The appointment procedures, orders requirements, and duties of personnel are described.
- Safety–This discusses requirements for grounding, handling, and storing COMSEC equipment.
- Power units.

- Rear battle responsibilities–This appendix discusses rear battle responsibilities, task organization, and support for reaction forces.

C-17. Annex F. (Chemical, Biological, Radiological, and Nuclear Operations.) This annex prescribes the policy, guidance, and procedures for CBRN defensive operations.

- Responsibilities.
- Chemical, biological, radiological, and nuclear reporting requirements and procedures.
  - Contamination avoidance.
  - Protection–Protection pertains to those measures each Soldier must take before, during, and after a CBRN attack to survive and continue the mission.
  - Decontamination–This discusses equipment requirements, procedures, and types of decontamination (such as hasty). This paragraph also provides guidance on deploying clean ambulances into a contaminated area.
  - Mission-oriented protective posture–This appendix provides guidance on the garments required for the different MOPP levels and identification procedures for personnel in MOPP.
  - Radiation exposure guidance–Establishes OEG. Discussion in this appendix includes determining what constitutes a radiological hazard, prescribing acceptable limits of potential casualty producing doses of radiation, minimizing exposure, and protecting against electromagnetic pulses.
  - Masking and unmasking procedures.
  - Radiological monitoring and survey operations.

C-18. Annex G. (Logistics.) This annex establishes logistics procedures for subordinate units when operating in a field environment.

- Nonmedical supply and services–A discussion of applicability, responsibilities, policy, classes of supply, requisition and delivery procedures, hours of operation, and other supply-relevant topics and available services (such as laundry and bath) can be addressed in this section.
- Medical logistics–This paragraph discusses supply and resupply procedures for Class VIII; policies and procedures for the backhaul of emergency Class VIII on medical vehicles; and hours of operation for the supporting medical logistics facility.
- Food service–This appendix discusses responsibilities, hours of operation, Class I supplies, sanitation requirements, layout of field kitchen, fuel storage, maintenance, safety precautions,

and administration, such as headcounts, required reports, shift schedules, meals, ready-to-eat (MRE), and inspections/visits of subordinate unit kitchens.

- Transportation/movement requirements–This appendix may cover the following areas: deployability data, applicability; responsibilities; policies on speed, vehicle markings, transporting flammable materials, transporting ammunition and weapons, and so on; convoy procedures; safety; and accident reporting.
- Fire prevention and protection–Guidance on the use of the tent stoves, flammable materials, use of cigarettes, matches, and lighters, electrical wiring and appliances, safety of tents and occupants, spacing of tents, stoves and ranges, installation of British thermal unit (BTU) heaters, and fire fighting equipment are presented in this appendix.
- Field hygiene and sanitation–This appendix provides uniform guidance and procedures for the performance of functions related to field hygiene and sanitation. It includes policies, communicable disease control, field water supply, water containers and cans, water purification bags, food sanitation, latrines, liquid waste disposal, and garbage and rubbish disposal.
- Conventional ammunition down/upload procedures–This appendix delineates responsibilities and provides guidance and procedures for the requisition, storage, and distribution of ammunition and weapons, reporting requirements, and safety.
- Petroleum, oils, and lubricants accounting.
- Medical logistics support. The medical logistics concept of operations, requisition and distribution procedures, accountability, and reports are provided in this appendix.
- Maintenance–This appendix includes information on the maintenance requirements of the battalion and the location and hours of operation of maintenance units and collection points. Maintenance for medical equipment, vehicles, aircraft, and communications and other categories of equipment are discussed.

C-19. Annex H. (Safety.) This annex establishes minimum essential safety guidance for commanders and units. It includes risk assessment, accident reporting, safety measures, emergency procedures, vehicle safety, ground guide procedures, fire prevention and protection, antennas, climate, survival training, animal and arthropods hazards, personal protective measures, hearing conservation, carbon monoxide poison, helicopter safety, and refueling operations.

C-20. Annex I. (Civil-Military Operations.) This annex discusses the participation in CMO. Medical elements are often involved in CMO, humanitarian and civic assistance, and disaster relief operations. The activities which may be covered include providing—

- Direct support medical evacuation for the operation.
- Guidance on developing a medical evacuation system in a HN.
- Training to a HN's medical infrastructure.

C-21. Annex J. (Mass Casualty Situations.) This annex discusses the procedures for providing medical evacuation support to MASCAL situations, to include coordination for nonmedical transportation assets and the augmentation of these assets with medical personnel to provide en route patient care. It discusses evacuation capabilities of US forces.

C-22. Annex K. (Minefield Operations.) This annex discusses notification procedures, training requirements, and techniques for clearing pathways to casualties, marking mines, and extracting personnel and equipment.

C-23. Annex L. (Religious Support.) This annex discusses procedures for processing unit requests for religious support and coordination of religious support to casualties, declared MASCALs, mortuary affairs support, unit memorial ceremonies/services, EPW, detainees, religious support logistics and supplies, and reports and administration.

Appendix D

Example of the Medical Evacuation Plan and of the Operations Order

SECTION I — EXAMPLE FORMAT OF THE MEDICAL EVACUATION PLAN

D-1. Figure D-1 provides an example of the medical evacuation plan.

(Classification)

Headquarters
Place
Date, time, and zone

Time Zone Used Throughout the OPLAN: *The time zone used throughout the OPLAN (including attachments) is the time zone applicable to the operation. Operations across several time zones use ZULU time. Place the classification and short title of the OPLAN at the top of the second and any subsequent pages of the base OPLAN.*

Task Organization: *List the number and coordinates of medical evacuation units here or in a trace or overlay. If you do not list units here, omit this heading.*

1. SITUATION. *(State the general medical evacuation factors affecting support of the operation. Include any information essential to understanding the current situation as it influences medical evacuation.)*

a. Enemy forces. *(Refer to the intelligence annex if it has been published or is to be published. List information about the composition, disposition, location, movements, estimated strengths, and identification of enemy forces. List enemy capabilities that could influence the medical evacuation mission.)*

(1) Strength and disposition. *(The size of the opposition force and its placement on the battlefield is important during the planning process for medical evacuation operations. When evacuation routes are selected, caution must be exercised to ensure medical evacuation assets are not compromised by going through enemy-held territory or by being ambushed by isolated pockets of resistance. Further, this information is vital in determining if medical evacuation assets [both ground and air] will require a security escort provided by movement and maneuvers forces before entering areas of the battlefield.)*

(2) Combat efficiency. *(Information on actual combat units or other forces, their training status, and their level of expertise and experience can be identified here. The level of AHS training and the development of a health care delivery system can also be included.)*

(3) Capabilities. *(This paragraph should discuss the potential capability to interfere with or disrupt medical evacuation operations.)*

(4) Logistics situation. *(This can include information on how well supplied the enemy/opposition force is with food, clothing, or other vital logistics factors. It may also include the financial backing and availability of future support from outside individuals/groups/nations.)*

(5) Weapons. *(This should include a discussion of the enemy's weapons that present the greatest threat to air and ground evacuation personnel and vehicles. Ground assets are more likely to face a small arms threat, rocket propelled grenades (RPGs), and improvised explosive devices (IEDs) while performing their mission; air assets are vulnerable to surface-to-air weapons and when on the ground to the same threats as ground elements.)*

b. Friendly forces. *(List pertinent information concerning friendly forces that might influence the medical evacuation mission.)*

(1) Strength and disposition. *(This should include all forces [US, allied, coalition, and HN] and should be maintained on overlays. It may also include liaison officers, interpreter support, and coordination requirements.)*

(2) Combat efficiency. *(The health of the command has a significant impact on this factor. Additionally, when there are significant numbers of DNBI casualties, the medical evacuation workload will increase.)*

(3) Present and projected operations. *(Medical evacuation planners must be familiar with plans for current and projected operations. A risk management assessment of current and projected operations is conducted. Casualty collecting points and AXPs must be designated during the planning process. Medical evacuation planners must also be able to anticipate changing requirements to ensure continuous medical evacuation is available to the supported force.)*

(4) Logistics situation. *(Since medical evacuation vehicles conduct emergency resupply of forward deployed medical units, the medical evacuation planner must maintain visibility of the current medical logistics situation.)*

(5) Sustainment area protection plan. *(Medical units play an important role in the sustainment area protection plan and should be included in the planning process.)*

(6) Weapons. *(Medical units only have small arms for self-defense and defense of their patients. Medical vehicles and aircraft should not carry any automatic or crew-served weapons; to do so would result in the loss of the protections afforded under the Geneva Conventions for the unit, medical personnel, and patients under its care.)*

(7) Health of the command.

(a) Acclimation of troops. *(Medical evacuation crews require acclimation when introduced into an AO where the temperature range and elevation are different from their home station. Medical evacuation operations involve heavy lifting and may require extended periods of time to complete. To reduce the risk of injury to litters bearers and to facilitate the evacuation effort, medical and nonmedical personnel engaged in these operations should be acclimated to the AO and a work/rest schedule should be developed and implemented. Air ambulance crew rest requirements impact aircraft availability and must also be considered.)*

(b) Presence of disease. *(The presence of disease impacts medical evacuation operations in two areas. The presence of disease in the AO [usually at subclinical levels in the native population] contributes to the incidence of disease manifesting itself within the supported force. As DNBI rates increase, so do the requirements for medical evacuation. Medical evacuation personnel are also susceptible to the endemic and epidemic diseases within the AO and/or multinational force. High rates of DNBI for medical personnel will adversely impact the medical evacuation capability.)*

(c) Status of immunizations and/or chemoprophylaxis. *(Commanders must ensure appropriate measures are taken to protect their Soldiers from DNBI. The records of replacement personnel need to be screened to ensure all required immunizations have been received and appropriate chemoprophylaxis/barrier creams for the AO initiated/provided.)*

(d) Status of nutrition. *(The nutrition status of the troops involved impacts on the susceptibility to disease and environmental injuries, morale, and fatigue. The medical evacuation commander must also ensure that his evacuation crews carry sufficient supplies of MREs to sustain themselves and their patients should delays in evacuation be experienced [such as delays resulting from shifts in the battle or inclement weather].)*

(e) Clothing and equipment. *(Special requirements for clothing and equipment to support a particular operation should be obtained prior to the beginning of the operation [for example, cold weather clothing for extreme cold weather operations; additional blankets for patients being transported (either to add padding or warmth); or block and tackle equipment for medical evacuation operations conducted in mountainous terrain or in urban areas].)*

(f) Fatigue. *(Mandatory work/rest schedules and sleep plans for air crews and ground evacuation personnel must be developed and implemented.)*

(g) Morale.

(h) Status of training. *(This can include any specialized training required for the conduct of a specific operation [for example: helicopter crews require deck-landing qualifications to perform shore-to-ship evacuation with US Navy ships or orientation to the social, political, economic, religious, and ethnic issues of a HN or supported population, or training nonmedical personnel in the proper techniques for carrying litters].)*

c. Environment.

(1) Terrain. *(This paragraph should discuss any aspects of the terrain that will either hinder or enhance the execution of the evacuation mission. It should discuss both natural and man-made terrain, as medical evacuation urban areas can pose significant challenges not found on a natural battlefield. Medical personnel must be able to recover injured and wounded Soldiers from below ground and from upper levels of man-made structures. If UO are planned, armor ambulances provide added protection for medical personnel and their patients; augmentation should be requested if armor ambulances are not organic to the unit. If additional resources will be required to accomplish evacuation due to impassable or difficult terrain [such as in mountain operations where additional litter bearers and medical personnel are required], it should also be addressed here. Further, the type of terrain to be traversed, such as rugged mountain or jungle swamps may require a patient [who would otherwise be ambulatory] to be transported on a litter until easier terrain is encountered.)*

(2) Weather. *(This should include a discussion of current weather conditions and seasonal variants. Weather conditions impact both ground and air evacuation operations; however, the most significant impact may be on AAs as severely inclement weather can ground all aircraft. It should also discuss the impact that the weather has on the terrain [such as rivers being frozen in winter or tundra becoming impassable in spring]. The climate may pose problems with acclimation as well as place additional requirements to sustain personnel during evacuation on evacuation assets [litter evacuation in the mountains in extreme cold weather operations may require warming tents and the capability to sustain the patient during nighttime when evacuation is difficult].)*

d. Civil considerations. Dislocated civilian population, EPW, and detainees. *(Dislocated civilians fleeing an area of hostilities can clog roadways leading away from the conflict area. This congestion on road networks may make evacuation along these routes almost impossible. The establishment of camps to sustain these categories of personnel may interrupt the road network requiring detours and lengthened evacuation times. Injured, ill, or wounded EPW or detainees are evacuated using the same evacuation means but are segregated from US, allied, or coalition patients. Coordination for nonmedical guards for EPW/detainee* *patients being evacuated through medical channels must be accomplished with the echelon commander. A determination of eligible beneficiaries for evacuation should also be developed and disseminated.)*

2. **MISSION.** *(Statement of the medical evacuation mission.)*

3. **EXECUTION.**

a. **Concept of operations.** *(The concept of operations describes how the commander sees the actions of subordinate units fitting together to accomplish the mission. Commanders ensure that their concept of operations is consistent with their intent and that of the next two higher headquarters. The concept of operations describes any other details the commander considers appropriate to clarify the concept of operations and ensure unity of effort. If the integration and coordination are too lengthy for this paragraph, they are addressed in the appropriate annexes. When an operation involves two or more clearly distinct and separate phases, the concept of operations may be prepared in subparagraphs describing each phase.)*

b. **Coordinating instructions.** *(List only instructions applicable to two or more units and not covered in unit SOPs.)*

c. **Patient estimates.** *(Indicate rates and numbers by type unit/division/corps/EAC.)*

(1) **Number of patients anticipated.** *(The anticipated number of patients affects the number and type of medical evacuation resources required.)*

(2) **Distribution within the area of operations.** *(Distribution within the AO is an important consideration because varying types of terrain will generate different requirements for support. Operations conducted in mountainous terrain rely heavily on manual and litter evacuation techniques, are labor intensive, and require more time to complete. Differences in road composition, traffic density, and abundance/absence of paved roads will impact differently on the type of support required. Further, if the distance between Role 2 and Role 3 MTFs exceeds the capability of US Army AAs, the use of the USAF aircraft should be anticipated and preplanned. Coordination with the USAF is required.)*

(3) **Distribution in time during the operation (evacuation time).** *(As the battle progresses over time and space, medical evacuation assets may be reallocated to support the units in contact. The medical evacuation plan must be sufficiently flexible to enable the commander to shift resources as required.)*

(4) **Areas of patient density.** *(Areas with a high patient density may require augmentation of the medical evacuation assets supporting that location.)*

(5) **Possible mass casualties.** *(MASCAL situations should be anticipated when possible. Evaluation of the general threat and health threat in the AO can facilitate the medical evacuation planner in forecasting evacuation requirements. When dealing with MASCAL situations, the use of nonmedical transportation platforms should be included. Augmentation of these vehicles by medical personnel to provide en route care or CLS to provide first aid will assist in reducing the deterioration of the patient's medical condition until arrival at an MTF.)*

(6) **Lines of patient drift.** *(This indicates what routes injured Soldiers are most likely to take from the battlefield. It is usually the most direct route over the least demanding terrain, such as at the base of the hill rather than over the hill where climbing would be required.)*

(7) **Evacuation routes/corridors.** *(Evacuation routes should be preplanned, indicated on the medical evacuation overlay, and reconnaissance accomplished. Routes that provide lucrative targets, have significant obstacles to circumvent, or will be unduly congested due to fleeing refugees or other displaced persons should only be used if no other routes are available. Air corridors are provided by the supporting A2C2 element. Due the placement of hospitalization resources within the JOA, AA may be required to operate out of fixed sectors and coordination for air escorts must be coordinated.)*

(8) **Support requirements.** *(This paragraph discusses the type of direct support, general support, or operational control required for the operation and its command relationship. Specific guidance on trigger points and release points from a particular category of support should be included.)*

(9) **Medical evacuation procedures.**

(a) **Evacuation overlays.** *(Evacuation overlays must be developed to facilitate the evacuation effort. Both supporting and supported units must maintain and update [as required] overlays throughout the operation.)*

(b) **Communications.** *(Medical evacuation frequencies must be designated at the outset of the operation. Requirements for liaison officers, A2C2, and interpreters must be anticipated.)*

(c) **Casualty collecting points.** *(Location, staffing, and activation trigger [such as crossing a phase line] must be known to supported and supporting units.)*

(d) **Ambulance exchange points.** *(Location, staffing, and activation trigger [such as crossing a phase line] must be known to supported and supporting units).*

Note: If an AXP is staffed with a treatment element, it is no longer considered to be an AXP but rather it becomes the forward element of a BAS or Role 2 MTF. The AXP is only a location on the ground where the patient is transferred from one evacuation platform to another. If it is staffed, it is usually staffed by one member of the evacuation platoon to direct returning ambulances to the area where the last ambulance departed from. The element designating the AXP is responsible for staffing it. Ambulance exchange points may be rendezvous points to be used only once during the operation.

Note: Ambulance exchange points should not be used as resupply points for Class VIII. (Planning on pushing Class VIII to these points may result in the loss of the Class VIII materiel, as there may not be anyone at the point to receive the supplies.)

(e) **Ambulance shuttle system.** *(The ambulance [or litter] shuttle system is a management tool to facilitate the medical evacuation of forward areas.)*

(10) **Casualty evacuation (transportation).**

(a) **Medical augmentation.** *(Units and/or facilities capable of providing medical augmentation support should be identified and tasked as appropriate.)*

(b) **Manual evacuation.**

(c) **Litter evacuation.** *(A trauma specialist or CLS should be included as a member of the litter team when there is a significant distance to be covered.)*

(d) **Pack animals.** *(In some remote locations, the use of pack animals may be the most feasible form of transportation from the POI to a place where vehicles, aircraft, or watercraft become accessible. Whenever possible, medical personnel should accompany patients evacuated in this manner.)*

(e) **Nonmedical vehicle.** *(Nonmedical vehicles which could be preplanned for use for casualty transport should be identified. Further, guidance on the use of vehicles of opportunity should also be discussed. When possible nonmedical vehicles used for these purposes should be augmented with medical personnel or CLS.)*

(f) **Nonmedical aircraft.** *(Nonmedical aircraft which could be preplanned for use for casualty transport should be identified. Guidance on the use of aircraft of opportunity should also be discussed. When possible nonmedical vehicles used for these purposes should be augmented with medical personnel or CLS.)*

(g) **Nonmedical watercraft.** *(Nonmedical boats, ships, or other watercraft which could be preplanned for use for casualty transport should be identified. Further, guidance on the use of these platforms of opportunity should also be discussed. When possible nonmedical vehicles used for these purposes should be augmented with medical personnel or CLS.)*

(11) **Resources available.**

(a) **Mode of transportation.**

(b) **Personnel.** *(This includes a discussion on personnel strengths and the impact shortages have on the medical evacuation operations.)*

(c) **Supplies and equipment.** *(This includes both medical and nonmedical supplies and equipment. Litter and blanket exchanges should be covered here. If PMI are an issue, they should be discussed here.)*

(12) **Use of smoke and obscurants.** *(In most cases, medical units will not have a high priority for the use of smoke and obscurants; however, if the supported units have planned for its use, medical evacuation units can take advantage of the situation to clear the battlefield of casualties. Guidance on the use of colored smoke to identify the pickup location for patients can also be included.)*

d. **Warfighting functions.**

(1) **Medical planners must take the following warfighting functions into consideration.**

(a) **Intelligence.** (Known or suspected enemy situation. Terrain and weather. What kind of road network is in the AO? How will the predicted weather impact the traffic capability on and off the roads? Are there LZs along the axis of advance and in or around the urban area? How will the weather impact on the availability of aircraft?)

(b) **Movement and maneuver.** *(What is the combat scheme of maneuver and are there medical assets readily available to care for potential casualties? Particular attention should be paid to the employment of the armored ambulances.)*

(c) **Fire support.** *(Are there any buildings specifically targeted? If so, medical planners should ensure CCPs, evacuation routes, and treatment elements remain at a safe distance away from these buildings to prevent potential fratricide incidents.)*

(d) **Protection.** *(Preserving the force includes protecting personnel (combatant and noncombatant), physical assets, and information of the US and multinational partners.)*

(e) **Sustainment.** *(Typically, the BCT moves a forward logistics element toward the battle area to provide support. If this is the case, then how are medical assets integrated into the plan, and, more importantly, where are they positioned in the convoy and how are they secured?)*

(f) **Command and control.** *(The senior military on-site or the senior medical person at the scene, if available, assigns the evacuation precedence and determines if the patient should be evacuated by ground or air resources. What are the primary and secondary frequencies? This element is the central point for prioritizing assets and medical evacuation requests. It is imperative that this element is capable of continuous situational awareness through accurate battle tracking.) From information provided by the S2, medical planners must understand the tactics of the enemy and be prepared for casualties at all possible areas of contact. The medical planners should be able to gain insight into this matter during the war gaming process. The medical planners should provide input into the counterreaction portion of the action, reaction, counterreaction drill.)*

(2) **Additional planning information.**

(a) *Availability and numbers of ground and AA assets, evacuation routes and flight windows.*

(b) *Nonstandard ground and aircraft availability (for MASCAL situations) and how to trigger their involvement.*

(c) *Additional medical assets from higher levels such as treatment teams, ambulances, and combat configured loads of Class VIII.*

(d) *Existing and proposed locations of all available MTFs.*

(e) *Class VIII resupply plan.*

(f) *The need for litter bearer teams and special equipment for extracting casualties from buildings and rubble.*

e. **Casualty evacuation synchronization.**

(1) *Medical evacuation and CASEVAC assets must be incorporated into the evacuation plan. Ideally there should be a single dedicated controlling authority. This prevents duplication of effort and confusion and enables the battlefield to be cleared faster and more efficiently.*

(2) *In MASCAL situations CASEVAC assets once put into operation should be dedicated versus designated (lifts of opportunity). Potential problems that arise arising from relying on designated versus dedicated CASEVAC include excessive time spent acquiring the information needed to transport casualties. Medical regulating to higher roles of care facilities may have to be accomplished by blocks of beds rather than on a by name basis. If a facility is designated for certain specialty care [such as head trauma] the evacuator must know which facilities have been identified and where they are located.).*

(3) *Another consideration is aircraft configuration. Casualty evacuation may be a seats-out operation. By dedicating aircraft and crews, they have time to prepare for the mission. Casualty evacuation crews must be briefed on the details of their mission including LZs, PZs, air corridors, communications architecture, and supported units.*

(4) *Any CASEVAC plan should clearly articulate the following information.*

(a) *Who is responsible (such as the company 1SG) for coordinating the company CASEVAC mission?*

(b) *Combat lifesavers role.*

(c) *Class VIII and blood resupply procedures.*

(d) *Personnel to act as litter bearers.*

(e) *Locations of CCPs, forward treatment teams, BAS, and AXPs.*

(f) *Location and designation of MTFs.*

(g) *Medical regulating officer coordination.*

(h) *Medical treatment facility and MRO contact frequencies.*

4. **SERVICE SUPPORT.**

a. **Supply.** *(Refer to TSOP or another annex whenever practical.)*

(1) **General supply.** *(Provide special instructions applicable to medical evacuation units.)*

(2) *Consider supply levels for all classes of supply, in the event of mission requirements in an austere environment and at extended distances from the full compliment of logistics and sustainment resources.*

b. **Medical logistics.** *(Provide special procedures applicable to the operations.)*

(1) **Requirements.** *(For sustaining the ground and AA operations and personnel.)*

(2) **Distribution.** *(This should include the method of distribution and any limitations or restrictions that are applicable. Additionally, if special transportation requirements exist, they should also be noted.)*

(3) **Medical logistic activities.** *(This includes the location of the medical supply activity supporting the AO and the means of communicating requests for resupply.)*

c. **Supplies required for ground and AA operations and personnel.** *(Includes estimates of the number of patients anticipated to be evacuated.)*

d. **Health service support.** *(Include information and instructions for supported units. Prescribe the plan for ground and air medical evacuation.)*

(1) **Evacuation.** *(List CCPs and AXPs. Establish ambulances shuttles, routes, means, and schedules of evacuation and en route treatment policies for the use of nonmedical transportation assets; policies for evacuation by ground and air; and evacuation of CBRN-contaminated patients. Include information about medical evacuation request procedures and channels, and holding policies.)*

(2) **Treatment.**

(a) **Treatment of supported forces.**

(b) **Policies.** *(Should address treatment and hospitalization policies.)*

(c) **Units.** *(This includes information concerning the location, capabilities, and communications means of units providing support.)*

(d) **Other.** *(This can be information on other medical assets that may be used in emergencies.)*

e. **Personnel.**

(1) Current subordinate unit manning levels/critical MOS shortages (consider all manning influences—Task Organization, R&R flow, Boots On Ground data, and so forth).

(2) New personnel requirements resulting from the operation (language skills, additional skill identifier [ASI], and so forth).

(3) Casualty estimates (as developed).

(4) Forecasted replacement availability.

(5) Evacuation policy for the operation.

(6) Supporting HR organizations (location, command/support relationship, controlling element).

(7) Supporting medical elements

(8) Replacement fill priority (Coordinate w/S3/G3).

(9) Crew/key leader replacement.

(10) Projected postal flow/limitations.

(11) Change to established PASR reporting flow/times.

(12) Change to casualty reporting flow (CLT change, reporting changes).

(13) RSOI Reception HR impacts.

(14) R&R schedule/flow operational impact (during operations conducted during sustained operations).

(15) Civilian and JIM manning impacts on the operation focusing on strengths and personnel service requirements.

(16) Specific Army G1 personnel planning guidance impacts on the operation.

(17) Status of other personnel service or personnel support factors (postal, MWR, legal support, military pay support).

f. Coordinating instructions. *(List instructions applicable to two or more units and not routinely covered in unit SOPs. This is always the last subparagraph in paragraph 4. Place complex instruction in an annex.)*

(1) Time or condition when a plan becomes effective. *(Include the time or the conditions under which the plan is to be placed in effect.)*

(2) Commander's critical information requirements. *(List only in coordinating instructions and not in annexes.)*

(3) Risk mitigation control measures. *(These are measures unique to this operation and not included in unit SOPs. They must include MOPP, OEG, troop safety, vehicle recognition signals, and fratricide prevention measures.)*

(4) Rules of engagement. *(If required.)*

(5) Boundaries. *(Address measures established to protect units. Specify which tactical units are to provide protection.)*

g. Special reports. *(List reports requiring special emphasis that is not addressed elsewhere.)*

5. COMMAND AND SIGNAL.

a. Command. *(State the map coordinates for the CP locations and at least one future location for each CP. Identify the chain of command if not addressed in unit SOPs.)*

b. Signal. *(Include the headquarters location and movements, liaison arrangements, recognition and identification instructions, and general rules concerning the use of communications and other equipment, if necessary. Use an annex when appropriate.)*

(Classification)

Figure D-1. Example of a medical evacuation plan

SECTION II — EXAMPLE FORMAT FOR AN OPERATIONS ORDER

D-2. Figure D-2 provides an example of the medical operations order.

(Classification)

Place the classification at the top and bottom of every page of the OPLAN.

(Change from verbal orders, if any)

The phrases "No change from verbal orders" or "No change from verbal orders except paragraph #" are required. This statement is only applicable if the commander issues a verbal order.

Copy ## of ## copies
Issuing headquarters
Place of issue

Show the name of the town or place in capital letters, coordinates in parentheses, and the country in capital letters. You may encode both.

Date-time group of signature

(The effective time for implementing the plan or order is the same as the date-time group of signature unless the coordinating instructions state otherwise. Use time zone Z unless the order states otherwise. When orders apply to units in different time zones, use Z time. When an OPORD does not specify the actual time to begin an operation, state that time in terms of an event.)

OPERATION ORDER 00-001, Operation Liberation
References: OPORD 00-000
Time zone: Zulu
Task Organization

Ground Ambulance Units participating in Operation Liberation
123 MED CO (GA) assigned to 1st MMB and provides emergency support to the 1st Marine Expeditionary Force (MEF)
234 MED CO (GA) assigned to 1st MMB
Ambulance Platoons assigned to:
C CO BSMC
C CO BSMC
C CO BSMC
C CO BSMC

Air Ambulance Units participating in Operation Liberation
C 2-1 MED CO (AA) CAB assigned to 1st Infantry Division
C 2-2 MED CO (AA) CAB assigned to 2nd Infantry Division
C 2-3 MED CO (AA) DS to the 1st MEF
C 2-4 MED CO (AA) CAB assigned to 1st Corps support

Air Force aeromedical evacuation assets
2d AF Combat Search and Rescue (CSAR), 1st AF Squad

1. **SITUATION.**
   a. **Enemy forces.** *As addressed in OPORD 00-000, the ground-to-air threat is significant in the all enemy sectors. The air-to-air threat is minimal, in all sectors. All ground planned evacuation routes will be targets for ambush by enemy forces and sympathetic host nationals.*
   b. **Friendly forces.** *A six (6) aircraft (HH-60) Air Force CSAR unit will be staged from the APOD. They will respond to downed aircrew evacuation recovery requests from the Air Force Combat Control Center.*
   c. **Attachments and detachments.** *None.*
   d. **Assumptions.** *None.*
2. **MISSION.** *On order provide tactical and operational ground and air intratheater medical evacuation or patient movement in support of Operation Liberation and continue coverage of all combined joint task force (CJTF) assets until relieved.*
3. **EXECUTION. Intent: State the commander's intent.**
   a. **Concept of support operations.** *This is a three phase operation.*
   Phase 1—Will be prior to initiation of the attack. *All US forces will be in the staging areas and in preattack positions. There will be no or very limited direct contact with enemy forces. C 2-1, C 2-2, and C 2-3 will not be required to*

provide evacuation support until D-3 days. Units in the staging area will be directed to call C 2-4 for evacuation at all times prior to D-3 days. All other evacuation units will concentrate on training and preparation for hostilities as directed by their commanders.

Phase 2—Will be the attack. *C 2-1, C 2-2, C 2-3 will provide DS to their assigned, GS to the corps sustainment area and the backhaul mission.*

Phase 3—Will be from the conclusion of the attack until stability operations commence. *Evacuation assets will begin transferring to an area support mission during this phase.*

Note. The concept of support operations takes into account both the tactical OPORD and the medical Annex to the OPORD. Information in both documents will shape the evacuation plan. Simultaneous planning should therefore occur.

---

b. Tasks to maneuver units: *Not applicable.*

c. Tasks to movement and maneuvers units:

(1) Evacuation responsibilities.

(a) Phase 1.

C 2-1, C 2-2, C 2-3 will provide direct support to their division assets from D-3 on. C 2-1, C 2-2, C 2-3 will respond to any request for evacuation that is geographically close to their positions even if not in their division area.

C 2-4 will provide support to the entire staging area (AO Buffalo). C 2-4 will perform the ship-to - shore mission as directed by the 11th MEDCOM TPMC

The 123 MED CO (GA) will operate in the 1st MEF sector in a direct support role. 1st MMB has will cover the 1st and 2nd Division footprint.

The 234th MED CO (GA) will provide support to Camps Jones, Smith, and Franks and perform patient transfers from camps to the APOD.

(b) Phase 2.

C 2-1, C 2-2, C 2-3 will provide direct support to their division assets. C 2-1, C 2-2, C 2-3 will be responsible for moving patients from within the division area back to the staging AO Buffalo until FLOT is beyond phase line alpha.

After FLOT is beyond phase line alpha (at which time one way flight time for the H-60 back to AO Buffalo will exceed 2 hours), C 2-1 will designate 1 FSMT to stage at XYZ Airfield to perform the backhaul and the area support mission until relieved. United States AF aeromedical transports will be coordinated for by 11th MEDCOM flying from XYZ airfield.

Beyond phase line Alpha C 2-2 will provide one FSMT to support the greater casualties expected in C 2-2 sector. Coordinate internally for link-up.

C 2-4 will provide support to the entire staging area (AO Buffalo). C 2-4 will perform the ship-to-shore mission as directed by the 11th MEDCOM MRO. C 2-4 will designate 4 aircraft to go forward as required to backhaul patients to AO Buffalo.

(c) Phase 3.

C 2-1, C 2-2, C 2-3 will provide direct support to their Divisions' assets.

The 1st MEF will identify areas of high casualty density conducive to ground evacuation (urban areas) and divide the resources of the 123rd MED CO (GA) to cover these areas efficiently, if required.

The 234th MED CO (GA) will conduct convoy support and provide coverage for AO Buffalo.

d. Evacuation mission guidelines.

(1) Phase 1, 2, and 3. *The senior military person advised by medical personnel on-scene will assign the evacuation precedence and determine the mode of evacuation (ground or air). Launch authority, a separate issue, based on risk will be according to the aviation SOP.*

(a) *Weather below acceptable minimums.*

(b) *No acceptable LZ in the immediate vicinity, except for hoist missions. Ground evacuation may be used for part of the evacuation.*

(c) *Chemical, biological, radiological and nuclear conditions at LZ would contaminate the aircraft. Ground evacuation may be used for part of the evacuation, followed by an air pickup after decontamination is complete.*

(d) *Patients are located at a LZ in direct enemy contact. Ground evacuation may be used for part of the evacuation, followed by an air pickup once clear of the enemy.*

(e) *Air evacuation assets are unavailable.*

(f) *Mission flight route requires an armed escort and one is not available.*

(2) The following are additional guidelines for evacuation:

(a) *Ground evacuation is the preferred method to clear casualties out of hostile urban areas. Air ambulance pick-up at a secured LZ outside of the urban area is preferred. For large urban areas, this may not be possible.*

(b) *If a mission requires the use of a hoist (such as thick forest or jungle) and the pick-up area has suspected enemy contact, ground evacuation is advised to a secure pick-up site.*

(c) *Prescheduled movement of ROUTINE or CONVENIENCE patients along secure routes of evacuation should be conducted using ground evacuation if the distance of evacuation is short.*

(3) Coordinating instructions:

(a) Medical regulating. *The 11th MEDCOM will oversee medical regulating during the operation.*

(b) Tentative locations of MTFs.

Phase 1

- *2nd CSH AB 1234 5678*

- *1st CSH AB 2345 6789*

Phase 2
- *2nd CSH AB 1234 5678*
- *1st CSH AB 2345 6789*

Phase 3
- *2nd CSH AB 1234 5678*
- *1st CSH AB 2345 6789*

(c) Reporting. *Evacuation units, through division G3 if assigned to a division, will submit a daily report of evacuation missions by 1200 the following day.*

(d) *If a MASCAL develops which overwhelms the resources of a single AA CO, contact GSAB for assistance.*

(e) Casualty collection. *Military treatment facilities can call 9 line missions to any medical evacuation asset for URGENT and PRIORITY missions however MTFs that need to clear beds will coordinate mass patient movements through CJTF G3 air for possible USAF support.*
- *1st CSH will call C 2-4 for evacuation requests during Phase 1.*
- *2nd CSH will call C 2-4 for evacuation requests during Phase 1.*
- *1st CSH will call C 2-1 for evacuation requests during Phase 2/3.*
- *2nd CSH will call C 2-2 for evacuation requests during Phase 2/3.*

(f) Ground evacuation. *Seasonal monsoons are expected to washout many of the main planned MSRs and cause flooding.*

(g) Air evacuation. *If air evacuation routes become so long as to seriously degrade the capabilities of an AA company, contact CJTF G3 Air for additional support and possible reallocation of evacuation assets. An AA CO is considered to be operating beyond its capability or inefficiently if:*
- *More than 50 per cent of missions require a one way trip time exceeding 2 hours.*
- *On more than 50 per cent of missions flown, routes of evacuation are flown through another AA company's area of direct support coverage.*

(h) Medical considerations for evacuation.
- US Army evacuation assets will be used to evacuate all injured personnel on the battlefield without prejudice.
- Human immunodeficiency virus is at epidemic proportions in Simula population. Take appropriate precautions during treatment and when dealing with biologicals.
- A significant number EPW and displaced civilians and detainees are predicted to be evacuated. Estimated to be a much as 75 per cent of total patient load.

(i) United States Air Force aeromedical evacuation.
- Combined joint task force G3 Air will coordinate for USAF intratheater aeromedical missions.

4. SERVICE SUPPORT.

a. Casualty estimates. Reference the Medical Annex for specific casualty estimates.

b. Aviation maintenance support. C 2-3 will receive AVUM/AVIM support from the 2-4 CAB.

c. Divisions G3 will report the following CCIR to CJTF G3 Air. Use the CJTF CCIR format.
- Any MASCAL.
- Any evacuation company with degraded capability for—
- Evacuation distance(defined above)
- Maintenance: below 70 percent operational readiness rate
- Personnel: number of crews less than available aircraft
- Any aircraft Class A accidents/events or shoot downs.

5. COMMAND AND SIGNAL.

a. Fragmentary orders (FRAGOs) to this order will be published to adjust the corps evacuation plan as necessary.

b. 11th MEDCOM MIRC Chat net name, Phone xxx-xxxx
- 1st Division Surgeon (G3), MIRC Chat net name
- 2-1 CAB MEDEVAC Cell, MIRC Chat net name
- 2nd Division Surgeon (G3), MIRC Chat net name
- 2-2 CAB MEDEVAC Cell, MIRC Chat net name
- 1st MEF, MEDEVAC Cell, MIRC Chat net name
- 1st Corps CAB MEDEVAC cell, MIRC Chat net name, phone xxx-xxxx

ACKNOWLEDGE:
PATRICK, LTG

Figure D-2. Example of a medical operations order

Appendix E

Use of Smoke and Obscurants in Medical Evacuation Operations

This appendix provides guidance for AMEDD personnel in the use of smoke and obscurants in medical evacuation operations.

TACTICAL COMMANDER

E-1. Throughout the battlefield, forces acquire and engage targets based on visual, laser, and microwave technologies. Friendly and enemy units use smoke and obscurants across the battlefield as a combat multiplier. The use of smoke and obscurants to mask combat operations is dictated by the tactical commander. He normally provides the operational guidance for units or elements operating in an area requiring obscuration. Permission to employ smoke and obscurants solely to mask medical evacuation operations may not be approved. However, if the tactical commander's plan indicates that smoke operations are to be employed in a given AO, the medical planner should consider both the advantages and disadvantages posed by their employment.

FACTORS TO CONSIDER

E-2. The medical planner should consider the factors which might impact the use of smoke and obscurants in medical evacuation operations. Factors to consider are the—

- Phase of the tactical operation in which smoke and obscurants will be employed.
- Effect on ground and air evacuation routes when operating in an obscured environment (such as limited hours of use, checkpoint or convoy requirements, or the elimination of NOE approaches).
- Potential for exploiting the use of the cover and concealment provided for clearing the battlefield of casualties, especially in defensive operations.
- Potential requirements for smoke generation to perform the medical evacuation mission which would not detract from the tactical capability and requirements.

E-3. Smoke can also be used to identify unit areas or LZs for which a medical evacuation request has been received. Further, smoke can indicate wind direction at a landing site for AA operations.

EMPLOYMENT OF SMOKE AND OBSCURANTS

E-4. Smoke and obscurants are employed to protect friendly forces from attack during the offense or defense. Smoke and obscurants disrupt enemy combat operations throughout the depth of the battlefield and across the operational continuum. They—

- Disrupt the ability to communicate.
- Conceal friendly forces.
- Deceive the enemy.
- Identify and signal.
- Degrade the effects of directed-energy weapons.

E-5. The benefit to AHS forces is derived through the tactical commander's use of smoke to obscure friendly tactical maneuvers. This obscuration—

- Prohibits the enemy from knowing how many casualties have been inflicted.
- Aids the movement of medical units and equipment.

- Enhances the ability to resupply forward deployed AHS elements.
- Aids in the tactical deception plan.

GENEVA CONVENTIONS AND THE USE OF SMOKE AND OBSCURANTS IN MEDICAL EVACUATION OPERATIONS

E-6. As discussed above, the 1949 Geneva Conventions for the Amelioration of the Condition of the Wounded and Sick in Armed Forces in the Field (GWS) provides protection of medical personnel and units from intentional attack so long as they carry out no duties harmful to the enemy (Article 21, GWS). In order to facilitate their identification so as to prevent their intentional attack, medical units, equipment, and personnel are authorized to display the distinctive emblem of the Red Cross (Article 41 and Article 42, GWS). Under tactical conditions, when requirements for concealment outweigh those for recognition, all distinctive emblems may be obscured or removed from medical equipment if ordered by competent military authority and authorized by Army regulations. Display of the distinctive emblem is not required to afford the right against intentional attack; attack of medical units, equipment, and personnel not displaying the distinctive emblem is prohibited if opposing forces realize that the forces about to be attacked are medical units performing humanitarian duties.

E-7. The use of smoke or obscurants in medical evacuation operations does not differ from the use of camouflage techniques and is not prohibited by the GWS. Its only effect will be to obscure the identity of units as they perform their humanitarian mission. Given the lethality of the modern battlefield, however, it would be difficult, if not impossible, to say that such obscuration of these units, equipment, and personnel would necessarily increase their risk from unintentional attack.

E-8. It is recognized that, with the advent of precision-guided munitions and electro-optical or laser target acquisition devices, there will be a substantial use of smoke and other obscurants on the modern battlefield as a result of normal combat operations. The legitimate use of obscurants by combatants to thwart the accuracy of precision guided munitions may increase the risk to units and equipment not employing obscurants. This may possibly place medical units and equipment at greater risk if they fail to employ them. Further, medical evacuation operations will have to be carried out on the battlefield as medical personnel find it, which will include obscurants employed for normal combat operations.

USE OF SMOKE IN AEROMEDICAL EVACUATION AND HOIST RESCUE OPERATIONS

EFFECTIVE USE

E-9. Smoke can be used effectively in aeromedical evacuation and overland hoist rescue operations to—

- Identify the landing site. Colored smoke is an excellent daytime marking method. The smoke generated from a smoke grenade is difficult to detect more than 2 to 3 miles away, but an aircraft in the area should have little difficulty in noting its location. As more than one unit may be operating in a given area, it is important that the unit requesting an aeromedical evacuation mission be able to signal the aircraft as to the correct landing site to use. Radio communications produce an electronic signature. The electronic signature created from a prolonged transmission to guide an AA to the landing site may not be an acceptable tactical risk.
- Ensure the LZ is controlled by friendly forces. When a unit employs colored smoke to mark a landing site, the aircrew should identify the color and confirm it with the ground personnel. The transmission time required for this procedure is limited, thereby reducing the electronic signature.
- Determine surface wind direction. The employment of smoke at the landing site also enables the aircrew to determine the wind direction.
- Provide cover and concealment. In some environmental conditions (such as desert operations), the phenomenon of inversion occurs. When this occurs, the smoke and obscurants used in

normal combat operations may provide an upper layer of smoke under which the AA can operate.

DISADVANTAGE

E-10. The use of smoke on aeromedical evacuation operations can be a disadvantage if incorrectly employed or if the smoke generated in the tactical operation interferes with the medical evacuation mission. Smoke can obscure the landing site and make NOE approaches unusable. Further, smoke on the battlefield can force aircraft to fly at higher than planned heights. This increases the risk of being acquired by the enemy.

OVERWATER HOIST OPERATIONS

E-11. In overwater hoist rescue operations, the employment of smoke for marking the patient pickup area, for determining surface wind conditions, and for spatial orientation is essential. The smoke employed by the aircrew must not interfere with the conduct of the operation or mask the location of the individual to be rescued.

EMPLOYMENT OF SMOKE IN GROUND MEDICAL EVACUATION OPERATIONS

E-12. The employment of smoke during ground evacuation operations must be in consonance with the tactical commander's plan. Smoke can mask medical evacuation operations on the battlefield, but must not interfere with the tactical mission. In all combat operations, but especially in UO, smoke can be employed to cover and conceal—

- Movement across open areas.
- Extraction of casualties from vehicles and/or buildings.
- Entry and exit into/out of structures.

Appendix F

Evacuation Capabilities of United States Forces

This appendix provides the evacuation capabilities of US forces and aircraft.

EVACUATION CAPABILITIES OF UNITED STATES AIR FORCE AIRCRAFT

F-1. Evacuation capabilities of USAF aircraft are provided in Table F-1. As the majority of USAF aircraft are not dedicated medical evacuation platforms, when used for medical evacuation the crew must be augmented with medical personnel to provide in-flight care.

F-2. The evacuation capability of Civil Reserve Air Fleet aircraft is provided in Table F-2.

Table F-1. Evacuation capabilities of United States Air Force aircraft

Transport Aircraft	*Litter*	*Ambulatory*	*Combination*
C-130 Hercules	74	92	Variety
C-141 Starlifter	48	170	Variety
C-5 Galaxy	70		
C-17A	*36	*54	
KC-135 and KC-10	8	24	
*Normal configuration			

Table F-2. Civil Reserve Air Fleet capabilities

Transport Aircraft	*Litter*	*Ambulatory*	*Combination*
Boeing B-767	*111	22	87 Litter 22 Ambulatory
* When activated, these aircraft are preconfigured for medical evacuation. The crew must be augmented with medical personnel			

EVACUATION CAPABILITIES OF UNITED STATES ARMY VEHICLES AND AIRCRAFT

F-3. The evacuation capabilities of US Army vehicles and aircraft are provided in Table F-3.

Table F-3. Evacuation capabilities of United States Army vehicles and aircraft

Vehicle/Aircraft	*Litter*	*Ambulatory*	*Combination*
Ground Vehicles			
M996 Truck, Ambulance	2	6	1 Litter 3 Ambulatory
M997 Truck, Ambulance	4	8	2 Litter 4 Ambulatory

M1010 Truck, Ambulance	4	8	
M792 Truck, Ambulance	3	6	2 Litter 3 Ambulatory
M1133, Stryker MEV	4	6	
Bus, Motor, 44 Passenger	18	37	
M113 Carrier, Personnel	4**	10	
M998 Truck, Cargo/Troop Carrier (Four Man)	3	4	
M998 Truck, Cargo/Troop Carrier (Two Man)	5		
Truck Cargo, 2 1/2 Ton, 5 Ton	12	16	
M997 Heavy Expanded Mobility Tactical Truck	9		
M871 Semitrailer, Cargo	16		
M1085 Truck, Cargo, Medium Tactical Vehicle, Long Wheel Base, 5 Ton	12	22	
M1093 Truck, Cargo, Medium Tactical Vehicle Air Drop/Air Delivery, 5 Ton	8	14	
M1081 Truck, Cargo, Medium Tactical Vehicle Light Air Drop/Air Delivery, 2 1/2 Ton	7	12	
Rotary-Wing Aircraft			
***UH-60 Blackhawk	6	7	*4 Litter 1 Ambulatory
****UH-60A Blackhawk	3	4	4 Litter 1 Ambulatory
UH-1H/V Iroquois	6	9	*3 Litter 4 Ambulatory
CH-47 Chinook	24	31	#
Fixed-Wing Aircraft			
C-12 Huron		8	

LEGEND
* Normal configuration
** Spall liner must be removed
*** without internal hoist installed, requires 6 man litter configuration kit
**** With internal hoist installed
# See Table F-5

RAILWAY CAR CAPABILITIES

F-4. Although railway cars are not available within the US Army inventory, it is important to know the approximate capacities of rail transport in the event they become available and/or required in civil support operations or through wartime HN support agreements. The approximate capabilities of railway cars are provided in Table F-4.

Table F-4. Capabilities of railway cars

Rail Transport	*Litter*	*Ambulatory*	*Combination*
Pullman Car (US)	32	48	
Sleeping Car (NATO/HN)	32	32	
Ambulance, Railway Car (NATO/HN)	24	30	
Ambulance, Railway Car, Personnel	21	21	
Railbus			40 Litter 16 Ambulatory

EVACUATION CAPABILITIES OF UNITED STATES NAVY SHIPS, WATERCRAFT, AND ROTARY-WING AIRCRAFT

F-5. The evacuation capabilities of USN ships, watercraft, and aircraft are provided in Table F-5. The entries for ambulatory and litter patients on ships are the same because all patients require a bunk.

Table F-5. Evacuation capabilities of United States Navy ships, watercraft, and aircraft

Ships/Watercraft/Aircraft	*Litter*	*Ambulatory*
Hospital Ships		
T-AH 19, US Naval Ship Mercy	1000	1000
T-AH 20, US Naval Ship Comfort	1000	1000
Casualty Receiving And Treatment Ships		
Amphibious Assault Ship (LHD)(Multipurpose)	604	604
Amphibious Assault Ship (LHA)(General Purpose)	367	367
Amphibious Assault Ship (LPH)	222	222
Amphibious Transport Dock (LPD)	14	14
Dock Landing Ship (LSD)	108	108
Limited Medical Capability Watercraft		
Amphibious Cargo Ship (LKA)	12	12
Amphibious Command Ship (LCC)	24	24
Rotary-Wing Aircraft		
CH-46 Sea Knight	15	25
CH-53 Sea Stallion	24	*55
V22 Osprey	12	24
* With centerline seating installed		

Appendix G

Selection of Patients for Aeromedical Evacuation and Patient Classification Codes and Precedence

Patient classification codes provide information to evacuators and treatment personnel in an abbreviated form. They can be used to complete administrative reporting requirements pertaining to the evacuation of patients.

Paragraphs G-2 through G-5 implement STANAG 3204.

BRIEFING OF PATIENTS PRIOR TO AEROMEDICAL EVACUATION

G-1. When patients are evacuated by aircraft (routinely in Roles 2, 3, 4, and 5), they should be briefed on the following points—

- A number of ambulatory patients will be detailed to assist with the evacuation of litter patients in any emergency.
- Safety belts and litter straps are to be properly fastened according to orders given by the pilot.
- Patients are instructed on the proper position to assume in preparing for an emergency. Flight crews and CCATT ensure that seat safety harnesses have been tightened.
- Ambulatory patients, with the exception of those designated to assist litter patients, are the first to leave a downed aircraft.
- Immobilized litter patients are freed from litters and assisted in leaving the aircraft. Litters will not normally be removed from their fastenings in view of the limited time available to evacuate the aircraft.
- Mentally disturbed patients should be quieted so that the orderly removal of other patients will not be jeopardized.

INTERNATIONAL STANDARDIZATION EVACUATION AGREEMENT

G-2. Table G-1 provides the patient classification codes defined in international standardization agreements.

Table G-1. Patient classification codes

Code	*Classification*
Class 1. Psychiatric Category	
1A	Stretcher patients who are frankly disturbed and inaccessible, requiring close supervision and sedation, or the use of restraint equipment.
1B	Patients who are not at the time of request grossly disturbed, but who may react badly to air travel, or who may commit acts likely to endanger themselves or the safety of the aircraft and its occupants.
1C	Sitting patients who are cooperative and have proved themselves to be reliable under preflight observation.
Class 2. Stretcher Patients Other Than Psychiatric	
2A	Patients, who in an emergency, would be unable to leave the aircraft without assistance.
2B	Patients, who in an emergency, would be able to leave the aircraft unaided.
2C	Patients who can walk onto an aircraft but on a long flight would benefit from a litter.
Class 3. Sitting Patients Other Than Psychiatric	
3A	Sitting patients, including handicapped persons, who may need medical or nursing attention en route and who, in an emergency, would require assistance to escape.
3B	Sitting patients who may need medical or nursing attention en route and who would be able to escape unassisted in an emergency.
Class 4. Sitting Patients Other Then Psychiatric	
4	Patients who will not need medical or nursing attention en route and are capable of traveling unescorted. Class 4 patients are normally required to make their own way from the deplaning airfield to their destination.

INTERNATIONAL STANDARDIZATION EVACUATION PRECEDENCE

G-3. Patients for AE will be given appropriate degrees of priority so that, if aircraft space is limited, the more urgent patients may be evacuated before those whose conditions are less serious. The degrees of priority are depicted in Table G-2. (The evacuation precedence used by the USAF is essentially the same as this listing. It contains a few word changes and introduces specific time limits. It does not contain Priority 4.)

Table G-2. Patient priorities as designated in STANAG 3204

PRIORITY 1/URGENT	Emergency patients for whom speedy evacuation is necessary to save life or limb, to prevent complication of serious illness, or to avoid serious permanent disability.
PRIORITY 2/PRIORITY	Patients who require specialized treatment not available locally and who are liable to suffer unnecessary pain or disability unless evacuated with the least possible delay.
PRIORITY 3/ROUTINE	Patients whose immediate treatment requirements are available locally but whose prognosis would definitely benefit by air evacuation on routine scheduled flights.
PRIORITY 4	Patients for whom air evacuation is a matter of medical convenience rather than necessity.

PATIENT CLASSIFICATION

G-4. Table G-3 provides the patient classification codes used aboard USAF aircraft and that can be used in completing DD Form 601 (Patient Evacuation Manifest). These codes are expanded to include categories of patients and other personnel which may or may not apply on the battlefield (such as infants, relatives, or friends).

UNITED STATES AIR FORCE EVACUATION PRECEDENCE

G-5. The evacuation precedence used by the USAF is dramatically different than that employed by the US Army medical evacuation system. These precedence should not be confused. Table G-4 (page G-5) provides the evacuation precedence and time frames used by the USAF.

Table G-3. Patient classification codes

Code	*Classification*
Class 1.	***Neuropsychiatric Patients***
1A	Severe psychiatric litter patients requiring the use of restraining apparatus, sedation, and close supervision at all times.
1B	Psychiatric litter patients of intermediate severity requiring tranquilizing medication or sedation, not normally requiring the use of restraining apparatus, but who react badly to air travel or who may commit acts likely to endanger themselves or the safety of the aircraft. Restraining apparatus should be available for use.
1C	Psychiatric walking patients of moderate severity who are cooperative and who have proved reliable under observation.
Class 2.	***Litter Patients (Other Than Psychiatric)***
2A	Immobile litter patients unable to move about of their own volition under any circumstances.
2B	Mobile litter patients able to move about of their own volition in an emergency.
Class 3.	***Walking Patients (Other Than Psychiatric)***
3A	Nonpsychiatric and nonsubstance abuse patients who require medical treatment, assistance, or observation en route.
3B	Recovered patients who are returning to their units and require no medical attention en route.
3C	Ambulatory drug or alcohol substance abuse patients.
Class 4.	***Infant Category***
4A	Infants under three years, occupying a seat or in a bassinet or a car seat secured in an ambulatory seat.
4B	Recovered infants under three years, occupying a seat or in a bassinet or car seat secured in an ambulatory seat.
4C	Infants in an incubator.
4D	Infants under three years on a litter.
4E	Outpatients under three years on a litter for comfort.
Class 5.	***Outpatient Category***
5A	Ambulatory, nonpsychiatric, and nonsubstance abuse outpatients who are traveling for an outpatient visit and do not require a litter or medical assistance in flight.
5B	Ambulatory drug or substance abuse outpatients going for treatment.
5C	Psychiatric outpatients going for treatment.
5D	Outpatients on a litter for comfort or safety.
5E	Returning outpatients on a litter for comfort or safety.
5F	Other returning outpatients.
Class 6.	***Attendant Category***
6A	Medical attendants, either physician, nurse, or technician, who are assigned to give specialized medical treatment or nursing care to a particular patient.
6B	Nonmedical attendants, either relatives or friends, who may assist with the patient's care and who may also require support.

Table G-4. Evacuation precedence used by the United States Air Force

Evacuation Precedence	*Definition And Time Frames*
URGENT PRECEDENCE	Applies only to the need for immediate life, limb or eyesight saving. These patients should be picked up and delivered to the designation facility as soon as possible.
PRIORITY PRECEDENCE	Applies to the need for prompt medical care not available locally. These patients should be picked up within 24 hours and delivered to the destination facility with the least possible delay.
ROUTINE PRECEDENCE	Applies to all other patients. Routine patients will be picked up and delivered on regularly scheduled flights.

Appendix H

Troop Leading Procedures, Precombat Checks/Precombat Inspections, and Leader Checklists

Troop leading procedures (TLPs) are universal in their application. They are used by leaders at all levels regardless of branch or MOS. For purposes of this appendix the leader titles of ambulance platoon leader and ambulance platoon sergeant are used for familiarity.

SECTION I — TROOP LEADING PROCEDURES

H-1. Leaders must conduct detailed planning and preparation before each mission in order to be successful. That is why the Army developed an eight-step process of TLP for leaders to follow.

H-2. Each step of the TLP process is designed to give leaders a framework on which to thoroughly develop their plans and orders. If these steps are not followed the resulting plans and orders may be missing information which may be critical to successfully execute the mission.

THE EIGHT STEPS OF TROOP LEADING PROCEDURES

STEP ONE: RECEIVE THE MISSION

H-3. Once staff planning is initiated the ambulance platoon leader gathers information concerning the operation from the company commander and disseminates to his subordinate leaders. Examples of this information include but are not limited to intelligence preparation of the battlefield (IPB) and METT-TC. If the company commander has not completed a full plan, he may still issue a warning order (WARNO) to the ambulance platoon leader.

- In keeping with the rules of time management, the ambulance platoon leader uses one-third of his overall time for planning the operation. The reverse planning sequence is recommended.
- The ambulance platoon leader's OPORD must also be understandable to subordinate personnel.
- Throughout the planning and development process the ambulance platoon leader may find it necessary to change or refine the plan as the tactical situation changes, as long as it follows the commander's intent.

STEP TWO: ISSUE A WARNING ORDER

H-4. The ambulance platoon leader issues a WARNO based on information received from the company commander's WARNO or OPORD.

- The ambulance platoon leader can present the order in any format.
- As a minimum, the platoon should use the company or battalion tactical standing operating procedures (TACSOP). The best method is to use already developed and implemented platoon level TACSOP that cover preparation for offensive or defensive missions. These TACSOPs should be thoroughly trained and rehearsed.
- Whatever the mission, the platoon can begin work on its preparation by assessing equipment needs through the various military supply classes (such as Class I, III, IV, V, or VIII).
- Every Soldier in the platoon should have a specific task and purpose (identified in the platoon SOP) once the ambulance platoon leader has issued a WARNO.

Step Three: Make a Tentative Plan

H-5. The ambulance platoon leader considers the following while preparing the tentative plan:

- Mission analysis, IPB, METT-TC considerations, higher mission and intent (two levels up).
- The ambulance platoon leader must know what systems are available for his use and how to employ them. He must fully understand his own equipment and capabilities as well as those of his enemy.
- Consider **Observation** and Fields of Fire, **Avenues** of Approach, **Key** Terrain, **Obstacles** and **Cover** and Concealment, (man-made and natural), (OAKOC), weather, morale, and flexibility to react to last-minute changes. As the commander updates the ambulance platoon leader, the ambulance platoon leader, in turn, continually updates the platoon with FRAGOs.
- Experience is a valuable asset for mental preparation. Knowing how Soldiers are going to react to situations, based on past experiences, can help any officer or NCO develop his plan.
- During the planning process, the leader should have a *vision* of the enemy and understand the threat using available intelligence. The better the ambulance platoon leader understands the enemy, the greater his opportunity for success.

Step Four: Start Necessary Movement

H-6. Even though listed as Step Four, necessary movement can begin as soon as the mission is received.

- If movement is required, the ambulance platoon sergeant, with the assistance of squad leaders, can move the platoon to a new tactical assembly area (TAA). This allows the ambulance platoon leader more time to prepare his order and graphics.
- At the new TAA, preparations continue with precombat checks (PCCs). This is a good opportunity to conduct PCCs on squad equipment by using the platoon SOP checklists and to certify squad rehearsals. If the ambulance platoon sergeant is unable to conduct squad inspections, the platoon SOP should designate someone else, such as the senior squad leader, for the task.
- If platoon linkups are time driven, movements and time management become critical. Failure to plan or allow for ample movement time creates late linkups and results in late rehearsals, late PCCs, and various other events that will not be accomplished on time or to standard.
- If movements were conducted late or linkups were late, the mission timeline will suffer.

Note. Time management is mission essential!

Step Five: Reconnoiter (Conduct Reconnaissance)

H-7. Although listed fifth, this step may be implemented at any time and as often as necessary during TLP. When the risk of enemy contact is too great, or establishing additional orders for reconnaissance parties is time consuming leaders may have to rely on map reconnaissance and or photographic imagery to reconnoiter an AO. Ambulance platoon leaders must ensure that ground evacuation routes are reconnoitered and strip maps developed for security escort vehicle drivers and ambulance drivers.

- A leader's reconnaissance must be performed! Leaders who see more of the AO are better able to develop more detailed and useful plans.
- Information gained by the leader's reconnaissance allows more accurate terrain models to be built and greater detail can be provided when the ambulance platoon leader is briefing the OPORD.

Step Six: Complete the Plan

H-8. Complete the plan—

- With reconnaissance completed, movement accomplished, and the initial plan established, the ambulance platoon leader should have most of the information that is needed to finish writing the OPORD.
- Any additional resources needed should be en route or already on-hand so the ambulance platoon leader, when briefing the OPORD, has positive control of additional equipment or manpower.
- Graphics must be finished before the OPORD is briefed to the platoon and quickly handed down to squad leader level. Logistics and sustainment graphics must include CBRN decontamination points, MTFs locations, and supported unit CCP and AXP.

STEP SEVEN: ISSUE THE ORDER

H-9. Avoid rushing through the OPORD briefing as this may result in pertinent information being left out of the brief.

- When completed, distribute the OPORD with graphics to each squad and team leader. Also include a terrain model or sand table to enhance visualization of the terrain.
- Minimize distractions. Having subordinate leaders in the right mindset to receive the order is critical to allow them to collect and retain all of the information presented at the OPORD brief.
- If possible, present the OPORD with a representation of the terrain in the background to enhance visualization of the mission.
- After the OPORD has been presented, the ambulance platoon leader ensures that the information has been disseminated to every Soldier. Backbriefs from Soldiers and NCOs to the ambulance platoon leader should reveal any portion or portions of the OPORD that were missed or not fully understood. Everyone leaving the OPORD brief should have a clear understanding of the mission.
- Contingencies need special consideration. If, during the OPORD brief, a potential activity is discussed that has no clear plan, subordinates can help develop a contingency plan, unless the activity is already specified in the platoon SOP. Defer making a decision on topics discussed during the OPORD brief which require more thought and possibly guidance from higher headquarters. If at all possible, allow a follow-up period to develop clarity on the situation and the techniques needed to accomplish the mission.

STEP EIGHT: REHEARSALS

H-10. Supervise rehearsals and plan to run as many rehearsals as time permits. After each rehearsal, conduct an after-action review (AAR). When conducting platoon rehearsals with attachments, ensure that the ambulance platoon leader/sergeant supervises the AAR with attachments. When conducting the AAR, ensure Soldier involvement. By this time all Soldiers understand the concept of the operation.

- When rehearsing the mission, ensure that the rehearsal lane coincides with what was briefed during the OPORD brief. Based on the latest intelligence, the rehearsal lane should match (as much as possible) what will be encountered during the actual mission. Do not allow for complacency. If there is information about the terrain and the obstacle that can be placed in the rehearsal lane to add realism, ensure that obstacle information is built into the lane. A company-level rehearsal is of limited value if the enemy obstacle does not resemble what was reconnoitered or templated.
- If time does not allow for a full rehearsal, backbriefs must be conducted. A successful understanding of the mission can be accomplished with key personnel by walking troops through the mission on the terrain model and asking questions. The backbrief is never a substitute for a rehearsal, but is an acceptable recourse when time is limited.
- Conduct final precombat inspections (PCI).
- If there are any last-minute concerns, they should be satisfied during this time.
- Conclude the brief.

H-11. Some steps of the TLPs require more development than others. Regardless of the mission, the ambulance platoon leader must not, for any reason, omit any step from the TLP.

SECTION II — PRECOMBAT CHECKS AND PRECOMBAT INSPECTIONS

H-12. Precombat checks and PCIs are tools which leaders have to ensure that their Soldiers—

- Fully understand the mission and their role.
- Have the equipment that is necessary to accomplish the mission on hand and that it is serviceable.

H-13. Leaders must conduct PCCs and PCIs for every mission. Failure to perform these vital checks could result in unnecessary casualties, loss of equipment, and failure to accomplish the mission.

H-14. When conducting PCCs/PCIs, as deficiencies are identified corrections must be made immediately.

H-15. Who performs PCCs and PCIs are both intended to accomplish the same thing, which is to have a leader ensure that personnel and equipment are ready. The only difference is who is doing the inspecting.

- Commanders conduct PCIs. Ambulance platoon leaders, ambulance platoon sergeants, and squad leaders conduct PCCs.
- Why do we identify a commander's inspection separately? Equipment readiness is essential for mission success. Squad leaders and ambulance platoon leaders are certainly capable of ensuring equipment readiness, but the commander's first-hand knowledge of mission requirements cannot be substituted.

H-16. Units need to establish standard PCC/PCI checklists based on their units mission profile. They should be included in the TACSOP and issued to all leaders. They should include routine daily checks and checks that are mission specific. They do not need to be extremely detailed; a list of items to check and the leader's knowledge of the equipment and the appropriate manuals should suffice. The PCCs checklists provided below are examples that units may use to develop their own or they can be modified as appropriate to meet their own requirements.

SECTION III — LEADER CHECKLISTS

H-17. This section provides sample checklists which may be used as is, or modified, to meet unit specific requirements. As with any checklist they are only effective if they are incorporated into a unit's TACSOPs and then used routinely.

GENERAL CONSIDERATIONS

H-18. Leader receives the mission from the next higher headquarters.

___ Clarifies any questions.

___ Coordinates with next higher headquarters as required.

___ Conducts mission analysis.

___ Produces the estimate.

___ Revises mission statement.

___ Produces tentative time schedule.

___ Issues a WARNO.

H-19. Unit members perform readiness, maintenance, and functional checks under the supervision of unit leaders which includes:

___ Medical equipment sets are present and all perishable and dated items are current and on-hand.

___ Vehicles/aircraft/generators.

___ Night vision devices.

___ Communications equipment.

___ Weapons and ammunition.

___ Field sanitation equipment and supplies.

___ Special equipment (such as hoist and forest penetrator).

___ Common table of allowances equipment.

H-20. Leader makes a tentative plan.

___ Uses estimate of the situation to depict plan of support.

___ Develops courses of action (COA).

___ War games COA.

___ Determines best COA.

___ Completes his plan.

H-21. Leader issues OPORD.

H-22. Leader uses sand table or sketches to depict plan of support.

H-23. Leader affects coordination for the mission.

___ Support requirements.

___ Current intelligence (to include health threat) update.

___ Control measures.

___ Communications and signal information.

___ Time schedule.

H-24. Leader receives attachments/augmentation (personnel), if appropriate.

___ Attachments are oriented to unit.

___ Attachments are briefed on the mission.

H-25. Leader supervises AHS mission preparation.

___ Key leaders brief back unit leader.

___ Key personnel rehearsals are conducted.

H-26. Unit leaders:

___ Supervise.

___ Inspect.

___ Ensure adequate security.

___ Conduct brief backs.

___ Rehearse.

___ Continue coordination.

H-27. Unit plans for support of combat operations.

___ Analyze mission.

___ Patient acquisition and medical evacuation requirements.

___ Area support requirements.

___ Requirements for water, Class VIII, and other supply classes.

___ Requirements for CBRN defense.

___ Transportation requirements.

H-28. Unit leaders:

___ Execute a work/rest plan based on work priorities and statutory (crew rest) requirements.

___ Monitor current situation.

___ Issues appropriate FRAGOs based on intelligence or operational updates.

___ React to messages or orders from higher headquarters.

___ Execute any actions and coordination resulting from change.

H-29. Unit headquarters remains current on positions and missions of higher, adjacent, and subordinate units.

Ground Ambulances Inspected For

___ Authorized MES are on hand.

___ Medical equipment is complete and serviceable.

___ Oxygen is on hand and serviceable.

___ Authorized medications are on hand and current.

___ Packing list is available.

___ Strip maps and/or road maps are available (with overlays).

___ On-vehicle equipment is on hand.

___ Log book is present and current.

___ All drivers are licensed.

___ Situational awareness equipment (position locator) is on hand and serviceable.

___ Communications equipment is on hand and serviceable and set to correct frequency.

___ Medical unit identification markers (according to the Geneva Conventions) are displayed.

Note. Markers are red on a white background only; camouflaged or subdued markers are not authorized.

Chemical, Biological, Radiological, and Nuclear Equipment

___ Individual protective equipment is on hand and serviceable. (One set is issued; the other is maintained in support.)

___ Protective masks are issued and serviceable.

___ Nerve Agent Antidote Kit is available and distributed, if required.

___ Convulsant antidote for nerve agent is available and distributed, if required.

___ Decontamination apparatus is complete and serviceable.

___ Basic load of decontamination supplies is on hand.

___ Chemical agent alarms are on hand and serviceable.

___ M256A1 detector kits are issued.

___ Chemical, biological, radiological, and nuclear contamination marking kits are distributed.

___ Chemical agent monitors are on hand, if authorized.

___ Replacement filters for protective masks are on hand.

___ Biological and chemical warfare agents prophylaxis, immunizations, and barrier creams have been accomplished, if appropriate.

___ Nerve agent pretreatment packets are available.

___ Radiac sets are on hand.

MISCELLANEOUS EQUIPMENT

___ Inspect binoculars.

___ Inspect camouflage nets and support systems, if appropriate.

___ Inspect night vision devices.

___ Ensure batteries are on hand and serviceable.

___ Inspect tentage.

___ Inspect global positioning systems, if available.

PERSONNEL

___ Ensure Soldiers are in the correct uniform.

___ Ask questions to ensure that Soldiers have been briefed on mission and situation.

___ Implement appropriate MOPP level.

___ Check for drivers licenses.

___ Brief Soldiers on operations safety and environmental injuries.

___ Individual equipment is on hand and stowed properly.

___ Soldier has eaten and is briefed on future field feeding.

___ Identification cards and tags are on hand and serviceable.

___ Camouflage self and equipment, when directed. (If medical vehicles are camouflaged, the Geneva Convention emblem must be removed.) (Refer to STANAG 2931.)

___ Work/rest plan (to include crew rest) is implemented.

___ Water discipline plan is implemented, if appropriate.

COMMUNICATIONS EQUIPMENT

___ Radios are operational (communications check completed).

___ Telemedicine equipment is available and operational, if available.

___ Speech security equipment functions, if available.

___ Frequencies are set.

___ Matching units are operational.

___ Antennas are tied down properly.

___ Connectors are clean and serviceable.

___ Field telephones (TA-312) are on hand and serviceable, if appropriate.

___ Batteries are on hand and charged.

___ Man-pack radio sets are complete.

___ Switchboard is on hand and serviceable.

___ Communications wire (WD-1) is on hand and serviceable.

___ Antennas and remotes are present and serviceable.

___ Signal operating instructions are available and secured.

___ Call signs, frequencies, and challenge passwords have been disseminated.

___ Perform communications check again.

VEHICLES

___ Loads are according to load plan; load plan is posted in the vehicle.

___ Hazardous cargo is properly identified and stored toward the rear of the vehicle for easy access and inspection.

___ Ammunition is issued and properly stored.

___ Vehicle fuel tank is topped off.

___ Package POL products and small arms lubricant are present.

___ Water cans are full.

___ Meals, read-to-eat are issued and stowed.

___ First-aid kits are present and complete.

___ Operator's manuals and lubrication orders are present for the vehicle, radios, and associated equipment.

___ Critical on-vehicle equipment and basic issue items are present.

___ Vehicle dispatch is complete.

___ Department of the Army (DA) Form 2404 (Equipment Inspection and Maintenance Worksheet) is complete.

___ No deadline deficiencies exist.

___ Before operation PMCS has been completed.

INDIVIDUAL WEAPONS

___ Clean and functional.

___ Cleaning tools/kits, bolts, and ruptured cartridge extractors are present.

___ Range cards are on hand.

___ Ammunition is issued, accounted for, and secured.

___ Magazines are issued.

___ Blank adapter installed, if appropriate.

___ Function check performed.

MEDICAL EVACUATION SUPPORT PLAN

___ Mission statement/commander's intent/task organization.

___ Intelligence preparation of the battlefield considerations.

___ Patient estimates by phase.

___ Identify expected areas of patient density.

___ Anticipate and plan for likely patient conditions.

___ Formulate/evaluate/implement/reevaluate COA.

___ Location of supported units.

___ Location of supporting MTFs/FSTs/hospitals.

___ Location of supporting evacuation units.

___ Augmentation of personnel and vehicles coordinated supporting MTFs.

___ Class VIII supply/resupply coordinated and available.

___ Location of supported unit CCPs and AXPs.

___ Evacuation routes primary and alternate.

___ Trafficability (enemy activity, hazards, condition, obstacles).

___ Security (gunship escort, MP escort, convoy support).

___ Strip maps/overlays/maps.

___ Landing zones designated for rotary-wing evacuation aircraft (primary and alternate)

___ Ensure that platoon/company/battalion/treatment facility litter teams are available to assist ambulance crews in loading patients.

___ Available litters and types (NATO standard, folding nonrigid, SKED®, and so forth).

___ Ground ambulances available (type/number/carrying capacity).

___ Air ambulances available (type/number/carrying capacity).

___ Supporting evacuation assets available (requested/ prepositioned).

___ Nonmedical vehicles for CASEVAC (type/number/carrying capacity).

___ Available aircraft (UH-60, CH-47).

___ Medical evacuation support plan for units without evacuation resources.

___ Treatment and evacuation of EPW and detainees.

___ Chemical, biological, radiological, nuclear considerations (contaminated evacuation vehicle plan, patient decontamination site, decontamination sites, contaminated aircraft).

___ Other.

CASUALTY CARE CONSIDERATIONS

___ Self-aid/buddy aid/CLS/trauma specialist.

___ Combat lifesavers trained and CLS bags stocked and issued.

___ Emergency care specialists and flight medics are rehearsed, aid bags fully stocked.

___ Trauma specialist Class VIII resupply sets fully stocked and pushed forward to company command post.

___ Locations of platoon/company/battalion/treatment facility CCPs are known.

___ Other.

Mass Casualty Plan (Resourced/Rehearsed/Coordinated)

___ Litter teams.

___ Litters (all).

___ Medical evacuation/CASEVAC vehicles (all).

___ Class VIII supplies (on hand).

___ Class VIII resupply (Prepackaged and prepositioned).

___ Company/battalion command net.

___ Brigade support medical company command net.

___ Personal effects, weapons, equipment, and property exchange.

___ Chemical, biological, radiological, and nuclear casualty/evacuation plan.

___ Location of decontamination site.

LANDING ZONE OPERATIONS

Safety

___ At all times take all the necessary safety precautions.

___ Platoon that is responsible for running the LZ provides all required safety gear. Included are—

  ___ Safety goggles.

  ___ Visual Signal, VS-17 panels/chemical lights.

  ___ Smoke (Appendix E).

___ Is the LZ secure.

Marking the LZ

___ Use colored smoke to identify the LZ and wind direction.

___ Use an inverted "Y" to mark the LZ.

___ During daylight, camouflage the LZ so the VS-17 panels cannot be seen until evacuation aircraft are confirmed and inbound.

___ At night, emplace chemical lights only when an AA is confirmed and is inbound.

Landing Requirements for Light Helicopters

___ Minimum real estate required is a cleared area 30 meters in diameter.

___ Area must have an approach and departure zone clear of obstructions and not over the established company area.

Appendix I
Safety on the Battlefield

The tactical environment provides ever-changing demands and unpredictable problems, often under stressful conditions. The interface of man, machine, and environment is constantly shifting. In this environment, mission accomplishment requires continuous leader involvement and flexible decision making. Not surprisingly, accidents and injuries increase during tactical operations. Safety in the tactical environment depends upon compliance with established standards. However, due to fluid conditions in the tactical environment, safe mission accomplishment relies heavily on complete integration of risk management into both planning and execution phases. Risk management assists commanders in anticipating and controlling hazards in the planning phase and in dealing with unexpected hazards as they arise in the execution phase.

SECTION I — GENERAL

COMMON ACCIDENTS

I-1. During all operations, it is critical commanders and leaders minimize risks through appropriate risk management procedures to protect the force and preserve the unit's war-fighting capability. Over half of all Army accidents during tactical operations are accredited to the following operations.

VEHICLE OPERATIONS

I-2. Most accidents in this category are caused by excessive speed for conditions (weather, traffic). Other causes are recklessness, fatigue, unfamiliarity with roads, and untrained and inexperienced drivers. A lack of equipment/vehicle handling characteristics knowledge also contributes to accidents. Strict enforcement of standards is needed for all vehicle operations. The senior occupant must be responsible for this enforcement. Only trained, licensed personnel should be assigned to operate vehicles or equipment. Ground guides are mandatory during movement in bivouac and assembly areas, when backing, and during periods of reduced visibility. Operators must be familiar with proper operation and maintenance of commercial equipment.

MATERIEL HANDLING

I-3. These accidents occur when an object is too large or too heavy to handle for those attempting to move it. As a result, someone may sustain a muscle or back injury, or may be crushed. Overconfidence in one's ability, a lack of planning, and fatigue are common factors in such accidents.

MAINTENANCE

I-4. These accidents are often caused by failure to follow procedures, using the wrong tools, and/or fatigue.

SEVERE ACCIDENTS

I-5. The following activities produce fewer accidents than those listed above; however, when they do occur, they often result in catastrophic damage or death. Commanders and leaders must pay special attention to unit missions which involve these activities or put Soldiers in situations or locations where they will come in contact with associated hazards and ensure they properly brief their personnel and mitigate risk to the fullest extent.

AMMUNITION AND EXPLOSIVES HANDLING

I-6. Horseplay, mishandling, disassembly, unauthorized use, and improper storage of ammunition and explosives account for many personnel injury accidents. It is essential to enforce accountability and security procedures for unexpended ammunition and explosives and to comply with explosive storage safety standards in AR 385–64 and DA Pam 385–64.

EXPLOSIVE SOUVENIRS

I-7. Educate Soldiers to dangers involved and the serious consequences of collecting unexploded ordnance on the battlefield or ranges. Post-tactical training shakedown inspections for this type of materiel are a must. Amnesty boxes are also useful. Platoon sergeants and squad leaders policing their Soldiers can prevent most of these accidents through proper training.

FIELD EXPEDIENTS

I-8. Be suspicious of shortcuts. Although tactical operations frequently involve employment of field expedients, risks and benefits must be carefully weighed. In many cases, field expedients are the result of a weak supply system or inadequate planning.

FIELD HEATERS AND STOVES

I-9. Operators of all types of heaters and stoves should be trained and licensed in advance. Equipment should be maintained and operated in accordance with operating instructions including use of proper fuel. Ensure combustible material is kept well away from heaters and stoves and fire fighting equipment is available for each heater and stove. Heaters or stoves with self-contained fuel supplies (small space heater) should not be refilled while the heater is on or still warm. Do not use heaters or stoves in tents or other confined spaces without use of proper ventilation, such as tent vent flaps, doors, or windows.

PETROLEUM, OILS, LUBRICANTS STORAGE AND HANDLING

I-10. The POL handlers must know and practice safety rules and procedures. Frequent inspections ensure safe storage and transfer of POL products. Proper grounding procedures must be followed. Field manual 10–67–1 describes use of protective equipment to prevent personnel exposure.

SOLDIER FATIGUE

I-11. When a Soldier's sleep time is dependent upon the tactical situation, debilitating fatigue can occur. Soldiers suffering from sleep loss experience various symptoms of fatigue including decreased coordination, narrowed attention span, and reduced performance to standard. Anticipate fatigue-related errors and take action to prevent them.

TACTICAL SLEEP PLAN

I-12. Commanders need to develop and enforce sleep plans. Control sleeping areas to prevent Soldiers from being crushed by moving vehicles in and out of the area.

Water Operations

I-13. Plan water operations carefully, as risks of drowning and equipment loss are high during water operations. Pair strong swimmers with weak ones to protect personnel. Secure equipment and float it across rather than requiring individuals to carry their equipment. Use safety lines and personal floatation devices.

Weapons

I-14. Most accidents occur when cleaning or clearing individual weapons, entering or exiting vehicles, or running with loaded weapons. Provide guidance for weapons handling, loading, and clearing, and ensure it is strictly enforced. Do not load weapons that are not essential for the current mission.

Weather-Related Casualties

I-15. Consider the effects of weather during planning. Unit effectiveness is quickly lost through weather-related casualties such as frostbite, heat stroke, lightning strikes, and falls. Instruct Soldiers in awareness, prevention, and first aid for weather-related injuries and when these conditions can be expected.

VEHICLE CONVOY OPERATIONS

I-16. Convoy operations can be very dangerous if not properly planned. Control of convoy speed and proper separation between vehicles is critical to reducing risk of an accident. See FM 4-01.45 for in-depth information on tactical convoy operations.

Convoy Commander Responsibilities

I-17. Convoy accidents are most commonly caused by a leader failing to perform his or her duties as the commander of a vehicle movement as required in FM 55-30. This includes failure to:

- Control the group's movement.
- Ensure vehicles maintain proper march speed.
- Properly mark unit vehicles.

I-18. Convoy commanders are responsible for ensuring the safe movement of the convoy. To do this, they must positively control the convoy's movement by using communications equipment among the vehicles. This includes ensuring each vehicle has a properly trained, equipped, and supervised crew leading from the front in the absence of radios and/or other means of managing the march. Control of the movement includes enforcing speed limits, march intervals, and seat belt usage. It also requires the ability to stop the march if an unexpected hazard is encountered along the route.

I-19. Preparing vehicles and Soldiers for movement is a leader responsibility. Inexperienced Soldiers, personnel turbulence, and ever increasing training requirements have caused some units to become complacent in managing risks associated with vehicle movements. According to the risk management process, leaders must identify hazards associated with the mission and develop, implement, and supervise control measures to mitigate those risks. These control measures include marking the vehicles in accordance with local SOPs, briefing crews on hazardous conditions expected along the route, precombat checks of personnel and equipment, and developing preaccident emergency contingency plans.

I-20. Unit SOPs should address leader responsibilities during movements of any number of vehicles and identify means to implement common controls.

Driver Skills

I-21. Operators should be taught the specific skills needed for tactical vehicle operations, which, in addition to requirements outlined in AR 600-55, include—

- Pulling and backing trailers.
- Vehicle recovery operations.

- Loading and lashing of cargo.
- Methods of negotiating difficult terrain such as sand dunes, rice paddies, mountainous terrain, icy roads.
- Ground-guide procedures and signals.
- Methods and procedures for retrieving vehicles stuck in snow, mud, sand, or other restrictive terrain.
- Proper parking and use of the proper-sized chock blocks.

Use of Safety Equipment

I-22. When the tactical situation allows, flashers should be turned on immediately if a vehicle is disabled or impedes traffic, and every effort made to move the disabled vehicle from the roadway. Highway warning kits should be provided for each vehicle in a convoy. When a vehicle is disabled, place the warning triangle a minimum of 100 meters to the rear of the vehicle and ensure personnel remain clear of the road and rear of the vehicle.

Night Operations

I-23. Personnel who are required to operate motor vehicles while wearing NVG must be trained and tested on use and operation of such devices, and the training recorded on the individual's driver training records. Other factors include:

- Ground guides should be used when moving vehicles at night in areas where troops are present.
- Blackout driving should be prohibited on roads open to the public.

FIRE PREVENTION

I-24. The risk of fire is high in areas where a large number of Soldiers are in tents. The following guidance will reduce the risk of fires:

- Establish a fire prevention and protection plan including procedures for inspecting and recharging fire extinguishers during tactical operations.
- Appoint a fire marshal for each bivouac area and train them in their duties. Train Soldiers in fire prevention techniques as well as emergency procedures in the event of a fire.
- Establish safe distances between tents to reduce risk of multiple losses from one fire.
- Provide available fire fighting equipment (portable extinguishers, sand, water buckets, and shovels) to contain small fires. Ensure personnel are trained on their use.
- Establish procedures for sounding fire alarms.
- Ensure no-smoking areas are established and enforced.
- Establish an inspection system ensuring compliance with fire prevention standards.
- Ensure flammable materials are stored according to appropriate directives and checklists.
- Ensure vehicle fire extinguishing/suppression systems are operational and crews are proficient in using the systems.
- Provide a designated fire plan, equipment, and trained personnel for POL storage, ammunition supply points, motor pools, hospitals, and hangars.

SECTION II — ASSEMBLY AREA OPERATIONS

SAFETY OFFICER/NONCOMMISSIONED OFFICER

I-25. The safety officer and NCO must establish, in coordination with the 1SG, platoon leaders, and platoon sergeants, procedures for ground guides and incorporation of these procedures into the unit SOP. Company commanders must weigh mission requirements and personnel available while in combat operations and use risk management analysis to validate procedures for implementation into the SOP.

Operating within the modular force structure may require assistance from support units to meet necessary personnel and vehicle requirements to safely ground guide and move aircraft.

GROUND VEHICLES

I-26. Commanders and leaders must be aware of hazards from ground vehicles in AAs. Vehicles moving in the AA should always have a ground guide leading the vehicle until out of the AA. This prevents injury to personnel or damage to equipment. Use of a one-way path through the unit area will assist greatly in preventing accidents. Additionally units may be operating in the vicinity of armored vehicles and Soldiers must take additional precautions due to the inability of armored vehicle crews to see personnel close to their vehicles.

I-27. During initial entry into the AA, platoon ground guides lead convoy parties to their platoon locations with each vehicle dismounting one person to ground guide their vehicle. During this period of movement into the AA, there is a higher probability of accident or injury due to the amount of traffic and unfamiliarity of the area. Commanders must ensure movement into the AA is thoroughly briefed.

AIRCRAFT

I-28. Aircraft in the AA present movement hazards and contribute to an additional hazard of blowing debris. Blowing debris is a hazard to personnel, aircraft, and equipment. The variation in environments in which a unit may operate determines the type and amount of debris. Desert environments, for instance, produce brownout conditions which are a hazard to personnel, equipment, aircraft, and crews. Preventive measures should be incorporated into the unit SOP mitigating this risk. An example of risk mitigation for aircraft operations, in and around the AA parking areas, is brownout mats. These mats reduce the brownout effect of operating in a desert environment. Commanders must ensure SOPs are implemented for each of the environments the unit may operate within. This will allow the unit to establish ground guide procedures for aircraft, whether while hovering into and out of the AA or during ground movement of the aircraft within the AA.

I-29. Ground guide procedures for flight into or out of the AA will be based on the distance between aircraft and proximity to hazards (natural or manmade). In environments where the AA allows for wide scatter of aircraft, the need for ground guides while hovering into and out of the parking area is negligible. In contrast, when the AA parking areas require close interval spacing of aircraft, the need for guides is critical to safety. Additionally, if the AA requires aircraft to park in close proximity to natural or manmade hazards, ground guides are required. The need for ground guides is more critical with operations under night or with NVG devices.

I-30. Upon initial entry into the AA, ground guides should meet incoming aircraft. The situation may dictate crews entering the AA know limited information on where they will park their aircraft. The ground guides guide the aircraft into their assigned parking spot. In desert environments, the ability of personnel to guide aircraft into the parking location is decreased. Brownout upon landing requires a thorough prebrief to the flight crew on the AA parking plan and as exact a location as possible in relation to a ground marking, such as an inverted Y, of each aircraft parking location. Certain situations may require aircraft to land directly on a parking spot. The company safety officer and NCO must be involved in the design and procedures planned for aircraft operations into and out of the AA.

WING WALKING

I-31. Ground guides will also be utilized while moving aircraft with ground handling equipment. Ground guiding aircraft while moving with a vehicle requires additional personnel to ensure rotor blades and aircraft do not come in contact with another aircraft or object. Commanders must ensure procedures designed by the safety officer/NCO are integrated into the SOP. The plan should address guides necessary for various settings (such as define how many are necessary for close proximity parking and scattered parking not in close proximity to another aircraft or hazard). Ground movement of aircraft is normally associated with aviation maintenance operations. The maintenance officer should be involved in the design of ground guide/wing walking procedures for the unit SOP. A critical component of the aircraft ground movement plan is communication. Personnel conducting these operations must conduct a team prebrief and ensure that the team leader has a device that will signal all members to immediately halt the operation (survival vest whistle).

GROUND GUIDES

I-32. Commanders and leaders must ensure ground guide operations, vehicle and aircraft, are incorporated into their company SOPs. During AA operations, ground guides perform a key role in preventing personnel injury, and vehicle and equipment damage. Ground guide operations are critical to maintaining combat readiness for the unit.

PYROTECHNICS

I-33. Aviation units normally use pyrotechnics while in an AA. Units utilize pyrotechnics along likely enemy avenues of advance within their security sector as a warning of activity along these avenues. Additionally, aircrews utilize pyrotechnics as signaling devices in downed aircrew rescue operations. Commanders, unit leaders, and the ALSE officer must train personnel in proper use of pyrotechnics and its associated hazards. These items will be addressed in the unit SOP.

I-34. Pyrotechnics require protection against moisture, dampness, and high temperature. Pyrotechnic items must be given high priority for the best available protection because of their sensitivity. Wet pyrotechnic material is hazardous to store; consequently, any boxes showing signs of dampness will be removed from a storage site and inspected. If the pyrotechnic materiel is wet, it will be destroyed. Certain kinds of materiel deteriorate with age and have expiration dates on the containers. Loose pyrotechnic tracer composition, flare composition, and similar mixtures spilled from broken containers should be carefully taken up and covered completely with SAE 10 (EO–10) engine oil and removed for appropriate disposal.

I-35. Pyrotechnics are hazard division 1.3 which includes items that burn vigorously and cannot usually be extinguished in storage situations. Explosions are confined to pressure ruptures of containers and do not produce propagating shock waves or damaging blast overpressure beyond the magazine distance specified in DA Pam 385-64, Table 5–16. Tossing about of burning container materials, propellant, or other flaming debris may cause a severe hazard of spreading fire.

I-36. The fire fighting extinguishing agent for pyrotechnic materials, Class D (combustible metals [magnesium, potassium, and so forth]) hazard, should be a dry powder agent. Pyrotechnics are a mass fire hazard due to their chemical agents. The mass fire hazard can be more hazardous based on other likely types of ammunition stored with pyrotechnics. These fires may be fought if:

I-37. Explosives are not directly involved.

- White phosphorous munitions are involved, there may be an explosion (immerse/spray with water continuously).
- Pyrotechnics and magnesium incendiaries do not use carbon dioxide, Halon extinguishers, or water on or near munitions. Allow magnesium to cool unless upon flammable materiel. In this case, use a 2-inch layer of dry sand or powder on the floor and rake the burning material onto this layer and re-smother.

WARNING

DO NOT DISCHARGE PYROTECHNICS IN THE VICINITY OF AIRCRAFT FLYING IN THE AREA.

SECTION III — PROTECTING THE SOLDIER

ENVIRONMENTAL CONSIDERATIONS

I-38. As the military looks into the next century, and even today, military units should try to avoid unnecessary environmental damage, not only in training, but also across the spectrum of operational missions. A mission's success may be determined by political or socio-economic stability, both of which are affected by environmental factors and resources. The Army must be able to identify ways to protect the natural environment while executing full range of their missions:

- Considering the environment in planning and decision making in conjunction with other essential considerations of national policy.
- Protecting the environment of home stations and training areas as a means of retaining resources for mission purposes.
- Using environmental risk assessment and management principles to integrate environmental considerations into mission performance.
- Instilling an environmental ethic in Soldiers.
- Understanding linkages between environmental protection issues and their associated impact on safety, force protection, and AHS.

I-39. Soldiers must be prepared to respond across full spectrum operations in any component (offense, defense, stability or civil support operations), sometimes within the same operation. Deployed forces must be able to conform to environmental protection requirements of the theater commander without impairing combat effectiveness. For in-depth information on environmental considerations and Soldier protection, see FM 3-100.4.

I-40. When deployed, company commanders will often deal with base camps. Base camps, while not installations, are comparable to small towns and require many of the considerations applied to installations. A mayor (often the HHC commander) assists base camp commander with control of base operations. A base camp coordination agency will provide expertise and support to the commander, largely through its subordinate base camp assistance/assessment team. Environmental expertise is resident or aligned with this team and available to support the base camp commander and designated mayor of the base camp. They provide technical recommendations and maintain appropriate standards. More information about this phenomenon is provided in the Center for Army Lessons Learned Newsletter 99-9, Integrating Military Environmental Protection.

I-41. Army Regulation 200-1 and the Army's Commander's Guide to Environmental Management specify commanders' environmental responsibilities. To carry out these responsibilities, commanders:

- Comply with an installation's environmental policies and legally applicable and appropriate federal, state, and local laws and regulations or country-specific final governing standards if outside the continental United States.
- Demonstrate a positive and proactive commitment to environmental stewardship and protection.
- Provide environmental training required by law, regulation, or command policy.
- Ensure all personnel perform their duties in compliance with environmental laws and regulations, and respond properly to emergencies.
- Promote proactive environmental measures and pollution prevention.
- Supervise compliance with environmental laws and regulations during operational, training, and administrative activities.
- Include environmental considerations in mission planning, briefings, meetings, execution, and AARs.
- Understand the requirements of Army environmental programs.
- Identify and assess environmental risks of proposed programs and activities.
- Coordinate unit activities with higher headquarters environmental elements.
- Appoint and train an environmental compliance officer (ECO) coordinator for the unit.

- Ensure SOPs contain all environmental considerations and regulatory requirements appropriate for the level of command.
- Conduct environmental self-assessment or internal environmental compliance assessments.
- Understand linkages between environmental considerations and their associated impact on safety, force protection, and AHS.

ENVIRONMENTAL DUTIES

I-42. The key to fulfilling environmental requirements successfully at unit level is the ECO. Army Regulation 200-1 directs all Army unit commanders "appoint and train ECOs at appropriate levels to ensure compliance actions take place." In company-sized units, this duty will generally translate into an extra duty. The ECO manages environmental issues at unit level and ensures environmental compliance. He also coordinates, through the respective chain of command, with the supporting installation environmental staff to clarify requirements and obtain assistance.

I-43. The ECO accomplishes environmental compliance requirements on behalf of the commander. He also coordinates with supporting installation environmental staff to clarify requirements and obtain assistance. While this position of responsibility is not a formal staff position, the ECO is critical to the commander's environmental program. The ECO:

- Advises the unit on environmental compliance during training, operations, and logistics functions.
- Serves as commander's eyes and ears for environmental matters.
- Coordinates between unit and higher/installation headquarters environmental staffs.
- Manages information concerning unit's environmental training and certification requirements.
- Performs unit environmental self-assessment inspections.
- Performs environmental risk assessments.

UNIT PLANNING

I-44. Staffs integrate environmental protection into planning for larger units. Unit leaders integrate environmental protection into unit planning for battalion- and company-level units. Soldier safety should be addressed in SOPs, OPORDS, and training plans.

STANDING OPERATING PROCEDURES

I-45. Unit leaders develop SOPs reflecting environmental protection considerations for routine tasks and activities. The SOPs provide information to Soldiers on how to accomplish routine tasks in an environmentally sound manner. The SOPs incorporate local requirements. As local requirements change, unit leaders update their SOPs. The SOPs also help define environmental protection requirements for all unit activities—facility operations, field operations, deployment, and combat. Unit leaders ensure SOPs comply with local requirements by coordinating with higher headquarters staff—usually the environmental office, the surgeon and his staff, preventive medicine personnel, and the SJA or engineer coordinator.

I-46. Unit leaders conduct environmental risk assessments when planning operations or activities. Risk assessment is a standard element of TLP. Unit leaders perform environmental risk assessments for activities not addressed in the SOP or when conditions differ significantly from those described in the SOP. A maintenance unit does not perform a risk assessment every time it performs a lubrication or service. Rather, the SOP describes the correct manner to perform these actions. Risk assessments apply to garrison operations as well as field operations.

OPERATION ORDERS

I-47. Unit leaders address environmental protection in their plans and orders including: WARNOs, OPORDs, OPLANs, concept of operation plans (CONPLANs), and FRAGOs. The higher headquarters staff develops an environmental appendix/annex to its OPORD/OPLAN/CONPLAN. Subordinate unit leaders draw environmental information from the environmental appendix to the OPORD/OPLAN/

CONPLAN, or from Annex L in a joint operations planning and execution system document. Field Manual 5-0 directs the inclusion of Appendix 2 (Environmental Considerations) to Annex F (Engineer) of the OPLAN/OPORD/CONPLAN (battalion level and above) and specifies that lower-level unit leaders/staffs include environmental information in coordinating instructions and service and support paragraphs.

Risk Management Plan

I-48. Risk management is an integral part of the military decision-making process. The risk management process provides the framework for making risk management a routine part of planning, preparing, and executing operational missions and everyday tasks. Assessing environmentally-related risks is part of the total risk management process.

I-49. Knowledge of environmental factors is a key to planning and decision making. With this knowledge, leaders quantify risks, detect problem areas, reduce risk of injury or death, reduce property damage, and ensure compliance with environmental laws and regulations. Unit leaders should conduct risk assessments before conducting any training, operations, or logistical activities.

Tactical Risk/Accident Risk

I-50. When assessing the risk of hazards in operations, the commander and leaders must look at two types of risk.

- Tactical risk is risk concerned with hazards existing because of the presence of either the enemy or an adversary, thus involving the considerations of force protection. It applies to all levels of war and across the spectrum of operations. For example, during the Gulf War, the enemy's demolition of oil fields created a significant health and environmental hazard to the surrounding countryside and those units maneuvering through the area.
- Accident risk includes all operational risk considerations other than tactical risk. It includes risk to friendly forces and risk posed to civilians by an operation, as well as the impact of operations on the environment. It can include activities associated with hazards concerning friendly personnel, civilians, equipment readiness, and environmental conditions. Examples of environmentally related accident risk are improper disposal of hazardous waste, personnel not properly trained to clean up a spill, and units maneuvering in ecologically sensitive terrain. Preventive medicine considerations also fall into this area of risk.

Note: Tactical risk and accident risk may be diametrically opposed. The commander may accept a high level of environmentally-related accident risk on reducing the overall tactical risk. For example, a commander may decide to destroy an enemy's petroleum storage area to reduce its overall tactical risk.

Training Plans

I-51. Formal planning for training culminates with publication of the training schedule. Informal planning and coordination (preexecution checks) continue until training is performed. During rehearsals, leaders ensure all safety and environmental considerations are met.

I-52. To conduct effective, meaningful training for Soldiers, leaders, and units, thorough preparation is essential. This means well prepared trainers, Soldiers, and support personnel are ready to participate, and their facilities, equipment, and materials are ready to use.

I-53. A unit executes training the same way it executes a combat mission. The chain of command is present, in charge, and responsible. During operations, leaders ensure environmental practices and preventive measures are being employed. Once Soldiers understand what is expected of them, these practices become merely another measure of unit proficiency and level of unit discipline.

FIGHTER MANAGEMENT

I-54. Fighter management describes a process essential to effective and efficient battle staff operations. At company level, commanders do not have staffs. The battle staff becomes those personnel in the unit who operate the CP. Successful continuous operations at the company level are more demanding than higher level organizations. The unit requires a TACSOP allowing rest, especially for critical personnel. At company level, the commander and 1SG will find it very demanding to try to rest. Their rest plan must be a priority for the organization to be lead effectively and their units to be successful on the battlefield.

I-55. The cycle of recurring events within a CP focuses the CP officer in charge/noncommissioned officer in charge (OIC/NCOIC) on meeting information and action requirements. Company personnel are normally required to attend battalion/company recurring meetings. This has an impact on company CP operations and schedules. These recurring events include—

- Shift changes.
- Battle update briefings to the commander.
- Battalion targeting meetings.
- Battalion reports.
- Battalion battle updates without the commander.
- Battalion commanders' collaborative sessions.
- Company CP OIC/NCOIC shift change briefing/collaborative sessions.

I-56. The company CP OIC/NCOIC must achieve battle rhythm for updating and viewing information and understand how to use it to affect operations. A well-established battle rhythm aids the commander and leaders with CP organization, information management and display, decision making, and fighting the battle. Battle rhythm demands careful planning and design. Many competing demands must be deconflicted. Even subordinate platoons and sections affect a company battle rhythm based on their needs and unit procedures.

I-57. Command discipline is required to enforce sleep cycles and create an environment where sound sleep can be achieved. Battle rhythm and associated planning is a proactive, not reactive, exercise.

SLEEP PLANS

I-58. Units must develop detailed rest plans and enforce them. Leaders must rest to maintain their effectiveness; however, some leaders attempt to get involved in every aspect of planning and execution. An integral part of the planning process is determining when senior leader presence is required. It is just as important to identify when a leader's presence is not required. The planning process should include the following supporting techniques:

- Include a sleep plan in the METT-TC analysis.
- Ensure leaders have confidence in second and third echelons of leadership and their ability to make routine decisions.
- Instill trust and confidence in the officers, junior NCOs, and specialists by effective training and SOPs.
- Consider contingencies and establish criteria for waking leaders.
- Post sleep plans in CPs, platoon, and section areas; everyone needs to know the plan.
- Synchronize sleep plans with higher headquarters and subordinate platoons/sections.

CONTINUOUS OPERATIONS

I-59. Continuous operations are combat operations continuing around the clock at a high pace, requiring Soldiers to fight without relief for extended periods. Opportunities for sleep are scattered throughout the day and night.

Sustained Operations

I-60. Sustained operations are operations conducted 24 hours a day with little or no opportunity for sleep. Sustained operations are when the same Soldiers or small units engage in combat operations with no opportunity for the unit to stand down and little time for Soldiers to sleep. Aviation units routinely conduct sustained operations to an extent even greater than the infantry due to aviation's maneuver, maneuver support and sustainment roles. To combat the negative effects of sustained operations, commanders and leaders must enforce fighter management principles on their organizations. A fighter management policy should be addressed in the unit SOP, and higher headquarters policy should correlate with company policy.

Degradation of Combat Capability

I-61. As sustained operations continue, all Soldiers begin to show effects of general fatigue and lack of sleep. Unless counteracted, unit performance of combat tasks decline. Recent studies indicate performance is degraded by 25 percent for each 24-hour period without sleep. After 96 hours, performance can be expected to be near zero. Determination to endure must be supplemented by countering the adverse effects to slow the rate of decline. It becomes more difficult to perform assigned tasks to the required standard. Leaders need to recognize signs of serious sleep deprivation in their subordinates. Studies show performance in all duty positions does not degrade the same. Performance in a duty position where there is a heavy load of mental tasks, such as aviation duties (determining, calculating, thinking, decision-making), degrades faster than the performance in a position whose tasks are mainly physical (firing, running, lifting, digging).

I-62. In addition to the degradation caused by fear, fatigue, and loss of sleep, there is a significant loss of effectiveness caused by operation in MOPP 4. When units are using full CBRN protective equipment, judgment is degraded, communications are less effective, and information flow between units is reduced.

FATIGUE IN FLYING OPERATIONS

I-63. Aviation operations are inherently dangerous. Commanders and leaders must be aware of increased dangers present in combat aviation continuous and sustained operations. Additionally, leaders must be able to recognize the symptoms of fatigue and how to deal with fatigued aircrews. Field Manual 3-04.301 provides an in-depth review of aeromedical factors associated with aviation.

OPERATIONAL TEMPO AND BATTLE RHYTHM

I-64. The aviation company should be staffed for 24-hour operations; however, it also conducts cyclical missions. Standard operating procedures establish methods of ensuring the right personnel are available for either cyclical or 24-hour operations. Regardless of methods used, practice during exercises must determine strengths and weaknesses of headquarters personnel for CP operations and training for additional personnel who may be utilized to staff the CP during continuous or sustained operations. Such knowledge allows leaders to focus on critical areas and personnel requiring additional training.

I-65. In planning, the battalion staff must consider battle rhythm requirements of subordinate companies and platoons. Depending on the situation, the battalion may schedule missions allowing company or platoon rotations to maintain their battle rhythm.

Absence of Battle Rhythm

I-66. Without procedures establishing battle rhythm, leaders and units reach a point of diminished returns. This typically occurs between 72 and 96 hours of operations. As leader fatigue sets in, information flow, planning process, execution, and sustainment suffer—often greatly. Symptoms of diminished battle rhythm include the following:

- Leader fatigue.
- Leaders not fully aware of critical decision points (DPs).
- Leaders not available at critical DPs.
- Disjointed time lines between various levels of command.

Presence of Battle Rhythm

I-67. Battle rhythm allows units and leaders to function at a sustained level of efficiency for extended periods. Effective battle rhythm permits an acceptable level of leadership at all times. It can focus leadership at critical points in the fight or during particular events. Procedures and processes facilitating efficient decision making and parallel planning are critical to achieving battle rhythm. Every component of battle rhythm contributes uniquely to sustained operations.

Training

I-68. It is difficult, if not impossible, to establish battle rhythm while simultaneously conducting operations. Preplanning makes it happen. Planning, preparing, and training before deployment lays a solid foundation for viable battle rhythm during operations. Commanders must ensure all company personnel are trained on TOC set-up and operations, and they understand the team concept required to conduct CP operations during combat operations for extended periods. Additionally, company commanders must coordinate with battalion staffs during training to deconflict staff operations and requirements which are counterproductive to the company battle rhythm.

Battle Rhythm Elements

I-69. Battle rhythm is a multifaceted concept including the following elements:

- Sleep/rest plans.
- Trained second- and third-tier leadership in CPs.
- Synchronized upward multiechelon timelines.
- Parallel planning.
- Established processes and SOPs.

Command Post Personnel Depth

I-70. Established processes and SOPs relieve many antagonistic effects of extended operations. Standard operation procedures establishing and maintaining battle rhythm by facilitating routine decisions and operations are a step in the right direction. Soldiers trained to act appropriately in the absence of leaders or orders can relieve commanders and leaders of many of the time-consuming tasks robbing them of essential rest. Examples of areas NCOs and junior officers can accomplish for the commander include—

- Battle summaries and updates during a fight.
- Intelligence updates before, during, and after a battle.
- Sustainment updates before, during, and after a battle.
- Updates to the next higher commander.
- Shift change briefings.

Challenges to Battle Rhythm

I-71. Challenges to battle rhythm include enlisted, NCO, and junior officer duties; and synchronization of planning, execution, rehearsal timelines, equipment maintenance, and sleep plans.

Noncommissioned Officer and Junior Officer Responsibilities

I-72. At the company level, all personnel are critical to the success of the operation and provide valuable contributions. It is imperative commanders make certain each Soldier understands the importance of even the most menial tasks, such as CP security and TOC setup and teardown. Given the amount of personnel in a company, 24-hour operations and force protection requirements, commanders must utilize all personnel available to successfully accomplish CP operations and continuous missions. This becomes even more critical when a company has platoons or sections staged forward in support of BCTs. The improper use of personnel produces the following results:

- Key leaders become exhausted.

- CP operations trained personnel become exhausted.
- The initiative of trained subordinates is stifled, and the incentive to train is diminished.

I-73. The following techniques ensure proper use of personnel:

- Appropriate tasks are assigned to junior NCOs and specialists.
- Effective training and SOPs instill trust in officers and confidence in junior NCOs and specialists.
- Effective command guidance to the company that it is a team effort and all personnel available contribute to the effort.

CONTINUOUS OPERATIONS AND TIMELINES SYNCHRONIZATION

I-74. Timelines for the operation at hand must allow for not only the next operation, but also extended continuous operations. Synchronized, multiechelon timelines assist units in achieving battle rhythm. If units do not address critical events at least one level up and down, disruption results. An example of an unsynchronized timeline is a battalion rehearsal that conflicts with platoon precombat inspections or other events in their internal timeline. Lower echelon units seldom recover from a poor timeline directed by a higher headquarters, as well as platoons and sections at company level. Company commanders must coordinate with their battalion staffs on development of SOPs that include planning, rehearsal, and execution timelines one level below battalion to prevent these conflicts.

STANDING OPERATING PROCEDURES UTILIZATION

I-75. SOPs must be practiced and reviewed during professional development and sergeants' time. The existence of an SOP will not resolve troop-leading challenges unless the SOP is practiced often and internalized by unit members. Checklists are critical, as many leaders will often find themselves rushed, physically fatigued, distracted, and deprived of sleep. Checklists ensure each step is considered even when leaders are exhausted.

SECTION IV — ACCIDENT/INCIDENT REPORTING

I-76. Persons involved in, or aware of, an accident must immediately report it to the commander or supervisor directly responsible for the operation, materiel, or person(s) involved. Table I-1 and Table I-2 identify reporting requirements for aviation and ground accidents/incidents (peacetime and combat). See AR 385-40, DA Pam 385-40, and AR 385-95 for in-depth instructions on Army accidents. Company commanders ensure unit SOPs cover accident/incident reporting, and it is fully addressed by the company aviation safety officer.

Table I-1. Aviation accident notification and reporting requirements and suspenses

Peacetime				***Combat ****	
Accident Class	***Notification Telephonic Worksheet***	***DA Form 2397***	***Reporting AAA report***	***Notification Telephonic Worksheet***	***Reporting AAA report***
A	Immediate–to USASC (telephonic notification–no hardcopy notification required (DSN 558-2660/2539/3410 or Commercial (334) 255-2660/2539/3410	(CAI/IAI) 90 calendar days	Aircraft ground ACDTS only-30 calendar days	Same as peacetime to USASC or Safety Rep, forward	(Only when CDR determines DA Form 2397 investigation/ report not feasible) submit as soon as conditions, situation permit-Do not exceed 30 calendar days.
B	Immediate–to	(CAI/IAI) 90	Aircraft ground	Same as	(Only when CDR

Peacetime				*Combat **	
Accident Class	***Notification Telephonic Worksheet***	***DA Form 2397***	***Reporting AAA report***	***Notification Telephonic Worksheet***	***Reporting AAA report***
	USASC (telephonic notification–no hardcopy notification required (DSN 558-2660/2539/3410 or Commercial (334) 255-2660/2539/3410	calendar days	ACDTS only 30 calendar days	peacetime to USASC or Safety Rep, forward	determines DA Form 2397 investigation/ report not feasible) submit as soon as conditions, situation permit- Do not exceed 30 calendar days.
C	Immediate–to USASC (telephonic notification–no hardcopy notification required (DSN 558-2660/2539/ 3410 or Commercial (334) 255-2660/2539/3410	N/A	30 calendar days	Same as peacetime to USASC or Safety Rep, forward	Same as peacetime
D	N/A (Unless SOF issue involved/ suspected)	N/A	10 calendar days	Same as peacetime	Same as peacetime
E	N/A (Unless SOF issue involved/ suspected)	N/A	10 calendar days	Same as peacetime	Same as peacetime
F	N/A (Unless SOF issue involved/ suspected)	N/A	10 calendar days	Same as peacetime	Same as peacetime
Submis-sion Methods	Class A-C telephonic (immediate) Class D, E, F-IF SOF	Mail	Typed or hand printed AAA reports by mail/FAX or courier/ message/ format/elec-tronic submission, Include attachments as required	Same as peacetime	Same as peacetime
Notes * Only when the senior tactical commander determines that the situation, conditions, and/or time does not permit normal peacetime investigating and reporting.					

Table I-2. Ground accidents notification and reporting requirements and suspenses (3)

Peacetime				*Combat (2)*	
Accident	***Telephonic Notification***	***AGAR***	***DA Form 285***	***Telephonic Notification***	***AGAR only*** ***By any means possible***
Class	***Worksheet***			***Worksheet***	***(Message, Electronic, FAX, Phone, Hand Carry, Mail)***
On-Duty					
A	Immediately (1)	Not required	IAI/CAI-90 days	Immediately (1)	As time permits (Not to exceed 30 days)
B	Immediately (1)	Not required	IAI/CAI-90 days	Immediately (1)	As time permits (Not to exceed 30 days)
C	Not required	Within 30 days	Not required	Not required	As time permits (Not to exceed 30 days)
D	Not required	Within 30 days	Not required	Not required	As time permits (Not to exceed 30 days)
Off-Duty					
A	Immediately (1)	Within 30 days	Not required	Immediately (1)	As time permits (Not to exceed 30 days)
B	Immediately (1)	Within 30 days	Not required	Immediately (1)	As time permits (Not to exceed 30 days)
C	Not required	Within 30 days	Not required	Not required	As time permits (Not to exceed 30 days)
D	Not required	Within 30 days	Not required	Not required	As time permits (Not to exceed 30 days)

Notes:
(1) USASC must be notified IMMEDIATELY by phone at DSN 558-2660/2539-3410 or Commercial (334) 255-2660/2539/3410 or notify USASC Safety Rep forward (during combat).
(2) ONLY when the senior tactical commander determines that the situation, conditions, and/or time does not permit normal peacetime investigating and reporting.
(3) Army civilian injury only accidents should be reported on appropriate Department of Labor Forms according to this regulation.

Glossary

SECTION I — ACRONYMS AND ABBREVIATIONS

A2C2	Army airspace command and control
AA	air ambulance/active Army
AAA	abbreviated aviation accidents
AAR	after-action review
AATF	air assault task force
ABCA	American, British, Canadian, and Australian
ACDTS	accidents
ACR	armor cavalry regiment
AD	air defense
AE	aeromedical evacuation
AECT	aeromedical evacuation control team
AELT	aeromedical evacuation liaison team
AF	Air Force
AGAR	abbreviated ground accident report
AGPU	aviation ground power units
AHS	Army Health System
ALCT	airlift control team
ALSE	aviation life support equipment
amb	ambulance
AMEDD	Army Medical Department
AO	area of operations
AOC	air operations center
AOE	Army of Excellence
APOD	aerial port(s) of debarkation
AR	Army regulation
ASI	additional skill identifier
ARSOF	Army special operations forces
ARNG	Army National Guard
ASCC	Army service component command
ASF	aeromedical staging facility
ASMC	area support medical company
ASOS	Army support to other Services
ASTS	aeromedical staging squadron
ATM	advanced trauma management
ATS	air traffic service

attn	attention
aug	augmentation
avail	available
AVIM	aviation intermediate maintenance
AVN	aviation
AVUM	aviation unit maintenance
AXP	ambulance exchange point
BAE	brigade aviation element
BAS	battalion aid station
BCT	brigade combat team
bde	brigade
BH	behavioral health
BIFV	Bradley infantry fighting vehicle
bn	battalion
BOA	basis of allocation
BSA	brigade support area
BSB	brigade support battalion
BSMC	brigade support medical company
BSS	brigade surgeon section
BTU	British thermal unit
C2	command and control
CA	civil affairs
CAB	combat aviation brigade
CAI	centralized accident investigation
CASEVAC	casualty evacuation
CASF	contingency aeromedical staging facility
CBRN	chemical, biological, radiological, and nuclear
CBRNE	chemical, biological, radiological, nuclear, and high yield explosives
CCATT	critical care air transport team
CCIR	commander's critical information requirements
CCP	casualty collecting point
CE	communications-electronic
CJA	Command Judge Advocate
CJTF	combined joint task force
CLS	combat lifesaver
CLT	casualty liaision team
CM	campaign module
CMCC	corps movement control center
CMD	command
CMO	civil-military operations

CO	company
COA	course(s) of action
COCOM	combatant command
COMSEC	communications security
CONPLAN	concept of operation plan
CONUS	continental United States
COSC	combat and operational stress control
COSR	combat and operational stress reaction
CP	command post
CRTS	casualty receiving and treatment ships
CSH	combat support hospital
CSAR	combat search and rescue
CTC	Combat Training Center
CZ	combat zone
DA	Department of the Army
DCO	defense coordination officer
DCSCOMPT	Deputy Chief of Staff, Comptroller, US Army
DCSIM	Deputy Chief of Staff for Information Managment, US Army
DCSLOG	Deputy Chief of Staff for Logistics, US Army
DCSPER	Deputy Chief of Staff for Personnel, US Army
DCSSPO	Deputy Chief of Staff for Support Operation, US Army
DD	Department of Defense
DHS	Department of Homeland Security
div	division
DMOC	division medical operations center
DNBI	disease and nonbattle injuries
DOD	Department of Defense
DODD	Department of Defense Directive
DP	decision point
DS	direct support
DSN	defense switched network
DSS	division surgeon section
EAB	echelons above brigade
EAC	echelons above corps
EAD	echelons above division
ECO	environmental compliance officer
EEE	early entry element(s)
EEH	early entry hospital
EEM	early entry module
EMEDS	expeditionary medical support

EMT	emergency medical treatment
EPS	essential personnel services
EPW	enemy prisoner(s) of war
1SG	first sergeant
F	Fahrenheit
FARP	forward arming and refueling point
FAX	facsimile
FEMA	Federal Emergency Management Agency
FHP	force health protection
FM	field manual/frequency modulated
FMC	field medical card
FMI	field manual interim
FRAGO	fragmentary order
FRSS	forward resuscitation
FSB	forward support battalion
FSMC	forward support medical company
FSMT	forward support MEDEVAC team
FST	forward surgical team
G3	Assistant Chief of Staff (Operations and Plans)
GA	ground ambubance
GAMC	ground ambulance medical company
GCC	geographic combatant commander
GRREG	graves registration
GS	general support
GSAB	general support aviation battalion
GWS	Geneva Convention for the Amelioration of the Condition of the Wounded and Sick in Armed Forces in Field
HBCT	heavy brigade combat team
HD	homeland defense
HHD	headquarters and headquarters detachment
HMMWV	high mobility multipurpose wheeled vehicle
HN	host nation
hosp	hospital
HQ	headquarters
HR	human resources
HS	homeland security
HSS	health service support
IAI	installation-level accident investigation
IED	improvised explosive device
IG	Inspector General
IM	information management

IPB	intelligence preparation of the battlefield
IPMC	intratheater patient movement center
ISB	intermediate staging base
IT	information technology
JFC	joint force commander
JHSS	joint health service support
JMD	joint manning document
JMTB	Joint Military Transportation Board
JOA	joint operations area
JP	joint publications
JPMRC	joint patient movement requirements center
JTF	joint task force
LHA	amphibious helicopter assault carrier
LHD	amphibious helicopter assault carrier dock
LIN	line item number
LOC	lines of communication
LOTS	logistics over-the-shore
LPD	amphibious transport dock
LPH	amphibious assault helicopter carrier–aviation
LSD	dock landing ship
LST	tank landing ship
LZ	landing zone
MACA	military assistance to civil authorities
MASCAL	mass casualty
MASF	mobile aeromedical staging facility
MBA	main battle area
MC3T	medical command, control, and communication telemedicine
MC4	medical communications for combat casualty care
MCB	movement control battalion
MCO	major combat operation
MCP	main command post
med	medical
MEDBDE	medical brigade
MEDCOM	medical command
MEDEVAC	medical evacuation
MEDLOG	medical logistics
MEF	Marine expeditionary force
MES	medical equipment sets
METT-TC	mission, enemy, terrain and weather, troops and support available, time available, and civil considerations

MEV	medical evacuation vehicle
MH	mental health
MHS	military health system
MIA	missing in action
MIJI	meaconing, intrusion, jamming, and interference
MMB	multifunctional medical battalion
MOC	medical operations cell
MOPP	mission-oriented protective posture
MOS	military occupational specialty
MP	military police
MRE	meals, ready-to-eat
MRI	medical reengineering initiative
MRO	medical regulating office
MSB	main support battalion
MSC	Military Sealift Command
MTF	medical treatment facility
MTOE	modification table of organization and equipment
MWD	military working dogs
MWR	morale, welfare, and recreation
N/A	not applicable
NATO	North Atlantic Treaty Organization
NAVAIDS	navigational aids
NBC	nuclear, biological, and chemical
NCO	noncommisioned officer
NOE	nap-of-the-earth
NP	neuropsychiatric
NSHS	National Strategy for Homeland Security
NVG	night vision goggles
OCP	operational command post
OEG	operational exposure guidance
OEH	occupational and environmental health
OMF	originating medical facility
OPCON	operational control
OPLAN	operation plan
OPORD	operation order
OPS	operations
OPSEC	operations security
OPTEMPO	operational tempo
OR	operating room
OVE	on vehicle equipment

PASR	personnel accountability and strength reporting
PAD	patient administrator
PCC	precombat checks
PCI	precombat inspections
PDS	personnel daily summary
PERS	personnel
PIM	personnel information management
PIR	priority intelligence requipment
PMCS	preventive maintenance checks and services
PMI	patient movement items
PMM	preventive medicine measures
PMRC	patient movement requestment center
POI	point of injury
POL	petroleum, oils, and lubricants
PR	personnel recovery
prep	preparation
PRM	personnel readiness management
PT	platoon
PSYOP	psychological operations
PZ	pickup zone
PVNTMED	preventive medicine
QSTAG	Quadripartite Standardization Agreement
R5	reception, replacement, return to duty, rest and recuperation, redeployment
R&R	rest and recuperation
RC	reserve component
rep	representative
RPG	rocket propelled grenade
RSTA	reconnaissance and target acquisition
RSOI	reception, staging, onward movement, and integration
RTD	return to duty
S&RO	stability and reconstruction operations
S1	Adjutant (US Army)
S2	Intelligence Officer (US Army)
S3	Operations and Training Officer (US Army)
S4	Supply Officer (US Army)
S6	Information Management (US Army)
SBCT	Stryker Bridage Combat Team
SEC	section
SF	special forces
SIMLM	single integrated medical logistics manager

SINCGARS	Single Channel Ground/Airborne Radio System
SJA	Staff Judge Advocate
SOA	special operations aviation
SOF	special operations forces
SOP	standing operating procedures
SPT	support
SRC	standard requirements code
STANAG	standardization agreement
SURG	surgical
SUST	sustainment
TAA	tactical assembly area
TACC	tanker airlift control center
TAC CP	tactical command post
TAC SOP	tactical standing operating order
TAES	Theater Aeromedical Evacuation System
TC	training circular
tm	team
TMIP	Theater Medical Information Program
tmt	treatment
tng	training
TO	theater operations
TOC	tactical operations center
TOE	tables of organization and equipment
TPFDL	time-phased force deployment list
TPMC	theater patient movement center
TPMRC	theater patient movement requirements center
TSOC	Theater Special Operations Command
TSOP	tactical standing operating procedures
UO	urban operations
US	United States
USAF	United States Air Force
USAMEDDC&S	United States Army Medical Department Center and School
USASC	United States Army Safety Center
USN	United States Navy
USNS	United States Navy Ship
USNORTHCOM	United States Northern Command
USTRANSCOM	United States Transportation Command
vet	veterinary
WARNO	warning order
WIA	wounded in action

WMD	weapons of mass destruction
XO	executive officer
ZULU	universal time

References

SOURCES USED

These are the sources quoted or paraphrased in this publication.

NATO STANAGs

2040. *Stretchers, Bearing Brackets, and Attachment Supports*, 23 September 1982 (Latest Amendment, 6 October 2004)

2087. *Medical Employment of Air Transport in the Forward Area*, 29 June 1983 (Latest Amendment, 10 September 1999)

2128. *Medical and Dental Supply Procedures*, 15 November 1982 (Latest Amendment, 21 November 1991)

2132. *Documentation Relative to Medical Evacuation, Treatment, and Cause of Death of Patients*, 7 August 1974 (Latest Amendment, 15 September 1986)

2454. *Road Movements and Movement Control*, 27 January 2005

2931. *Orders for the Camouflage of the Red Cross and Red Crescent on Land in Tactical Operations*, 18 October 1984 (Latest Amendment, 3 April 1998)

3204. *Aeromedical Evacuation*, 26 September 1973 (Latest Amendment, 10 September 2001)

ABCA QSTAGs

436. *Minimum Labeling Requirements for Medical Materiel* (Amendment 1), 8 October 1980

470. *Documentation Relative to Medical Evacuation, Treatment, and Cause of Death of Patients*, 23 February 1979 (Latest Amendment, 14 August 1989)

PRESIDENTIAL DIRECTIVE

Homeland Security Presidential Directive-5, *Management of Domestic Incidents*, February 28, 2003

DODD

DoD Directive 3025.15, *Military Assistance to Civil Authorities*, February 18, 1997

DoD Directive 5200.31E, *DoD Military Working Dog (MWD) Program*, March 29, 2006

JOINT AND MULTISERVICE PUBLICATIONS

Joint Pub 1-02, *DOD Dictionary of Military and Associated Terms*, 23 March 1994
Joint Pub 4-02, *Doctrine for Health Service Support in Joint Operations*, 26 April 1995
Joint Pub 5-0, *Joint Operation Planning*, 26 December 2006
Joint Pub 5-00.1, *Joint Doctrine for Campaign Planning*, 25 January 2002
Joint Pub 5-00.2, *Joint Task Force (JTF) Planning Guidance and Procedures*, 13 January 1999

ARMY PUBLICATIONS

AR 40-5, *Preventive Medicine*, 22 July 2005

AR 71-32, Force Development and Documentation-Consolidated Policies, 3 March 1997

AR 200-1, *Environmental Protection and Enhancement*, 21 February 1997

AR 385-40, *Accident Reporting and Records*, 1 November 1994

AR 385-64, *U.S. Army Explosives Safety Program*, 1 February 2000

AR 385-95, *Army Aviation Accident Prevention*, 12 December 1999

AR 600-55, *The Army Driver and Operator Standardization Program (Selection, Training, Testing, and Licensing)*, 31 December 1993

DA Pam 385-40, *Army Accident Investigating Reporting*, 1 November 1994

DA Pam 385-40, *Ammunition and Explosives Safety Standards*, 15 December 1999

FM 1-202, *Environmental Flight*, 23 February 1983

FM 1-564, *Shipboard Operations*, 29 June 1007

FM 3-0, *Operations*, 14 June 2001

FM 3-04.301, *Aeromedical Training for Flight Personnel*, 29 September 2000

FM 3-05, *Army Special Operations Forces*, 20 September 2006

FM 3-06.11, *Combined Arms Operations in Urban Terrain*, 28 February 2002

FM 3-07, *Stability Operations and Support Operations*, 20 February 2003

FM 3-11.5, *Multiservice Tactics, Techniques, and Procedures for Chemical, Biological, Radiological, and Nuclear Decontamination*, 4 April 2006

FM 3-34.2, *Combined-Arms Breaching Operations (Including Changes 1—3)*, 31 August 2000

FM 3-97.6, *Mountain Operations*, 28 November 2000

FM 3-100.4, *Environmental Considerations in Military Operations (Including Change 1)*, 15 June 2000

FM 4-01.45, *Multi-Service Tactics, Techniques, and Procedures for Convoy Operations*, 24 March 2005

FM 4-02, *Force Health Protection in a Global Environment*, 13 February 2003

FM 4-02.1, *Combat Health Logistics*, 28 September 2001

FM 4-02.4, *Medical Platoon Leader's Handbook—Tactics, Techniques, and Procedures (Including Change 1)*, 24 August 2001

FM 4-02.6, *The Medical Company—Tactics, Techniques, and Procedures (Including Change 1)*, 4 September 2004

FM 4-02.7, *Health Service Support in a Nuclear, Biological, and Chemical Environment—Tactics, Techniques, and Procedures*, 10 January 2002

FM 4-02.12, *Health Service Support in Corps and Echelons above Corps*, 2 February 2004

FM 4-02.17, *Preventive Medicine Services*, 28 August 2000

FM 4-02.18, *Veterinary Services—Tactics, Techniques, and Procedures*, 30 December 2004

FM 4-02.21, *Division and Brigade Surgeon's Handbook (Digitized)—Tactics, Techniques, and Procedures*, 15 November 2000

FM 4-02.24, *Area Support Medical Battalion—Tactics, Techniques, and Procedures*, 28 August 2000

FM 4-02.25, *Employment of Forward Surgical Teams—Tactics, Techniques, and Procedures*, 28 March 2003

FM 4-25.12, *Unit Field Sanitation Team*, 25 January 2002

FM 4-02.43, *Force Health Protection Support for Army Special Operations Forces*, 28 November 2006

FM 5-0, *Army Planning and Orders Production*, 20 January 2005

FM 8-42, *Combat Health Support in Stability Operations and Support Operations*, 27 October 1997

FM 8-55, *Planning for Health Service Support*, 9 September 1994

FM 8-250, *Preventive Medicine Specialist (Including Change 1)*, 27 January 1986
FM 10-67.1, *Concepts and Equipment of Petroleum Operations*, 2 April 1998
FM 21-10, *Field Hygiene and Sanitation*, 21 June 2000
FM 31-70, *Basic Cold Weather Manual*, 12 April 1968
FM 55-30, *Army Motor Transport Units and Operations (Including Change 1)*, 15 September 1999
FM 90-3, *Desert Operations*, 24 August 1993
FM 90-13, *River Crossing Operations*, 26 January 1998
ST 4-02.121, *Multifunctional Medical Battalion*, October 2005
TC 1-400, *Brigade Aviation Element Handbook*, 27 April 2006
TB MED 507, *Heat Stress Control and Heat Casualty Management*, 7 March 2003
CALL Newsletter 99-9, *Integrating Military Environmental Protection*, 1999
FMI 4-90.1, *Heavy Brigade Combat Team Logistics*, 15 March 2005
FMI 5-0.1, *The Operations Process*, 31 March 2006

DOCUMENTS NEEDED

These documents must be available to the intended users of the publication.

DA Form 285 Series, *U.S. Army Accident Report*
DA Form 1156, *Casualty Feeder Card*
DA Form 2397 Series, *Technical Report of U.S. Army Aircraft Accidents*
DA Form 2404, *Equipment Inspection and Maintenance Worksheet*
DD Form 601, *Patient Evacuation Manifest*
DD Form 1380, *U.S. Field Medical Card*

Index

References are to paragraph numbers unless otherwise stated.

S

T